Lecture Notes in Computer Science   2272

Edited by G. Goos, J. Hartmanis, and J. van Leeuwen

Springer
Berlin
Heidelberg
New York
Barcelona
Hong Kong
London
Milan
Paris
Tokyo

Didier Bert   Jonathan P. Bowen
Martin C. Henson   Ken Robinson (Eds.)

ZB 2002: Formal Specification and Development in Z and B

2nd International Conference of B and Z Users
Grenoble, France, January 23-25, 2002
Proceedings

Springer

Volume Editors

Didier Bert
CNRS, Laboratoire LSR, IMAG
681, rue de la Passerelle
38402 Saint Martin d'Hères Cedex, France
E-mail: didier.bert@imag.fr

Jonathan P. Bowen
South Bank University, SCISM, Centre for Applied Fromal Methods
Borough Road, London SE1 0AA, UK
E-mail: jonathan.bowen@sbu.ac.uk

Martin C. Henson
University of Essex, Department of Computer Science
Wivenhoe Park, Colchester CO4 3SQ, UK
E-mail: hensm@essex.ac.uk

Ken Robinson
The University of New South Wales, UNSW
CAESER, The School of Computer Science and Engineering
Sydney NSW 2052, Australia
E-mail: k.robinson@unsw.edu.au

Cataloging-in-Publication Data applied for

Die Deutsche Bibliothek - CIP-Einheitsaufnahme

ZB 2002: formal specification and development in Z and B : proceedings /
2nd International Conference of B and Z Users, Grenoble, France, January
23 - 25, 2002. Didier Bert ... (ed.). - Berlin ; Heidelberg ; New York ;
Barcelona ; Hong Kong ; London ; Milan ; Paris ; Tokyo : Springer, 2002
   (Lecture notes in computer science ; Vol. 2272)
   ISBN 3-540-43166-7

CR Subject Classification (1998): D.2.1, D.2.2, D.2.4, F.3.1, F.4.2, F.4.3

ISSN 0302-9743
ISBN 3-540-43166-7 Springer-Verlag Berlin Heidelberg New York

Springer-Verlag Berlin Heidelberg New York
a member of BertelsmannSpringer Science+Business Media GmbH

http://www.springer.de

© Springer-Verlag Berlin Heidelberg 2002
Printed in Germany

Typesetting: Camera-ready by author, data conversion by PTP-Berlin, Stefan Sossna
Printed on acid-free paper   SPIN 10846165   06/3142   5 4 3 2 1 0

Preface

These proceedings record the papers presented at the second International Conference of B and Z Users (ZB 2002), held on 23–25 January 2002 in the city of Grenoble in the heart of the French Alps. This conference built on the success of the first conference in this series, ZB 2000, held at the University of York in the UK. The location of ZB 2002 in Grenoble reflects the important work in the area of formal methods carried out at the *Laboratoire Logiciels Systèmes Réseaux* within the *Institut d'Informatique et Mathématiques Appliquées de Grenoble* (LSR-IMAG), especially involving the B method.

B and Z are two important formal methods that share a common conceptual origin; each are leading approaches applied in industry and academia for the specification and development (using formal refinement) of computer-based systems. At ZB 2002 the B and Z communities were brought together to hold a second joint conference that simultaneously incorporated the 13th International Z User Meeting and the 4th International Conference on the B method. Although organized logistically as an integral event, editorial control of the joint conference remained vested in two separate but cooperating program committees that respectively determined its B and Z content, but in a coordinated manner.

All the submitted papers in these proceedings were peer reviewed by at least three reviewers drawn from the B or Z committee depending on the subject matter of the paper. Reviewing and initial selection were undertaken electronically. The Z committee met at South Bank University in London on 27th September 2001 to determine the final selection of Z papers. The B committee met on the morning of 28th September 2001 at the Conservatoire National des Arts et Métiers (CNAM) in Paris to select B papers. A joint committee meeting was held at the same location in the afternoon to resolve the final paper selection and to draft a program for the conference. Sergiy Vilkomir of the Centre for Applied Formal Methods (CAFM) at South Bank University aided in the local organization of the Z meeting. Véronique Viguié Donzeau-Gouge helped in the organization of the meetings at CNAM.

The conference featured a range of contributions by distinguished invited speakers drawn from both industry and academia. The invited speakers addressed significant recent industrial applications of formal methods, as well as important academic advances serving to enhance their potency and widen their applicability. Our invited speakers for ZB 2002 were drawn from Finland, France, and Canada. Ralph-Johan Back, Professor of Computer Science at Åbo Akademi University and Director of the Turku Centre for Computer Science (TUCS) has made important contributions in the development of the refinement calculus, influential and relevant to many formal methods, including B and Z. Pierre Chartier of RATP (Régie Autonome des Transports Parisiens), central in rail transport for Paris, is a leading expert in the industrial application of the B method. Eric C.R. Hehner, Professor of Computer Science at the University of Toronto, has always presented his novel ideas for formal methods using an elegant simplicity.

Besides its formal sessions, the conference included tool demonstrations, exhibitions, and tutorials. In particular, a workshop on *Refinement of Critical Systems: Methods, Tools, and Experience* (RCS 2002) was organized on 22 January 2001 with the support of the EU IST-RTD Project *MATISSE: Methodologies and Associated Technologies for Industrial Strength Systems Engineering*, in association with the ZB 2002 meeting. Other conference sessions included a presentation on the status of the international Z Standard, in its final stages of acceptance. In addition, the International B Conference Steering Committee (APCB) and the Z User Group (ZUG) used the conference as a convenient venue for open meetings intended for those interested in the B and Z communities respectively.

The topics of interest to the conference included: Industrial applications and case studies using Z or using B; Integration of model-based specification methods in the software development lifecycle; Derivation of hardware-software architecture from model-based specifications; Expressing and validating requirements through formal models; Theoretical issues in formal development (e.g., issues in refinement, proof process, or proof validation, etc.); Software testing versus proof-oriented development; Tools supporting tools for the Z notation and the B method; Development by composition of specifications; Validation of assembly of COTS by model-based specification methods; Z and B extensions and/or standardization.

The ZB 2002 conference was jointly initiated by the Z User Group (ZUG) and the International B Conference Steering Committee (APCB). LSR-IMAG provided all local organization and financial backing for the conference. Without the great support from many local staff at LSR-IMAG and others in Grenoble, ZB 2002 would not have been possible. In particular, we would like to thank the Local Committee Chair, Marie-Laure Potet. ZB 2002 was supported by CNRS (Centre National de la Recherche Scientifique), INPG (Institut National Polytechnique de Grenoble), Université Joseph Fourier (Grenoble), and IMAG. ClearSy System Engineering, Gemplus, the Institut National de Recherche sur les Transports et leur Securité (INRETS), and RATP provided sponsorship. We are grateful to all those who contributed to the success of the conference.

On-line information concerning the conference is available under the following Uniform Resource Locator (URL):

http://www-lsr.imag.fr/zb2002/

This also provides links to further on-line resources concerning the B method and Z notation.

We hope that all participants and other interested readers benefit scientifically from these proceedings and also find them stimulating in the process.

November 2001   Didier Bert
                                                                             Jonathan Bowen
                                                                              Martin Henson
                                                                             Ken Robinson

Program and Organizing Committees

The following people were members of the ZB 2002 Z program committee:

Conference Chair: Jonathan Bowen, South Bank University, London, UK
Program Chair: Martin Henson, University of Essex, UK

Ali Abdallah, South Bank University, London, UK
Rob Arthan, Lemma 1, Reading, UK
Paolo Ciancarini, University of Bologna, Italy
Neville Dean, Anglia Polytechnic University, UK
John Derrick, The University of Kent at Canterbury, UK
Mark d'Inverno, University of Westminster, UK
Wolfgang Grieskamp, Microsoft Research, USA
Henri Habrias, University of Nantes, France
Jonathan Hammond, Praxis Critical Systems, UK
Ian Hayes, University of Queensland, Australia
Jonathan Jacky, University of Washington, USA
Randolph Johnson, National Security Agency, USA
Steve King, University of York, UK
Kevin Lano, Kings College London, UK
Yves Ledru, LSR-IMAG, Grenoble, France
Jean-Francois Monin, France Telecom R&D, France
Fiona Polack, University of York, UK
Norah Power, University of Limerick, Ireland
Steve Reeves, University of Waikato, New Zealand
Mark Saaltink, ORA, Ottawa, Canada
Thomas Santen, Technical University of Berlin, Germany
Graeme Smith, University of Queensland, Australia
Susan Stepney, Logica Cambridge, UK
Sam Valentine, LiveDevices, York, UK
John Wordsworth, The University of Reading, UK

The following served on the ZB 2002 B program committee:

Program Chair: Didier Bert, CNRS, LSR-IMAG, Grenoble, France
Co-chair: Ken Robinson, The University of New South Wales, Australia

Christian Attiogbé, University of Nantes, France
Richard Banach, University of Manchester, UK
Juan Bicarregui, CLRC, Oxfordshire, UK
Pierre Bieber, CERT, Toulouse, France
Egon Börger, University of Pisa, Italy
Michael Butler, University of Southampton, UK
Dominique Cansell, LORIA, University of Metz, France
Pierre Chartier, RATP, Paris, France
Steve Dunne, University of Teesside, UK
Mark Frappier, University of Sherbrooke, Canada
Andy Galloway, University of York, UK
Jacques Julliand, University of Besançon, France
Jean-Louis Lanet, GemPlus Research Lab, France
Brian Matthews, CLRC, Oxfordshire, UK
Luis-Fernando Mejia, Alstom Transport Signalisation, France
Jean-Marc Meynadier, Matra Transport, France
Louis Mussat, DCSSI, France
Marie-Laure Potet, LSR-IMAG, Grenoble, France
Emil Sekerinski, McMaster University, Canada
Bill Stoddart, University of Teesside, UK
Helen Treharne, Royal Holloway, UK
Véronique Viguié Donzeau-Gouge, CNAM, Paris, France
Marina Walden, Åbo Akademi, Finland

The following people helped with the organization of the conference in various capacities:

B submissions:	Ken Robinson, The University of New South Wales Didier Bert, LSR-IMAG, Grenoble
Z submissions:	Martin Henson, University of Essex Sonia Oakden, University of Essex
Invited speakers:	Ken Robinson, The University of New South Wales
Tool demonstrations:	Mark d'Inverno, University of Westminster Yves Ledru, LSR-IMAG, Grenoble
Tutorials:	Henri Habrias, University of Nantes
Proceedings:	Didier Bert, LSR-IMAG, Grenoble
Local committee:	Marie-Laure Potet (chair), LSR-IMAG, Grenoble Pierre Berlioux, Jean-Claude Reynaud

We are especially grateful to the above for their efforts in ensuring the success of the conference.

External Referees

We are grateful to the following people who aided the program committees in the reviewing of papers, providing additional specialist expertise:

Yamine Ait Ameur, ENSAE/Aérospatiale and ONERA-CERT Toulouse, France
Françoise Bellegarde, Université de Franche-Comté, France
Eerke Boiten, The University of Kent at Canterbury, UK
Lilian Burdy, Laboratoire CEDRIC, CNAM, France
Alessandra Cavarra, Oxford University Computing Laboratory, UK
Fabien Combret, GemPlus, France
Axel Dold, University of Ulm, Germany
Benoit Fraikin, University of Sherbrooke, Canada
Lindsay Groves, Victoria University, New Zealand
Paul Howells, University of Westminster, UK
Olga Kouchnarenko, Université de Franche-Comté, France
Leonid Mikhailov, University of Southampton, UK
Pascal Poizat, Université d'Évry, France
Mike Poppleton, Open University, UK
Antoine Requet, GemPlus, France
Hector Ruiz Barradas, Universidad Autónoma Metropolitana de México
Marianne Simonot, Laboratoire CEDRIC, CNAM, France
Carsten Sühl, GMD, Berlin, Germany
Bruno Tatibouet, Université de Franche-Comté, France
Ray Turner, University of Essex, UK
Mark Utting, University of Waikato, New Zealand
Norbert Völker, University of Essex, UK
Jim Woodcock, The University of Kent at Canterbury, UK

Support

ZB 2002 greatly benefited from the support of the following organizations:

CNRS
IMAG
INP Grenoble
Université Joseph Fourier, Grenoble
Ministère français des Affaires Etrangères

and sponsorship from:

ClearSy System Engineering
GemPlus
INRETS
RATP

Table of Contents

Theories, Implementations, and Transformations

Eric Hehner and Ioannis T. Kassios

Department of Computer Science, University of Toronto,
Toronto ON M5S 3G4 Canada
{hehner, ykass}@cs.utoronto.ca

Abstract. The purpose of this paper is to try to put theory presentation and structuring in the simplest possible logical setting in order to improve our understanding of it. We look at how theories can be combined, and compared for strength. We look at theory refinement and implementation, and what constitutes proof of correctness. Our examples come from both the functional style and imperative (state-changing) style of theory. Finally, we explore how one implementation can be transformed to another.

1 Introduction

A classic paper by Burstall and Goguen in 1977 [2] taught us to think about data types used in computer programs as logical theories, presented by axioms, whose properties can be explored by logical deduction. The following year, a paper by Guttag and Horning [4] developed the idea further, showing us the algebraic properties of data types presented as theories. Another important contribution came from Abrial [8] in the design of Z, and more recently B [1]. He brought to theory design all the structuring and scoping that programming languages provide, enabling us to build large theories by composing smaller ones. With the work of the Z and B community, and a change of terminology, theory design became an important part of software development.

The purpose of this paper is to try to put theory presentation and structuring in the simplest possible logical setting in order to improve our understanding of it. It is not the purpose of this paper to provide a notation or language for practical engineering use; for that task the Z and B community are the leaders.

2 Notation

Notation is not the point of this paper; as much as possible, we will use standard, or at least familiar, notations. The two booleans are $\top$ and $\bot$, and the boolean operators are $\neg \wedge \vee = \neq \Rightarrow \Leftarrow$. The same equality $=$ and unequality $\neq$ will be used with any type. we also use a large version $= \Rightarrow \Leftarrow$ of equality and implication that are identical to the small version except for their precedence; the only purpose is to save

D. Bert et al. (Eds.): ZB 2002, LNCS 2272, pp. 1-21, 2002.

a clutter of parentheses. The empty bunch is *null* . The comma (,) is bunch union, which is commutative, idempotent, and associative. The colon (:) is bunch inclusion. For example,

$$2, 9 : 0, 2, 5, 9$$

is a boolean expression with value T because the left operand of colon is included in the right operand. We use the asymmetric notation $x,..y$ for the bunch of integers from and including x up to but excluding y . The empty list is $[nil]$, and the list [2; 6; 4; 8] contains four items. The notation $[x;..y]$ is used for the list of integers from and including x up to but excluding y . Lists are indexed from 0 . List formation distributes over bunch union, so if *nat* is the natural numbers, then $[nat]$ is the list whose one item is the bunch of natural numbers, or equally, the bunch of all lists whose one item is a natural number. A star denotes repetition of an item, so $[*nat]$ is all lists of natural numbers. We use # for list length. We use a standard lambda notation $\lambda x: D \cdot fx$ for functions, and juxtaposition for function application. We use $A{\rightarrow}B$ for the bunch of all functions with domain at least A and range at most B . Quantifiers $\forall \exists$ apply to functions, but for the sake of familiarity they replace the lambda.

Here are all the notations of the paper in a precedence table.

0.	T ⊥ ( ) [ ] numbers names	(true, false, precedence, list brackets)
1.	juxtaposition	(function application) right-to-left
2.	# * →	(list length, item repetition, function space) right-to-left
3.	+ − +	(addition, subtraction, catenation) left-to-right
4.	; ;..	(sequencing of list items) associative
5.	, ,.. |	(bunch union, function selection) associative
6.	= ≠ < > ≤ ≥ :	(equality, unequality, order, inclusion) continuing
7.	¬	(negation) right-to-left
8.	∧	(conjunction) associative
9.	∨	(disjunction) associative
10.	⇒ ⇐	(implication) continuing
11.	:=	(assignment)
12.	**if then else**	(if then else)
13.	;	(sequential composition) associative
14.	λ· ∀· ∃·	(function, quantifiers)
15.	**= ⇒ ⇐**	(equality, implication) continuing

To say that = is continuing is to say that $a = b = c$ neither associates to the left nor associates to the right, but means $a = b \wedge b = c$. A mixture of continuing operators can be used; for example, $a \leq b < c$ means $a \leq b \wedge b < c$. For further details on notation and basic theories please consult [5] or [6].

3 Theories

Here is a little theory presented in a style similar to [2] and [4].

Theory0:   names: *chain, start, link, isStart*
            signatures: *start: chain*
            link: chain→chain
            isStart: chain→bool
            axioms: *isStart start*
            ∀*c: chain·* ¬ *isStart* (*link c*)

Theory0 introduces four new names into our vocabulary. The signatures section tells us something about the role these names will play in the theory. Then the axioms tell us what can be proven, what are the theorems, in this theory.

The first problem with this presentation of Theory0 is that names cannot be attached to theories. For example, this theory uses the name *bool* , and many others do too, and each of them is telling us something about *bool*. And when we build large theories by composing smaller ones, no particular theory in the composition can claim a name as its own. And it isn't just names that get introduced by theories; symbols like ≤ , or in our example ∀ and ¬ , and even = , are used in many theories, and each of them is telling us something more about the use of those symbols. Names and symbols are defined by their use in all theories where they appear; and we can always add more theories to the collection. As part of a library of theories, we need a linked, browsable dictionary of names and symbols, telling us which theories use them. This dictionary should be generated automatically from the library of theories, so that it is always up-to-date. The first change to theory presentation is to remove the list of names.

The next change to theory presentation is to consider a signature to be a kind of boolean expression. One of the uses of Bunch Theory is as a fine-grained type theory. The boolean expression

$$5:\ 0, 3, 5, 8$$

has value T and says, " 5 is included among 0, 3, 5, 8 ". But we can also read it as " 5 has type 0, 3, 5, 8 ". Defining *nat* as the bunch of all natural numbers, the boolean expression 5: *nat* has value T . And so *x: nat* can be given as an axiom about *x* . So too *x, y: nat* can be an axiom, just as 3, 5: 0, 3, 5, 8 has value T . The expression *A→B* consists of all functions with domain at least *A* and range at most *B* . For example,

$$(\lambda n:\ nat\cdot\ n+1):\ nat\rightarrow nat$$

has value T . And so *f: nat→nat* can be an axiom about *f* . By "currying", *A→B→C* consists of two-variable functions, and so on.

The final change to theory presentation is just to write all the axioms as one big axiom by taking their conjunction. Now a theory consists of one single axiom, so there is now no difference between a theory and an axiom. *Theory0* can be written as follows.

$$
\begin{array}{lll}
\textit{Theory0} & = & \textit{start: chain} \\
& \wedge & \textit{link: chain}{\rightarrow}\textit{chain} \\
& \wedge & \textit{isStart: chain}{\rightarrow}\textit{bool} \\
& \wedge & \textit{isStart start} \\
& \wedge & \forall c\text{: chain} \cdot \neg\, \textit{isStart (link c)}
\end{array}
$$

4 Composition

The original paper by Burstall and Goguen [2] presents four operations on theories: combination, enrichment, induction, and derivation. To illustrate theory combination, here is a second theory.

$$
\begin{array}{lll}
\textit{Theory1} & = & \textit{start: chain} \\
& \wedge & \textit{link: chain}{\rightarrow}\textit{chain} \\
& \wedge & \forall c\text{: chain} \cdot \textit{start} \neq \textit{link c} \\
& \wedge & \forall c, d\text{: chain} \cdot (c{=}d) = (\textit{link c} = \textit{link d})
\end{array}
$$

Theory0 and *Theory1* have much in common, but also some differences; there are theorems in each that are not theorems in the other. With our form of theory presentation, we can combine the two theories with ordinary boolean conjunction.

$$
\textit{Theory2} \;=\; \textit{Theory0} \wedge \textit{Theory1}
$$

Burstall and Goguen's next theory operation, enrichment, is also just conjunction, but with further axioms rather than with a named theory. Here is an example.

$$
\begin{array}{lll}
\textit{Theory3} & = & \textit{Theory2} \\
& \wedge & \forall c\text{: chain} \cdot \textit{start} \leq c < \textit{link c}
\end{array}
$$

The next of Burstall and Goguen's theory operations adds a structural induction scheme over the generators of the new data type. For us, it is again just conjunction of another axiom.

$$
\begin{array}{lll}
\textit{Theory4} & = & \textit{Theory3} \\
& \wedge & \forall P\text{: (chain}{\rightarrow}\textit{bool)} \cdot \\
& & \quad P\ \textit{start} \wedge (\forall c\text{: chain} \cdot P\ c \Rightarrow P\ (\textit{link c})) \\
& \Rightarrow & \quad \forall c\text{: chain} \cdot P\ c
\end{array}
$$

That is the familiar form of induction; a neater, equivalent form is as follows.

$$Theory4 \quad = \quad Theory3$$
$$\wedge \quad \forall C \cdot start, link\ C\colon C \;\Rightarrow\; chain\colon C$$

To briefly explain this axiom, most operators and functions distribute over bunch union. For example,

$$(2, 5, 9) + 1 \;=\; (3, 6, 10)$$

So *link C* consists of all the results of applying *link* to things in *C* . The axiom says that if *start* and all the links of things in *C* are included in *C* , then *chain* is included in *C* . The antecedent can be rewritten as

$$start\colon C \quad \wedge \quad link\colon C{\rightarrow}C$$

and, regarding *C* as the unknown, *chain* is one solution. The axiom therefore says that *chain* is the smallest solution.

Burstall and Goguen's final operation on theories, derivation, allows part of a theory to be hidden from the theory users. For us, that's existential quantification.

$$Theory5 \quad = \quad \exists start\colon chain \cdot Theory4$$

Theory5 has all the same theorems as *Theory4* minus those that mention *start* . If we want to keep all the theorems of *Theory4* but rename *start* as *new* , define

$$Theory6 \quad = \quad \exists start\colon chain \cdot start{=}new \wedge Theory4$$

We can combine theories with other boolean operators too, such as disjunction and implication. For example,

$$Theory7 \quad = \quad (\forall c\colon chain \cdot new \le c) \Rightarrow Theory6$$

This makes *Theory7* such that if we had the axiom $\forall c\colon chain \cdot new \le c$ then we would have *Theory6* . In a vague sense, *Theory7* is *Theory6* without $\forall c\colon chain \cdot new \le c$. To be precise, if we take *Theory7* and add the axiom $\forall c\colon chain \cdot new \le c$, we get back *Theory6* .

$$Theory6 \quad = \quad Theory7 \wedge \forall c\colon chain \cdot new \le c$$

New theories are not always built by additions to old theories; sometimes they are built by deletions. One of the problems with object-orientation is that, although subclassing allows us to add attributes, there is no way to delete attributes and make a superclass, nor to make an interclass between two existing classes.

These examples illustrate that our theory presentation is both a simplification and a generalization of the early work. By reducing theories to boolean expressions we understand them in the simplest possible way, and we allow all combinations that make logical sense.

5 Refinement and Implementation

A theory can serve as a specification of a data type, and of computation in general. Specifications can be refined, usually by resolving nondeterminism. Specification A refines specification B if all computer behavior satisfying A also satisfies B . If theories are expressed as single boolean expressions,

> theory A refines theory B means $A \Rightarrow B$
> theory B is refined by theory A means $B \Leftarrow A$

Refinement is just implication. So far, we have

$$Theory6 \Rightarrow Theory7$$
$$Theory4 \Rightarrow Theory5$$
$$Theory4 \Rightarrow Theory3$$
$$Theory3 \Rightarrow Theory2$$
$$Theory2 \Rightarrow Theory1$$
$$Theory2 \Rightarrow Theory0$$

When we define a theory, and especially when we combine theories, there is always the danger of inconsistency. The only way to prove the consistency of a theory is to implement it. As software engineers, our goal is to design useful theories (they must be consistent to be useful), and to implement them.   A theory is said to be implemented when all names and symbols appearing in it have been implemented. A name or symbol is implemented by defining it in terms of other names and symbols that are implemented.   Ultimately, the computing machinery provides the ground theory on top of which all other theories are implemented.   (To logicians, an implementation is known as a "model", and the ultimate machinery is usually taken to be set theory, although they might claim that the model is the sets themselves and not set theory.)

An implementation can be expressed in exactly the same form as a theory: a boolean expression. Here is an example implementation of *Theory4* , assuming that *nat* is an implemented data type, and that functions are implemented.

$$
\begin{aligned}
Imp \quad = \quad & chain = nat \\
\wedge \quad & start = 0 \\
\wedge \quad & isStart = (\lambda c\text{: } nat\cdot\ c{=}0) \\
\wedge \quad & link = (\lambda c\text{: } nat\cdot\ c{+}1)
\end{aligned}
$$

An implementation is also a theory, but of a particular form. It is a conjunction of equations, and each equation has a left side consisting of one of the names needing an implementation, and a right side employing only names and symbols that are already implemented.

The benefit in expressing an implementation in the same form as a theory is that the proof of correctness of the implementation is now just a boolean implication. We prove that *Imp* correctly implements *Theory4* by proving

$$Imp \implies Theory4$$

Implementation is just refinement by an implemented theory. By the transitivity of implication we have immediately that *Imp* also implements *Theory5* , *Theory3* , *Theory2* , *Theory1* , and *Theory0* .

6 Functional Stack

From a typical mathematician's viewpoint, a stronger theory is a better theory because it allows us to prove more. But the theory must not be so strong as to be inconsistent, for then we can prove everything trivially. The game is to add axioms, approaching the brink of inconsistency as closely as possible without falling over. For example, here a strong but consistent theory of stacks.

$$
\begin{aligned}
Stack0 \;=\; \lambda X \cdot \quad & empity: stack \\
\wedge \quad & push: stack {\to} X {\to} stack \\
\wedge \quad & pop: stack {\to} stack \\
\wedge \quad & top: stack {\to} X \\
\wedge \quad & (\forall S \cdot empty,\, push\, S\, X: S \;\implies\; stack: S) \\
\wedge \quad & (\forall s: stack \cdot \forall x: X \cdot push\, s\, x \neq empty) \\
\wedge \quad & (\forall s, t: stack \cdot \forall x, y: X \cdot \\
& push\, s\, x = push\, t\, y \;=\; s{=}t \wedge x{=}y) \\
\wedge \quad & (\forall s: stack \cdot \forall x: X \cdot pop\,(push\, s\, x) = s) \\
\wedge \quad & (\forall s: stack \cdot \forall x: X \cdot top\,(push\, s\, x) = x)
\end{aligned}
$$

And here is an implementation, assuming lists, functions, and integers are already implemented.

$$
\begin{aligned}
Stack \;=\; \quad & stack \;=\; [*int] \\
\wedge \quad & empty \;=\; [nil] \\
\wedge \quad & push \;=\; (\lambda s: stack \cdot \lambda x: int \cdot s^{+}[x]) \\
\wedge \quad & pop \;=\; (\lambda s: stack \cdot \mathbf{if}\ s{=}empty\ \mathbf{then}\ empty\ \mathbf{else}\ s\,[0;..\#s{-}1]) \\
\wedge \quad & top \;=\; (\lambda s: stack \cdot \mathbf{if}\ s{=}empty\ \mathbf{then}\ 0\ \mathbf{else}\ s\,(\#s{-}1))
\end{aligned}
$$

where [*int] is all lists of integers, [nil] is the empty list, + is catenation, # is length, and s [0;..#s−1] is list s up to but not including its last item. To prove that Stack1 is an implementation of Stack0 we must prove

$$Stack1 \Rightarrow Stack0 \; int$$

but we won't spend the space here.

The only way to prove the consistency of a theory is to implement it. The only way to prove the incompleteness of a theory is to implement it twice such that some boolean expression is a theorem of one implementation, and its negation is a theorem of the other. In our example,

pop empty = empty
top empty = 0

are theorems of Stack1 . But here is another implementaion of Stack0 int :

Stack2 =	stack	=	[*int]
∧	empty	=	[nil]
∧	push	=	(λs: stack· λx: int· s+[x])
∧	pop	=	(λs: stack· if s=empty then push empty 0
			else s [0;..#s−1])
∧	top	=	(λs: stack· if s=empty then 1 else s (#s−1))

in which their negations are theorems. So Stack0 int is incomplete. That means we can find a stronger theory of stacks by saying what pop empty and top empty are. But do we want a stronger theory? What is the purpose of this theory?

In Stack0 , we have empty: stack and pop: stack→stack ; from them we can prove pop empty: stack . In other words, popping the empty stack gives a stack, though we do not know which one. An implementer is obliged to give a stack for pop empty , though it does not matter which one. If we never want to pop an empty stack, then the theory is too strong. We should weaken the conjunct pop: stack→stack and remove the implementer's obligation to provide something that is not wanted. The weaker conjunct

$$\forall s: stack· \; s{\neq}empty \; \Rightarrow \; pop \; s: stack$$

says that popping a nonempty stack yields a stack, but it is implied by the remaining conjuncts and is unnecessary. Similarly from empty: stack and top: stack→X we can prove top empty: X ; deleting top: stack→X removes an implementer's obligation to provide an unwanted result for top empty .

We may decide that we have no need to prove anything about all stacks, and can do without induction ∀S· empty, push S X: S ⇒ stack: S . After a little thought, we may realize that we never need an empty stack, nor to test if a stack is empty. We can always work on top of a given (possibly non-empty) stack, and in most uses we are required to do so, leaving the stack as we found it. We can delete *empty*: *stack* and all mention of *empty* . We must replace it with the weaker *stack* ≠ *null* so that we can still declare variables of type *stack* . If we do want to test whether a stack is empty, we can begin by pushing some special value, one that will not be pushed again, onto the stack; the empty test is then a test whether the top is the special value.

For most purposes, it is sufficient to be able to push items onto a stack, pop items off, and look at the top item. The theory we need is considerably simpler than the one presented previously.

$$
\begin{aligned}
Stack3 \quad = \quad \lambda X \cdot \qquad & stack \neq null \\
\wedge \quad & (\forall s: stack \cdot \ \forall x: X \cdot push\ s\ x: stack) \\
\wedge \quad & (\forall s: stack \cdot \ \forall x: X \cdot pop\ (push\ s\ x) = s) \\
\wedge \quad & (\forall s: stack \cdot \ \forall x: X \cdot top\ (push\ s\ x) = x)
\end{aligned}
$$

For the purpose of studying stacks, as a mathematical activity, we want a strong theory so that we can prove as much as possible. As an engineering activity, theory design is the art of excluding all unwanted implementations while allowing all the others. It is counter-productive to design a stronger theory than necessary;   it makes implementation harder, and it makes theory extension harder.

7 Imperative Stack

It is an accident of history that the usual stack specification is functional in style, while the usual stack implementation is imperative. Functions were familiar mathematics, suitable for formal specification, at a time when imperative programs were still understood only as commands for the operation of a computer. We now have a mathematical understanding of imperative, state-changing programs. We can equally well have specifications that are both mathematical and imperative.

In the simplest version of imperative stack theory, *push* is a procedure with parameter of type X , *pop* is a program, and *top* is an expression of type X . In this theory, *push* 3 is a program (assuming 3: X); it changes the state. Following this program, before any other pushes and pops, *print top* will print 3 . Here is the theory.

$$
\begin{aligned}
Stack4 \quad = \quad \forall x: X \cdot \qquad & (top' = x \ \Leftarrow \ push\ x) \\
\wedge \quad & (ok \ \Leftarrow \ push\ x;\ pop)
\end{aligned}
$$

The first conjunct says that following a push, the top tem is the item pushed. In the second conjunct, *ok* (sometimes called *skip*) is a program (which is a specification, which is a boolean expression) that says that all final values of variables equal the corresponding initial values (the identity relation on states). So the second conjunct says that a pop undoes a push. In fact, it says that any natural number of pushes are undone by the same number of pops.

	ok		use *ok* ⇐ *push x*; *pop*
⇐	*push x*; *pop*		*ok* is identity for sequential composition
=	*push x*; *ok*; *pop*	Reuse *ok* ⇐ *push x*; *pop* and ; is monotonic
⇐	*push x*; *push y*; *pop*; *pop*		

We can prove things like

$$top'=x ⇐ push\ x;\ push\ y;\ push\ z;\ pop;\ pop$$

which say that when we push something onto the stack, we find it there later at the appropriate time. That is all we really want from a stack.

If we need only one stack, we obtain an economy of expression and of execution by leaving it implicit, as in *Stack4* . There is no need to say which stack to push onto if there is only one. (If we need more than one stack, we can add an extra parameter to each operation.)

In imperative theories, the state is divided into two kinds of variables: the user's variables and the implementer's variables. A user of the theory enjoys full access to the user's variables, but cannot directly access (see or change) the implementer's variables. A user gets access to the implementer's variables only through the theory. On the other side, an implementer of the theory enjoys full access to the implementer's variables, but cannot directly access (see or change) the user's variables. An implementer gets access to the user's variables only through the theory.

To implement *Theory4* , we introduce an implementer's variable s: [*X] and now we define

Stack5	=		$(push = λx: X·\ s:= s+[x])$
		∧	$(pop = s:= s\ [0;..#s–1])$
		∧	$(top = s\ (#s–1))$

The proof that *Stack5* implements *Stack4* , as always, is just an implication.

$$Stack5 ⇒ Stack4$$

By implementing *Stack4* we prove that it is consistent. But it is incomplete. Incompleteness is a freedom for the implementer, who can trade economy against robustness. If we care how this trade will be made, we should strengthen the theory. For example, we could add

Stack6	=		*Stack4*
		∧	$(print\ "error" ⇐ mkempty;\ pop)$

A slightly fancier imperative stack theory tells us about *mkempty* (a program to make the stack empty) and *isempty* (a boolean to say whether the stack is empty). Letting $x: X$, the theory is

$$
\begin{array}{lll}
\textit{Stack7} & = & \textit{Stack4} \\
& \wedge & (\forall x: X \cdot \neg \textit{isempty}' \Leftarrow \textit{push } x) \\
& \wedge & (\textit{isempty}' \Leftarrow \textit{mkempty})
\end{array}
$$

The imperative stack theory we presented first, *Stack4* , can be weakened and still retain its stack character. We must keep

$$
\textit{top}'=x \ \Leftarrow \ \textit{push } x
$$

but we do not need the composition *push x; pop* to leave all variables unchanged. We do require that any natural number of pushes followed by the same number of pops gives back the original top. The theory is

$$
\begin{array}{lll}
\textit{Stack8} = \exists \textit{balance} \cdot & (\textit{top}'=x \ \Leftarrow \ \textit{push } x) \\
& \wedge & (\textit{top}'=\textit{top} \ \Leftarrow \ \textit{balance}) \\
& \wedge & (\textit{balance} \ = \ \textit{ok} \vee \exists x \cdot (\textit{push } x; \ \textit{balance}; \ \textit{pop}))
\end{array}
$$

This weaker theory allows an implementation in which popping does not restore the implementer's variable *s* to its pre-pushed value, but instead marks the last item as "garbage".

A weak theory can be extended in ways that are excluded by a strong theory. For example, we can add the names *count* (of type *nat*) and *start* (a program), as follows:

$$
\begin{array}{lll}
\textit{Stack9} & = & \textit{Stack8} \\
& \wedge & (\textit{count}'=0 \ \Leftarrow \ \textit{start}) \\
& \wedge & (\forall x: X \cdot \textit{count}'=\textit{count}+1 \ \Leftarrow \ \textit{push } x) \\
& \wedge & (\textit{count}'=\textit{count}+1 \ \Leftarrow \ \textit{pop})
\end{array}
$$

so that *count* counts the number of pushes and pops. From a software engineering point of view, the weakest theory is best.

8 Functional Tree

Here is a strong theory that is good for mathematicians who want to study trees.

$$
\begin{array}{lll}
\textit{Tree0} & = \lambda X \cdot & \textit{emptree}: \textit{tree} \\
& \wedge & \textit{graft}: \textit{tree} \rightarrow X \rightarrow \textit{tree} \rightarrow \textit{tree} \\
& \wedge & (\forall T \cdot \textit{emptree}, \textit{graft } T \ X \ T: T \ \Rightarrow \ \textit{tree}: T)
\end{array}
$$

$\land$ $(\forall t, u: tree \cdot \forall x: X \cdot graft\ t\ x\ u \neq emptree)$

$\land$ $(\forall t, u, v, w: tree \cdot \forall x, y: X \cdot$
$\qquad graft\ t\ x\ u = graft\ v\ y\ w\ =\ t{=}v \land x{=}y \land u{=}w)$

$\land$ $(\forall t, u: tree \cdot \forall x: X \cdot left\ (graft\ t\ x\ u) = t)$

$\land$ $(\forall t, u: tree \cdot \forall x: X \cdot root\ (graft\ t\ x\ u) = x)$

$\land$ $(\forall t, u: tree \cdot \forall x: X \cdot right\ (graft\ t\ x\ u) = u)$

For programming purposes, a simpler, weaker theory is sufficient. As with stacks, we don't really need to be given an empty tree. As long as we are given some tree, we can build a tree with a distinguished root that serves the same purpose. And we probably don't need tree induction.

$Tree1\ =\ \lambda X \cdot$ $\qquad\qquad tree \neq null$

$\land$ $(\forall t, u: tree \cdot \forall x: X \cdot graft\ t\ x\ u: tree)$

$\land$ $(\forall t, u: tree \cdot \forall x: X \cdot left\ (graft\ t\ x\ u) = t)$

$\land$ $(\forall t, u: tree \cdot \forall x: X \cdot root\ (graft\ t\ x\ u) = x)$

$\land$ $(\forall t, u: tree \cdot \forall x: X \cdot right\ (graft\ t\ x\ u) = u)$

If lists and recursive data definition are implemented, then we can implement a tree of integers by the following theory.

$Tree2\ =$ $\qquad\qquad\qquad tree\ =\ emptree,\ graft\ tree\ int\ tree$

$\land$ $emptree\ =\ [nil]$

$\land$ $(graft\ =\ \lambda t: tree \cdot \lambda x: int \cdot \lambda u: tree \cdot [t;\ x;\ u])$

$\land$ $(left\ =\ \lambda t: tree \cdot t\ 0)$

$\land$ $(right\ =\ \lambda t: tree \cdot t\ 2)$

$\land$ $(root\ =\ \lambda t: tree \cdot t\ 1)$

Here is another implementation.

$Tree3\ =$ $\qquad\qquad\qquad tree\ =\ emptree,\ graft\ tree\ int\ tree$

$\land$ $emptree\ =\ 0$

$\land$ $(graft\ =\ \lambda t: tree \cdot \lambda x: int \cdot \lambda u: tree \cdot$
$\qquad\qquad$ "left"$\rightarrow t\ |$ "root"$\rightarrow x\ |$ "right"$\rightarrow u)$

$\land$ $(left\ =\ \lambda t: tree \cdot t$ "left")

$\land$ $(right\ =\ \lambda t: tree \cdot t$ "right")

$\land$ $(root\ =\ \lambda t: tree \cdot t$ "root")

According to $Tree2$, the tree

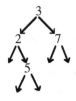

is

$$[[[nil]; 2; [[nil]; 5; [nil]]]; 3; [[nil]; 7; [nil]]]$$

and according to *Tree3* it is

"left" →   ("left" → 0
                    | "root" → 2
                    | "right" →   ("left" → 0
                                            | "root" → 5
                                            | "right" → 0))

        | "root" → 3
        | "right" →   ("left" → 0
                                | "root" → 7
                                | "right" → 0)

Both *Tree2* and *Tree3* implement *Tree0* , and therefore also *Tree1* .

$$Tree2 \vee Tree3 \ \Rightarrow \ Tree0 \ int \ \Rightarrow \ Tree1 \ int$$

9 Imperative Tree

Imagine a tree that is infinite in all directions; there are no leaves and no root. You are looking at one particular item in the tree. The name *node* (of type *X*) tells the value of that item; *change* (a procedure with parameter of type *X*) causes the value of that item to change; *goUp* (a program) changes your view to the parent of that item; *goLeft* changes your view to the left child of that item; *goRight* changes your view to the right child of that item. The theory uses six auxiliary definitions: *L* means "go down left, then do anything except go back up (ending in the same place), then go back up"; *R* means "go down right, then do anything except go back up (ending in the same place), then go back up"; *U* means "go up, then do anything except go back down the way you came (ending in the same place), then go back down the way you came"; *L* means "do anything except for going down left (ending where you start)"; *R* means "do anything except for going down right (ending where you start)"; *U* means "do anything except for going up (ending where you start)".

$$Tree4 \ = \ \exists L, R, U, \underline{L}, \underline{R}, \underline{U}\cdot$$
$$(\forall x: X\cdot node'=x \ \Leftarrow \ change \ x)$$

$$\wedge \quad (node'{=}node \iff L \vee R \vee U)$$
$$\wedge \quad (L \;=\; goLeft;\; \underline{U};\; goUp)$$
$$\wedge \quad (R \;=\; goRight;\; \underline{U};\; goUp)$$
$$\wedge \quad (goLeft;\; U \;=\; \underline{L};\; goLeft)$$
$$\wedge \quad (goRight;\; U \;=\; \underline{R};\; goRight)$$
$$\wedge \quad (\underline{L} \;=\; ok \vee (\exists x\cdot\ change\ x) \vee R \vee U \vee (\underline{L};\ \underline{L}))$$
$$\wedge \quad (\underline{R} \;=\; ok \vee (\exists x\cdot\ change\ x) \vee U \vee L \vee (\underline{R};\ \underline{R}))$$
$$\wedge \quad (\underline{U} \;=\; ok \vee (\exists x\cdot\ change\ x) \vee L \vee R \vee (\underline{U};\ \underline{U}))$$

10 Transformation

A program is a specification of computer behavior. Sometimes (but not always) a program is the clearest kind of specification. Sometimes it is the easiest kind of specification to write. If we write a specification as a program, there is no work to implement it.

Even though a specification may already be a program, we can, if we like, implement it differently. An imperative theory is presented in terms of user's variables and implementer's variables; the former provide the user's interface to the theory; the latter may be for implementation purposes or they may just be for explanatory purposes. Perhaps the implementer's variables were chosen to make the specification as clear as possible, but other implementer's variables might be more storage-efficient, or provide faster access on average. Since a theory user has no access to the implementer's variables except through the theory, an implementer is free to change them in any way that provides the same theory to the user.

Let the user's variables be u, and let the implementer's variables be v (u and v represent any number of variables). Now suppose we want to replace the implementer's variables by new implementer's variables w. We accomplish this transformation by means of a transformer, which is a boolean expression D relating v and w such that $\forall w\cdot \exists v\cdot D$. Let D' be the same as D but with primes on all the variables. Then each specification S in the theory is transformed to

$$\forall v\cdot D \Rightarrow \exists v'\cdot D' \wedge S$$

Specification S is in variables u and v, and the transformed specification is in variables u and w.

Transformation is invisible to the user. The user imagines that the implementer's variables are initially in state v, and then, according to specification S, they are finally in state v'. Actually, the implementer's variables will initially be in state w related to v by D; the user will be able to suppose they are in a state v because $\forall w\cdot \exists v\cdot D$. The implementer's variables will change state from w to w' according to the transformed specification $\forall v\cdot D \Rightarrow \exists v'\cdot D' \wedge S$. This says that whatever related initial state v the user was imagining, there is a related final state v' for the user to imagine as the result of S, and so the fiction is maintained. Here is a picture of it.

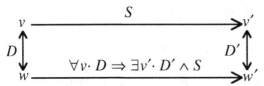

Implementability of S (in its variables v and v') becomes (via the transformer D and D') the new specification (in the new variables w and w'). This transformation is one form of data refinement.

11 Limited Queue

We illustrate theory transformation with the example of an imperative queue of limited size. Here's the theory we start with.

$$Queue0 \;=\;$$
$$\forall x\colon X\cdot \quad (mkemptyq \Rightarrow isemptyq')$$
$$\wedge \quad (isemptyq \wedge \neg isfullq \wedge join\ x \;\Rightarrow\; front'=x \wedge \neg isemptyq')$$
$$\wedge \quad (\neg isemptyq \wedge leave \;\Rightarrow\; \neg isfullq')$$
$$\wedge \quad (\neg isemptyq \wedge \neg isfullq \wedge join\ x \;\Rightarrow\; front'=front \wedge \neg isemptyq')$$
$$\wedge \quad (isemptyq \wedge \neg isfullq \;\Rightarrow\; (join\ x;\ leave \;=\; mkemptyq))$$
$$\wedge \quad (\neg isemptyq \wedge \neg isfullq \;\Rightarrow\; (join\ x;\ leave \;=\; leave;\ join\ x))$$

Let the limit be positive natural n, and let $Q\colon [n*X]$ and $p\colon nat$ be implementer's variables. Then here is a theory to implement $Queue0$.

$$Queue1 \quad = \qquad\qquad (mkemptyq \;=\; p:=0)$$
$$\wedge \qquad (isemptyq \;=\; p=0)$$
$$\wedge \qquad (isfullq \;=\; p=n)$$
$$\wedge \qquad (join \;=\; \lambda x\colon X\cdot Qp:=x;\ p:=p+1)$$
$$\wedge \qquad (leave \;=\; \mathbf{for}\ i:=1;..p\ \mathbf{do}\ Q(i-1):=Qi;\ p:=p-1)$$
$$\wedge \qquad (front \;=\; Q0)$$

A user of $Queue1$ would be well advised to precede any use of $join$ with the test $\neg isfullq$, and any use of $leave$ or $front$ with the test $\neg isemptyq$, but that's not our business at the moment. A new item joins the back of the queue at position p taking constant time to do so. The front item is always found instantly at position 0.

Unfortunately, removing the front item from the queue takes time $p-1$ to shift all remaining items down one index.

We want to transform the queue so that all operations are instant. Variables Q and p will be replaced by R: $[n*X]$ and f, b: $0,..n$ with f indicating the current front of the queue and b its back.

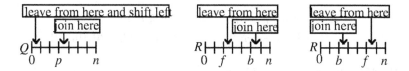

The idea is that b and f move cyclically around the list; when f is to the left of b the queue items are between them; when b is to the left of f the queue items are in the outside portions.

Here is the transformer D.

$$0 \le p = b{-}f < n \;\wedge\; Q[0;..p] = R[f;..b]$$
$$\vee \qquad 0 < p = n{-}f{+}b \le n \;\wedge\; Q[0;..p] = R[f;..n \;;\; 0;..b]$$

One great thing about theory transformation is that once we have stated the transformer, which is the relation between the old and new variables, there is no further invention required; the operations of the theory are transformed for us. First we transform $mkemptyq$.

$$\forall Q, p \cdot D \Rightarrow \exists Q', p' \cdot D' \wedge mkemptyq$$
$=\qquad \forall Q, p \cdot D \Rightarrow \exists Q', p' \cdot D' \wedge p'{=}0 \wedge Q'{=}Q$ \qquad by several omitted steps
$=\qquad f'{=}b'$
$\Leftarrow\qquad f:= 0;\; b:= 0$

The other great thing about theory transformation is that it never transforms incorrectly, even if we have an incorrect transformer! Next we transform $u:=$ $isemptyq$.

$$\forall Q, p \cdot D \Rightarrow \exists Q', p' \cdot D' \wedge (u:= isemptyq)$$
$=\qquad \forall Q, p \cdot D \Rightarrow \exists Q', p' \cdot D' \wedge u'{=}(p{=}0) \wedge p'{=}p \wedge Q'{=}Q$

by several omitted steps

$=\qquad f{<}b \;\wedge\; f'{<}b' \;\wedge\; b{-}f = b'{-}f' \;\wedge\; R[f;..b] = R'[f';..b'] \;\wedge\; \neg u'$
$\vee\qquad f{<}b \;\wedge\; f'{>}b' \;\wedge\; b{-}f = n{+}b'{-}f'$
$\wedge\qquad R[f;..b] = R'[(f';..n); (0;..b')] \;\wedge\; \neg u'$
$\vee\qquad f{>}b \;\wedge\; f'{<}b' \;\wedge\; n{+}b{-}f = b'{-}f'$
$\wedge\qquad R[(f;..n); (0;..b)] = R'[f';..b'] \;\wedge\; \neg u'$
$\vee\qquad f{>}b \;\wedge\; f'{>}b' \;\wedge\; b{-}f = b'{-}f'$
$\wedge\qquad R[(f;..n); (0;..b)] = R'[(f';..n); (0;..b')] \;\wedge\; \neg u'$

Initially R might be in the "inside" or "outside" configuration, and finally R' might be either way, so that gives us four disjuncts. Very suspiciously, we have $\neg u'$ in every case.   That's because $f{=}b$ is missing!   So the transformed operation is unimplementable.   That's the transformer's way of telling us that the new variables do not hold enough information to answer whether the queue is empty.   The problem occurs when $f{=}b$ because that could be either an empty queue or a full queue.   A solution is to add a new variable m: *bool*   to say whether we have the "inside" mode or "outside" mode.   We revise the transformer D as follows:

$$m \;\wedge\; 0 \leq p = b{-}f < n \;\wedge\; Q[0;..p] = R[f;..b]$$
$$\vee \qquad \neg m \;\wedge\; 0 < p = n{-}f{+}b \leq n \;\wedge\; Q[0;..p] = R[(f;..n); (0;..b)]$$

Now we have to retransform   *mkemptyq* .

$$\forall Q, p \cdot D \;\Rightarrow\; \exists Q', p' \cdot D' \wedge mkemptyq$$
$$= \qquad \forall Q, p \cdot D \;\Rightarrow\; \exists Q', p' \cdot D' \wedge p'{=}0 \wedge Q'{=}Q \qquad \text{by several omitted steps}$$
$$= \qquad m' \wedge f'{=}b'$$
$$\Leftarrow \qquad m{:=} \dagger; \; f{:=} 0; \; b{:=} 0$$

$$\forall Q, p \cdot D \;\Rightarrow\; \exists Q', p' \cdot D' \wedge (u{:=} isemptyq)$$
$$= \qquad \forall Q, p \cdot D \;\Rightarrow\; \exists Q', p' \cdot D' \wedge u'{=}(p{=}0) \wedge p'{=}p \wedge Q'{=}Q$$
$$\text{by several omitted steps}$$

$$
\begin{aligned}
= \quad & m \;\wedge\; f{<}b \;\wedge\; m' \;\wedge\; f'{<}b' \;\wedge\; b{-}f = b'{-}f' \\
& \wedge \quad R[f;..b] = R'[f';..b'] \;\wedge\; \neg u' \\
\vee \quad & m \;\wedge\; f{<}b \;\wedge\; \neg m' \;\wedge\; f'{>}b' \;\wedge\; b{-}f = n{+}b'{-}f' \\
& \wedge \quad R[f;..b] = R'[(f';..n); (0;..b')] \;\wedge\; \neg u' \\
\vee \quad & \neg m \;\wedge\; f{>}b \;\wedge\; m' \;\wedge\; f'{<}b' \;\wedge\; n{+}b{-}f = b'{-}f' \\
& \wedge \quad R[(f;..n); (0;..b)] = R'[f';..b'] \;\wedge\; \neg u' \\
\vee \quad & \neg m \;\wedge\; f{>}b \;\wedge\; \neg m' \;\wedge\; f'{>}b' \;\wedge\; b{-}f = b'{-}f' \\
& \wedge \quad R[(f;..n); (0;..b)] = R'[(f';..n); (0;..b')] \;\wedge\; \neg u' \\
\vee \quad & m \;\wedge\; f{=}b \;\wedge\; m' \;\wedge\; f'{=}b' \;\wedge\; u' \\
\vee \quad & \neg m \;\wedge\; f{=}b \;\wedge\; \neg m' \;\wedge\; f'{=}b' \\
& \wedge \quad R[(f;..n); (0;..b)]{=}R'[(f';..n); (0;..b')] \;\wedge\; \neg u'
\end{aligned}
$$
$$\Leftarrow \qquad u' = (m \wedge f{=}b) \wedge f'{=}f \wedge b'{=}b \wedge R'{=}R$$
$$= \qquad u{:=} \; m \wedge f{=}b$$

The transformed operation offered us the opportunity to rotate the queue within R , but we declined to do so.   Each of the remaining transformations offers the same useless opportunity, and we decline each time.

$$\forall Q, p \cdot D \;\Rightarrow\; \exists Q', p' \cdot D' \wedge (u{:=} isfullq)$$
$$= \qquad \forall Q, p \cdot D \;\Rightarrow\; \exists Q', p' \cdot D' \wedge u'{=}(p{=}n) \wedge p'{=}p \wedge Q'{=}Q$$
$$\text{by several omitted steps}$$

$$\Leftarrow \qquad u{:=} \; \neg m \wedge f{=}b$$
$$\forall Q, p \cdot D \;\Rightarrow\; \exists Q', p' \cdot D' \wedge join \; x$$

$=$ $\forall Q, p \cdot D \Rightarrow \exists Q', p' \cdot D' \land Q' = Q[0;..p] + x + Q[p+1;..n] \land p' = p+1$

by several omitted steps

$\Leftarrow$ $Rb := x;$ **if** $b+1=n$ **then** $(b := 0;\ m := f)$ **else** $b := b+1$

$\forall Q, p \cdot D \Rightarrow \exists Q', p' \cdot D' \land leave$

$=$ $\forall Q, p \cdot D \Rightarrow \exists Q', p' \cdot D' \land Q' = Q[(1;..p); (p;..n)] \land p' = p-1$

by several omitted steps

$\Leftarrow$ **if** $f+1=n$ **then** $(f := 0;\ m := \uparrow)$ **else** $f := f+1$

$\forall Q, p \cdot D \Rightarrow \exists Q', p' \cdot D' \land (u := front)$

$=$ $\forall Q, p \cdot D \Rightarrow \exists Q', p' \cdot D' \land u' = Q0 \land p' = p \land Q' = Q$

by several omitted steps

$\Leftarrow$ $u := R f$

$Queue2 =$ $(mkemptyq = m := \uparrow;\ f := 0;\ b := 0)$

$\land$ $(isemptyq = m \land f = b)$

$\land$ $(isfullq = \neg m \land f = b)$

$\land$ $(join = \lambda x : X \cdot Rb := x;$ **if** $b+1=n$ **then** $(b := 0;\ m := f)$

else $b := b+1)$

$\land$ $(leave = $ **if** $f+1=n$ **then** $(f := 0;\ m := \uparrow)$ **else** $f := f+1)$

$\land$ $(front = R f)$

A transformation can be done by steps, as a sequence of smaller transformations. A transformation can be done by parts, as a conjunction of smaller transformations. But we don't pursue the topic further.

12 Incompleteness

Transformation is sound in the sense that a user cannot tell that a transformation has been made; that was the criterion of its design. But it is possible to find two theories that behave identically from a user's view, but for which there is no transformer to transform one into the other. Transformation is therefore incomplete.

An example to illustrate incompleteness comes from Gardiner and Morgan [3]. The user's variable is i and the implementer's variable is j, both of type nat. The theory is

$GM0$ $=$ $(initialize = i' = 0 \leq j' < 3)$

$\land$ $(step = $ **if** $j > 0$ **then** $(i := i+1;\ j := j-1)$ **else** $ok)$

The user can look at i but not at j. The user can $initialize$, which starts i at 0 and starts j at any of 3 values. The user can then repeatedly $step$ and observe that i increases 0 or 1 or 2 times and then stops increasing, which effectively tells the user what value j started with.

If this were a practical problem, we would notice that *initialize* can be refined, resolving the nondeterminism. For example,

$$initialize \iff i:= 0; \; j:= 0$$

We could then transform *initialize* and *step* to get rid of j, replacing it with nothing. The transformer is $j=0$. It transforms the implementation of *initialize* as follows:

$$\forall j \cdot j=0 \Rightarrow \exists j' \cdot j'=0 \land i'=j'=0$$
$$= \qquad i:= 0$$

And it transforms *step* as follows:

$$\forall j \cdot j=0 \Rightarrow \exists j' \cdot j'=0 \land step$$
$$= \qquad \forall j \cdot j=0 \Rightarrow \exists j' \cdot j'=0 \land \textbf{if } j>0 \textbf{ then } (i:= i+1. \; j:= j-1) \textbf{ else } ok$$
$$= \qquad ok$$

The very simple transformed theory

$$GM1 \quad = \qquad\qquad (initialize \; = \; i:= 0)$$
$$\land \qquad (step \; = \; ok)$$

cannot be distinguished from the original by the user. If this were a practical problem, we would be done. But the theoretical problem is to replace j with boolean variable b without resolving the nondeterminism, producing the theory

$$GM2 \quad = \qquad\qquad (initialize \; = \; i'=0)$$
$$\land \qquad (step \; = \; \textbf{if } b \land i<2 \textbf{ then } i'= i+1 \textbf{ else } ok)$$

Now *initialize* starts b either at $\top$, meaning that i will be increased, or at f, meaning that i will not be increased. Each use of *step* tests b to see if we might increase i, and $i<2$ to ensure that i remains below 3. If i is increased, b is again assigned either of its two values. The user will see i start at 0 and increase 0 or 1 or 2 times and then stop increasing, exactly as in the original specification *GM0*. The nondeterminism is maintained. But there is no transformer in variables i, j, and b to do the job; transformation is an incomplete method.

Where there's a will, there's a way. The criterion for being a transformer D is $\forall new \cdot \exists old \cdot D$. This criterion is sufficient to guarantee that when the *old* variables are replaced by the *new*, the result will be correct, but it is not always necessary for correctness. First, we rewrite *GM0* by introducing variable k to stand for the nondeterministically chosen initial value of j.

$$GM0 \quad = \quad \exists k: 0,..3 \cdot \quad (initialize \; = \; i:= 0; \; j:= k)$$
$$\land \qquad (step \; = \; \textbf{if } j>0 \textbf{ then } (i:= i+1; \; j:= j-1) \textbf{ else } ok)$$

Next we replace j with b using $i+j=k \land b=(j>0)$. This does not meet the criterion for being a transformer, but it is still safe because $i+j=k$ is established by *initialize* and maintained invariant by *step* , and $b=(j>0)$ is a transformer (although it produces an unimplementable result). Now we transform.

$$\forall j\cdot\ i+j=k \land b=(j>0) \implies \exists j'.\ i'+j'=k \land b'=(j'>0) \land \textit{initialize}$$
$$= \quad \forall j\cdot\ i+j=k \land b=(j>0) \implies \exists j'.\ i'+j'=k \land b'=(j'>0) \land i'=0 \land j'=k$$
$$= \quad b=(i<k) \implies (i:= 0;\ b:= i<k)$$

$$\forall j\cdot\ i+j=k \land b=(j>0) \implies \exists j'\ i'+j'=k \land b'=(j'>0) \land \textit{step}$$
$$= \quad \forall j\cdot\ i+j=k \land b=(j>0) \implies \exists j'.\ i'+j'=k \land b'=(j'>0)$$
$$\land \textbf{if } j>0 \textbf{ then } i'=i+1 \land j'=j-1 \textbf{ else } i'=i \land j'=j$$
$$= \quad b=(i<k) \implies b'=(i'<k) \land \textbf{if } b \textbf{ then } i'=i+1 \textbf{ else } i'=i$$
$$= \quad b=(i<k) \implies \textbf{if } b \land i<2 \textbf{ then } (i:=i+1;\ b:= i<k) \textbf{ else } ok$$

The resulting theory is

$$GM3 = \exists k: 0,..3\cdot \quad (\textit{initialize} = b=(i<k) \implies (i:= 0;\ b:= i<k))$$
$$\land \quad (\textit{step} = b=(i<k) \implies$$
$$\textbf{if } b \land i<2 \textbf{ then } (i:=i+1;\ b:= i<k) \textbf{ else } ok)$$

Variable j has disappeared, and variable b has appeared, as desired. But we also have k , which we added just to help make the transformation, and now it is no longer wanted. Eliminating it produces *GM2* as desired.

The incompleteness of transformation, like the incompleteness of first-order logic, is demonstrated with an example carefully crafted to show the incompleteness, not one that would ever arise in practice. We should not switch to a more complicated rule, or combination of rules, that are complete. We should stay with the simple rule that is adequate for all transformations that will ever arise in any problem other than a demonstration of theoretcal incompleteness. And even then, all we need is to soften the criterion for being a transformer. For further reading, see [7].

13 Conclusion

A theory can be presented as a boolean expression. Theories can then be combined by ordinary conjunction, and by other boolean connectives, and compared for strength by ordinary implication. Strong theories serve mathematicians who want to prove a lot, but weak theories are better for software engineers who need to implement them. This kind of theory presentation is both a simplification and a generalization of the early work. By reducing theories to boolean expressions we understand them in the simplest possible way, and we allow all combinations that make logical sense. Theory refinement is just implication. Implementation can also be expressed as a theory in a particular form.   Then implementation is just a refinement, and the proof of correctness of the implementation is just a boolean implication.   This kind of theory presentation, as a single boolean expression, works for both the functional style and imperative (state-changing) style of theory.

Theory transformation is a safe and automatic way to reimplement a theory, once the transformer has been written.   Although the method of transformation is incomplete in a theoretical sense, it is complete enough for all practical purposes.

References

1. J.-R.Abrial: *the B book, Assigning Programs to Meanings*, Cambridge University Press, 1996
2. R.M.Burstall, J.A.Goguen: "Putting Theories Together to make Specifications", in R.Reddy (ed.): *Proceedings of the fifth International Joint Conference on Artificial Intelligence*, volume 6 pages 1045-1058, Morgan Kaufman , Cambridge MA, 1977
3. P.H.B.Gardiner, C.C.Morgan: "a Single Complete Rule for Data Refinement", *Formal Aspects of Computing*, volume 5 number 4 pages 367-382, 1993
4. J.V.Guttag, J.J.Horning: "the Algebraic Specification of Abstract Data Types", *Acta Informatica*, volume 10 pages 27-52, 1978
5. E.C.R.Hehner: *a Practical Theory of Programming*, second edition, Springer, 2002
6. I.T.Kassios: Theory Theory and an Attempt to Orient Objections to Object Orientation, MSc thesis, University of Toronto, 2001
7. W.-P.deRoever, K.Engelhardt: *Data Refinement: Model-Oriented Proof Methods and their Comparisons*, tracts in Theoretical Computer Science volume 47, Cambridge University Press, 1998
8. J.M.Spivey: *Introducing Z: a Specification Language and its Formal Semantics*, Cambridge University Press, 1988

Incremental Proof of the Producer/Consumer Property for the PCI Protocol

Dominique Cansell[1], Ganesh Gopalakrishnan[2], Mike Jones[3], Dominique Méry[1], and Airy Weinzoepflen[1] *

[1] LORIA, Nancy France{Dominique.Cansell, Dominique.Mery, Airy.Weinzoepflen}@loria.fr
[2] University of Utah, Salt Lake City, Utah, USA ganesh@cs.utah.edu
[3] presently at Brigham Young University, Provo, Utah, USA jones@cs.byu.edu

Abstract. We present an incremental proof of the producer/consumer property for the PCI protocol. In the incremental proof, a corrected model of the multi-bus PCI 2.1 protocol is shown to be a refinement of the producer/consumer property. Multi-bus PCI must be corrected because the original PCI specification violates the producer/consumer property. The final model of PCI includes transaction types and reordering along with the completion mechanism for delayed PCI transactions. Verification results include multiple concurrent sessions of the producer/consumer property in a family of topologically isomorphic network configurations. The remaining configurations are identified and left for future work. In contrast to previous case studies involving this problem [13,15], the incremental proof provides structure which simplifies otherwise difficult monolithic proof attempts.

1 Introduction

Overview. Modelling complex systems can be improved by the use of refinement techniques. A refinement technique allows one to gradually develop a system step by step, or to tackle complex problem like the PCI Transaction Ordering Problem. The B event-based method provides a framework for deriving abstract systems satisfying the producer consumer property. The paper is an illustration of the effectiveness of refinement for those systems. The work described in this paper is a joint effort between researchers at LORIA, Nancy and the University of Utah and was facilitated by several exchange visits between institutions.

Refining abstract systems. The essence of refinement is to separate abstract systems with respect to safety properties and to simplify the view of the system under development. The development process begins with a very simple and abstract system which states properties expressed in requirements. The development process continues by adding more and more details through the refinement relationship. The refinement relationship expresses the enrichment of the initial abstract system and is supported by the Atelier B tool[20]. When a new refined system is analyzed, the tool generates proof obligations (POs) which

---

* Supported in part by NSF grants CCR-9987516 and CCR-00814006 and in part by NSF/CNRS cooperation project 1998-2000 and PRST IL/QSL/DIXIT project

D. Bert et al. (Eds.): ZB 2002, LNCS 2272, pp. 22–41, 2002.

are proved either automatically, or using the Atelier B interactive prover. The refinement process is repeated for a chain of abstract systems and stops when the last model is suitably close to the implementation of system.

Understanding systems by proofs. The documentation of a non-trivial system is very tedious and complex, since it is often expressed using diagrams and English descriptions that are similar to automata. The main aspect of our approach is the use of a general theorem prover which helps in the proof developement process by identifying what is missing in the model under development. A chain of refinements allows us to get very close to the concrete model described in the documentation of the system. The prover derives obligations as invariants and safety properties which must then be proven interactively. The prover plays a central role in the refinement process and allows us to get a complete proof of the development and hence of the final protocol. The verification is *uniform* in the sense that no assumptions are made about the number of nodes in the system. The incremental process is based on refinement and we have to add step by step details of the document. The main problem is to start from a very abstract, yet significant, system expressing the goal of the protocol (namely the producer/consumer system in this case study). The methodology allows us to address problems in small steps because of the incremental process. It allows us to think of the properties of the specified system and to communicate together using precise formal definitions.

Outline. In our study, we build a model of the PCI protocol (*Peripheral Component Interconnect*, an I/O standard) using refinement, and prove that this protocol satisfies a crucial property, the Producer/Consumer property.

First, we briefly recall the PCI protocol, then present the B System methodology. Next, we use the B System to model a simple statement of the Producer/Consumer property. We then refine this statement into an abstraction of the PCI protocol at the bus/bridge level. We conclude with some observations about refining a model of PCI from the Producer/Consumer property and compare the refinement verification with previous efforts to verify the same property using theorem proving and model checking.

2   The PCI Transaction Ordering Problem

2.1   A Short Review of the PCI Protocol

A PCI network is a network of *agents* connected to *buses*, the *buses* being interconnected via *bridges*. The entire network forming an acyclic graph with exactly one path from any agent to every other agent as shown on figure 1. Routing is performed using a static routing table.

All agents have two waiting queues (or channels): one incoming (the target channel) and one outgoing (the master channel). Similarly, every bridge between buses b1 and b2 has two channels: one channel with b1 as its in-bus and b2 as its out-bus, and one channel with b2 as its in-bus and b1 as its out-bus. The PCI protocol (Revision 2.1)[16] uses two kinds of transactions: *Posted* transactions and *Delayed* transactions. For simplicity of explanation, consider a simple PCI network with two agents connected to two different buses linked via a bridge.

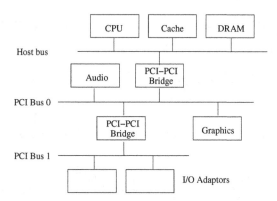

Fig. 1. A PCI acyclic network

First, suppose an agent, the *Consumer* emits a delayed read (DR) transaction addressed to another agent *Data*. The entire mechanism is shown on figure 2.

A routing table indicates that messages addressed to Data pass through bus b1, so the Consumer attempts to enqueue its transaction in the bridge channel going from b1 to b2 (1). After the first attempt, the DR transaction is marked as committed by the Consumer (2), the Consumer re-tries the transaction until the target bridge channel receives this transaction (3). The Consumer retains a copy of the delayed transaction in its master channel. Next, the bridge channel tries to relay this transaction toward the target channel of Data, after one try the transaction is also marked committed by the bridge channel (4). The bridge channel continues to re-send the transaction until it is received by the target (5). The bridge also keeps a copy of the delayed transaction. The copies of delayed transactions are kept and deleted by the completion, which is described next.

Finally the DR transaction reaches the target channel of Data and the completion mechanism begins. First, the delayed read transaction is erased from the target channel of Data and is replaced by a Completion transaction in the Data master channel (6). The completion containing the return value of Data and the ID of target agent of the original DR transaction (Data in this case). The Completion transaction then travels backward from Data to Consumer, deleting copies of the DR transaction in every channel it crosses, until it reaches the Consumer (7). The Completion carries the target of the original transaction (Data in this case) so it is easy to match a completion transaction with its corresponding DR transaction copy. When the completion reaches the Consumer, the transaction is complete and there are no more DR transaction copies pending since they have all been erased on the way back (8).

The difference between Posted transactions and Delayed transactions is that a Posted transaction is not acknowledged by a completion. A posted transaction completes on the originating bus before completing on the destination bus, while it is the exact opposite for a Delayed transaction. One advantage of a delayed transaction is that the bus is not held in a wait state while completing an access to a slow device. In order to avoid deadlock in a PCI network, re-orderings in the channels are allowed. Legal re-orderings are shown in figure 3. Also, uncommitted transactions may be discarded.

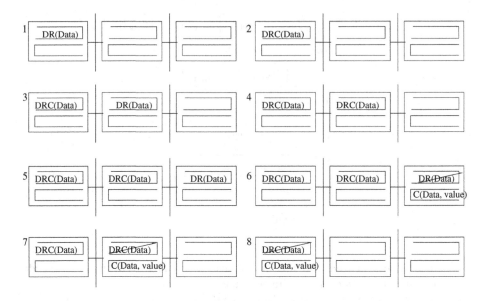

Fig. 2. Delayed read mechanism

2.2   A Bug in the Protocol

Despite the re-orderings allowed to avoid deadlock, the PCI protocol is intended to satisfy the Producer/Consumer transaction ordering property.

The Producer/Consumer Property. The Producer/Consumer property (P/C) states that if an agent (the Producer) writes a value to another agent (the Data), and then sets a flag at a third agent (the Flag), then any other agent (the Consumer) which reads the value written in Flag is guaranteed to read the value written to Data by Producer when reading Data— assuming no other agents overwrite the value written by the Producer. This property is intended to hold in every PCI network for any choice of Producer, Consumer, Data and Flag. Unfortunately, PCI violates this property due to a phenomenon called *completion stealing*.

Completion Stealing, Breach of P/C . Figure 4. shows a violation of Producer/Consumer involving completion stealing.

Suppose an agent, called Observer, is introduced, and that the Observer requests a value at agent Data. Next the transaction completion from Data, intended for the Observer, waits in the bridge whilst Producer writes to Data and Flag. Then Consumer reads the value written in Flag and emits a Delayed Read transaction toward Data, which is then committed.

The completion from the Observer transaction is still waiting in the bridge. The completion *for* the Observer stores the address of the target (Data) but not the source (Observer).

Row pass Column?	P	DR DW	CDR/ CDW
P	No	Yes	Yes
DR DW	No	Y/N	Y/N
CDR	No	Yes	Y/N
CDW	Y/N	Yes	Y/N

Fig. 3. The ordering rules in a bridge. C stands for *Completion*, P for *Posted*, D for *Delayed*, R for *Read* and W for *Write*. Yes means *Must be allowed to pass*, No means *Can not be allowed to pass*, Y/N means *the designer may choose either way*

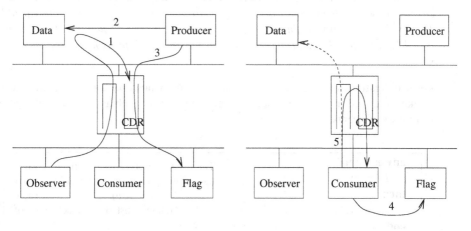

Fig. 4. Completion stealing in a PCI network

The violation happens when the completion waiting on the bridge completes with the Consumer instead of the Observer and the Consumer reads the *old* value of Data.

2.3 A Revision Proposition

Corella made a proposal to fix this bug [11] :

- implement local-master identifiers so that a delayed transaction knows where it originated from.
- require bridges and multi-function devices to have only one outstanding transaction at a time for a given address (because of some aliasing problems that might occur and that might enable completion-stealing).

Previously, we (Utah's group) have tried to verify the Producer/Consumer property with Corella's solution using a combination of model checking and theorem proving [15]. In this paper, we will propose abstract models which will be more and more precise and the last one will be close to the (concrete) model used in [15].

3   Proof-Based Development

The B system [2,3] approach is based on B notations of the B method [4] and extends the methodological scope of initial concepts as set-theoretical notations and generalized substitutions to take into account *abstract systems*. An *abstract system* is characterized by a (finite) list of variables possibly modified by a (finite) list of events; an invariant establishes properties satisfied by variables and maintained by the activation of events reacting to the environment, when guards are true. Abstract systems are close to actions systems of Back [6,5] and to UNITY programs [9]. We briefly recall definitions and principles of abstract systems and we explain how they can be managed by Atelier B [20].

Definition 1. *(event)*
An event has the following general form: yyy = ***any*** *x* ***where*** $P(x,v)$ ***then*** $S(v,x)$ ***end*** *where* **yyy** *is the name of the event, x are local variables, v are state variables, $P(x,v)$ is a condition defining the guard upto an exitencial quantification - namely $\exists x . P(x,v)$ - and $S(v,x)$ is a (generalized) substitution expressing transformations of states.*

Proof obligations are produced from substitutions, whose semantics is defined using weakest preconditions. For instance, we give two rules for obtaining proof obligations from texts of events.

any x **where**
    $P(x,v)$
 then
    $v := E(x,v)$
 end
	The event preserves the invariant I, when the following proof obligation holds: $I(v) \wedge P(x,v) \Rightarrow I(E(x,v))$; $E(x,v)$ is the value of the variable v just after the activation of the event.

The refinement of events defines a link between events and preserves properties of events; it maintains the invariant and there are several ways to refine events. We give an example of refinement and a list of proof obligations is generated to validate the refinement.

any x **where**
    $P(x,v)$
 then
    $v := E(x,v)$
 end
	is refined by, denoted $\sqsubseteq$,	**any** y **where**
		    $Q(y,w)$
		 then
		    $w := F(y,w)$
		 end

- $I(v)$ is an invariant of an abstract system S and $J(v,w)$ is an invariant of a refined abstract system S' for w and a gluing invariant linking v and w.
- $I(v) \wedge J(v,w) \wedge Q(y,w) \Rightarrow \exists x.(P(x,v) \wedge J(E(x,v),F(y,w)))$ is the proof obligation for checking the refinement.

Proof obligations of an event refinement contain informations that help in modifying invariant or event, when one or more are remaining unproved because they are unprovable; the latter case leads to a reformulation of the event or a new statement of the refined invariant. The proof engine is the tool supporting the validation of the refinement. Refinement of events leads to define refinement of abstract systems; refining an abstract system by another one is based on the refinement of events in the abstract system according to the relation defined above: $S \sqsubseteq S'$ (read S' refines S) means that S and S' have the same set of events and every abstract event (of S) is refined by the concrete event of S'. However, new events can be introduced in the new event system and refine the abstract *silent* event. A *silent* event is either a skip event (variables are not modified in the abstract model) or a keep event (modifying variables but maintaining abstract invariant). Rules for the refinement of abstract systems are simply defined as follows.

Definition 2. *(refinement of abstract systems)*
Let S and S' abstract systems
S is refined by $S' \cup U$ where $U = \{U_k : k \in K\}$, when

- $S \sqsubseteq S'$ and **silent** $\sqsubseteq U_k$, $(k \in K)$,
- $I(v) \Rightarrow grd(S') \vee grd(U)$ *(grd(U) (resp. grd(S') stands for the disjunction of U's guards (resp. the disjunction of S''s guards),*
- $I(v) \Rightarrow V \in \mathbb{N}$ *(V is the variant controling the convergence),*
- $I(v) \Rightarrow (V = n \Rightarrow [U_k](V < n))$, $(k \in K)$.

The refinement of abstract systems allows us to enrich system models in a *step by step* approach and in a gradual way. Refinement provides a way to construct stronger invariants of the system and to introduce communications, if needed. No assumptions are made regarding the size of the system, in contrast to model checking. The case study will show that refinement helps in understanding the system since the complexity of the system is distributed and proofs are made easier. Finally, the *B system* approach is a part of the validation process. Proof obligations ensure that properties are preserved when events are triggered, deadlocks are avoided and no new events take control of the entire machine, that is, the events of the previous machine still happen—but in less time. Atelier B is not yet updated to directly handle the *B system* approach, but it allows us to generate proofs obligations for abstract systems and refinements; the case study is developed to illustrate the adequacy of that approach when constructing distributed systems.

4   Modelling in Order to Prove

4.1   Refining from the Producer/Consumer Property toward an Abstraction of PCI Revision 2.1

The goal of our study is to formally prove that the *revised PCI protocol* satisfies the Producer/Consumer property, using abstraction and refinement techniques in the B event-based (system) method.

We shall start from a simple model of a system of the Producer/Consumer property, refine it gradually into a system abstracting the revised PCI protocol. The refinement will be validated at every step, thus we will have proven that the revised PCI protocol satisfies the Producer/Consumer property. The Atelier B [20] software was used in this study and abstract systems have been completely checked by the tool.

4.2   Initial Specification of the Producer/Consumer Property

The first abstract system describes the behavior of the Producer and Consumer. The Producer produces a series of time-stamped data: the event *produce* describes the production of the data. The Consumer consumes a datum in the set of available data by copying the datum from prod to cons: the *consume* event models this task.

In this abstraction, the Producer/Consumer property is expressed by the fact that all consumed datum has been produced previously (*cons* $\subseteq$ *prod*).

In our first model *pci1*, we define two sets *AD* and *VAL*. The variables of this model are *prod* and *cons* which are initialized to the empty set. The invariant of *pci1* is the following:

$$
\begin{array}{l}
prod \ \in \ AD \ \nrightarrow \ VAL \ \wedge \\
cons \ \in \ AD \ \nrightarrow \ VAL \ \wedge \\
cons \ \subseteq \ prod
\end{array}
$$

There are only two events in this abstract model which are :

$$
\begin{array}{l}
produce \ = \ \textbf{any} \ ad, data \ \textbf{where} \\
\qquad ad \ \in \ AD \ \wedge \\
\qquad data \ \in \ VAL \ \wedge \\
\qquad ad \ \notin \ \text{DOM}(prod) \\
\textbf{then} \\
\qquad prod \ := \ prod \cup \{ad \ \mapsto \ data\} \\
\textbf{end};
\end{array}
$$

$$
\begin{array}{l}
consume \ = \ \textbf{any} \ ad \ \textbf{where} \\
\qquad ad \ \in \ AD \ \wedge \\
\qquad ad \ \in \ \text{DOM}(prod) \ \wedge \\
\qquad ad \ \notin \ \text{DOM}(cons) \\
\textbf{then} \\
\qquad cons \ := \ cons \cup \{ad \ \mapsto \ prod(ad)\} \\
\textbf{end};
\end{array}
$$

4.3   Introduction of the Data and Flag Agent

The second model refines the first model. This model introduces agents Data (*mem*) and Flag (*flag*) which are intermediate agents between the Consumer and Producer. In the state of the Producer/Consumer property for the PCI protocol, the Consumer can not consume (read) the data before checking the value of the flag to determine if the data is ready to read. The value of the flag is updated by the Producer after producing a new data value (*produce*) and writing the value to the data agent (*datawrite.*) The new and refined observable events in this model are:

$$
\begin{aligned}
produce \; = \; &\textbf{any } ad, data \textbf{ where}\\
&\quad ad \; \in \; AD \; \wedge\\
&\quad data \; \in \; VAL \; \wedge\\
&\quad ad \; \notin \; \text{DOM}(prod)\\
&\textbf{then}\\
&\quad prod \; := \; prod \cup \{ad \mapsto data\}\\
&\textbf{end};
\end{aligned}
$$

$$
\begin{aligned}
datawrite \; = \; &\textbf{any } ad \textbf{ where}\\
&\quad ad \; \in \; AD \; \wedge\\
&\quad ad \; \notin \; \text{DOM}(mem) \; \wedge\\
&\quad ad \; \in \; \text{DOM}(prod)\\
&\textbf{then}\\
&\quad mem \; := \; mem \cup \{ad \mapsto prod(ad)\}\\
&\textbf{end};
\end{aligned}
$$

$$
\begin{aligned}
flagwrite \; = \; &\textbf{any } ad \textbf{ where}\\
&\quad ad \; \in \; AD \; \wedge\\
&\quad ad \; \notin \; \text{DOM}(flag) \; \wedge\\
&\quad ad \; \in \; \text{DOM}(mem)\\
&\textbf{then}\\
&\quad flag \; := \; flag \cup \{ad \mapsto mem(ad)\}\\
&\textbf{end};
\end{aligned}
$$

$$
\begin{aligned}
consume \; = \; &\textbf{any } ad \textbf{ where}\\
&\quad ad \; \in \; AD \; \wedge\\
&\quad ad \; \notin \; \text{DOM}(cons) \; \wedge\\
&\quad ad \; \in \; \text{DOM}(flag)\\
&\textbf{then}\\
&\quad cons \; := \; cons \cup \{ad \mapsto mem(ad)\}\\
&\textbf{end};
\end{aligned}
$$

The invariant still states that data is consumed after it is produced but adds requirements on when the Consumer may consume the data:

$$
cons \subseteq flag \subseteq mem \subseteq prod
$$

4.4   Transaction Abstractions

In this model, abstractions of the transactions are added using new set-variables : dw, fw, fr, cfr, dr, cdr. In this abstraction, we keep only the address of the data produced (ad) and not the corresponding value (*data* in method *produce* of the previous model).

The set-variable dw represents the set of completed *data write* transactions, fw represents the set of completed *flag write* transactions, fr represents the set of *flag read* transactions having reached the Flag agent, cfr represents the set of *flag read completion* transactions having completed at the Consumer agent, dr represents the set of *data read* transactions having reached the Data agent and cdr represents the set of *data read completion* transactions having completed at the Consumer agent.

We restrict our study to *posted* write transactions, and *delayed* read transactions because reads can not be posted and delayed writes allow only one write message to be in transit at a time for the Producer/Consumer property, while posted writes allow both write messages (*data write* and *flag write*) to be in transit at the same time. Thus the model with posted writes is of more interest.

In this model, transactions travel instantaneously. In later models, when topology, bridges and buses will be introduced (cf. 4.5), we will introduce a delay in transactions travel.

The Producer/Consumer property is expressed as: $fr \subseteq fw \land dr \subseteq dw$. The invariant is now:

$$
\begin{aligned}
&dr \subseteq dw \land \\
&cfr \subseteq fr \cap dw \land \\
&dr \subseteq cfr \land \\
&cdr \subseteq dr \land \\
&fr \subseteq fw \land \\
&dw = \text{DOM}(mem) \land \\
&fw = \text{DOM}(flag) \land \\
&\text{DOM}(cons) \subseteq cdr
\end{aligned}
$$

In order to abstract the transactions, we also need to update our old observable events (modify the guards) and add new intermediate observable events. We proved this refinement without much difficulty.(cf.figure 7)

$$
\begin{aligned}
consume = \;&\textbf{any } ad \textbf{ where} \\
&\quad ad \in AD \land ad \in cdr \land ad \notin \text{DOM}(cons) \\
&\textbf{then} \\
&\quad cons := cons \cup \{ad \mapsto mem(ad)\} \\
&\textbf{end};
\end{aligned}
$$

```
produce  =  any ad, data where
                ad ∈ AD ∧ data ∈ VAL ∧ ad ∉ DOM(prod)
            then
                prod := prod ∪ {ad ↦ data}
            end;
```

```
datawrite  =  any ad where
                  ad ∈ AD ∧ ad ∉ DOM(mem) ∧ ad ∈ DOM(prod)
              then
                  mem := mem ∪ {ad ↦ prod(ad)} ‖
                  dw := dw ∪ {ad}
              end;
```

```
flagwrite  =  any ad where
                  ad ∈ AD ∧ ad ∉ DOM(flag) ∧ ad ∈ DOM(mem)
              then
                  flag := flag ∪ {ad ↦ mem(ad)} ‖
                  fw := fw ∪ {ad}
              end;
```

```
flagread  =  any ad where
                 ad ∈ AD ∧ ad ∉ fr ∧ ad ∈ fw
             then
                 fr := fr ∪ {ad}
             end;
```

```
compflagread  =  any ad where
                     ad ∈ AD ∧ ad ∉ cfr ∧ ad ∈ fr
                 then
                     cfr := cfr ∪ {ad}
                 end;
```

```
dataread  =  any ad where
                 ad ∈ AD ∧ ad ∈ cfr ∧ ad ∉ dr
             then
                 dr := dr ∪ {ad}
             end;
```

$$\boxed{\begin{aligned}
&compdataread \;=\; \textbf{any } ad \textbf{ where}\\
&\qquad\qquad ad \;\in\; AD \wedge ad \;\in\; dr \wedge ad \;\notin\; cdr\\
&\qquad \textbf{then}\\
&\qquad\qquad cdr \;:=\; cdr \cup \{ad\}\\
&\qquad \textbf{end};
\end{aligned}}$$

4.5   Concrete Transactions and Topologies

In this section, we abstract the PCI protocol at the bus/bridge level by introducing topologies (equivalence classes for PCI networks), concrete transactions, bridges, …
We still do not allow re-ordering at this level of abstraction, but we prepare the way by modelling bridge channels. We proved that this model is a refinement of the previous model, and thus still satisfies the Producer/Consumer property. This refinement proof was considerably more difficult than the previous ones (see the discussion in the next subsection).

Topologies.   The Producer, Consumer, Data and Flag agents can be arranged into one of four distinct families of labeled configurations shown in figure 5. A fixed set of configuration families is obtained by distinguishing configurations only by their topology (or shape). This is done by ignoring the number of bridges in a sequence of bridges between agents or a common bridge (labeled B in the figure). As a result, each of the four families contain an unbounded number of unique configurations. For example, the box labeled P in figure 5 may include an unbounded number of bridges connecting the Producer to the common bridge labeled B.
In general, the set of topologies for configuration families over n agents can be enumerated by an algorithm based on full Steiner topologies [12]. A Steiner topology is a Steiner tree without the associated geometrical embedding. Steiner trees are used to connect a set of points with minimal edge length. New internal points, called Steiner points, can be added to a Steiner tree to mininize edge length. A full Steiner topology is a Steiner topology with the maximal number of Steiner points. The algorithm for enumerating all configuration families on n agents first enumerates all full Steiner topologies with n agents or less then considers all topologically unique ways to add the remaining agents to the full Steiner topology.
For properties defined on four agents (such as the producer/consumer property) the configuration enumeration algorithm generates two unique topologies. The remaining configuration families are generated by considering unique labelings of the topologies. Each configuration family must be verified separately.

PCI at the bus/bridge level.   For the purpose of our study, we restricted ourselves to one of the four topology equivalence classes, and completed the proof for this class, rather than simultaneously develop a model for every topology.
We choose the *most interesting* topology, configuration B in figure 5. This topology is the most interesting, because it allows the most transactions to cross the common bridge, thus allowing reordering rules to interfere with the Producer/Consumer property.

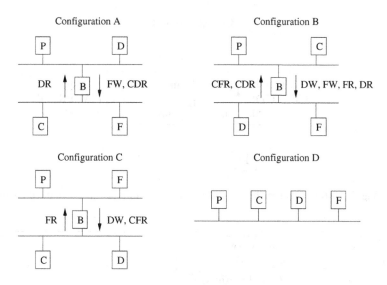

Fig. 5. Topologies equivalence classes

In the model of PCI at the bus-bridge level, transactions travel instantaneously on the bus, but not the bridge.

Thus, we modelled the two channels of the common bridge by two queues each : $inb1$ and $outb1$ for the channel from bus b1 to bus b2, $inb2$ and $outb2$ for the channel from bus b2 to bus b1 as shown in figure 6. We define this following set of states

$$STATES = \{DW, FW, FR, CFR, DR, CDR\}$$

Our queues are defined as partial injections $\mathbb{N} \rightarrowtail AD \times STATES$

Transactions going from bus b1 to bus b2, enter the queue $inb1$ at the first available position ($cinb1$), then exit the bridge by entering queue $outb1$, and finally they reach their targeted agent. One part of the gluing invariant is:

$$
\begin{aligned}
dw &= \text{DOM}(\text{RAN}(inb1) \rhd \{DW\}) \wedge \\
dw1 &= \text{DOM}(\text{RAN}(outb1) \rhd \{DW\})
\end{aligned}
$$

We introduced new set-variables $dw1, fw1, fr1, cfr1, dr1$ and $cdr1$, which are subsets of the previous dw, fw, fr, cfr, dr and cdr variables (cf.4.4) and indicate which transactions have already crossed the common bridge, and which have not.

In this model, the channel queues are strictly first in first out. (FIFO). That will not be the case later when we introduce re-ordering (cf.4.6).

Rather than giving the complete set of events, we give only the events which might be observable when a *flag write* transaction travels from the Producer to its corresponding Flag agent (and crosses the common bridge in the current configuration, cf. figure 6):

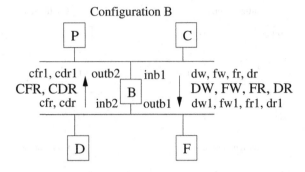

Fig. 6. Abstraction of the bridge and concretization of transactions, in the B configuration

---

flagwrite = **any** *ad* **where**
$\qquad ad \in AD \wedge$
$\qquad ad \notin \text{DOM}(\textit{flag}) \wedge$
$\qquad ad \in \text{DOM}(\textit{mem}) \wedge$
$\qquad (ad \mapsto FW) \notin \text{RAN}(\textit{inb1}) \wedge$
$\qquad (ad \mapsto DW) \in \text{RAN}(\textit{inb1})$
$\quad$ **then**
$\qquad \textit{flag} := \textit{flag} \cup \{ad \mapsto \textit{mem}(ad)\} \;\|$
$\qquad \textit{inb1} := \textit{inb1} \cup \{\textit{cinb1} \mapsto (ad \mapsto FW)\} \|$
$\qquad \textit{fw} := \textit{fw} \cup \{ad\} \;\|$
$\qquad \textit{cinb1} := \textit{cinb1} + 1$
$\quad$ **end**;

FWout = **any** *ad* **where**
$\qquad ad \in AD \wedge$
$\qquad \textit{coutb1} < \textit{cinb1} \wedge$
$\qquad \textit{coutb1} \mapsto (ad \mapsto FW) \in \textit{inb1} \wedge$
$\qquad (ad \mapsto FW) \notin \text{RAN}(\textit{outb1})$
$\quad$ **then**
$\qquad \textit{coutb1} := \textit{coutb1} + 1 \;\|$
$\qquad \textit{outb1} := \textit{outb1} \cup \{\textit{coutb1} \mapsto (ad \mapsto FW)\} \;\|$
$\qquad \textit{fw1} := \textit{fw1} \cup \{ad\}$
$\quad$ **end**;

Example of a Very Hard PO: compflagread.17 (1 p.):
"Local hypotheses $\cdots \wedge$
$\quad ad \in AD \wedge$
$\quad ad \in \textit{fw1\$1} \wedge$
$\quad \textit{not}(ad \in \textit{cfr\$1}) \wedge$
$\quad ad \in \textit{fr1\$1} \wedge$
$\quad \textit{not}(ad \mapsto CFR \in \text{RAN}(\textit{inb2\$1})) \wedge$

"*Check that the invariant cfr $\subseteq$ dw1 is preserved by the operation − ref 4.4, 5.5* $\cdots \wedge$

$\Rightarrow$

$ad \in dw1\$1$

This proof obligation was generated when refining method *compflagread* with concrete transactions. We were required to prove that if the *flag read* transaction is about to complete for a given address ($ad \in fr1\$1 \wedge ad \in fw1\$1 \wedge not(ad \in cfr\$1)$), then the corresponding *data write* transaction has previously exited the bridge ($ad \in dw1\$1$). This property seemed obvious when thinking about the protocol, yet it was difficult to prove (about one week of work). To do so, we had to introduce the following properties in the global invariant. Although these properties seemed obvious according to the protocol and our current model, we had to prove them for every observable event. The original PO, compflagread.17, helped us build a stronger invariant.

- $\forall ad.(ad \in AD \wedge ad \mapsto DW \in \text{RAN}(outb1)$
  $\Rightarrow ad \mapsto DW \in \text{RAN}(inb1) \wedge$
  $outb1^{-1}(ad \mapsto DW) \le inb1^{-1}(ad \mapsto DW))$

- $\forall ad.(ad \in AD \wedge ad \mapsto FW \in \text{RAN}(outb1)$
  $\Rightarrow ad \mapsto FW \in \text{RAN}(inb1) \wedge$
  $outb1^{-1}(ad \mapsto FW) \le inb1^{-1}(ad \mapsto FW))$

- $\forall ad.(ad \in AD \wedge ad \in fw$
  $\Rightarrow ad \mapsto DW \in \text{RAN}(inb1) \wedge$
  $0 \le -1+inb1^{-1}(ad \mapsto FW)-inb1^{-1}(ad \mapsto DW) \wedge$
  $inb1^{-1}(ad \mapsto DW)+1 \le inb1^{-1}(ad \mapsto FW))$
- $\forall ad.(ad \in dw \wedge inb1^{-1}(ad \mapsto DW) \in \text{DOM}(outb1)$
  $\Rightarrow ad \mapsto DW \in \text{RAN}(outb1))$

Note that we used $\le$ instead of = in the first two properties. This is because later on, when re-ordering will be introduced (cf.4.6), posted transactions FW and DW might exit the bridge at a position preceding the position they have entered, since they will be able to bypass delayed transactions waiting in the bridge. If we had kept the equality sign, we wouldn't have been able to prove refinement in our last model.

These properties will remain true in further models, when we introduce re-ordering, since *data write* is a posted transaction, it must have exited the bridge before the FW transaction did (DW was emitted before FW, from the same agent and can NOT be bypassed), and FW has exited the bridge since $ad \in fw1\$1$.

4.6   Last Step toward Abstract PCI

In the final step, we refine the previous models in an abstraction of PCI with re-ordering rules. We also add commitment and delays in transactions travel from the common bridge toward the target agent, to reach a complete PCI abstraction. After proving this refinement, we deduce that PCI satisfies the Producer/Consumer property.

From the Bridge to the Agents.   Adding a delay in transactions travelling from the common bridge toward the target agent is an easy but necessary refinement.

Easy because we just add new variable-sets ($dw2$, $fw2$, $fr2$, $cfr2$, $dr2$ and $cdr2$), subsets of the previous variable-sets ($dw1$, $fw1$, $fr1$, $cfr1$, $dr1$ and $cd1$) and new methods like :

$$
\begin{aligned}
FWbridge2agent \; = \; &\textbf{any } ad \textbf{ where} \\
&\quad ad \in AD \wedge \\
&\quad ad \in fw1 \wedge \\
&\quad ad \notin fw2 \\
&\textbf{then} \\
&\quad fw2 := fw2 \cup \{ad\} \\
&\textbf{end};
\end{aligned}
$$

$$
\begin{aligned}
FRbridge2agent \; = \; &\textbf{any } ad \textbf{ where} \\
&\quad ad \in AD \wedge \\
&\quad ad \in fr1 \wedge \\
&\quad ad \in fw2 \wedge \\
&\quad ad \notin fr2 \\
&\textbf{then} \\
&\quad fr2 := fr2 \cup \{ad\} \\
&\textbf{end};
\end{aligned}
$$

the guard $ad \in fw2$ is very important, it means that the FR transaction won't bypass the corresponding FW transaction on its way between the bridge and the Flag

Necessary, because even if $fw \subseteq dw$, we don't have $fw2 \subseteq dw2$. The *data write* transaction is still emitted by the Producer before its corresponding *flag write* transaction, but it doesn't necessarily reach its target (Data) before the *flag write* transaction, since the path between the bridge and Data and the path between the bridge and Flag are different, independent, and might have different lengths. (remember our study is based on topology equivalence classes, cf.4.5)

Commitment.   We modelled commitment over the bridge channels, in fact only channel 1 (queues $inb1$ and $outb1$) in our current configuration. Deletion was taken care of with completion transactions, crossing the common bridge through channel 2 in our configuration. Commited transactions are held in $AD \times STATES$ set-variables $commitinb1$ and $commitoutb1$.

The invariant gluing this abstraction to the previous one is :

$$
\begin{aligned}
commitinb1 \;\; &\subseteq \; \text{RAN}(inb1) \; \triangleright \; \{FR, DR\} \wedge \\
commitoutb1 \;\; &\subseteq \; \text{RAN}(outb1) \; \triangleright \; \{FR, DR\}
\end{aligned}
$$

Indeed, only delayed transactions are commited (so *not FW* and *DW*, cf. 4.4), and in the current configuration they both cross the common bridge through channel 1.

Deletion happens during completion transaction travels.

Here is an example of commitment, and an example of where deletion is added in this model :

$$flagread \ = \ \textbf{any} \ ad \ \textbf{where} \ ...$$
$$\textbf{then}$$
$$... \ \|$$
$$commitinb1 \ := \ commitinb1 \cup \{ad \mapsto FR\}$$
$$\textbf{end};$$

$$CFRout \ = \ \textbf{any} \ ad \ \textbf{where} \ ...$$
$$\textbf{then}$$
$$... \ \|$$
$$commitinb1 \ := \ commitinb1 - \{ad \mapsto FR\}$$
$$\textbf{end};$$

The main invariant to verify on the commitment variables is :

$$\forall(ad).(ad \in AD \wedge ad \in cdr2$$
$$\Rightarrow \{ad\} \lhd commitinb1 = \emptyset \wedge \{ad\} \lhd commitoutb1 = \emptyset)$$

This invariant means that at the end of every Producer/Consumer cycle ($ad \in cdr2$), all pending commited transactions relative to this cycle have been deleted.

Unfortunately, this invariant could not be proven directly. We had to decompose the Producer/Consumer cycle, and add an invariant for every step of the cycle. Each of these proofs were therefore made easier, and it also strengthened our invariant.

One such step-invariant was :

$$\forall(ad).(ad \in AD \wedge ad \in dr1 \wedge ad \notin dr2$$
$$\Rightarrow \{ad\} \lhd commitinb1 = \{(ad \mapsto DR)\} \wedge$$
$$\{ad\} \lhd commitoutb1 = \{(ad \mapsto DR)\})$$

These two first refinements were quite easy to prove with Atelier B, that was not the case for the next one.

Re-ordering Rules. In this final model, we introduced re-ordering rules.

In the current configuration, the only re-ordering rule to consider is the one where posted transactions are allowed to pass delayed read and delayed write transactions. For the Producer/Consumer property, other transaction types do not interact in the current configuration. Our study also does not include delayed write, so only these four re-ordering rules are to be considered :

- reordering1: DW is allowed to pass FR
- reordering2: DW is allowed to pass DR
- reordering3: FW is allowed to pass FR
- reordering4: FW is allowed to pass DR

We introduced them very easily in the model :

$reordering1 =$
    any $ad1, ad2$ **where**
       $ad1 \in AD \land ad2 \in AD \land$
       $ad1 \neq ad2 \land (ad1 \mapsto DW) \in \text{RAN}(inb1-outb1) \land$
       $(ad2 \mapsto FR) \in \text{RAN}(inb1-outb1) \land$
       $inb1^{-1}(ad2 \mapsto FR) \leq inb1^{-1}(ad1 \mapsto DW) \land$
       $inb1^{-1}(ad2 \mapsto FR)+1 = inb1^{-1}(ad1 \mapsto DW)$
    then
       $inb1 := inb1 \lhdminus (\ \{inb1^{-1}(ad2 \mapsto FR) \mapsto (ad1 \mapsto DW)\}$
                    $\cup \{inb1^{-1}(ad1 \mapsto DW) \mapsto (ad2 \mapsto FR)\})$
    end;

The rule is read : "if one DW transaction relative to one P/C cycle ($ad1$) has entered the common bridge and not yet exited it (($ad1 \mapsto DW$) $\in$ RAN($inb1-outb1$)), while a FR transaction relative to another P/C cycle ($ad2 \neq ad1$) has also entered the common bridge, before this DW transaction ($inb1^{-1}(ad2 \mapsto FR) \leq inb1^{-1}(ad1 \mapsto DW)$), and has not yet exited the common bridge either (($ad2 \mapsto FR$) $\in$ RAN($inb1-outb1$)), and if these two transactions are next to each other in the queue ($inb1^{-1}(ad2 \mapsto FR)+1 = inb1^{-1}(ad1 \mapsto DW)$), then, the re-ordering event reordering1 is likely to be observed. reordering1 consists of a simple position exchange in the queue inb1.

It took us almost one week, to prove this refinement, so we only introduced re-ordering rules reordering2, reordering3 and reordering4, checked the PO generated, and claim that the refinement proofs should be symetrically equivalent. So far we did not manage to do them.

5   Conclusion and Related Works

This study identified potential applications of the B-event method for proving simple properties about complex systems. In this case study, we refined a simple statement of the producer/consumer property into a model of the multi-bus PCI 2.1 protocol.

The proof method used in this case study is particularly interesting because the proof was completed gradually using refinement. Breaking the proof into smaller pieces through refinement reduced the difficulty of the intermediate proofs. In fact, most of the nonobvious PO generated by Atelier B were proved automatically (about 2/3 of the nPO column on figure 7), the others had to be proved interactively using Atelier B, which is especially designed for this purpose.

The final model, while still an abstraction, is very close to the PCI protocol, even. Having proved that it refines a model of the Producer/Consumer property enables us to say that PCI multi-bus networks (with the topology addressed in the proof, three topologies we left unproven) satisfy this property regardless of other traffic present in the network.

In this case study, parameterized system verification is performed using incremental proof. The parameterized verification problem has been studied in the context of model checking [8,14,1,17] and, more recently, logic program transformations [18] and theorem proving combined with model checking [7,13]. Incremental refinement proof requires interactive proof, unlike model checking, but avoids the state explosion and non-termination problems encountered by model checking solutions. While all interactive proof methods avoid state explosion and non-termination problems, incremental proofs provide a convenient framework for breaking an otherwise difficult proof into manageable pieces.

Previously, the co-authors from the University of Utah (in collaboration with Mokkedem and Hosabettu) attempted to prove the same property using a higher-order logic theory of PCI transitions [15]. Creating the proof monolithically in one step from property to protocol was prohibitively difficult. The one-step proof was abandoned, unfinished, after 18 months of effort. In contrast, the incremental proof was completed in less than three months. The incremental proof effort using Atelier B divided the proof into manageable steps that could be completed with comparatively minimal user effort. The PCI protocol has been considered in other case studies [19,10]. In each of the other case studies, the bus signalling properties of PCI, rather than multi-bus transaction ordering properties, were of interest. Both case studies rely on forms of model checking.

```
Project status
+-----------+----+-----+------+------+-----+-----+
| COMPONENT | TC | POG | Obv  | nPO  | nUn | %Pr |
+-----------+----+-----+------+------+-----+-----+
| pci1      | OK | OK  |   77 |    6 |   0 | 100 |
| pci2      | OK | OK  |  238 |   16 |   0 | 100 |
| pci3      | OK | OK  |  650 |   37 |   0 | 100 |
| pci4B     | OK | OK  | 1981 |  332 |   0 | 100 |
| pci5B     | OK | OK  | 3003 |  657 |   0 | 100 |
+-----------+----+-----+------+------+-----+-----+
| TOTAL     | OK | OK  | 5949 | 1048 |   0 | 100 |
+-----------+----+-----+------+------+-----+-----+
```

Fig. 7. Summary of proofs, all the Proof Obligations generated (Obv and nPO) have been proved (%Pr = 100).

References

1. P. A. Abdulla, A. Bouajjani, B. Jonsson, and M. Nilsson. Handling global conditions in parameterized system verification. In Nicolas Halbwachs and Doron Peled, editors, *Computer-Aided Verification, CAV '99*, volume 1633 of *Lecture Notes in Computer Science*, pages 134–145, Trento, Italy, July 1999. Springer-Verlag.
2. J.-R. Abrial. Extending b without changing it (for developing distributed systems). In H. Habrias, editor, *1st Conference on the B method*, pages 169–190, November 1996.
3. J.-R. Abrial and L. Mussat. Introducing dynamic constraints in B. In D. Bert, editor, *B'98: Recent Advances in the Development and Use of the B Method*, volume 1393 of *Lecture Notes in Computer Science*. Springer-Verlag, 1998.
4. Jean-Raymond Abrial. *The B Book - Assigning Programs to Meanings*. Cambridge University Press, 1996.

5. R. J. R. Back. On correct refinement of programs. *Journal of Computer and System Sciences*, 23(1):49–68, 1979.

6. R. J. R. Back and K. Sere. Stepwise refinement of action systems. In J. L. A van de Snep-scheut, editor, *Mathematics for Program Construction*, pages 113–138. Springer-Verlag, june 1989. LNCS 375.

7. Karthikeyan Bhargavan, Davor Obradovic, and Carl A. Gunter. Routing information protocol in HOL/SPIN. In *Theorem Provers in Higher-Order Logics 2000: TPHOLs00*, August 2000.

8. M.C. Browne, E.M. Clarke, and O. Grumberg. Reasoning about networks with many identical finite state processes. *Information and Computation*, 81:13–31, April 1989.

9. K. M. Chandy and J. Misra. *Parallel Program Design A Foundation*. Addison-Wesley Publishing Company, 1988. ISBN 0-201-05866-9.

10. Edmund Clarke, Somesh Jha, Yuan Lu, and Dong Wang. Abstract BDDs: A technique for using abstraction in model checking. In Laurence Pierre and Thomas Kropf, editors, *Correct Hardware Design and Verification Methods, CHARME '99*, volume 1703 of *Lecture Notes in Computer Science*, Bad Herrenalb, Germany, September 1999. Springer-Verlag.

11. Francisco Corella. Proposal to fix ordering problem in PCI 2.1, 1996. Accessed June 2001 www.pcisig.com/reflector/thrd8.html#00704.

12. F. K. Hwang, D. S. Richards, and P. Winter. *The Steiner Tree Problem*, volume 53 of *Annals of Discrete Mathematics*. North-Holland, Amsterdam, Netherlands, 1992.

13. Michael Jones and Ganesh Gopalakrishnan. Verifying transaction ordering properties in unbounded bus networks through combined deductive/algorithmic methods. In Warren A. Hunt Jr. and Steven D. Johnson, editors, *Formal Methods in Computer-Aided Design: FMCAD'00*, number 1954 in LNCS, page 505, Austin, Texas, November 2000.

14. Y. Kesten, O. Maler, M. Marcus, A Pnueli, and E. Shahar. symbolic model checking with rich assertional languages. In Orna Grumburg, editor, *Computer-Aided Verification, CAV '97*, volume 1254 of *Lecture Notes in Computer Science*, Haifa, Israel, June 1997. Springer-Verlag.

15. Abdel Mokkedem, Ravi Hosabettu, Michael D. Jones, and Ganesh Gopalakrishnan. Formalization and proof of a solution to the PCI 2.1 bus transaction ordering problem. *Formal Methods in Systems Design*, 16(1):93–119, January 2000.

16. PCISIG. PCI Special Interest Group–PCI Local Bus Specification, Revision 2.1, June 1995.

17. Amir Pnueli and Elad Shahar. Liveness and acceleration in parameterized verification. In E. Allen Emerson and A. Prasad Sistla, editors, *Computer-Aided Verification, CAV '00*, volume 1855 of *Lecture Notes in Computer Science*, pages 328–343, Chicago, IL, July 2000. Springer-Verlag.

18. A. Roychoudhurry, K. N. Kumar, C. R. Ramakrishnan, I.V. Ramakrishnan, and S. A. Smolka. Verification of parameterized systems using logic program transformations. In S. Graf and M. Schwartzbach, editors, *Tools and Algorithms for the Construction and Analysis of Systems (TACAS)*, volume 1785 of *Lecture Notes in Computer Science*, pages 172–187. Springer-Verlag, 2000.

19. Kanna Shimizu, David L. Dill, and Alan J. Hu. Monitor-based formal specification of PCI. In Warran A. Hunt Jr. and Steven D. Johnson, editors, *Formal Methods in Computer-Aided Design: FMCAD'00*, number 1954 in LNCS, page 335, Austin, Texas, November 2000.

20. STERIA - Technologies de l'Information, Aix-en-Provence (F). *Atelier B, Manuel Utilisateur*, 3.5 edition, 1998.

Controlling Control Systems: An Application of Evolving Retrenchment

Michael Poppleton[1] and Richard Banach[2]

[1] Department of Computing, Open University, Walton Hall,
Milton Keynes MK7 6AA, UK,
m.r.poppleton@open.ac.uk,
http://mcs.open.ac.uk/mp529
[2] Department of Computer Science, Manchester University,
Manchester M13 9PL, UK,
banach@cs.man.ac.uk,
http://www.cs.man.ac.uk/~banach/

Abstract. We review retrenchment as a liberalisation of refinement, for the description of applications too rich (e.g. using continuous and infinite types) for refinement. A specialisation of the notion, *evolving retrenchment*, is introduced, motivated by the need for an approximate, evolving notion of simulation. The focus of the paper is the case study, a substantial second-order linear control system. The design step from continuous to zero-order hold discrete system is expressible as an evolving retrenchment. Thus we demonstrate that the retrenchment approach can formalise the development of useful applications, which are outside the scope of refinement.

The work is presented in a data type-enriched language containing the B language of J.-R. Abrial.

1 Introduction

From early concerns about proving correctness of programs such as Hoare's [23] and Dijkstra's [15], a mature *refinement calculus* of specifications to programs has developed. Thorough contemporary discussion can be found in [14,2]. In particular, the *simulation* proof method, appearing in *inter alia* [34,32], is central.

In this context of model-based specifications the term "refinement" has a very precise meaning; according to Back and Butler [3] it is a "...correctness-preserving transformation...between (possibly abstract, non-executable) programs which is transitive, thus supporting stepwise refinement, and is monotonic with respect to program constructors, thus supporting piecewise refinement." Relationally, refinement is characterised as a development step requiring the concrete precondition to be weaker than the abstract (the applicability, or termination condition), and the concrete transition relation to be stronger, or less nondeterministic, than the abstract (the correctness, or transition condition). The most succinct characterisation of refinement is that of "operational indistinguishability", i.e. that every concrete behaviour be a possible abstract one.

D. Bert et al. (Eds.): ZB 2002, LNCS 2272, pp. 42–61, 2002.

Refinement is a strong technique in software development, both in descriptive power, and in delivering proof obligations that assert strongly coupled structure between levels of abstraction. So it is not surprising that refinement cannot be used, without simplifying or approximating assumptions and informal justifications, in many real-world system situations. Early work on clean termination of programs [11,10] shares our concern with refining specifications on abstract, infinite domains to finite computer-oriented domains. Another approach to this finiteness problem was Neilson's thesis [24], which proposed a notion of acceptably inadequate design, i.e. that refinement over an infinite domain could be regarded as a limit of finite refinements. Partial logic approaches have also been proposed [26].

This work develops and applies the *retrenchment* method, a liberalisation of refinement. We argued, when introducing the notion [5,6], for a weakening of the retrieve relation over the operation step, allowing concrete non-simulating behaviour in retrenchment. Concrete I/O may have different type to the abstract counterpart, and moreover the retrenchment relation may define fluidity between state and I/O components across the development step from abstract to concrete model. This early work made a more engineering-oriented than formal case for moving away from refinement.

[29] reported initial work, using transitivity and monotonicity arguments, on the development of a calculus of retrenchment in B. This calculus was completed in [31], which showed all primitive operators of the B GSL, including operation sequence, to be monotonic with respect to retrenchment. [8,9] gave a substantial theoretical presentation and development of the retrenchment framework, exploring the landscape between refinement, simulation and retrenchment. This was done by means both of B and automata-theoretic formalisms, using specialised (if widely applicable) further assumptions to nudge the retrenchment relation closer to refinement-like structure. [7] examined what can be said when a simulation is partial, or *punctured*, rather than full. This allows the description of breaks, or punctures (e.g. failure and reset) in the simulation. [4] addressed the integration of refinement and retrenchment from a methodological perspective. In [30] we presented a generalisation, *evolving retrenchment*.

This paper is squarely in the B setting (albeit with some data-type enrichment), and uses the generalisation (rather than specialisation) that is evolving retrenchment. The motivation for this variant of retrenchment is the need for an approximate, evolving notion of simulation. In simple retrenchment, a simulation fails at the point at which an abstract/concrete operation step fails to establish the retrieve relation. Here we propose that the retrieve relation be allowed to *evolve* in the sense that the precision of representation may change over time. We give a simple supporting example of the transformation of real to floating-point addition. As an evolving retrenchment, this transformation demonstrates the approximate nature of the simulation of real by FP addition. It also serves as a formal bound on error propagation, thus giving a link to classical numerical analysis. The introductory example is completed with a demonstration of how the monotonicity of evolving retrenchment with respect to operation sequence supports an evolving notion of simulation.

The focus of this paper is the case study, a substantial second-order linear control system. We summarise the classical control engineering design transformation from continuous- to discrete-time using the zero-order hold approximation. Two evolving retrenchment descriptions are given for this transformation, in a B framework enriched with appropriate data types.

The paper contents are as follows. In Sect. 2 we briefly recap syntactic and semantic definitions for retrenchment. Evolving retrenchment is defined and demonstrated in a small example. Section 3 introduces the case study, a second-order linear control system. This is done as two descriptions, a classical continuous control system, and the zero-order hold discrete approximation. Section 4 gives two possible design formulations in terms of evolving retrenchment. We discuss the limitations of a method such as B for such a formulation. Section 5 concludes.

2   From Simple to Evolving Retrenchment

The notion of an evolving relationship between system models emerges from the intrinsically approximate nature of modelling real-world systems which are described with continuous mathematics. For example, to model the motion of a projectile near to the earth we might use a linear differential equation relating the acceleration of the projectile to the forces of gravity and air friction. We might choose to omit, for practical reasons, the Coriolis effect, or the effect of time-varying crosswinds. So the best mathematical model available to us will always be approximate, for reasons of cost, practicality, or even lack of mathematical sophistication.

This work is, however, concerned with the relationship between mathematical models at adjacent levels of abstraction. In the projectile example we might (unwisely choosing a large reduction in abstraction level) choose to specify the next layer of model using some finite set of floating-point numbers appropriate to computer implementation. Clearly, a second level of approximation is introduced: before any system dynamics are considered, precision is limited to that determined by the floating-point set chosen. Moreover, when some abstract (continuous) operation is considered in relation to its concrete (floating point) counterpart, precision of representation is in general reduced. For simple addition, precision decays by a factor of two. For a more complex operation and its retrenchment, say exponentiation, the precision decay is far more complex.

From a refinement point of view this suggests that relaxing the invariant nature of the retrieve relation between abstract and concrete models, allowing it to *evolve*, might give a more expressive development step than either refinement or simple retrenchment. Although retrenchment allows breach of the invariant (via establishment of the concession), such breach effectively stops any simulation of the abstract operation sequence by the concrete. Specialisations of retrenchment, as indicated before, seek to combine the concession achieved with application domain structure to re-establish the retrieve relation where the simulation has failed to establish it. The approach here is more general in relaxing the immutability of the retrieve relation, and allowing it to evolve. That is, if one operation step results under retrenchment in achievement of a weaker, evolved retrieve relation, this latter relation can serve as a starting point for a subsequent retrenchment step.

2.1   Simple Retrenchment

Simple retrenchment is, loosely speaking, the strengthening of the precondition, the weakening of the postcondition, and the introduction of mutability between state and I/O at the two levels of abstraction.

Figure 1 defines the B syntax. Abstract machine M has parameter a, state variable u, and invariant predicate $I(u)$. Variable u is initialised by substitution $X(u)$, and is operated on by operation $OpName$, a syntactic wrapper for substitution $S(u, i, o)$, with input i and output o. Unlike a refinement, which in B is a construct derived from the refined machine, a retrenchment is an independent MACHINE. Thus N is a machine with parameter b (not necessarily related to a), state variable v, invariant $J(v)$, initialisation $Y(v)$, and operation $OpNameC$ as wrapper for $T(v, j, p)$, a substitution with input j and output p. Viewed as an independent machine, N cannot refer to M. So the new retrenchment syntax proposed here must in this case be regarded as null text.

Here we state that machine N RETRENCHES machine M; in general either a MACHINE or a REFINEMENT may be retrenched. The RETRENCHES clause (similarly to REFINES) makes visible the contents of the retrenched construct. We further assume that the name spaces of the retrenched and retrenching constructs are disjoint, but admit an injection of (retrenched to retrenching) operation names. An *injection* allows further, independent dynamic structure in the retrenching machine. In the retrenching machine N we have state variable v under invariant $J(v)$, and the RETRIEVES clause $G(u, v)$. The existence of N as an independent machine requires that its (local) invariant be stated independently of the retrieve relation, unlike in B, where the two take the form of a joint predicate.

The relationship between concrete and abstract state is fundamentally different *before* and *after* the operation. We model this by distinguishing between a strengthened before-relation between abstract and concrete states, and a weakened after-relation. Thus the syntax of the concrete operation $OpNameC$ in N is precisely as in B, with the addition of the *ramification*, a syntactic enclosure of the operation. We call this augmented syntax for the operation a *ramified generalised substitution*. Before, the relationship is constrained (precondition is strengthened) by the new *WITHIN* condition $P(u, v, i, j, A)$ which may change the balance of components between input and state, and may further define an optional "memory" variable LVAR A for reference in the after-state. The after-relation is weakened (postcondition is weakened) by the CONCEDES clause (the *concession*) $C(u, v, o, p, A)$, which specifies what the operation guarantees to achieve (in terms of after-state, output and logical variable A) if it cannot

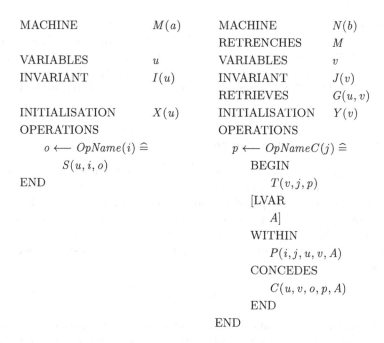

Fig. 1. Syntax of simple retrenchment

maintain the retrieve relation G. The concession will describe the weaker state of representation after the operation, or preserve exception information if representation fails completely.

The operation retrenchment POB is central in this theory since it serves as the semantic definition of retrenchment. Analogously to the operation refinement POB, although with more predicate information, concrete $OpNameC$ (substitution T) retrenches abstract operation $OpName$ (substitution S) if:

$$I(u) \wedge G(u,v) \wedge J(v) \wedge P(i,j,u,v,A) \wedge \mathsf{trm}(T(v,j,p))$$
$$\Rightarrow \mathsf{trm}(S(u,i,o)) \wedge [T(v,j,p)] \neg [S(u,i,o)] \neg (G(u,v) \vee C(u,v,o,p,A)) \quad (1)$$

The retrenchment initialisation POB is as for refinement, guaranteeing a joint starting state satisfying $G(u,v)$.

Definition (1) is justified by comparison with the refinement proof obligation. In refinement, "operational indistinguishability" of the refined operation is the central issue. When the abstract operation terminates, so does the concrete one: this is precondition weakening. In retrenchment we invert this relationship. The retrenching machine is different from the retrenched machine and more faithful to the system being built. It can contain features incompatible with those of the abstract machine. These are described explicitly in the CONCEDES clause, and implicitly in the WITHIN clause which restricts the portions of the two models being related. To the extent that the retrenching machine might not terminate in places that the retrenched machine is guaranteed to do so, we must consciously exclude such possibilities in the WITHIN clause from the relationship asserted between the two machines. Thus we strengthen the precondition by swapping the two termination clauses in the POB, and use the WITHIN clause to tailor the scope of the POB over and above what is already stated in the RETRIEVES clause.

2.2 Evolving Retrenchment

We now propose that the retrieve relation G becomes "variant" (i.e. evolves), in the sense that it becomes mediated by some "precision" parameters α and β. That is, $G_\alpha(u, v)$ is required to start the retrenchment step, and $G_\beta(u', v')$ is a possible outcome. The intuition behind this is that α, β belong to some ordered set, where increasing α denotes decreasing precision of the representation of abstract u by concrete v in $G_\alpha(u, v)$. That is, we will usually (but not always) expect that

$$\alpha \leq \beta \Rightarrow (G_\alpha \Rightarrow G_\beta)$$

This formulation describes a typical precision-decay situation over a simulation, for example in a sequence of arithmetic steps (real to floating point). We choose the notion of *evolving* rather than *decaying* retrenchment as this suggests, because it is quite possible for precision to *increase* over a simulation step. In a control system, sensor readings provide the software with the inputs necessary to model the real system state. In control systems with a large number of sensors, it is quite likely that, at a given point in time, some sensors have failed, and that the modelled state is representing the real system state in some degraded mode[1]. Some algorithm extrapolating the failed sensor value from neighbouring sensors may be applied, to make the best of things, but accuracy is nonetheless lost. Assuming it is possible for sensors to be repaired in-flight, and for repair status to be detectable, it is manifestly possible for precision of the system state representation to improve.

The extension to syntax for evolving retrenchment involves some subscripting of clauses G, P, C in Fig. 1. Semantically, *evolving retrenchment* is defined:

$$I(u) \wedge G_\alpha(u, v) \wedge J(v) \wedge P_\alpha(i, j, u, v, A) \wedge \mathsf{trm}(T)(v, j) \wedge \alpha \leq \beta$$
$$\Rightarrow \mathsf{trm}(S)(u, i) \wedge [T(v, j, p)] \neg [S(u, i, o)] \neg (G_\beta(u, v) \vee C_\beta(u, v, o, p, A)) \quad (2)$$

The following shorthand form will be more convenient:

$$S \lesssim T \ \ \text{w.r.t.} \ \ (G_\alpha, P_\alpha \longrightarrow G_\beta, C_\beta)$$

Thus the (u, v) relationship is now mediated through evolving precision parameters α and β. Notice that WITHIN and CONCEDES predicates are also parameterised: this is necessary since each of these further constrain the (u, v)

---

[1] This assumes reliable sensor failure determination, an issue we will not pursue here.

relationship, in context of inputs and outputs respectively. Generalising these predicates also allows these relations between abstract/concrete I/O and state to evolve. Note that the precision parameters are not free variables, they are more akin to B machine parameters. A relation of evolving retrenchment between two machines M and $N(\alpha, \beta)$ can be interpreted as a collection of simple retrenchments indexed on the carrier sets of precision parameters α, β. This collection describes the (usually decaying) evolution of the RETRIEVES relationships involved at each step.

2.3   From an Example to Simulation

A simple arithmetic example will demonstrate the use of evolving retrenchment. Floating point arithmetic is well understood, e.g. [19]. We choose the retrenchment of real to floating-point addition, simplifying the scenario to highlight the evolving approximation of representation that we wish to describe. Figure 2 shows the example: the upper line of numbers represents a sequence of three real additions and the lower line represents the corresponding floating point sequence. Δ represents the error bound at each step.

Abstract model: $\mathbb{R}$

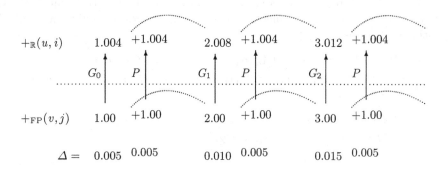

Concrete model: floating point $\mathrm{FP}(10, 4, e)$

Fig. 2. Evolving retrenchment of real to floating point addition

We restrict ourselves, using the WITHIN clause, to (abstract state and input variables u, i) the real interval $[1, 9.9995)^2$, and the corresponding set of base-ten floating point numbers with four decimal places (concrete state and input variables v, j). e is the limiting exponent value for FP representation. We use a

---

[2] This overloaded syntax is the classical real notation for a semi-open interval including 1 and excluding 9.9995

rounding regime for FP addition. We have (using some syntactic shorthand for clause P):

$$I(u) \mathrel{\hat{=}} u \in \mathbb{R} \qquad\qquad J(v) \mathrel{\hat{=}} v \in \mathrm{FP}(10, 4, e)$$
$$OpA(u, i) \mathrel{\hat{=}} u := +_{\mathbb{R}}(u, i) \qquad\qquad OpC(v, j) \mathrel{\hat{=}} v := +_{\mathrm{FP}}(v, j)$$
$$G_k(u, v) \mathrel{\hat{=}} \mid u - v \mid \,\le 5 * (k + 1) * 10^{-3} \text{ for } k \in \mathbb{N}$$
$$P(i, j, u, v) \mathrel{\hat{=}} u, i, u + i \in [1, 9.9995)$$
$$\wedge\; v, j \ge 1 \wedge v, j \le 9.999 \wedge \mid i - j \mid \,\le 5 * 10^{-3}$$
$$C_k(u, v) \mathrel{\hat{=}} G_{k+1}(u, v) \tag{3}$$

Because these are simple assignment operations, the operations terminate universally. Initial error G_0 is the standard state representation error $5 * 10^{-3}$ for these FP parameters; the evolving error results from the accumulation of the constant input representation error $5 * 10^{-3}$. So the evolving retrenchment POB is a succinct formal statement that the retrieve relation will degrade by at most $5 * 10^{-3}$ every step:

$$I(u) \wedge G_k(u, v) \wedge J(v) \wedge P(i, j, u, v)$$
$$\Rightarrow [+_{\mathrm{FP}}(v, j)]\neg\, [+_{\mathbb{R}}(u, i)]\neg\, (G_k(u, v) \vee C_k(u, v)) \tag{4}$$

The apparently curious concession definition is explained by the POB: the last subclause could more clearly be written $(G_k(u, v) \vee G_{k+1}(u, v))$.

From [30], we repeat the monotonicity statement for operation sequence with respect to evolving retrenchment. This is a simpler form for universally substituting operations:

$$S1 \precsim T1 \text{ w.r.t. } (G_\alpha, P1_\alpha(i_1, j_1, u, v) \longrightarrow G_\beta, C1_\beta(u, v, o_1, p_1))$$
$$\wedge\; S2 \precsim T2 \text{ w.r.t. } (G_\beta, P2_\beta(i_2, j_2, u, v) \longrightarrow G_\gamma, C2_\gamma(u, v, o_2, p_2))$$
$$\wedge\; I \wedge G_\alpha \wedge J$$
$$\wedge\; P1 \wedge [T1]\neg\, [S1]\neg\, (G_\beta \wedge P2_\beta)$$
$$\vdash [T1; T2]\neg\, [S1; S2]\neg\, (G_\gamma \vee C2_\gamma)$$
$$\text{...that is}$$
$$S1; S2 \precsim T1; T2 \text{ w.r.t. } (G_\alpha, P_{\beta^i} \longrightarrow G_\gamma, C2_\gamma) \tag{5}$$
$$\text{...where } P_{\beta^i} \text{ denotes last line of hypothesis}$$

The POB (4) shows that starting from G_k, we might achieve G_k, but certainly will achieve the weaker G_{k+1}. Hence the monotonicity property is directly applicable to guarantee a two-step transition from G_k to at least G_{k+2}, and so on, inductively.

This formalises the sequential composition of the OpA/OpC retrenchments into what we might call an *evolving simulation*. This precisely describes the evolution of the representation. This is a formal statement of error propagation, a subject thoroughly analysed in classical numerical analysis such as [1]. Thus evolving retrenchment points the way to the exploitation of numerical analysis in prover-supported verification of continuous system development.

3   Case Study: Second-Order Linear Control System

We now test the intuition of evolving retrenchment against the kind of continuous application it is intended for. The case study is based on a classical second-order linear control system. Typical control engineering practice [27] is to design a discrete, computer based system ("digital system", in control engineering parlance) by approximation to the continuous. We will describe this approximate design step as a retrenchment.

A continuous control system is analogue in structure, and has analogue connections to the plant under control. System dynamics and control vary continuously in time. On the other hand, a digital system emphasises the discrete computer and interfaces with the continuous plant through sensors and A-D converters. These interfaces are discrete by virtue of being sampled, say every T seconds.

The discrete system is inevitably an approximation of the continuous, losing all information about state and input dynamics between the sample points. The precision of the approximation is a function of time. The design step from continuous to discrete control is thus an obvious candidate for description as a retrenchment.

We will use the "zero-order hold", or ZOH approximation [27,18]. In practice this is the preferred choice over a number of approximation methods. It is named after the fact that the discrete controller output $y_d(kT)$, available only at discrete sample times, is converted into continuous $y(t)$ by the ZOH chip as follows:

$$y(t) \mathrel{\hat=} y_d(kT) \qquad \text{for } kT \le t < kT + T \tag{6}$$

The aim of the retrenchment to be presented is to give a precise formal description of the precision of approximation of a continuous by a discrete control system. Such a description is typical of the style sought by formal methods practitioners, i.e. capable of formal verification. Such a description cannot be provided by refinement.

Figure 3 shows a second-order linear system describing the control of a simple industrial process. A mass m is attached to a wall by a spring k_1/dashpot k_2 damper assembly. It is constrained to move horizontally, in one dimension only. The mass is also subject to external disturbance $d(t)$ and an operator input $x_i(t)$, representing the desired reference position of the mass. The actual position (of the centre of gravity of the mass) is the system output $x_o(t)$. The uncontrolled acceleration of m, by elementary physics, then depends on $k_1x_o(t)$, $k_2\dot{x}_o(t)$, and $d(t)$. A controller force is required to counteract the disturbance and to move the mass towards the reference position; a typical such force is defined by the equation:

$$\text{controller force } = k_3(x_o(t) - x_i(t)) + k_4\dot{x}_o(t) \tag{7}$$

The system is described by the equation:

$$m\ddot{x}_o(t) = -k_1x_o(t) - k_2\dot{x}_o(t) - k_3(x_o(t) - x_i(t)) - k_4\dot{x}_o(t) + d(t) \tag{8}$$

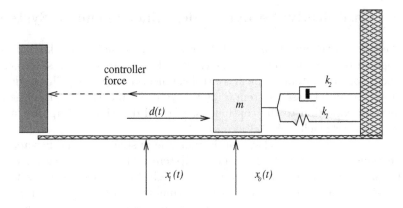

Fig. 3. A second-order linear control system

For reasons of space, only those results of the control-theoretic analysis necessary for understanding and for the retrenchment description are given; see [31] for a full analysis.

3.1 A Continuous State Space Model

We describe the case study using the state-space formulation of modern control theory. The (control-theoretic) state of a dynamic system is defined to be the vector of minimum dimensionality of variables $\mathbf{y}(t)$, such that knowledge of $\mathbf{y}(t_0)$ and all system inputs for all time after t_0, completely characterises the system behaviour.

To describe a model such as the above in state-space, we "flatten" an nth order linear differential equation into n first order equations (here $n=2$). Part of this process is renaming output and input variables and their derivatives to facilitate the matrix presentation below. For our example we assign state variable y_i to each derivative $x_o^{(i-1)}$. Thus y_1 represents mass position x_o and y_2 represents mass velocity $\dot{x}_o$. Inputs are renamed as components of a vector $\mathbf{u}$. Here, the following system of equations is generated:

$$y_1 = x_o$$
$$y_2 = \dot{x}_o = \dot{y}_1$$
$$\dot{y}_2 = \ddot{x}_o = -\frac{1}{m}\left((k_1 + k_3)y_1 + (k_2 + k_4)y_2\right) + \frac{1}{m}(k_3 u_1 + u_2) \qquad (9)$$
$$\dots \text{ where}$$
$$u_1(t) = x_i(t) \qquad u_2(t) = d(t)$$

Thus the system can be described by the following matrix equation:

$$\dot{\mathbf{y}} = \mathbf{A}\mathbf{y} + \mathbf{B}\mathbf{u} \qquad \dots \text{ that is}$$

$$\begin{bmatrix} \dot{y}_1 \\ \dot{y}_2 \end{bmatrix} = \begin{bmatrix} 0 & 1 \\ -\frac{1}{m}(k_1 + k_3) & -\frac{1}{m}(k_2 + k_4) \end{bmatrix} \begin{bmatrix} y_1 \\ y_2 \end{bmatrix} + \begin{bmatrix} 0 & 0 \\ \frac{k_3}{m} & \frac{1}{m} \end{bmatrix} \begin{bmatrix} u_1 \\ u_2 \end{bmatrix} \tag{10}$$

Laplace transformation, e.g. [13,16], gives

$$\mathbf{Y}(s) = \varPhi(s)\mathbf{y}(0) + \varPhi(s)\mathbf{B}\mathbf{U}(s) \qquad \text{... where} \tag{11}$$
$$\varPhi(s) = (s\mathbf{I} - \mathbf{A})^{-1} \tag{12}$$

$\mathbf{Y}(s), \varPhi(s), \mathbf{U}(s)$ are the Laplace transforms of $\mathbf{y}(t), \varPhi(t), \mathbf{u}(t)$ respectively, where $\varPhi(t)$ is called the *transition matrix* of the system. For all $t \geq 0$, it is convenient that $\varPhi(t) = e^{\mathbf{A}t}$. $\varPhi(t)$ is calculated by inverse Laplace transformation of $\varPhi(s)$, one element at a time, by recourse to a table of Laplace transforms such as [13].

The important point here is that output $\mathbf{y}(t)$ is the sum of an initial-conditions term independent of input, and an integral term over the input. The time-domain solution for the linear system (10) is

$$\mathbf{y}(t) = \varPhi(t)\mathbf{y}(0) + \int_0^t \varPhi(t - \tau)\mathbf{B}\mathbf{u}(\tau)\, d\tau \tag{13}$$

3.2   A Discrete State Space Model

The discrete model is conceptually closer to being implementable, since it assumes input $\mathbf{u}(t)$ is only available at sampling points of interval T. Information is only needed at these discrete time points: the state trace is now a sequence rather than a function of real-valued time.

The approximation lies in assuming the input to be piecewise constant, with value $\mathbf{u}(kT)$ across the sampling interval $[kT, kT+T)$. [27] shows that continuous system (10), with input piecewise constant in this way, can be described exactly as the following difference equation in state space. We call the discrete state $\mathbf{y}_d$ to distinguish it from continuous state $\mathbf{y}$. For our purposes $\mathbf{y}_d$ is only defined at sample times kT for $k \in \mathbb{N}$, although for practical purposes $\mathbf{y}_d$ is piecewise constant in continuous time by use of the ZOH chip.

$$\mathbf{y}_d(kT + T) = \mathbf{A}_d(T)\mathbf{y}_d(kT) + \mathbf{B}_d(T)\mathbf{u}(kT) \qquad \text{... where}$$
$$\mathbf{A}_d(T) = \varPhi(T) = e^{\mathbf{A}T} \qquad \mathbf{B}_d(T) = \int_0^T \varPhi(\tau)\mathbf{B}\, d\tau \tag{14}$$

This picture is "local" to the k'th sampling interval. A straightforward derivation gives an analogous local expression to (14) for the continuous model, over an interval of length T, from (13). We give the continuous and discrete state space equations over interval $[kT - T, kT)$ for comparison:

$$\mathbf{y}(kT) = e^{\mathbf{A}T}\mathbf{y}(kT - T) + \int_0^T e^{\mathbf{A}\tau}\mathbf{B}\mathbf{u}(kT - \tau)\, d\tau \tag{15}$$
$$\mathbf{y}_d(kT) = e^{\mathbf{A}T}\mathbf{y}_d(kT - T) + \mathbf{B}_d(T)\mathbf{u}(kT - T) \tag{16}$$

Given an input function $\mathbf{u}(t)$, then, we are interested in the approximation of continuous by discrete state, and its description as a retrenchment. That is, we seek an explicit expression for the difference function $\Delta(t) \mathrel{\hat{=}} \mathbf{y}(t) - \mathbf{y}_d(t)$, where from (15,16) we have

$$\Delta(kT) = e^{\mathbf{A}T} \Delta(kT - T)$$
$$+ \int_0^T e^{\mathbf{A}\tau} \mathbf{B} \mathbf{u}(kT - \tau)\, d\tau - \mathbf{B}_d(T)\mathbf{u}(kT - T) \qquad (17)$$

3.3   Making the Approximation Explicit

Provided $\mathbf{A}$ is nonsingular (as is required by a nontrivial system (10)), then we have

$$\mathbf{B}_d(T) = \int_0^T e^{\mathbf{A}\tau} \mathbf{B}\, d\tau = \mathbf{A}^{-1} e^{\mathbf{A}\tau} \mathbf{B}\Big|_0^T = \mathbf{A}^{-1}(e^{\mathbf{A}T} - \mathbf{I})\mathbf{B} \qquad (18)$$

Thus the discrete contribution to error $\Delta(kT)$ from the k'th interval can be calculated exactly, giving a matrix of exponential-sinusoidal expressions in T and model parameters. The matrix $\mathbf{B}_d(t)$ (which we do not give here) is independent of k. Judicious Taylor expansion gives

$$\int_0^T e^{\mathbf{A}\tau} \mathbf{B} \mathbf{u}(kT - \tau)\, d\tau - \mathbf{B}_d(T)\mathbf{u}(kT - T)$$
$$= \begin{bmatrix} O(T^3) \\ \frac{T^2}{2m}(k_3 \dot{u}_1(kT - T) + \dot{u}_2(kT - T)) + O(T^3) \end{bmatrix} \qquad (19)$$

To $O(T^3)$, we see that for step k the discrete model approximates the continuous y_1, i.e. the system output x_o, exactly. The step k contribution to the error in y_2, i.e. velocity $\dot{x}_o$, is a term in T^2 and the first derivatives of the inputs. If the expansion is performed to $O(T^4)$ then the derivatives of the inputs can be seen to contribute to error in x_o.

4   Control System Design as an Evolving Retrenchment

The basis of arguing that this is a case for retrenchment is now clear: (17) and (19) provide the means, given error of approximation $\Delta(kT - T)$ at start of step k, to describe the error $\Delta(kT)$ at the end of that step.

We describe the retrenchment in the B framework of this paper, but again need to imagine the language is further enriched. As a minimum we need the reals, differentiable functions on the reals, vectors and matrices, and the associated tools of real and complex analysis implicit in this work. The syntax and semantics of B predicates and formulas needs enrichment to make statements about such types, e.g. (10) and (15).

Note that in this section, as we discuss the composition of the difference vector $\Delta(kT)$ at time kT (17), we will refer to its two summands as the *history*

Δ-vector $e^{\mathbf{A}T}\Delta(kT - T)$ and the *previous* Δ-vector (19). We will call $\Delta(kT)$ the *current* Δ-vector. Also note that from here we will use the term "state" in the sense of state variables in a retrenchment model, rather than in the control-theoretic sense.

The abstract, extended-B model is the continuous state-space model (10). State variables are given by the 2-vector $\mathbf{y}(t) = \{y_1(t), y_2(t)\}^3$ of real-valued functions, which describe system output $x_o(t)$ and its derivative. Operationally, $\mathbf{y}$ will be defined in terms of the solution (13) to the linear system (10). The retrenchment relation will represent that continuous behaviour at the discrete sample points (multiples kT of sample time T) of the concrete model[4]. The abstract model respects the granularity of the discrete in the sense that, for step k, $\mathbf{y}$ is only defined over time interval $[kT - T, kT]$. Abstract inputs for step k are given by the input vector $\mathbf{u} = \{u_1(t), u_2(t)\}$ and its derivative $\dot{\mathbf{u}} = \{\dot{u}_1(t), \dot{u}_2(t)\}$ over $[(k-1)T, kT]$.

Since the system behaviour is defined in terms of the solution (13) to differential equation (10), the latter is clearly appropriate as abstract invariant, conjoined with a suitable typing statement:

$$I(\mathbf{y}) \,\hat{=}\, \dot{\mathbf{y}} = \mathbf{A}\mathbf{y} + \mathbf{B}\mathbf{u} \land \mathbf{y} \in [kT - T, kT] \to \mathbb{R}^2 \tag{20}$$

The abstract operation $OpA(k)$ for step k is defined in terms of the solution[5] (13). The abstract initialisation is simply $OpA(1)$:

$$OpA(k) \,\hat{=}\, \mathbf{y} := \lambda\, t : [kT - T, kT] \bullet e^{\mathbf{A}(t - kT + T)}\mathbf{y}(kT - T)$$
$$+ \int_0^{t - kT + T} e^{\mathbf{A}\tau}\mathbf{B}\mathbf{u}(t - \tau)\, d\tau \tag{21}$$

$$InitA \,\hat{=}\, OpA(1) \tag{22}$$

Operations OpA and its concrete counterpart OpC (26) are assignments and therefore terminate universally. In retrenchments like this we can thus dispense with concern about termination.

The usual operation consistency obligation for OpA is

$$I(\mathbf{y}) \Rightarrow [OpA(k)]I(\mathbf{y}) \tag{23}$$

This is straightforwardly discharged by differentiating the OpA definition (21) with respect to t by use of the Leibnitz rule [22] for differentiating across an integral:

$$\ldots \text{If} \qquad F(t) = \int_{A(t)}^{B(t)} f(t, \lambda)\, d\lambda$$

---

[3] We enclose horizontally displayed vectors in parentheses {}.

[4] As in Sect. 2.2, we really mean a collection of k-indexed simple retrenchments.

[5] Note the logical close coupling that differentiation gives to invariant and behaviour; this is unusual in B models.

$$\ldots \text{ then } \quad \frac{\mathrm{d}}{\mathrm{d}t} \int_{A(t)}^{B(t)} f(t,\lambda)\,\mathrm{d}\lambda = \int_{A}^{B} \frac{\delta f(t,\lambda)}{\delta t}\,\mathrm{d}\lambda + f(t,B)\frac{\mathrm{d}B}{\mathrm{d}t} - f(t,A)\frac{\mathrm{d}A}{\mathrm{d}t}$$

(24)

The rule is valid over an interval (a, b) for t provided $f, \frac{\delta f}{\delta t}$ are continuous on $[a, b]$, $A'(t), B'(t)$ are continuous on (a, b), and $A(t) \leq \lambda \leq B(t)$. $f(t, \lambda)$ is the integrand term in (21), including the input term $\mathbf{u}(t)$. This rule imposes conditions on the input, but these are no more onerous than those imposed by our use of Taylor's theorem in Sect. 3.3.

The concrete model is given by the discrete state-space model (16). We call the state vector $\mathbf{y}_d = \{y_{d1}, y_{d2}\}$. Concrete inputs are given by the input vector $\mathbf{u}$; assume for now these are identical to the abstract input values at times kT.

It is a feature of the ZOH design method that (10) is approximately satisfied by the discrete model (16). The discrete world is not rich enough to state this in the discrete invariant; indeed, to attempt to do so would breach separation of concerns in design. The retrenchment will describe this approximation. For the concrete invariant a simple bounding constraint (for some $L_1, U_1, L_2, U_2 \in \mathbb{R}$), which serves as a typing predicate, will suffice:

$$J(\mathbf{y}_d) \mathrel{\hat{=}} y_{d1} \in [L_1, U_1] \wedge y_{d2} \in [L_2, U_2]$$

(25)

The concrete operation is defined in the obvious way from (16):

$$OpC(k) \mathrel{\hat{=}} \mathbf{y}_d := e^{\mathbf{A}T}\mathbf{y}_d + \mathbf{B}_d(T)\mathbf{u}(kT - T)$$

(26)

Initialisation is in terms of user-defined initial conditions $\mathbf{y}_{dINIT}$ at time T, not necessarily the same as $\mathbf{y}(T)$. However, the refinement of initialisations POB (30) dictates how close $\mathbf{y}(T)$ and initial-state $\mathbf{y}_d$ should be.

$$InitC \mathrel{\hat{=}} \mathbf{y}_d := \mathbf{y}_{dINIT}$$

(27)

The operation consistency obligation for OpC is established trivially from the OpC definition (26):

$$J(\mathbf{y}_d) \Rightarrow [OpC(k)]J(\mathbf{y}_d)$$

(28)

The retrieve relation G decomposes the current Δ-vector $\Delta(kT)$ into its two elements, and does not refer to $\Delta(kT-T)$. G bounds $\Delta(kT)$ in terms of constants $\varepsilon_k, \dot{\varepsilon}_k{}^6$: the designer chooses bounds $\varepsilon_1, \dot{\varepsilon}_1$, and we will shortly see how the evolving bounds are defined inductively. For $k \in \mathbb{N}_1$:

$$G(\mathbf{y}, \mathbf{y}_d, k, \varepsilon_k, \dot{\varepsilon}_k) \mathrel{\hat{=}} |y_1(kT) - y_{d1}| \leq \varepsilon_k \wedge |y_2(kT) - y_{d2}| \leq \dot{\varepsilon}_k$$

(29)

We require refinement of initialisations to state the precision of approximation at start time T:

$$[InitC]\neg\,[InitA]\neg\,G(\mathbf{y}, \mathbf{y}_d, 1, \varepsilon_1, \dot{\varepsilon}_1)$$

(30)

---

[6] We use a little lexical licence in naming the error bound on the derivative term $\dot{\varepsilon}_k$.

Next we seek the remaining syntactic components of the evolving retrenchment of $OpA(k)$ by $OpC(k)$, as defined by (2) – WITHIN clause P and CONCEDES clause C:

$$I(\mathbf{y}) \wedge G(\mathbf{y}, \mathbf{y}_d, k - 1, \varepsilon_{k-1}, \dot{\varepsilon}_{k-1}) \wedge J(\mathbf{y}_d) \wedge P(\mathbf{y}, \mathbf{y}_d, \mathbf{u}, \dot{\mathbf{u}}, k - 1)$$
$$\Rightarrow [OpC(k)] \neg [OpA(k)]$$
$$\neg (G(\mathbf{y}, \mathbf{y}_d, k - 1, \varepsilon_{k-1}, \dot{\varepsilon}_{k-1}) \vee C(\mathbf{y}, \mathbf{y}_d, \mathbf{u}, \dot{\mathbf{u}}, k, \varepsilon_k, \dot{\varepsilon}_k)) \qquad (31)$$

We will not use the logical variable A and expect that the concession will state different error bounds $\varepsilon_k, \dot{\varepsilon}_k$ to those of $G(k - 1)$. For the WITHIN clause it suffices to define:

$$P(\mathbf{y}, \mathbf{y}_d, \mathbf{u}, \dot{\mathbf{u}}, k - 1) \,\hat{=}\, \text{true} \qquad (32)$$

Given (29), at time $kT - T$ we have for each element of the history Δ-vector

$$\left| \{e^{\mathbf{A}T} \Delta(kT - T)\}_i \right| \leq \left| e_{i1}^{\mathbf{A}T} \right| |\Delta(kT - T)_1| + \left| e_{i2}^{\mathbf{A}T} \right| |\Delta(kT - T)_2|$$
$$\leq \left| e_{i1}^{\mathbf{A}T} \right| \varepsilon_{k-1} + \left| e_{i2}^{\mathbf{A}T} \right| \dot{\varepsilon}_{k-1} \qquad (33)$$

For the previous Δ-vector we have, from (19), for some constant M_1, M_2:

$$|\{\text{previous } \Delta\text{-vector}\}_1| \leq M_1 T^3$$

$$|\{\text{previous } \Delta\text{-vector}\}_2| \leq \frac{1}{2m} |k_3 \dot{u}_1(kT - T) + \dot{u}_2(kT - T)| T^2 + M_2 T^3 \quad (34)$$

The CONCEDES clause C will bound the current difference vector $\Delta(kT)$ according to (33, 34). That is, C has effectively been derived in the (implicit, in this paper) error analysis (19) of difference function (17). Hence C will always be derivable by a prover sufficiently rich in theory and strategy to reproduce this reasoning.

C is free in all state variables and abstract input derivatives $\dot{\mathbf{u}}$:

$$C(\mathbf{y}, \mathbf{y}_d, \dot{\mathbf{u}}, k) \,\hat{=}\,$$
$$|y_1(kT) - y_{d1}| \leq \left| e_{11}^{\mathbf{A}T} \right| \varepsilon_{k-1} + \left| e_{12}^{\mathbf{A}T} \right| \dot{\varepsilon}_{k-1} + M_1 T^3$$
$$\wedge |y_2(kT) - y_{d2}| \leq \left| e_{21}^{\mathbf{A}T} \right| \varepsilon_{k-1} + \left| e_{22}^{\mathbf{A}T} \right| \dot{\varepsilon}_{k-1}$$
$$+ \frac{1}{2m} |k_3 \dot{u}_1(kT - T) + \dot{u}_2(kT - T)| T^2 + M_2 T^3 \qquad (35)$$

The RHS's of the two inequalities in (35) are renamed to $\varepsilon_k, \dot{\varepsilon}_k$ respectively. Any concern about defining ε_k and thus G_k in terms of input values can be addressed by reading the last line of (35) in terms of the maximum absolute values, over all time of interest, of $\dot{u}_1(t), \dot{u}_2(t)$[7]. The concession is then in the right form to be read as retrieve relation G for step k. This yields a neat description in terms of *evolving retrenchment*. Since the two bounding expressions $\varepsilon_k, \dot{\varepsilon}_k$ of (35) are functions only of $\varepsilon_{k-1}, \dot{\varepsilon}_{k-1}$ (and fixed model parameters) respectively, we can define for all $k > 1$:

$$G(\mathbf{y}, \mathbf{y}_d, k, \varepsilon_k, \dot{\varepsilon}_k) \,\hat{=}\, C(\mathbf{y}, \mathbf{y}_d, \dot{\mathbf{u}}_{\max}, k - 1) \qquad (36)$$

---

[7] The assumption of such maxima is typical control engineering practice.

and this gives the inductive definition of $\varepsilon_k, \dot{\varepsilon}_k$. Therefore the retrenchment statement (31) becomes:

$$I(\mathbf{y}) \wedge G(\mathbf{y}, \mathbf{y}_d, k - 1, \varepsilon_{k-1}, \dot{\varepsilon}_{k-1}) \wedge J(\mathbf{y}_d) \tag{37}$$
$$\Rightarrow [OpC(k)] \neg [OpA(k)] \neg (G(\mathbf{y}, \mathbf{y}_d, k - 1, \varepsilon_{k-1}, \dot{\varepsilon}_{k-1}) \vee G(\mathbf{y}, \mathbf{y}_d, k, \varepsilon_k, \dot{\varepsilon}_k))$$

This is precisely the type of scenario that the evolving variant of retrenchment was devised to describe. Retrieve relation $G(k - 1)$ of predetermined error may be preserved by step k; the concession $G(k)$ certainly will be. Thus (37) recasts the concession of (31) as an *evolved* retrieve relation[8], which is available for use as the basis for the step $k + 1$ retrenchment.

We briefly analyse the various sources of error contributed by step k, which the concession describes. Taylor expansion error is small at $O(T^3)$. The error from abstract input derivatives $\dot{u}_1, \dot{u}_2$, having a T^2-factor, is only significant for rapidly varying inputs. In the real world of the model, this is only possible for the disturbance force $d(t) = u_2(t)$. Usual engineering practice is to assume disturbance inputs have bounded (if large) derivatives.

The significant factors which multiply start-of-step error $G(k - 1)$ are the elements of transition matrix $e^{\mathbf{A}T}$, i.e. constant expressions in the model parameters. It is thus the transition matrix, independent of external inputs, which will usually determine improvement or decay in the error evolution $\Delta(kT)$. A lower bound on sample time T is determined by the target hardware, but the designer has some freedom in fixing the other model parameters. There are standard control engineering design constraints on parameter-setting, concerned with required system response to representative input signals [25,27]. It may be the case that sufficient design freedom exists to enable model parameters to be chosen such that every $\left| e_{ij}^{\mathbf{A}T} \right|$ is sufficiently less than 1. In this situation the approximation error of the retrenchment can be guaranteed to improve.

4.1   Another Retrenchment

An appealing fact about this ZOH discrete approximation to a continuous control system is that the difference vector $\Delta(kT)$ (17, 19) depends only on derivatives of system inputs (at least to $O(T^3)$). This fact makes it easy to generalise the example to distinguish between abstract and concrete inputs $\mathbf{u}$ and $\mathbf{u}_d$. Typically, precision would be lost in moving from an analogue specification in continuous variables and inputs, to digital hardware-based input sensors with finite precision. Revisiting (19) with input approximation, we have a bigger expression in u_{d1}, u_{d2}. Defining difference vectors for the input representations $\Delta u_i(kT) \hat{=} u_i(kT) - u_{di}(kT)$ for $i \in \{1, 2\}$, we have[9]:

$$\int_0^T e^{\mathbf{A}\tau} \mathbf{B} \mathbf{u}(kT - \tau) \, d\tau - \mathbf{B}_d(T) \mathbf{u}(kT - T)$$

---

[8] This retrenchment suggests a definition of *evolving refinement*.
[9] ρ, ω are derived parameters from the model (8)

$$= \begin{bmatrix} \frac{T^2}{2m}(k_3\Delta u_1(kT-T) + \Delta u_2(kT-T)) + O(T^3) \\ \left(\begin{array}{c} \frac{(T-\rho\omega T^2)}{m}(k_3\Delta u_1(kT-T) + \Delta u_2(kT-T)) \\ +\frac{T^2}{2m}(k_3\dot{u}_1(kT-T) + \dot{u}_2(kT-T)) + O(T^3) \end{array} \right) \end{bmatrix} \quad (38)$$

The extra error introduced by input approximation is a linear combination of the $\Delta u_i(kT-T)$ terms, with factors in powers of T. We can therefore choose WITHIN bounds to input error $\Delta u_i(kT-T)$ that make the overall extra error contributed insignificant. For example, to make

$$|\{\text{previous } \Delta\text{-vector}\}_1| \le \frac{\varepsilon_k}{10} + \frac{\dot{\varepsilon}_k}{10} + M_1 T^3 \quad (39)$$

$$\wedge |\{\text{previous } \Delta\text{-vector}\}_2| \le \frac{\varepsilon_k}{10} + \frac{\dot{\varepsilon}_k}{10} \quad (40)$$

$$+ \frac{1}{2m}|k_3\dot{u}_1(kT-T) + \dot{u}_2(kT-T)| T^2 + M_2 T^3$$

bound the input error to reduce the new error terms in (38) as follows:

$$|\Delta u_1(kT-T)| \le \min\left\{ \frac{\varepsilon_k m}{5k_3 T^2}, \frac{\varepsilon_k m}{5k_3(T-\rho\omega T^2)} \right\}$$

$$\wedge |\Delta u_2(kT-T)| \le \min\left\{ \frac{\dot{\varepsilon}_k m}{5 T^2}, \frac{\dot{\varepsilon}_k m}{5(T-\rho\omega T^2)} \right\} \quad (41)$$

The retrenchment description of this approximated-input scenario is that already given, with P strengthened to specify the input error bounds (41), and C weakened in its error bounds by $\frac{\varepsilon_k}{10} + \frac{\dot{\varepsilon}_k}{10}$ as per (39, 40).

5   Conclusion

Retrenchment was devised as a method to support the formal description of design transformations too rich for refinement. In this paper we introduce a generalisation, evolving retrenchment, to enable description of simulation-like behaviour through an evolving representation relation.

The retrenchment approach, like refinement, takes the abstract system model as given. The case study retrenchment in this paper has been developed on that assumption. The suggestion following the retrenchment (37) suggested another, more intricate and ambitious use for retrenchment. We indicated the possibility of exploiting design freedom (i.e. freedom of choice of model parameters) in the abstract description in order to "improve" the retrenchment. Improvement in the case study constituted a decreasing (as opposed to increasing) sequence of error bounds, i.e. an increasingly precise retrieve relation $G(k)$ over the k-indexed evolving retrenchment. Of course, the notion of improvement of retrenchment must be application-specific, and in any event about getting logically closer to some ideal refinement.

A more ambitious use of retrenchment in formal development, then, is one that exploits underspecification or nondeterminism in the abstract specification, in order to reduce the logical distance from pure refinement. Such a usage would contribute to two obvious methodological questions about retrenchment:

"How closely does the program approximate the requirement? If you can't refine, how close can you get to refinement?".

These questions relate to nontrivial developments involving the composition of retrenchment and refinement steps, from abstract model to compilable code.

The first question is answered by the transitivity property of retrenchment [5,29]: the composite concession of two composed retrenchments can be calculated mechanically. A more sophisticated answer is given by *maximally abstract retrenchment* [4] to the related question "Given a retrenchment from abstract model M_A to concrete model M_C, what is the corresponding abstract model M_U which is refinable to M_C?". In a theoretical sense, this gives requirements validation by constructing the abstract counterpart to the implemented system, for comparison with original requirements.

The second question is addressed, but not answered, by two pieces of work. The notion of *weakest retrenchment* [31] derives from the intuition that a weaker WITHIN clause is preferred where possible. The dual notion of *strongest retrenchment*, in terms of logical proximity of the CONCESSION and RE-TRIEVES clauses has been proposed [31], and is more relevant to the second question.

The focus of the paper is the description of the design step from continuous to ZOH-discrete linear system as an evolving retrenchment. This is set in a version of B much enriched with appropriate types, axioms and analysis. For practical work supported by theorem provers, it is straightforward to describe the models and their retrenchment in a language such as PVS [12] or HOL [20]. These provide some real analysis support: typically, prover support outside the usual discrete types and logics is scarce. However, current interest in the integration of theorem prover and computer algebra technologies such as [21,17, 33,28], therefore holds great potential for support of applications such as ours.

References

[1] K.E. Atkinson. *An Introduction to Numerical Analysis.* Wiley, 1989.

[2] R. J. R. Back and J. von Wright. *Refinement Calculus: A Systematic Introduction.* Springer, 1998.

[3] R.J.R. Back and M. Butler. Fusion and simultaneous execution in the refinement calculus. *Acta Informatica*, 35:921–949, 1998.

[4] R. Banach. Maximally abstract retrenchments. In *Proc. IEEE ICFEM2000*, pages 133–142, York, August 2000. IEEE Computer Society Press.

[5] R. Banach and M. Poppleton. Retrenchment: An engineering variation on refinement. In D. Bert, editor, *2nd International B Conference*, volume 1393 of *LNCS*, pages 129–147, Montpellier, France, April 1998. Springer.

[6] R. Banach and M. Poppleton. Retrenchment: An engineering variation on refinement. Technical Report Report UMCS-99-3-2, University of Manchester Computer Science Department, 1999.

[7] R. Banach and M. Poppleton. Retrenchment and punctured simulation. In K. Araki, A. Galloway, and K. Taguchi, editors, *Proc. IFM'99:Integrated Formal Methods 1999*, pages 457–476, University of York, June 1999. Springer.

[8] R. Banach and M. Poppleton. Sharp retrenchment, modulated refinement and simulation. *Formal Aspects of Computing*, 11:498–540, 1999.

[9] R. Banach and M. Poppleton. Retrenchment, refinement and simulation. In J. Bowen, S. King, S. Dunne, and A. Galloway, editors, *Proc. ZB2000*, volume 1878 of *LNCS*, York, September 2000. Springer.

[10] A. Blikle. The clean termination of iterative programs. *Acta Informatica*, 16:199–217, 1981.

[11] D. Coleman and J.W. Hughes. The clean termination of pascal programs. *Acta Informatica*, 11:195–210, 1979.

[12] J. Crow, S. Owre, J. Rushby, N. Shankar, and M. Srivas. A tutorial introduction to PVS. In R. France, S. Gerhart, and M. Larrondo-Petrie, editors, *WIFT'95: Workshop on Industrial-Strength Formal Specification Techniques*, Boca Raton, Florida, April 1995. IEEE Computer Society Press.

[13] J.J. D'Azzo and C.H. Houpis. *Linear Control System Analysis and Design*. McGraw-Hill, 4 edition, 1995.

[14] W.-P. de Roever and K. Engelhardt. *Data Refinement: Model-Oriented Proof Methods and their Comparison*. Cambridge University Press, 1998.

[15] E.W. Dijkstra. *A Discipline of Programming*. Prentice-Hall, 1976.

[16] R.C. Dorf and R.H. Bishop. *Modern Control Systems*. Addison-Wesley, 1998.

[17] M. Dunstan, T. Kelsey, U. Martin, and S. Linton. Lightweight formal methods for computer algebra systems. In *ISSAC'98: Proceedings of the 1998 International Symposium on Symbolic and Algebraic Computation*, Rostock, 1998. ACM Press.

[18] G.F. Franklin, J.D. Powell, and M.L. Workman. *Digital Control of Dynamic Systems*. Addison-Wesley Longman, 3rd edition, 1998.

[19] D. Goldberg. What every computer scientist should know about floating-point arithmetic. *ACM Computing Surveys*, 1991.

[20] M.J.C. Gordon and T.F. Melham. *Introduction to HOL: A theorem proving environment for higher order logic*. Cambridge University Press, 1993.

[21] John Harrison and Laurent Théry. A skeptic's approach to combining HOL and Maple. *Journal of Automated Reasoning*, 21:279–294, 1998.

[22] F.B. Hildebrand. *Advanced Calculus for Applications*. Prentice-Hall, 1962.

[23] C.A.R. Hoare. An axiomatic basis for computer programming. *Communications of the ACM*, 12(10):576–583, October 1969.

[24] D.S. Neilson. *From Z to C: Illustration of a Rigorous Development Method*. PhD thesis, Oxford University Programming Research Group, 1990. Technical Monograph PRG-101.

[25] K. Ogata. *Modern Control Engineering*. Prentice-Hall, 1997.

[26] O. Owe. Partial logics reconsidered: A conservative approach. *Formal Aspects of Computing*, 3:1–16, 1993.

[27] P.N. Paraskevopoulos. *Digital Control Systems*. Prentice-Hall, 1996.

[28] E. Poll and S. Thompson. Adding the axioms to Axiom: Towards a system of automated reasoning in Aldor. Technical Report 6-98, Computing Laboratory, University of Kent, May 1998.

[29] M. Poppleton and R. Banach. Retrenchment: extending the reach of refinement. In *ASE'99: 14th IEEE International Conference on Automated Software Engineering*, pages 158–165, Cocoa Beach, Florida, October 1999. IEEE Computer Society Press.

[30] M. Poppleton and R. Banach. Retrenchment: Extending refinement for continuous and control systems. In *Proc. IWFM'00*, Springer Electronic Workshop
in Computer Science Series http://ewic.org.uk/ewic, NUI Maynooth, July 2000.
Springer.

[31] M.R. Poppleton. *Formal Methods for Continuous Systems: Liberalising Refinement in B*. PhD thesis, Department of Computer Science, University of Manchester, 2001.

[32] S. Stepney, D. Cooper, and J. Woodcock. More powerful Z data refinement:
Pushing the state of the art in industrial refinement. In J.P. Bowen, A. Fett, and
M.G. Hinchey, editors, *11th International Conference of Z Users*, volume 1493 of
LNCS, pages 284–307, Berlin, Germany, September 1998. Springer.

[33] S. Thompson. Integrating computer algebra and reasoning through the type system of Aldor. In H. Kirchner and C. Ringeissen, editors, *Frontiers of Combining
Systems: Frocos 2000*, volume 1794 of *LNCS*, pages 136–150. Springer, March
2000.

[34] J. Woodcock and J. Davies. *Using Z: Specification, Refinement and Proof*.
Prentice-Hall, 1996.

Checking Z Data Refinements Using an Animation Tool

Neil J. Robinson

Software Verification Research Centre,
The University of Queensland, Brisbane, Australia

Abstract. We describe how a Z animation tool can be used to check Z data refinements. We illustrate two approaches. In the first approach the tool is used to interactively step through operations of the abstract and concrete specifications, checking whether the refinement relationship holds. In the second approach the tool is used to automatically check refinements and to provide counter-examples should the refinement fail. We envisage these techniques being used in order to improve understanding of refinements and to help validate their correctness.

1 Introduction

The concept of refinement is considered to be well understood in the formal methods community. However most practicing systems engineers and software engineers are only dimly aware of the concept. Where refinement is applied in practice and formal methods experts need to communicate with non-experts, then lack of understanding is a severe handicap to achieving proper validation of refinements. Even experts struggle to verify non-trivial data refinements. We believe that this situation can be remedied by improving tool support, novel use of existing tool support and by techniques for visualisation of refinement.

Earlier [16], we showed two different approaches to visualising refinements. The second of these involved the use of an animation tool, Possum [7], to simultaneously animate an abstract specification, and some concrete specification that is intended to be a refinement of the abstract specification. The motivation was to explore the specifications together, in order to both improve understanding of the refinement and, at each step, visually confirm that the refinement relationship still holds. In that work we did not define how to step through the operations and we left the comparison between abstract and concrete states to the user of the animation tool. Here we progress that work by defining a suitable framework for stepping through operations of the abstract and concrete specifications together. At each step, the tool checks whether or not the refinement relation still holds.

We also define a means of automatically checking a refinement using an animation tool. This approach allows proof of a refinement and also provides counter-examples (particular cases in which the refinement relationship does not hold). This is a novel use of an animation tool, since previous work such as that reported by Miller and Strooper [14], has equated the use of animation tools with testing only.

D. Bert et al. (Eds.): ZB 2002, LNCS 2272, pp. 62–81, 2002.

We begin with a summary of data refinement in both the relational world and for the specification language Z. We then describe the two approaches we have developed for checking refinements using an animator, and discuss their advantages and limitations. We finish with a review of similar work and our conclusions.

2   Refinement

A *specification* defines the required properties of a system. It includes a definition of expected inputs and a definition of the required outputs, usually based on the inputs and the internal *state* of the system. A specification may be *nondeterministic*, in that it may permit many possible outputs for a given input. Also, a specification may not define the required behaviour for all possible inputs. When an input occurs that is not allowed for in the specification, we assume that the specification allows nondeterministic behaviour (i.e., the outputs and internal state can be set arbitrarily).

If we have a system which conforms to an abstract specification A and it is replaced by a system which conforms to a concrete specification C, then if it could never be detected in any single run that the replacement has occurred, we say that specification C is a *refinement* of specification A [4].

2.1   Relational Model of Refinement

The Z approach to refinement has its roots in the relational model of refinement [8], in which refinement is defined simply as a subset relationship between total relations. A *concrete* relation CO is a refinement of an *abstract* relation AO if $CO \subseteq AO$.

Data refinement allows for refinements which include changes to the type of a relation. A *retrieve* relation is introduced, which translates between the two types. A retrieve relation R can be defined in the *downward* direction from abstract to concrete, or a relation S can be defined in the *upward* direction from concrete to abstract. Woodcock and Davies [19] show how the data refinement rules[1] can be visualised using a commuting diagram, as shown in Figure 1. For brevity we omit initialisation and finalisation, and we assume the relations AO and CO are total, the relations R and S propagate undefinedness, and $S = R^{-1}$. The small circles represent states, the arrows in the horizontal direction represent applications of the abstract and concrete relations and the arrows in the vertical direction represent applications of the retrieve relation.

From the diagram, it is clear that a downward simulation can be checked by the rule $R \mathbin{\mathrm{\S}} CO \subseteq AO \mathbin{\mathrm{\S}} R$, where '$\mathbin{\mathrm{\S}}$' denotes forward relational composition, and AO and CO are total relations. Similarly, for upward simulation, the rule is $CO \mathbin{\mathrm{\S}} S \subseteq S \mathbin{\mathrm{\S}} AO$. If either of these rules hold then the relation CO is a refinement of the relation AO.

---

[1] We use the word *downward* instead of Woodcock and Davies' *forward* and, similarly, *upward* instead of *backward*.

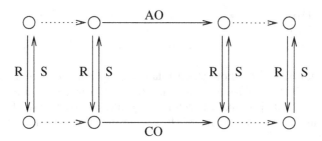

Fig. 1. Upward and Downward simulation

As described by Woodcock and Davies [19], these rules can be *relaxed* so that they apply for untotalised partial relations and then be *unwound* so that they allow for input and output.

Although this relational view of data refinement appears to be intuitive, in practice it is difficult to apply. In common with other data refinement approaches, a strong enough retrieve relation is often hard to find, and is often discovered only through trial and error. Techniques which can help us understand data refinements in practice are therefore highly desirable.

2.2   Data Refinement in Z

In the specification language Z, specifications are conventionally expressed in terms of a *state* defining variables and invariant properties of those variables, an *initialisation* defining the initial properties of the state variables, and a number of *operations* on the state. The precondition of an operation specifies the situations in which the operation is applicable and may be explicit (i.e. appearing in the written specification) or implicit (i.e. implied by the specification, but not actually written down). As stated earlier, if we use an operation in a state in which the precondition is false, then the operation may change the state arbitrarily. Operations are not ordered in any way and we make no assumptions about what has happened previously when we look at an individual operation — we do not even assume that initialisation has taken place. Data refinement rules allow transformation of the variables in the state (e.g., change of type), but usually assume that inputs and outputs are not changed. Data refinement also involves changing all the operations that can access the transformed variables.

Woodcock and Davies [19] translate the relational refinement rules into predicates on Z operations. To apply the Z rules, it is necessary to define a retrieve schema R relating the abstract state A with the concrete state C, and to index the abstract and concrete operations (pair them up). The indexing is introduced to convey the assumption that whenever an abstract operation would have been used, it is the matching concrete operation that is used instead. This may involve pairing up an operation in one specification, with a *no change* operation in the other specification. For each indexed pair of abstract operation AO and concrete operation CO, the downward simulation rules are as follows.

Dinit $\forall\,C' \bullet CI \Rightarrow (\exists\,A' \bullet AI \wedge R')$
Dapp $\forall\,A;\ C \bullet \text{pre}\,AO \wedge R \Rightarrow \text{pre}\,CO$
Dcorr $\forall\,A;\ C;\ C' \bullet \text{pre}\,AO \wedge R \wedge CO \Rightarrow (\exists\,A' \bullet AO \wedge R')$

We can inuitively see how these relate to the relational refinement rules. The subsetting in the relational rules has been translated to implications in the Z rules. Note that, just like the refinement rules for partial relations [9], both the upward and downward forms of the Z rules are needed for completeness [19].

3   Animation of Z Specifications

We now return to our main focus, the use of a Z animation tool for checking refinements. The animation tool we use is Possum [7], a part of the Cogito [5] toolset. Possum can accept a variety of Z style notations. However, we use the Sum language [11], which provides various extensions to Z, including modularisation.

   As stated by Hazel et al. [7], Possum interprets queries and responds with simplifications of those queries. It can be used either to step through consecutive states of a state machine by "executing" the operations of that machine, or to evaluate arbitrary expressions and predicates.

   For example, consider the operation *increase* as shown in Figure 2. If we

Fig. 2. Operation *increase*

call *increase* when x is bound to 0, then Possum responds with:

$$[x := 0, x' := 1]$$

Note that the response given by Possum in the above example is just one of the possibilities allowed by the specification. Where there is nondeterminism in a specification, Possum picks the first binding it can find that satisfies the specification. It does this predictably, i.e., it will always respond as above when x is bound to 0.

   We can examine all behaviours permitted by the specification using the query:

$$\{increase \bullet x \mapsto x'\}$$

to which Possum responds with:

$$\{0 \mapsto 1, 0 \mapsto 2, 0 \mapsto 3, 0 \mapsto 4, 1 \mapsto 2, 1 \mapsto 3, 1 \mapsto 4, 2 \mapsto 3, 2 \mapsto 4, 3 \mapsto 4\}$$

4   Stepping through Operations of the Abstract and Concrete Specifications

For each pair of abstract and concrete operations, we define a combined operation which attempts to step forward both the abstract and concrete operations together, whilst checking that the refinement relation still holds.

There are a number of possible outcomes at each step, which we classify as follows:

Fail_init. The initialisation check failed. The refinement relation does not hold (rule **Dinit** fails).

No_conc_init. It is not possible to initialise the concrete specification. Whilst this is not desirable, it does not violate the refinement rules.

Fail_app. The abstract operation can step forward, but the concrete operation cannot. The refinement relation does not hold (rule **Dapp** fails).

Fail_cor. The concrete operation can step forward to a state which cannot be matched by a corresponding abstract operation. The refinement does not hold (rule **Dcorr** fails).

Ok. Both the abstract and concrete operations successfully stepped forward to a matching state. Assuming all previous steps were also OK, the refinement relation still holds.

No_match. The current abstract and concrete states do not match. There is no point in stepping forward from non-matching states, as this will not tell us anything about the refinement relation.

Stuck. The abstract precondition is false. There is no point in stepping forward the concrete operation, as this will not tell us anything about the refinement relation. Note that if we call an operation whose precondition is currently *false* in Possum, the tool returns *no solution* and leaves the state unchanged — hence our use of the word *stuck*. However, the usual interpretation of Z is to assume that an operation allows chaotic behaviour when its precondition is *false*.

We now define a template for exploring downward simulations using the Possum animation tool. We use the modular structure of Sum [11] to structure our template. The abstract and concrete specifications are defined within modules A and C. Each module contains a *state* schema, an initialisation schema *init*, and a number of indexed operation schemas $AO1, AO2, ...$ for the abstract, and $CO1, CO2, ...$ for the concrete. The indices must be matched appropriately.

Next we define a module *down* that imports the abstract and concrete modules. This module contains a *state* schema which includes the abstract and concrete states, a *retrieve* schema relating the abstract and concrete states, an *init* schema which performs the initialisation check, and a number of operation schemas $dnchk\_o1, dnchk\_o2, ..,$ one per abstract/concrete operation pair.

4.1   Template for Initialisation Check

The initialisation template is shown in Figure 3. This check is straightforward and is derived from the Z refinement rule for initialisation **Dinit**, as provided by Woodcock and Davies [19], and repeated here in Section 2.2.

```
┌─ init ─────────────────────────────────────────────────────────────────
│ s! : Success
├────────────────────────────────────────────────────────────────────────
│ if ∃ C.state' • C.init then
│     if ∃ C.state' • C.init ∧ (∀ A.state' • A.init ⇒ ¬ retrieve')
│     then
│             C.init ∧ (∀ A.state' • A.init ⇒ ¬ retrieve') ∧ A.init ∧
│             s! = fail_init
│     else
│             C.init ∧ A.init ∧ retrieve' ∧ s! = ok
│     fi
│ else changes_only { } ∧ s! = no_conc_init
│ fi
└────────────────────────────────────────────────────────────────────────
```

Fig. 3. Template for initialisation check (downward simulation)

The declaration part of the *init* operation includes an output $s!$, which records the success of the operation in terms of the outcomes we introduced earlier. Note that, in Sum, it is not necessary to explicitly include $\Delta State$ in an *init* schema.

First, the tool checks that it is possible to perform the concrete initialisation $C.init$. If this is not possible then the tool responds with *no_conc_init* and leaves the state unchanged. Note that the Sum construct changes_only{} means that all variables in the declaration of the schema are unchanged.

Next, the tool attempts to find a concrete initialisation for which there is no matching abstract initialisation. If such an initialisation exists, then the tool performs it, and reports a *fail_init*. If there is no such initialisation, then the tool initialises the specifications to a matching state and reports *ok*.

4.2 Template for Checking Downward Simulation of a Pair of Operations

The template for checking the downward simulation of a pair of operations is shown in Figure 4. The template is derived from the downward simulation rules as provided by Woodcock and Davies [19], and repeated here in Section 2.2.

The operation is intended to step forward the abstract and concrete operations, $A.AO1$ and $C.CO1$ and check the applicability and correctness rules as it goes.

The Possum user calls the operation by typing:

$$dnchk\_o1 \gg$$

The piping operator $\gg$ causes Possum to show outputs separately from the state bindings. It also prevents the output variables binding to a value after the operation, so that next time the operation is called the output variable can take a new value. For convenience, we simplify the treatment of inputs and outputs in the abstract and concrete operations, by embedding them in the states. Note that

---
dnchk_o1
$\Delta state$
$s! : Success$

---
if *retrieve* then
    if $\exists A.state' \bullet A.AO1$ then
        if $\exists C.state' \bullet C.CO1$ then
            if $\exists C.state'$
                $\bullet\ C.CO1 \wedge (\forall A.state' \bullet A.AO1 \Rightarrow \neg retrieve')$
            then
                $C.CO1 \wedge (\forall A.state' \bullet A.AO1 \Rightarrow \neg retrieve') \wedge$
                $A.AO1 \wedge s! = fail\_cor$
            else
                $C.CO1 \wedge A.AO1 \wedge retrieve' \wedge s! = ok$
            fi
        else changes_only $\{\} \wedge s! = fail\_app$
        fi
    else changes_only $\{\} \wedge s! = stuck$
    fi
else changes_only $\{\} \wedge s! = no\_match$
fi

Fig. 4. Template for downward simulation check

in the Z rules presented by Woodcock and Davies [19], variables (including inputs and ouputs) with the same names in both the abstract and concrete specifications are unified. Our modular structure prevents unification, as variable names are prefixed by the module name (e.g. $A.x$ is not unified with $C.x$). However, when necessary, corresponding variables from the abstract and concrete modules can be explicitly equated in the *retrieve* schema.

First, the animation tool checks that the current abstract and concrete states correspond, by checking the *retrieve* schema (evaluating the retrieve schema's predicate in the current state). If this check fails then there is no change to the state and the tool responds with *no_match*.

Otherwise, the tool checks the abstract precondition. If the abstract precondition is *false* then no change of state occurs and the tool responds with *stuck*. Note that because our outputs are embedded in the state, we can use $\exists state' \bullet OP$, to check the precondition of an operation OP (we do not need to explicitly existentially quantify the outputs).

If the abstract precondition is true, then the tool next checks that the concrete precondition is also true. If this applicability check fails, then the refinement relation does not hold. Again, the state does not change and the tool responds with *fail_app*.

If the applicability check does not fail, then the tool moves on to check correctness. At this point it is necessary to deal with a potentially nondeterministic choice of how to proceed. The concrete operation may be able to step forward to a number of different states. Once it has stepped forward, it is necessary to check that the abstract operation can step forward to a matching state. Recall that part of the aim of this process is to check the refinement holds. With this aim in mind, the tool looks for a concrete after-state which cannot be matched by the abstract operation. If it can find such a state, then the tool steps forward the concrete operation to that state, steps forward the abstract operation arbitrarily, and reports a *fail_cor*. After such a failure, the abstract and concrete states will not match.

If the correctness check passes, then the tools steps forward both the abstract and concrete operations to an arbitrary matching state and reports *ok*.

4.3   Controlling the Exploration of the Concrete State Space

When the user checks the refinement using the above operations, choices of which direction to take in the concrete specification are left to the tool to choose arbitrarily. However this is not likely to be very thorough, as with this approach, the user can only explore one path through the concrete state space.

Possum includes the ability to *undo* operations. As queries are performed, the tool stacks the old states, so that the user can go back to any previous state. Therefore, the user can explore a number of different paths, by allowing the tool to choose arbitrarily, then undoing back to the desired state. The user can then step forward again, but this time control the path that is taken.

The control can be achieved by constraining the after-state of an operation, using standard Z syntax. For example, to check the operation and move to a concrete state where, say $cx = 10$, the user types:

$$[dnchk\_o1 \mid C.cx' = 10] \gg$$

We now consider a simple example to illustrate the operation of the refinement checker.

4.4   Unique Number Allocator

The Unique Number Allocator example, as shown in Figure 5, is due to Hayes [6]. Its task is to, each time it is called, return a natural number that it has not returned on any previous call. As discussed earlier, we include the output as part of the state. The abstract operation $AO1$ includes an implicit precondition requiring that $as \neq T$, since if as were equal to T, then it would not be possible to choose an out which is in T but not in as.

The concrete specification uses a single natural number cx which is initially set to 0. On each call, cx is increased and the output is set to the new value.

The module *down*, which performs the downward simulation check is shown in Figure 6. The schemas *init*, as shown in Figure 3, and *dnchk_o1*, as shown in Figure 4, are included unchanged. The relationship between abstract and concrete states is given in the *retrieve* schema. Note that for any abstract set $A.as$ there are many possible values of $C.cx$ and for each value of $C.cx$ there may be many possible values of $A.as$ [6].

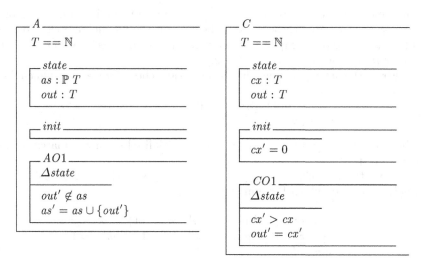

Fig. 5. Abstract and Concrete specifications of the unique number allocator

```
┌─ down ─────────────────────────────────────────────────────
│ import A
│ import C
│ Success ::= ok │ no_match │ fail_init │
│               no_conc_init │ fail_app │ fail_cor │ stuck
│ ┌─ state ─────────────────────────────────────────────────
│ │ A.state
│ │ C.state
│ └─────────────────────────────────────────────────────────
│ ┌─ retrieve ──────────────────────────────────────────────
│ │ state
│ │ ────────
│ │ A.as ⊆ 1 .. C.cx
│ │ A.out = C.out
│ └─────────────────────────────────────────────────────────
│ Check initialisation ..
│ Check operation ..
└─────────────────────────────────────────────────────────────
```

Fig. 6. Downward simulation check for unique number allocator

Recall that, for each step, our template for the downward refinement check performs a correctness check, by searching for a concrete after-state without a corresponding abstract after-state. Clearly, for steps in which the correctness rule holds, this is not possible when there are an infinite number of possible after-states to search through, as is the case for the unique number allocator. To allow us to use our template, we naively change the specifications by replacing the definition of T with $T == 0 .. 2$.

First, the user performs the initialisation check $init \gg$, to which Possum responds:

$$[\ A.as' := \{\}, A.out' := 0, C.cx' := 0, C.out' := 0\ ]$$
$$s! := ok$$

The result is ok, and the abstract state is initialised so that it matches with the concrete state.

Next, the user checks a step of the operations $AO1$ and $CO1$ using $dnchk\_o1 \gg$, to which Possum responds:

$$[\ A.as := \{\}, A.out := 0, A.as' := \{1\}, A.out' := 1,$$
$$C.cx := 0, C.out := 0, C.cx' := 1, C.out' := 1\ ]$$
$$s! := ok$$

The result is ok, and the abstract operation steps forward to match the concrete operation. The refinement relation appears to hold until we perform the check when $C.cx$ is equal to the largest value in T (in our case 2). This time when the user performs the query $dnchk\_o1 \gg$, Possum responds with:

$$[\ A.as := \{1, 2\}, A.out := 2, A.as' := \{1, 2\}, A.out' := 2,$$
$$C.cx := 2, C.out := 2, C.cx' := 2, C.out' := 2\ ]$$
$$s! := fail\_app$$

The applicability check has failed. The abstract precondition is *true*, but the concrete precondition is *false*. It is clear why the concrete precondition is *false* — there is no number in T that is greater than 2. However, the abstract operation $AO1$ is still able to output the number 0. This problem does not arise if T is the natural numbers, since every natural number has a successor, and so the concrete precondition is always *true*.

One way to correct this problem, and other applicability problems (see Section 5.1), is to change the concrete type T to $-1..2$ and change the first predicate of $CO1$ to:

$$cx' = cx + 1$$

We can obtain a suitable retrieve schema by changing the first predicate in the schema *retrieve* to:

$$A.as = 0 .. C.cx$$

For completeness, we can also change the concrete initialisation to $cx' = -1$. With these changes the refinement holds.

4.5   Discussion

Our example shows how the animation tool can be used to reveal errors in a refinement. The example also shows how even in the simplest examples, there can be subtle errors. Our use of the animation tool in this approach is analogous to testing, in that we are exploring the state space with the aim of revealing the presence of errors. An advantage over black-box testing is that the tool is able to examine the internal state of the specification under test. However, in another respect, we have a slightly harder task than conventional testing of sequential programs in a controlled environment, since we must allow for nondeterminism in the concrete specification. It is in these cases, where there is nondeterminism in the concrete specification, that we anticipate this approach being most useful.

Note that whilst our approach does not necessarily achieve an exhaustive check of all possible states, for each before-state of the concrete operation which the user examines, the tool searches through all possible after-states to look for a case where the correctness rule fails. This achieves better coverage than just checking a single path through the concrete state space. However the tool achieves this at a price — the search of all the possible after-states is sometimes not achievable in practice. Note that, in general, it is the degree of nondeterminism in the operations that will affect the feasibility of this approach, not the size of the abstract and concrete states. For the unique number allocator, there is unbounded nondeterminism in both the abstract and concrete operations, and so we were forced to convert the infinite states into finite states. Abstraction techniques, as used in model checking [3], could be used to perform the conversion more systematically.

An alternative is to leave the choice to the user (using the approach described in Section 4.3). In either case it is useful to be able to view the history of states that have been explored during the check, and the mappings between them. This is achieved by constructing trees which show the abstract and concrete states that have been explored. Each node of each tree is mapped to the corresponding nodes in the other tree. Similar visualisations have been used in other papers on data refinement, for example by Stepney et al [17]. Implementing this in an animator is a possible direction for future work.

Recall that the downward simulation rules we have based our template on are not complete. The advantage with the downward simulation rules for our purposes, is that they are quantified over before-states of the abstract and concrete operations. We can therefore naturally use the animation tool to step through the operations one step at a time, looking forward and exploring the possible after-states. In principle it is also possible to define a similar template for exploring upward simulations. However this will not be so intuitive in practice since, in order to check the correctness condition, we need to be able to step forward the concrete operation and check that for all matching after-states of the abstract operation, there is an abstract before-state that matches the concrete before-state. This does not fit so well with our model of stepping through the abstract and concrete states simultaneously.

5   Checking Refinements Automatically

The guided exploration of the abstract and concrete state spaces is very effective for improving understanding of the refinement and may also reveal errors in the refinement. However, it is also desirable to be able to automatically check the refinement and find counter-examples if the refinement relationship does not hold.

This is simple to achieve using our existing template. Figure 7 shows two additional schemas which define operations in which refinement rules fail. The first is an operation where the applicability check fails and the other is one in which the correctness check fails.

To perform an automatic check, we use set comprehensions. For each of the schemas in Figure 7, we perform a query which causes Possum to construct the set of all cases in which the refinement rule does not hold. Clearly, if the check passes, the query returns the empty set. This is illustrated below.

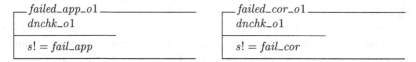

Fig. 7. Operations in which the refinement check fails

5.1   Automatic Check for the Unique Number Allocator

For the (uncorrected) unique number allocator, the user performs the following query in order to check the applicability condition:

$$\{\mathit{failed\_app\_o1} \bullet (A.as,\ C.cx)\}$$

Possum responds with:

$$\{(\{\}, 2), (\{1\}, 2), (\{2\}, 2), (\{1, 2\}, 2)\}$$

This set of counter-examples includes the case which we discovered in Section 4.4, and reveals further cases in which the abstract operation precondition is *true* but a matching concrete operation precondition is *false*.

To check the correctness condition, the user enters the following query:

$$\{\mathit{failed\_cor\_o1} \bullet ((A.as, (A.as', A.out')), (C.cx, (C.cx', C.out')))\}$$

Possum responds with the empty set $\{\}$, showing that the correctness rule passes.

Note that the user must provide the required format of the counter-examples, as the format will depend on the particular operation that is being examined. Generally, for checking applicability, it is helpful to see the before-states of the operations in which the rule does not hold. For checking correctness, it is helpful to see both the before and after-states of the operations — if counter-examples are produced, the after-states of the operations will not match.

5.2   Discussion

Our automatic approach is useful for showing the correctness of a refinement, but less useful for achieving an understanding of a refinement since we do not interactively explore the specifications as we did with the first approach.

We have also experimented with a similar automatic approach which uses the relational refinement rules directly. In order to make this work in the animation tool, we use set comprehensions to convert schemas back into relations. This approach is more efficient in the animator and works better for achieving an understanding of a refinement, since it allows the user to examine the possible behaviours of each of the operations, by inspecting the set comprehensions.

The approach described above performs an exhaustive search through the possible state spaces, and in cases with very small state spaces, will find all the counter-examples in which the refinement relationship does not hold. Clearly this technique is analogous to model checking, and in just the same way that users of model checkers are forced to use abstraction techniques [3], we are forced to reduce the state space of specifications in order to make the approach work in practice.

It is a weakness of model checking that in general it only works for finite state spaces. One of the main advantages of reasoning in Z is that it allows us to specify and reason with *predicates*, meaning that we can deal with infinite state spaces. Animation tools like Possum only make crude use of Z, in that as they step through operations, the current state, as stored in the animator, is just one of the possible current states given the operations that have happened up to that point. This is effective for validating specifications and producing visualisations that a customer can understand, but is not an effective approach for reasoning about the specifications.

An alternative approach to animation is to utilise *symbolic execution*. As stated by Waeselynck and Behnia [18], such animators (e.g., the Atelier B animator) evaluate predicates and set theoretic expressions using rewriting rules. The final effect of an operation may be displayed in symbolic form, or the user can select values for variables from a range of possible values. This approach to animation makes better use of the reasoning power of Z-style languages, and may be better suited to the purpose of checking refinements. This is another possible direction for future work.

6   The Phoenix and Apollo Cinemas

We now consider one further example from the literature. This case study is due to Woodcock and Davies and appears in the book "Using Z" [19] on pages 270–273. It explores a refinement between two cinema booking specifications. The specifications are shown in Figure 8. Note that we have included the outputs in the state, as explained in Section 4.2, and have also added initialisation schemas.

The specifications are intended to describe the required behaviour of a ticket booking system. They focus on the issuing of tickets to a single customer, and specify two operations - one for booking a ticket and one for arrival at the cinema.

In the Phoenix cinema, the system keeps a pool of tickets, *ppool*, and a record, *bkd*, of whether the customer has booked a ticket or not. When a customer books a ticket (operation $AO1$), *bkd* is set to *yes*, and then when the customer arrives (operation $AO2$), a ticket is allocated from the pool.

In the Apollo cinema, the system also keeps a pool of tickets, *apool*, but allocates a ticket, *tkt*, when the customer books (operation $CO1$). When the customer arrives (operation $CO2$), they are given the ticket that has already been allocated. When there is no ticket booked, *tkt* is set to *null*. Note that we have simplified the type *ATicket* from the version given by Woodcock and Davies, as Possum does not support embedded types.

When the operations are used as expected, the customer cannot tell the difference between the different systems used by the two cinemas. The two specifications are actually equivalent (the refinement works both ways), but we will focus on the refinement from Phoenix to Apollo, which can be treated using downward simulation.

Woodcock and Davies [19] provide the retrieve schema shown in Figure 9, in order to prove the data refinement.

Using the templates described in Section 4, the user can animate the Phoenix and Apollo cinemas together, checking the refinement relationship at each step. The user begins by entering *init*$\gg$, to which Possum responds with a successful initialisation:

$$[\ A.bkd' := no, A.ppool' := \{t1, t2\}, A.tout' := t1, C.apool' := \{t1, t2\},$$
$$C.tkt' := null, C.tout' := t1\ ]$$
$$s! := ok$$

Next, the user animates the booking of a ticket, by entering *dnchk_o1*$\gg$, to which Possum responds:

$$[\ A.bkd := no, A.ppool := \{t1, t2\}, A.tout := t1, A.bkd' := yes,$$
$$A.ppool' := \{t1, t2\}, A.tout' := t1, C.apool := \{t1, t2\}, C.tkt := null,$$
$$C.tout := t1, C.apool' := \{t2\}, C.tkt' := t1, C.tout' := t1\ ]$$
$$s! := ok$$

In the abstract specification, Phoenix, a ticket is now booked but the choice of ticket has not been made. In Apollo, the ticket has been chosen, and has been taken out of the pool. Note that, although the outputs, *A.tout* and *C.tout*, appear in the state, they are irrelevant for this operation.

Now, the user animates the arrival of a customer, by entering *dnchk_o2*$\gg$, to which Possum responds:

$$[\ A.bkd := yes, A.ppool := \{t1, t2\}, A.tout := t1, A.bkd' := no,$$
$$A.ppool' := \{t2\}, A.tout' := t1, C.apool := \{t2\}, C.tkt := t1,$$
$$C.tout := t1, C.apool' := \{t2\}, C.tkt' := null, C.tout' := t1\ ]$$
$$s! := ok$$

Both Phoenix and Apollo output the same ticket $t1$ to the customer. Phoenix now removes the ticket from the pool, and sets *bkd* back to *no*. Apollo sets *tkt* back to *null*. Both cinemas are now ready for another arrival.

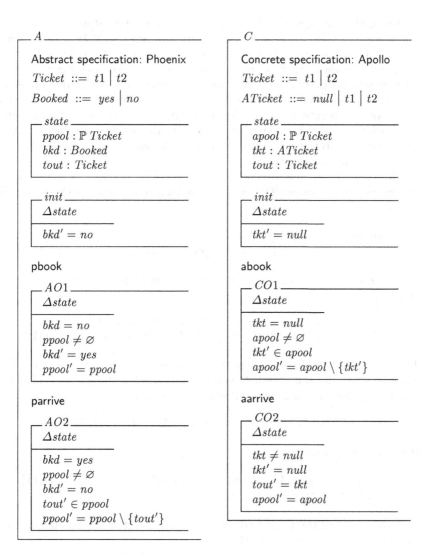

Fig. 8. The Phoenix and Apollo cinemas

$\underline{\quad retrieve\quad}$
state
___
$A.bkd = no \Rightarrow C.tkt = null \wedge A.ppool = C.apool$
$A.bkd = yes \Rightarrow C.tkt \neq null \wedge A.ppool = C.apool \cup \{C.tkt\}$
$A.tout = C.tout$

Fig. 9. Cinemas - Retrieve schema

Using the animation tool, the user can successfully step through the operations until both cinemas have exhausted their pool of tickets, at which point they both become *stuck*. However, that does not mean that the refinement has been fully checked. In fact, in the case when a ticket has been booked, the retrieve schema provided by Woodcock and Davies is not strong enough. The retrieve schema should exclude mappings to states where the ticket reserved in the Apollo (concrete) state is also in the Apollo ticket pool. These states are not reachable when simultaneously stepping through the operations, as the user only examines states which can be reached by executing the operations when their preconditions are true.

However, the automatic check considers *all* matching before-states, and so can reveal the problem. The user can run an automatic check on the arrival operation, by entering:

$$\{failed\_corr\_o2 \bullet (\ ((A.bkd, A.ppool), (A.bkd', A.ppool', A.tout')),$$
$$((C.tkt, C.apool), (C.tkt', C.apool', C.tout'))\ )\}$$

Possum responds with all the cases which fail the correctness check, two of which are:

$$\{\ (((yes, \{t1\}), (no, \{\}, t1)), ((t1, \{t1\}), (null, \{t1\}, t1))), ...$$
$$(((yes, \{t1, t2\}), (no, \{t1\}, t2)), ((t1, \{t1, t2\}), (null, \{t1, t2\}, t1))), ...\ \}$$

The user can examine one of these cases using the interactive approach — for example when the concrete before-state is $tkt = t1$ and $apool = \{t1\}$. The user begins by changing the state in Possum, as follows

$$[state' \mid C.apool' = \{t1\} \wedge C.tkt' = t1 \wedge retrieve']$$

Possum responds with:

$$[\ A.bkd' := yes, A.ppool' := \{t1\}, A.tout' := t1, C.apool' := \{t1\},$$
$$C.tkt' := t1, C.tout' := t1\ ]$$

This shows one of the undesired matching states — the Apollo cinema has allocated a ticket which is also in the pool.

Now, the user can check the arrival operations from this state, by entering $dnchk\_o2\gg$, to which Possum responds:

$$[\ A.bkd := yes, A.ppool := \{t1\}, A.tout := t1, A.bkd' := no,$$
$$A.ppool' := \{\}, A.tout' := t1, C.apool := \{t1\}, C.tkt := t1,$$
$$C.tout := t1, C.apool' := \{t1\}, C.tkt' := null, C.tout' := t1\ ]$$
$$s! := fail\_cor$$

The Apollo operation fails the correctness test, since there is no matching after-state for the Phoenix operation ($A.ppool' \neq C.apool'$).

This can be easily fixed by strengthening the retrieve schema as shown in Figure 10. An additional constraint has been added in the second predicate which ensures that, when a booking has been made, the ticket allocated in the Apollo

```
┌─ retrieve ─────────────────────────────────────────────
│ state
│ ─────────
│ A.bkd = no ⇒ C.tkt = null ∧ A.ppool = C.apool
│ A.bkd = yes ⇒
│       C.tkt ≠ null ∧ A.ppool = C.apool ∪ {C.tkt} ∧ C.tkt ∉ C.apool
│ A.tout = C.tout
└────────────────────────────────────────────────────────
```

Fig. 10. Cinemas - amended Retrieve schema

cinema is not also in the Apollo ticket pool ($C.tkt \notin C.apool$). With this change, the refinement holds. This can be confirmed using automatic checks, which now return empty sets. The user can also confirm that the case examined previously is now eliminated. If the user enters:

$$[state' \mid C.apool' = \{t1\} \wedge C.tkt' = t1 \wedge retrieve']$$

Possum responds with *no solution*, meaning there is no such state.

7   Related Work

Miller and Strooper describe the use of the animation tool, Possum, for testing a specification systematically [14]. An animation plan is produced, part of which identifies a case selection strategy, for testing the specification against user requirements. The strategy used for case selection follows the principles of conventional test case selection. This is related to our work in that they are using an animation tool for validating a specification. However, they define a method for systematically deriving test cases from the specification under test, whilst we consider verifying and validating specifications against more abstract specifications.

In another paper, Miller and Strooper show how the results of animating a specification can be used to test an implementation of the specification [15]. They use the animation test cases to derive test cases for the implementation. The derivation process includes the use of data refinement techniques. This is related to our work in that they are using an animator to check the refinement to code. However, we allow for nondeterminism in the concrete specification. Also our focus is on exploring the specifications to get a better understanding of the refinement, and interactively observing the behaviour of both the abstract and concrete specification simultaneously.

The Nitpick tool [10] checks properties of specifications written in a subset of Z. It specialises in producing counter-examples that show cases where the property does not hold and has been optimised to perform this task efficiently. We expect that Nitpick could be used to perform automatic checks of Z data refinements far more efficiently than an animation tool. However, our primary motivation for using an animation tool is to permit the interactive check of the refinement, as described in Section 4. This is intended to increase the user's understanding of the refinement, and is not possible with a model checking type tool, such as Nitpick.

Waeselynck and Behnia examine the relationship between formal refinement and testing via animation [18]. They propose a process in which an animator is used to test a specification against user requirements. Test cases and oracles are determined from user requirements. At a certain stage in the development, when the user is satisfied with the animated specification, further testing of more concrete formal models would not bring any new information [18], since refined models will pass any tests that a more abstract model has passed. From that point on formal refinement steps are used to develop the specification towards code. Their work differs from ours in that they propose the use of an animator for testing a specification against test cases and test oracles, which are derived from an understanding of user requirements. We use the animator to check refinement steps.

The authors go on to discuss the desirable properties of B [1] animation tools for the purposes of testing, e.g., how the tools should treat nondeterminism. They suggest, that animation tools could deal with nondeterminism by returning all possible results, rather than just a single arbitrary result. For example, suppose we wished to check that a nondeterministic operation can return the values 0..2. When testing we would call the operation, and the animator would return all possible results, which could then be compared with the desired results. They observe that there are no B tools that currently meet their criteria. Recall from our discussion in Section 4.3 that Possum [7] *can* meet their requirements, since we can form many test oracles, each of which checks that a particular result is allowed by the specification. Using the above example, we would use multiple test oracles, some that check for values within the required range (e.g. $op\{1/o!\}$, and, if necessary, some that check that the operation cannot return values outside that range (e.g. $op\{3/o!\}$). This is a similar concept to *operation checkers*, part of the *passive test oracle* approach described by McDonald and Strooper [13]. Alternatively, Possum can evaluate set comprehensions, as shown in Section 5, in order to generate all possible outputs from an operation.

Other refinement checking tools exist, such as the Zeta tool [12] and the Refinement Calculator [2]. However these tools are aimed at supporting formal refinement *proofs*. The tools support the user by providing libraries of theorems that are useful in proving refinements and by automating certain routine tasks, including the documentation of the proof process.

8   Conclusion

We have shown two approaches to checking Z data refinements using an animation tool. The first involves interactively exploring the state spaces of the abstract and concrete specifications, whilst the animation tool checks that the refinement relationship still holds. The second automatically checks the refinement relationship and produces counter-examples in cases where the refinement does not hold. We have demonstrated the approaches on a small example.

The approaches are useful for improving understanding of refinements and checking their correctness, and we have argued that this is important since even experts find non-trivial refinements hard to verify. The fact that there are mistakes in many refinement examples we have examined, points to the need for extending the current tool support for checking refinements. For critical systems, model checking can be effective. However, as we have demonstrated, there is also a role for more interactive tool support, for which specification animation tools are ideal.

Acknowledgements. I would like to thank Colin Fidge and Graeme Smith for their help and guidance in performing this work and for their comments on drafts of this paper. I would also like to thank John Derrick and the anonymous ZB2002 reviewers, for their insightful comments on the draft submission of the paper.

References

1. J.-R. Abrial. *The B-book : assigning programs to meanings*. Cambridge University Press, Cambridge, UK, 1996.
2. M. Butler, J. Grundy, T. Långbacka, R. Rukšėnas, and J. von Wright. The refinement calculator: Proof support for program refinement. In L. Groves and S. Reeves, editors, *Formal Methods Pacific '97*, pages 40–61. Springer-Verlag, 1997.
3. E. Clarke, O. Grumberg, and D. Long. Model-checking and abstraction. In *Proc. of the 19th ACM Symposium on Principles of Programming Languages*, pages 343–354. ACM Press, 1992.
4. P. H. B. Gardiner and C. Morgan. A single complete rule for data refinement. *Formal Aspects of Computing*, 5:367–382, 1993.
5. N. Hamilton, D. Hazel, P. Kearney, O. Traynor, and L. Wildman. A complete formal development using Cogito. In C. McDonald, editor, *Computer Science '98: Proc. 21st Australasian Computer Science Conference*, pages 319–330. Springer-Verlag, 1998.
6. I. J. Hayes. Correctness of data representations. In *Proceedings of the 2nd Australian Software Engineering Conference (ASWEC-87), Canberra*, pages 75–86. IREE (Australia), May 1987.
7. D. Hazel, P. Strooper, and O. Traynor. Possum: An animator for the Sum specification language. In W. Wong and K. Leung, editors, *Proceedings Asia-Pacific Software Engineering Conference and International Computer Science Conference*, pages 42–51. IEEE Computer Society Press, December 1997.
8. C. A. R. Hoare and Jifeng He. *Unified Theories of Programming*. Prentice Hall International Series in Computer Science, 1998.
9. C. A. R. Hoare, Jifeng He, and J. W. Sanders. Prespecification in data refinement. *Information Processing Letters*, 25(2):71–76, May 1987.
10. D. Jackson and C. A. Damon. Elements of style: Analyzing a software design feature with a counterexample detector. *IEEE Transactions on Software Engineering*, 22(7):484–495, July 1996.
11. W. Johnston and L. Wildman. The Sum reference manual. Technical Report 99-21, Software Verification Research Centre, School of Information Technology, The University of Queensland, Brisbane 4072, Australia, November 1999.

12. D. T. Jordan, C. J. Locke, J. A. McDermid, C. E. Parker, B. A. P. Sharp, and I. Toyn. Literate formal development of Ada from Z for safety critical applications. In *Proceedings of SafeComp'94*. ISA, 1994.

13. J. McDonald and P. Strooper. Translating Object-Z specifications to passive test oracles. In John Staples, Michael Hinchey, and Shaoying Liu, editors, *Second International Conference on Formal Engineering Methods, Brisbane, Australia, December 9-11, 1998*, pages 165–174, 1998.

14. T. Miller and P. Strooper. Animation can show only the presence of errors, never their absence. In D. D. Grant and L. Sterling, editors, *Proceedings of the Australian Software Engineering Conference, ASWEC 2001, Canberra, Australia*, pages 76–85. IEEE Computer Society Press, 2001.

15. T. Miller and P. Strooper. Combining the animation and testing of abstract data types. In *Proceedings of the Second Asia-Pacific Conference on Quality Software, APAQS 2001, Hong Kong*. IEEE Computer Society Press, December 2001. (to appear).

16. N. J. Robinson and C. Fidge. Visualisation of refinements. In D. D. Grant and L. Sterling, editors, *Proceedings of the Australian Software Engineering Conference, ASWEC 2001, Canberra, Australia*, pages 244–251. IEEE Computer Society Press, 2001.

17. S. Stepney, D. Cooper, and J. Woodcock. More powerful Z data refinement: Pushing the state of the art in industrial refinement. In J. Bowen, A. Fett, and M. Hinchey, editors, *ZUM'98: The Z Formal Specification Notation, 11th International Conference of Z Users*, Lecture Notes in Computer Science, pages 284–307. Springer-Verlag, 1998.

18. H. Waeselynck and S. Behnia. B model animation for external verification. In J. Staples, M. Hinchey, and S. Liu, editors, *Second International Conference on Formal Engineering Methods, Brisbane, Australia, December 9-11, 1998*, pages 36–45. IEEE Computer Society Press, 1998.

19. J. Woodcock and J. Davies. *Using Z: Specification, refinement, and proof*. International Series in Computer Science. Prentice Hall, 1996.

Encoding Object-Z in Isabelle/HOL

Graeme Smith[1], Florian Kammüller[2], and Thomas Santen[2]

[1] Software Verification Research Centre
University of Queensland 4072, Australia
smith@svrc.uq.edu.au
[2] Technische Universität Berlin, Softwaretechnik, FR 5-6
Franklinstr. 28/29, 10587 Berlin, Germany
{flokam,santen}@cs.tu-berlin.de

Abstract. In this paper, we present a formalisation of the reference semantics of Object-Z in the higher-order logic (HOL) instantiation of the generic theorem prover Isabelle, Isabelle/HOL. This formalisation has the effect of both clarifying the semantics and providing the basis for a theorem prover for Object-Z. The work builds on an earlier encoding of a value semantics for object-oriented Z in Isabelle/HOL and a denotational semantics of Object-Z based on separating the internal and external effects of class methods.

Keywords: Object-Z, reference semantics, higher-order logic, Isabelle

1 Introduction

Isabelle/HOL is an instantiation of the generic theorem prover Isabelle [8] with a classical higher-order logic based on that of the HOL System [3]. It supports a large library of definitions and rules including those of set theory, and advanced facilities for constructing recursive datatypes and inductive and co-inductive definitions[1]. It has proven to be an ideal basis for theorem prover support for formal specification languages. Existing encodings of formal specification languages include those of Z [7] and CSP [15].

The Z encoding in Isabelle/HOL, referred to as $\mathcal{HOL}\text{-}\mathcal{Z}$, has been extended by Santen [9,12] to support notions of classes and objects similar to those in Object-Z [13]. The main difference between the language encoded by Santen's approach and Object-Z, however, is that the former adopts a *value semantics*: values representing objects are used directly in definitions; in particular, in the definitions of classes of other objects. Object-Z, on the other hand, has a *reference semantics*: values of objects are only referenced from definitions, not used directly. The inclusion of object references in Object-Z facilitates the refinement of specifications to code in object-oriented programming languages, which also have reference semantics.

---

[1] An inductive definition specifies the smallest set consistent with a given set of rules. A co-inductive definition specifies the greatest set.

D. Bert et al. (Eds.): ZB 2002, LNCS 2272, pp. 82–99, 2002.

Object references also have a profound influence on the structuring of specifications. When an object is merely referenced by another object, it is not encapsulated in any way by the referencing object. This enables the possibility of self and mutually recursive structures. While these can be very useful in specifications, reasoning about them is not always straightforward. For example, when an object calls a method (i.e., invokes an operation) of another object, the called object may in turn call a method on an object, and so on. In specifications involving recursive structures, such sequences of method calls may repeat (when one of the calls is identical to an earlier call in the sequence). The semantics of method calls needs to account for this possibility and hence is most easily defined in terms of fixed points [14].

In this paper, we build on the work of Santen, modifying and extending it to support Object-Z's reference semantics. Our approach is based on a notion of "messages" which define an object's interaction with other objects in the specification. This approach, inspired by the denotational semantics of Object-Z defined by Griffiths [5,4], supports a modular approach to reasoning about encoded specifications. It also enables us to utilise Isabelle/HOL's inductive definition facility in order to avoid the need for explicitly calculating fixed points of recursively defined method calls.

In Section 2, we outline the general approach and discuss how object references and recursion are handled. In Section 3, we discuss the Isabelle/HOL encoding of classes and objects and, in Section 4, we show how collections of these are used to define specifications and (object) environments respectively. Section 6 sketches some technicalities of the encoding that, for the sake of readability and conciseness, we ignore throughout the rest of the paper. In Section 7, we conclude with a brief discussion of future work.

2 References and Recursion: A Message-Based Approach

An Object-Z class typically includes a state schema, initial state schema and one or more methods (i.e., operations) defined by schemas or operation expressions (which are similar to schema expressions in Z). An example is class C shown in Fig. 1.

The state schema of class C declares a variable n of type naturals and a reference a to an object of class A. It constrains the variable x of the object referenced by a to be less than or equal to n. Initially, n is zero and the object referenced by a satisfies the initial condition of class A. The class has two methods: Inc_n increases the state variable n of an object of class C by an amount input as $n?$; $Inc\_n\_and\_x$ increases n in the same way and simultaneously applies the method Inc_x of class A to the object referenced by a.

The method Inc_n affects the state of an object of class C only. The method $Inc\_n\_and\_x$, on the other hand, may have a wider effect: the method Inc_x of A may affect the state of the object referenced by a, or may call further methods. Since methods in other classes may have references to objects of class C, there is potential for recursion.

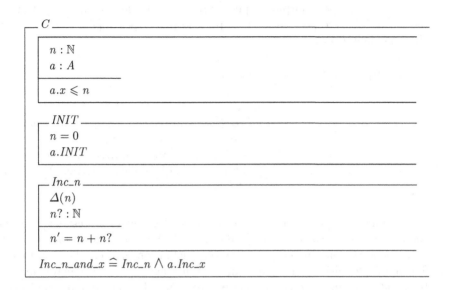

<figure>
C

$n : \mathbb{N}$
$a : A$

$a.x \leqslant n$

INIT

$n = 0$
$a.INIT$

Inc_n

$\Delta(n)$
$n? : \mathbb{N}$

$n' = n + n?$

$Inc\_n\_and\_x \mathrel{\widehat{=}} Inc\_n \wedge a.Inc\_x$
</figure>

Fig. 1. Example class specification.

The obvious approach to handling this recursion in Object-Z is through fixed points. The schemas and expressions modelling the state, initial state and methods of objects in a specification can be defined as the fixed point of characteristic functions based on their syntactic definitions [14]. A specification can be defined as a fixed point on the classes of the specification where classes are ordered according to the orders on their constituent schemas.

This approach, however, ignores the modularity in the specification: the meaning of a schema is given in terms of its effect on the whole system and not just the state of the class to which it belongs. This is not ideal for reasoning about specifications where we would prefer to take advantage of the specification's modularity. Instead, we want an approach where the effect of a schema is separated into its effect on the state of its class and its effect on the rest of the system.

Following a similar argument, Griffiths [5] developed a denotational semantics for Object-Z in which operation schemas are modelled as having two parts: a relation on the state of their class, and a "message" which defines their interaction with the rest of the system. Our approach to encoding Object-Z in Isabelle/HOL adopts this approach and extends it to state and initial state schemas, which may also affect the rest of the system (as in the example above).

A message encodes an Object-Z operation expression. Syntactically, such an expression comprises names (of methods and object references) and various operation operators. We encode a message in Isabelle/HOL using the recursive datatype facility as follows. (α is a polymorphic type representing method names.

ι, ω and υ are polymorphic types representing tuples of inputs, outputs and hidden variables respectively. These polymorphic types appear as parameters to the type definition.)

$$
\begin{aligned}
\mathit{datatype} \quad & (\alpha, \iota, \omega, \upsilon) \; \mathit{Message} = \\
& \quad \mathit{name} \; \alpha \\
& \mid \; \mathit{call} \; \alpha \; \alpha \\
& \mid \; \mathit{and} \; (\alpha, \iota, \omega, \upsilon) \; \mathit{Message} \; (\alpha, \iota, \omega, \upsilon) \; \mathit{Message} \\
& \qquad ([\iota * \omega, \iota * \omega] \rightarrow \iota * \omega * \upsilon) \\
& \mid \; \mathit{choice} \; (\alpha, \iota, \omega, \upsilon) \; \mathit{Message} \; (\alpha, \iota, \omega, \upsilon) \; \mathit{Message} \\
& \mid \; \mathit{sequence} \; (\alpha, \iota, \omega, \upsilon) \; \mathit{Message} \; (\alpha, \iota, \omega, \upsilon) \; \mathit{Message} \\
& \qquad ([\iota * \omega, \iota * \omega] \rightarrow \iota * \omega * \upsilon)
\end{aligned}
$$

Such a datatype is similar to a free-type in Z, but more restrictive, because the parameters to the definition are types, which correspond to maximal sets in Z, whereas the parameters to a free-type definition in Z can be arbitrary sets. The above definition defines five kinds of messages. The constructor *and* serves to encode the Object-Z operation operators $\wedge$ and $\parallel$. The operators *choice* and *sequence* correspond to the Object-Z operation operators $[\!]$ and $\S$ respectively. Each combine two messages. The functions of type $[\iota * \omega, \iota * \omega] \rightarrow \iota * \omega * \upsilon$ associated with the *and* and *sequence* messages are isomorphisms [11] used to combine the input and output tuples of component messages to form those of the composite message. For $\parallel$ and $\S$, the third result type υ represents those inputs and outputs hidden in the composite message. The use of these isomorphisms is explained in Section 4 and illustrated by the example in Section 5. Other uses of isomorphisms in our embedding, which for reasons of conciseness are largely ignored in this paper, are briefly discussed in Section 6.

The other two kinds of messages, *name* and *call* represent the base cases of the recursive definition. A *name* message is used to refer to a method within the class in which the message occurs. This method may be another message or a schema defining an effect on the state of the class. A *call* message is used to refer to a method of another object. It comprises two names: the first is of the object reference of the object, and the second of the method from that object's class. For example, the message corresponding to the method $Inc\_n\_and\_x$ above is *and* (*name* Inc_n) (*call* a Inc_x) IO, where the function IO determines the inputs and outputs of $Inc\_n\_and\_x$ from those of Inc_n and $a.Inc\_x$.

3   Classes and Objects

Building on the approach of Santen [9,12], we encode an Object-Z class in Isabelle/HOL as a tuple. (γ is a polymorphic type representing *object identifiers* which are used to reference objects. κ and σ are polymorphic types representing tuples of constants and state variables respectively.)

$$
\begin{aligned}
typedef \quad & (\gamma, \alpha, \kappa, \sigma, \iota, \omega, \upsilon) \; classschema \\
= \; & \{(C, S, I, Mths, Msgs, Refs) \mid \\
& (C :: \kappa \; constschema) \\
& (S :: (\kappa, \sigma) \; stateschema) \\
& (I :: (\kappa, \sigma) \; initschema) \\
& (Mths :: (\alpha, (\kappa, \sigma, \iota, \omega, \upsilon) \; methodschema) \; finmap) \\
& (Msgs :: (\alpha, (\alpha, \iota, \omega, \upsilon) \; Message) \; finmap) \\
& (Refs :: (\alpha, \omega \rightarrow \gamma) \; finmap). \\
& dom_m \; Refs \subseteq dom_m \; Mths\}
\end{aligned}
$$

The first three elements of a class tuple, C, S and I, are $\mathcal{HOL}\text{-}\mathcal{Z}$ schemas, i.e., functions from tuples to Booleans [7], defining the allowable tuples of constants, state variables and initial state variables respectively[2]. They represent the specified conditions on the state of an object of the class only. That is, for the example class of Section 2, the predicates S and I represent the following schemas.

$$
\begin{array}{|l}
\hline
n : \mathbb{N} \\
a : A \\
\hline
\exists a\_x : \mathbb{N} \bullet a\_x \leqslant n \\
\hline
\end{array}
$$

$$
\begin{array}{|l}
\hline
\_INIT \underline{\hspace{3cm}} \\
\exists a\_init : \mathbb{B} \bullet \\
\quad n = 0 \\
\quad a\_init \\
\hline
\end{array}
$$

These schemas facilitate a modular approach to reasoning. They enable reasoning about objects of the class in isolation from any specified system in which they may occur. For reasoning about complete specifications, however, the global effects implied by a specified class's constant definitions, state schema and initial state schema need also to be captured by the class encoding. For the constants and state schema, the effect is captured by a message *State*. For the example class, this message corresponds to the operation expression $a.X \parallel St$ where St is a method schema

$$
\begin{array}{|l}
\hline
\_St \underline{\hspace{3cm}} \\
x? : \mathbb{N} \\
\hline
x? \leqslant n \\
\hline
\end{array}
$$

and X is an *observer*, i.e., a method schema which outputs the value of a state variable,

---

[2] The actual definitions of the types *constschema*, *stateschema*, *initschema*, as well as that of *methodschema*, can be found in Santen [12, Chapter 5]

$$\boxed{\begin{array}{l} X \\ \hline x! : \mathbb{N} \\ \hline x! = x \end{array}}$$

in class A. In general, the message $State$ includes one observer for each state variable of a referenced object referred to in the constant definitions or state schema.

The global effect of the initial state schema is captured in a similar way by a message $Init$. For the example class, this message corresponds to the operation expression $a.INIT \parallel In$ where In is a method schema

$$\boxed{\begin{array}{l} In \\ \hline init? : \mathbb{B} \\ \hline n = 0 \\ init? \end{array}}$$

and $INIT$ is a method schema in A which outputs $init! : \mathbb{B}$ whose value is true precisely when the initial condition of A is true.

The next two elements of a class tuple, $Mths$ and $Msgs$, are finite partial functions (encoded using the type $finmap$ of finite mappings [12, Appendix B.1]) between names and method schemas, and names and messages respectively. The method schemas capture all the ways in which the class can affect the state of its objects. They include an observer for each state variable and constant (including the implicit constant $self$ which is the identifier of a given object) and the method schemas St and In. The messages capture the ways in which objects of the class can interact with other objects in a specified system. They include the messages $State$ and $Init$.

The final element of a class tuple is a function $Refs$ that maps all observers which output an object identifier to the identifier they output. This is necessary as the strong typing of Isabelle/HOL will not allow an output of generic type ω to be identified with an object identifier of generic type γ. The only constraint on a class tuple ensures that the domain of $Refs$ is a subset of that of $Mths$, i.e., all observers which output an object identifier are also methods. (This is encoded using the function dom_m which returns the domain of a finite mapping [12, Appendix B.1]).

Given this encoding of a class, we can encode the notion of objects in a way suggested by Santen [12]. Firstly, we encode an object state as a cross product of two tuples, corresponding to the values of the object's constants and variables respectively.

$$types \quad (\kappa, \sigma) \; objstate = \kappa * \sigma$$

Then, the object states belonging to objects of a given class are returned by a function $ObjsS$ (cma and sma return the Boolean-valued functions representing the constant and state schemas of a class respectively).

constdefs $ObjsS :: (\gamma, \alpha, \kappa, \sigma, \iota, \omega, \upsilon)$ *classschema* $\rightarrow (\kappa, \sigma)$ *objstate set*
$ObjsS\ Cls == \{(c, s) \mid c\ s.\ (cma\ Cls)\ c \wedge (sma\ Cls)\ c\ s\}$

Given this definition, an object is encoded as an ordered pair where the first element is the class of the object and the second element is the object's state, which must be an object state of its class.

typedef $(\gamma, \alpha, \kappa, \sigma, \iota, \omega, \upsilon)$ *object*
$= \{(Cls, obj) \mid obj \in ObjsS\ Cls\}$

Based on the definition in Santen [12, Chapter 6], we define method invocation as follows. ($\underline{obj}$ denotes the ordered pair representing an object *obj*. *mths* returns the set of methods associated with a class. $\diamond_c^1$ is the method selector for classes, i.e., $Cls \diamond_c^1 n$ returns the method of class *Cls* named *n*. *Meth* returns the boolean-valued function associated with a method schema. This function takes as arguments a constant tuple *c*, two variable tuples *s* and *s'* denoting the pre- and post-states, an input tuple *i* and an output tuple *o*.)

constdefs

$\xrightarrow{(\_)} :: [(\gamma, \alpha, \kappa, \sigma, \iota, \omega, \upsilon)$ *object* $* \iota, \alpha, (\gamma, \alpha, \kappa, \sigma, \iota, \omega, \upsilon)$ *object* $* \omega] \rightarrow$ *bool*
$oi \xrightarrow{n} oo == ($**let** $(obj, i) = oi;$
$(Cls, obj_{st}) = \underline{obj};$
$(c, s) = obj_{st};$
$(obj', o) = oo;$
$(Cls', obj'_{st}) = \underline{obj'};$
$(c', s') = obj'_{st}$
in $c' = c \wedge Cls' = Cls \wedge n \in mths\ Cls \wedge$
$(Meth\ (Cls \diamond_c^1 n))\ c\ s\ s'\ i\ o)$

4   Specifications and Environments

An Object-Z specification comprises a set of classes each with a unique name and a set of unique object identifiers. We encode it as a tuple comprising two finite mappings as follows. (β is a polymorphic type representing class names).

typedef $(\beta, \gamma, \alpha, \kappa, \sigma, \iota, \omega, \upsilon)$ *spec*
$= \{(CMap, IdMap) \mid$
$(CMap :: (\beta, (\gamma, \alpha, \kappa, \sigma, \iota, \omega, \upsilon)$ *classschema*) *finmap*)
$(IdMap :: (\beta, \gamma\ set)$ *finmap*).
$dom_m\ CMap = dom_m\ IdMap \wedge$
$(\forall c_1\ c_2.$
$c_1 \in dom_m\ CMap \wedge c_2 \in dom_m\ CMap \wedge c_1 \neq c_2$
$\Rightarrow IdMap\ c_1 \cap IdMap\ c_2 = \varnothing)\}$

The first mapping *CMap* relates class names in the specification with classes. The second *IdMap* associates class names with sets of identifiers. The constraint ensures that each class in the specification is associated with a set of unique identifiers.

To reason about specifications using the message-based approach, we need to introduce the notion of an *environment*. An environment is an instance of a specification (in much the same way that an object is an instance of a class). It associates each object identifier with an object in a way which satisfies the specification, i.e., each identifier belonging to a class in the specification maps to an object of that class.

We encode the notion of an environment as a function from identifiers to objects. The identifier mapping to a given object must be the same as the constant *self* of that object. (*cls_of* is a function which returns the class component of an object. *refs* is a function which returns the finite mapping *Refs* of a class. $\diamond_m$ is the application operator for finite mappings.)

$$
\begin{aligned}
typedef \quad &(\gamma, \alpha, \kappa, \sigma, \iota, \omega, \upsilon) \ Env \\
= \{&e :: \gamma \to (\gamma, \alpha, \kappa, \sigma, \iota, \omega, \upsilon) \ object. \\
&\forall id. \ (\exists \, out. \ (refs \ (cls\_of \ (e \ id))) \diamond_m self) \ out = id)\}
\end{aligned}
$$

The environments which are valid for a specification are given by the function *spec_envs*. (*classes_of* is a function which returns the finite mapping *CMap* of a specification. *ids_of* is a function which returns the finite mapping *IdMap* of a specification).

$$
\begin{aligned}
constdefs \quad spec\_envs &:: (\beta, \gamma, \alpha, \kappa, \sigma, \iota, \omega, \upsilon) \ spec \to (\gamma, \alpha, \kappa, \sigma, \iota, \omega, \upsilon) \ Env \ set \\
spec\_envs &== (\lambda S. \\
&\{e. \ (\forall cn : dom_m \ (classes\_of \ S). \ (\forall id : (ids\_of \ S) \diamond_m cn. \\
&\quad cls\_of \ (e \ id) = (classes\_of \ S) \diamond_m cn))\})
\end{aligned}
$$

To determine the effect on an environment of a particular event, we introduce an *effect* as a tuple comprising a pre-environment, a post-environment, an object identifier, a message and a tuple of inputs and output parameters.

$$
\begin{aligned}
types \quad &(\gamma, \alpha, \kappa, \sigma, \iota, \omega, \upsilon) \ Effect \\
&= ((\gamma, \alpha, \kappa, \sigma, \iota, \omega, \upsilon) \ Env * (\gamma, \alpha, \kappa, \sigma, \iota, \omega, \upsilon) \ Env \\
&\quad *\gamma * (\alpha, \iota, \omega, \upsilon) \ Message * \iota * \omega)
\end{aligned}
$$

The set *Effects* denotes all possible effects, i.e., all changes to environments caused by sending a message to one of their constituent objects.

$$
consts \quad Effects :: ((\gamma, \alpha, \kappa, \sigma, \iota, \omega, \upsilon) \ Effect) \ set
$$

It can be defined using Isabelle/HOL's inductive definition facility. It is the smallest set satisfying the following rules. (*msgs* returns the finite mapping *Msgs* of a class. $\underline{e}$ denotes the function representing an environment e. $\diamond_c^2$ is the message selector for classes, i.e., $Cls \diamond_c^2 n$ returns the message of class Cls named n.)

inductive Effects

intrs

$$\frac{\begin{array}{l} n \in dom_m \; (mths \; (cls\_of \; (\underline{e} \; id))) \\ ((\underline{e} \; id), i) \xrightarrow{n} ((\underline{e'} \; id), o) \end{array}}{(e, e', id, name \; n, i, o) \in Effects} \quad [\; nameImth \;]$$

$$\frac{\begin{array}{l} (refs \; (cls\_of \; (\underline{e} \; id)) \diamond_m n) \; o' = id' \\ (e, e', id', name \; m, i, o) \in Effects \end{array}}{(e, e', id, call \; n \; m, i, o) \in Effects} \quad [\; callI \;]$$

$$\frac{\begin{array}{l} (e, e', id, m_1, i_1, o_1) \in Effects \\ (e, e', id, m_2, i_2, o_2) \in Effects \\ (i, o, h) = IO \; (i_1, o_1) \; (i_2, o_2) \end{array}}{(e, e', id, and \; m_1 \; m_2 \; IO, i, o) \in Effects} \quad [\; andI \;]$$

$$\frac{(e, e', id, m_1, i, o) \in Effects}{(e, e', id, choice \; m_1 \; m_2, i, o) \in Effects} \quad [\; choiceIleft \;]$$

$$\frac{(e, e', id, m_2, i, o) \in Effects}{(e, e', id, choice \; m_1 \; m_2, i, o) \in Effects} \quad [\; choiceIright \;]$$

$$\frac{\begin{array}{l} (e, e'', id, m_1, i_1, o_1) \in Effects \\ (e'', e', id, m_2, i_2, o_2) \in Effects \\ (i, o, h) = IO \; (i_1, o_1) \; (i_2, o_2) \end{array}}{(e, e', id, sequence \; m_1 \; m_2 \; IO, i, o) \in Effects} \quad [\; sequenceI \;]$$

$$\frac{\begin{array}{l} n \in dom_m \; (msgs \; (cls\_of \; (\underline{e} \; id))) \\ m = (cls\_of \; (\underline{e} \; id)) \diamond_c^2 n \\ (e, e', id, m, i, o) \in Effects \end{array}}{(e, e', id, name \; n, i, o) \in Effects} \quad [\; nameImsg \;]$$

The rule *nameImth* has two conditions which must be met for an effect $(e, e', id, name \; n, i, o)$ to be in *Effects*. The first condition is that n is the name of a method schema of the object identified by id in environment e. The second is that the invocation of this method schema with inputs i and outputs o transforms the object to that identified by id in environment e'.

From the base set of effects generated by *nameImth*, the other rules define the set of all possible effects. For example, the rule *callI* states that the effect $(e, e', id, call\ n\ m, i, o)$ is in *Effects* whenever n is an observer of the class of the object identified by id in e that outputs id', and $(e, e', id', name\ m, i, o)$ is a member of *Effects*. Similarly, the rules *andI*, *choiceIleft* and *choiceIright*, and *sequenceI* state when *and*, *choice* and *sequence* messages, respectively, are in *Effects*.

The parameter functions *IO* of *and* and *sequence* messages map the inputs and outputs of the constituent messages to those of the composite messages and those that are hidden in the composite messages. They model the effect of building the union of schema signatures and subtracting the hidden part of the signatures in the schema calculus. The third component h of the result of applying *IO* is effectively hidden by the rules *andI* and *sequenceI*, because h does not appear in the conclusion of those rules. Similarly, the environment e'' representing the "intermediate state" of a method sequencing is hidden by *sequenceI*.

The final rule *nameImsg* states that the effect $(e, e', id, name\ n, i, o)$ is in *Effects* whenever n is the name of a message of the class of the object identified by id in e, and (e, e', id, m, i, o) is in *Effects* where m is the message associated with name n.

Each Object-Z specification has a distinguished *system class* which specifies the structure of the system and possible interactions between its objects. The effects which are valid for a specification are those that correspond to methods and messages of the system class. They are defined by the function *spec_effects* which takes as parameters a specification and the identifier of its system object (an object of the system class). This function also ensures that the state invariants of all objects are met before and after the effect. This is specified by showing that the effect corresponding to the message *State* (see Section 3), is allowed in the pre- and post-environments of the effect. (Note that *State* is simply a condition on the environment and does not change it.)

constdefs

$$spec\_effects :: [(\beta, \gamma, \alpha, \kappa, \sigma, \iota, \omega, \upsilon)\ spec, \gamma] \rightarrow (\gamma, \alpha, \kappa, \sigma, \iota, \omega, \upsilon)\ Effect\ set$$
$$spec\_effects == (\lambda S\ id.$$
$$\{(e, e', id, name\ n, i, o).\ e \in spec\_envs\ S \land e' \in spec\_envs\ S \land$$
$$(e, e, id, name\ State, (), ()) \in Effects \land$$
$$(e', e', id, name\ State, (), ()) \in Effects\})$$

To reason in a modular fashion, we can treat any class in the specification as the system class of a *sub-specification*. The sub-specification corresponding to a given class comprises that class and the classes of all objects which it references.

When reasoning about the effect of a particular message or sequence of messages on an environment, we often wish to limit our attention to those environments which are *initial environments* of the specification. That is, those environments which satisfy the initial state schema of the system class. The set of such environments for a given specification and system identifier is encoded in terms of the *Init* message (see Section 3) of the system class as follows.

constdefs

> $initial\_envs :: [(\beta, \gamma, \alpha, \kappa, \sigma, \iota, \omega, \upsilon) \; spec, \gamma] \rightarrow (\gamma, \alpha, \kappa, \sigma, \iota, \omega, \upsilon) \; Env \; set$
> $initial\_envs == (\lambda S \; id.$
> $\{e. \; e \in (spec\_env \; S) \wedge (e, e, id, name \; Init, (), ()) \in (spec\_effects \; S \; id)\})$

In other cases, we need to limit our attention to *reachable environments* of the specification. That is, those environments which result from the application of zero or more valid effects to an initial environment. The set of such environments for a given specification and system identifier is encoded as an inductive definition as follows.

consts

> $reachable\_envs :: [(\beta, \gamma, \alpha, \kappa, \sigma, \iota, \omega, \upsilon) \; spec, \gamma] \rightarrow (\gamma, \alpha, \kappa, \sigma, \iota, \omega, \upsilon) \; Env \; set$

inductive reachable_envs S id

intrs

$$\frac{e \in (initial\_envs \; S \; id)}{e \in (reachable\_envs \; S \; id)} \quad [\; initI \;]$$

$$\frac{e \in (reachable\_envs \; S \; id)}{\quad (e, e', id, m, i, o) \in (spec\_effects \; S \; id)}{e' \in (reachable\_envs \; S \; id)} \quad [\; effectI \;]$$

The rule *initI* states that all initial environments are reachable environments. The rule *effectI* states that all environments which result from the application of a valid effect to a reachable environment are also reachable environments.

5 Example Encoding

In this section, we present an example encoding of an Object-Z specification. The specification is of a simple multiplexer based on that of Smith [13, Chapter 1]. The specification comprises two classes modelling a generic queue and a multiplexer comprising two input queues and an output queue of messages.

The generic queue in Fig. 2 is modelled with three state variables: *items*, denoting the sequence of items in the queue, *in* denoting the total number of items which have ever joined the queue, and *out* denoting the total number of items which have ever left the queue. Initially, the queue is empty and no items

$Queue[Item]$

$items : \text{seq } Item$
$in, out : \mathbb{N}$

$INIT$
$items = \langle\,\rangle$
$in = out = 0$

$Join$
$\Delta(items, in)$
$item? : Item$

$items' = items \frown \langle item? \rangle$
$in' = in + 1$

$Leave$
$\Delta(items, out)$
$item! : Item$

$items = \langle item! \rangle \frown items'$
$out' = out + 1$

Fig. 2. Class $Queue$.

have joined or left it. Operations $Join$ and $Leave$ allow items to join and leave the queue respectively.

In Fig. 3, the multiplexer is modelled as having two input queues, $input_1$ and $input_2$, of a given type $Message$ and an output queue, $output$, also of the type $Message$. The class's state invariant ensures that the queues are distinct, i.e., not aliases for the same queue object, and that the number of items that have joined the output queue is equal to the sum of the numbers of items that have left the input queues.

Initially, each queue is in its initial state as defined in class $Queue$. Operations $Join_1$ and $Join_2$ allow messages to be joined to queues $input_1$ and $input_2$ respectively. The operation $Transfer$ allows a message from one of the input queues to be transferred to the output queue and the operation $Leave$ allows a message to leave the output queue.

To illustrate our encoding, we represent this example of an Object-Z specification in HOL. Since the class $Queue$ does not contain any messages, its representation is similar to a representation in Santen's earlier encoding [12]. We just show the observer $INIT$ of $Queue$ below; the new aspects of encoding a class are illustrated by sketching the representation of the class $Multiplexer$.

```
┌─ Multiplexer ─────────────────────────────────────────────────
│ ┌──────────────────────────────────────────────────────────
│ │ input₁, input₂, output : Queue[Message]
│ ├──────────────────────────────────────────────────────────
│ │ input₁ ≠ input₂ ∧ input₁ ≠ output ∧ input₂ ≠ output
│ │ output.in = input₁.out + input₂.out
│ └──────────────────────────────────────────────────────────
│ ┌─ INIT ───────────────────────────────────────────────────
│ │ input₁.INIT ∧ input₂.INIT ∧ output.INIT
│ └──────────────────────────────────────────────────────────
│ Leave₁ ≙ input₁.Leave
│ Leave₂ ≙ input₂.Leave
│ Transfer ≙ (Leave₁ ∏ Leave₂) ∥ output.Join
│ Leave ≙ output.Leave
└──────────────────────────────────────────────────────────────
```

The *Multiplexer* schema contains:

$input_1, input_2, output : Queue[Message]$

$input_1 \neq input_2 \wedge input_1 \neq output \wedge input_2 \neq output$
$output.in = input_1.out + input_2.out$

INIT:
$input_1.INIT \wedge input_2.INIT \wedge output.INIT$

$Leave_1 \mathrel{\widehat{=}} input_1.Leave$
$Leave_2 \mathrel{\widehat{=}} input_2.Leave$
$Transfer \mathrel{\widehat{=}} (Leave_1 \mathbin{[\!]} Leave_2) \parallel output.Join$
$Leave \mathrel{\widehat{=}} output.Leave$

Fig. 3. Class *Multiplexer*.

Assuming the definition of the methods and observers that are needed, we build the messages *State* and *Init* (see Section 3) by composing the corresponding constituents. For example, the message *Init* is built using the *INIT* observer of class *Queue*. This is a method schema which returns a boolean value $init! : \mathbb{B}$ representing whether the initial condition of *Queue* holds. It is thus defined as:

```
┌─ INIT ───────────────────────────────
│ Δ()
│ init! : 𝔹
├──────────────────────────────────────
│ items = ⟨⟩
│ in = out = 0
└──────────────────────────────────────
```

For each of the three aggregated objects, $input_1$, $input_2$, and *output*, a call to this method is issued in the initialisation of *Multiplexer*. For illustration purposes, the output variables $init!$ are all made distinct by renaming them accordingly. These init observers are composed in parallel with a method schema *In* of the class *Multiplexer*.

```
┌─ In ─────────────────────────────────
│ init₁?, init₂?, init₃? : 𝔹
├──────────────────────────────────────
│ init₁? ∧ init₂? ∧ init₃?
└──────────────────────────────────────
```

Thus the message *Init* for the class *Multiplexer* can be constructed as

$input_1.INIT \wedge input_2.INIT \wedge output.INIT \parallel In.$

In our encoding, this *Init* message is represented as

constdefs

> $MultiplexerInit ==$
> and (and (and (call "$input_1$" "INIT")(call "$input_2$" "INIT")...)
>   ("output" "INIT")...)
>   (name "In")
> $\lambda_{iso}((i_1, o_1), (i_2, o_2)).(i_1, o_2, o_1) \mid_{\{((i_1,o_1),(i_2,o_2)).o_1=i_2\}}$.

Using *and* and hiding, we can model the parallel composition operator $\parallel$: we build the conjunction of the first three init observers and the method *In* and use the function that is a further argument to the constructor of an *and* message to express the amalgamation and hiding of the parameters. The λ_{iso} term identifies o_1 and i_2 by using a domain restriction, and maps the result to the third position in the image — the hiding position. Evaluating the effects of a specification (see Section 4, rule *andI*) discards the element of an *and* message. For clarity, we omit the functions to compose the inputs and outputs of the first two *and* submessages. They are similar to the last function, but simpler as they neither identify parameters nor hide them.

The *State* message that comprises observers and methods for the internal and external effects of the state schema of the class *Multiplexer* is built in a very similar fashion. The messages for *Leave$_1$*, *Leave$_2$*, and *Leave* are left out here as they are simple *call* messages to the respective objects.

The message *Transfer* is constructed as

constdefs

> $MultiplexerTransfer ==$
> and (choice (name "$Leave_1$")(name "$Leave_2$"))
>   (call "output" "Join")
> $(\lambda_{iso}((i_1, o_1), (i_2, o_2)).(i_1, o_2, o_1)) \mid_{\{((i_1,o_1),(i_2,o_2))\mid o_1=i_2\}}$

Similar to the *Init* message, the parallel composition is achieved by a corresponding function that identifies the inputs and outputs of the submessages and maps elements that have to be hidden to the third position of the resulting triple.

After these preparations, we can build the representation of the class *Multiplexer* by adding all the constructed observers, methods and messages to a starting element — *basicclass*. The operators $\boxdot_r$ and $\boxdot$ add observers (the former, observers which return an object reference), $\boxplus$ adds methods, and $\boxplus_m$ adds messages to a class.

constdefs

$MultiplexerCls ==$
   $((basicclass\ MultiplexerConstSchema$
                    $MultiplexerStateSchema$
                    $MultiplexerInitSchema)$
   $\square_r\ ("self", MultiplexerselfObs)$
   $\square_r\ ("input_1", Multiplexerinput_1 Obs)$
   $\square_r\ ("input_2", Multiplexerinput_2 Obs)$
   $\square\quad ("INIT", MultiplexerinitObs)$
   $\boxplus\quad ("State", MultiplexerStateOp)$
   $\boxplus\quad ("In", MultiplexerInitOp)$
   $\boxplus_m\ ("Leave1", MultiplexerLeave1)$
   $\boxplus_m\ ("Leave2", MultiplexerLeave2)$
   $\boxplus_m\ ("Leave", MultiplexerLeave)$
   $\boxplus_m\ ("Init", MultiplexerInit)$
   $\boxplus_m\ ("Transfer", MultiplexerTransfer))$

In a final step, the representations for the classes *Queue* and *Multiplexer* are
used to form a specification which is accessible as a single HOL object.

6   A Note on HOL Technicalities

In order to concentrate on the semantic issues that form the core of the encoding
of the reference semantics of Object-Z outlined in this paper, we left out some
technical details concerning the types of methods and messages. However, having
made the essentials clear, we will now discuss a few specialities of the encoding
that are particularly interesting from the theorem proving perspective of this
work. Naturally, they are relevant for the value of our approach as well, as they
determine the applicability of the encoding.

As a major design decision for the mechanical support of Object-Z in Isabelle,
we have adopted the approach to formalisation that has previously been taken
by Santen [12]: our encoding uses a so-called *shallow embedding* of Z and a *deep
embedding* of Object-Z on top of it. In this final section, we will briefly describe
these technical terms in order to lead on to an informal discussion of the problems
we have encountered with the way we chose to encode Object-Z.

6.1   Shallow versus Deep Embedding

The terms shallow and deep [2] refer to the style chosen for the formalisation of a
formal language, say Object-Z, in the logic of a theorem prover. The depth of an
embedding describes to what extent the encoded language is made a first class
citizen of the logic. That is, are the objects of the language all explicitly described
as terms in the logic — then we have deep embedding — or are some concepts
of the language identified with concepts of the logic — a shallow embedding?
For example, in the shallow embedding of Z the input and output types of

operations are identified with types of higher-order logic (HOL). This is sound as the semantics of Z and HOL are very similar in particular with respect to the type system [10].

In general, for reasoning about a language it is advisable to have a deep embedding when the aim is to reason *about* a language rather than *in* the language. In particular, in cases where the encoded language has abstract structures, like Object-Z classes, it has been illustrated that in order to support modular reasoning in a general fashion one needs to represent structures as first class citizens [6]. When it comes to reasoning about a concrete sentence of the embedded language, like the encoding of the class *Multiplexer*, a deep embedding is not very efficient. Therefore, the art of designing a feasible and practical embedding lies in deciding how deep to embed each feature of the language.

6.2   Input and Output Types

Although the shallow embedding of Z is very useful for dealing with concrete specifications, the level of detail provided by the concrete input and output types of operations gets in the way when it comes to the level of Object-Z. A class in Object-Z includes a set of methods, that may in general have arbitrary input and output types. Since a type of classes is needed to be able to express concepts like specifications and relations between them, like refinement or behavioral conformance, it is necessary to unify all the different types of inputs and outputs of methods of a class. The unified input and output types are then used in the definition of the class type (cf. Section 3). In other words, at the Object-Z level, we need a deep embedding (in which we explicitly describe the unified input and output types of methods) to be able to achieve the right level of abstraction.

The type system of Isabelle/HOL has been chosen in such a way that type checking and typeability are decidable. In terms of the λ-cube [1], the type system of HOL corresponds to $\lambda \to$-Church. More powerful type systems of this cube, like $\lambda 2$, have universally quantified type variables, e.g. $\forall \alpha.\alpha$, that could be used to express unified types for inputs and outputs of class methods. For example, $(nat * nat) \to bool$ and $nat \to nat$ are both instances of $(\forall \alpha.\alpha) \to (\forall \alpha.\alpha)$.

However, in HOL, i.e. $\lambda \to$-Church, a polymorphic type expression, e.g. $\alpha \to \alpha$ may refer to an arbitrary α but all occurrences of α in instances have to be the same. In other words, the universal quantification of the type variable α is always at the beginning of a type expression. Therefore, it is necessary to build unified input and output types of all class methods in HOL.

One way to do this is to use the binary sum type constructor $+$ that constructs the sum of two types, i.e., a new type containing two distinguishable copies of both parameter types. Applying this type constructor in an iterated way to all the input and output types of methods, a most general type of all inputs and outputs is created. This provides a means of expressing the type of a class in HOL in which methods also have a most general type.

However, in order to apply the methods unified in this most general type, we need to be able to retrieve the original input and output types of the methods. To that end, we have to administer injections that record the positions where the concrete input and output types are embedded in the general sum types of the method types.

These injections have been developed in a general way by Santen [11] as so-called *isomorphisms*, as they build bijections on the general sum types that respect the term structure. In our encoding, not only methods but also the messages of Object-Z have input and output types. Therefore, we integrate the existing concept of isomorphisms for methods with the recursive datatype for messages. The concept of isomorphisms is also used for capturing the internal relationship of inputs and outputs of the composite *and* and *sequence* messages.

The handling of the types of inputs and outputs is manageable in our encoding, but it creates a considerable formal overhead. We consider the actual encoding of the input and output types with isomorphisms an implementation detail that may be interesting from the theorem proving perspective, but has to be hidden from the user. We need to implement specially tailored tactics that exploit internal term structure information of the theorem prover representation to hide such implementation detail from the user of our encoding. Fortunately, it is possible to create such additional tactic support in Isabelle. Although we have not implemented it yet, we will do so in the near future.

7   Conclusions

We have presented a new approach of encoding object-oriented specifications with reference semantics in higher-order logic. Central to our encoding is the distinction between the internal state changes and the external effects of method invocations that Griffiths has proposed. This makes the encoding more modular than a straightforward fixed point construction of a global environment of objects would.

Technically, we have built a theory to support our encoding in Isabelle/HOL. It remains to implement tactics that hide the technicalities of the encoding, in particular type conversions by isomorphisms, from the users' view. With those tactics at hand, we will be able provide a proof environment for Object-Z and investigate how the modularity of our encoding helps to modularise reasoning about Object-Z specifications. We feel that only modularisation of proofs will enable us to verify properties of specifications of realistic size and complexity.

Acknowledgements. This research was carried out while Graeme Smith was visiting Berlin on a Research Fellowship generously provided by the Alexander von Humboldt Foundation (AvH), Germany. As well as thanking the AvH, Graeme would like to thank GMD-FIRST, Berlin, for the use of an office and computing facilities during his stay.

References

1. H. Barendregt. Lambda calculi with types. In *Handbook of Logic in Computer Science, Vol. 2*. Oxford University Press, 1992.

2. J. Bowen and M. Gordon. A shallow embedding of Z in HOL. *Information and Software Technology*, 37(5–6):269–276, 1995.
3. M.J.C. Gordon and T.F. Melham, editors. *Introduction to HOL: A theorem proving environment for higher order logic.* Cambridge University Press, 1993.
4. A. Griffiths. *A Formal Semantics to Support Modular Reasoning in Object-Z.* PhD thesis, University of Queensland, 1997.
5. A. Griffiths. Object-oriented operations have two parts. In D.J. Duke and A.S. Evans, editors, *2nd BCS-FACS Northern Formal Methods Workshop*, Electronic Workshops in Computing. Springer-Verlag, 1997.
6. F. Kammüller. *Modular Reasoning in Isabelle.* PhD thesis, Computer Laboratory, University of Cambridge, 1999. Technical Report 470.
7. Kolyang, T. Santen, and B. Wolff. A structure preserving encoding of Z in Isabelle/HOL. In J. von Wright, J. Grundy, and J. Harrison, editors, *Theorem Proving in Higher Order Logics (TPHOLs 96)*, volume 1125 of *Lecture Notes in Computer Science*, pages 283–298. Springer-Verlag, 1996.
8. L.C. Paulson. *Isabelle: A Generic Theorem Prover*, volume 828 of *Lecture Notes in Computer Science*. Springer-Verlag, 1994.
9. T. Santen. A theory of structured model-based specifications in Isabelle/HOL. In E.L. Gunter and A. Felty, editors, *Theorem Proving in Higher-Order Logics (TPHOLs 97)*, volume 1275 of *Lecture Notes in Computer Science*, pages 243–258. Springer-Verlag, 1997.
10. T. Santen. On the semantic relation of Z and HOL. In J. Bowen and A. Fett, editors, *ZUM'98: The Z Formal Specification Notation*, LNCS 1493, pages 96–115. Springer-Verlag, 1998.
11. T. Santen. Isomorphisms – a link between the shallow and the deep. In Y. Bertot, G. Dowek, A. Hirschowitz, C. Paulin, and L. Théry, editors, *Theorem Proving in Higher Order Logics*, LNCS 1690, pages 37–54. Springer-Verlag, 1999.
12. T. Santen. *A Mechanized Logical Model of Z and Object-Oriented Specification.* Shaker-Verlag, 2000. Dissertation, Fachbereich Informatik, Technische Universität Berlin, (1999).
13. G. Smith. *The Object-Z Specification Language.* Kluwer Academic Publishers, 2000.
14. G. Smith. Recursive schema definitions in Object-Z. In A. Galloway J. Bowen, S. Dunne and S. King, editors, *International Conference of B and Z Users (ZB 2000)*, volume 1878 of *Lecture Notes in Computer Science*, pages 42–58. Springer-Verlag, 2000.
15. H. Tej and B. Wolff. A corrected failure-divergence model for CSP in Isabelle/HOL. In J. Fitzgerald, C.B. Jones, and P. Lucas, editors, *Formal Methods Europe (FME 97)*, volume 1313 of *Lecture Notes in Computer Science*, pages 318–337. Springer-Verlag, 1997.

Characters + Mark-up = Z Lexis

Ian Toyn[1] and Susan Stepney[2]

[1] Department of Computer Science, University of York,
Heslington, York, YO10 5DD, UK.
ian@cs.york.ac.uk
[2] Logica UK Ltd,
Betjeman House, 104 Hills Road, Cambridge, CB2 1LQ, UK.
stepneys@logica.com

Abstract. The mathematical symbols in Z have caused problems for users and tool builders in the past—precisely what is allowed? ISO Standard Z answers this question. This paper considers the Z notation at the level of the individual characters that make up a specification. For Z authors: it reviews the internationalisation of Z, discusses what characters can be used in forming names, and summarises the changes made to LaTeX mark-up in ISO Standard Z. For Z tool builders: it explains the sequence of processing that is prerequisite to the lexing of a Standard Z specification, and considers in detail the processing of LaTeX mark-up.

1 Introduction

Consider a typical paragraph from a Z specification.

$$
\begin{array}{l}
\Phi Update \\
\hline
\Delta System \\
\Delta File \\
f? : ID \\
\hline
f? \mapsto \theta \ File \in fs \\
fs' = fs \oplus \{f? \mapsto \theta File \ '\}
\end{array}
$$

It exhibits several characteristic features: the outline around the mathematics, the use of Greek letters (Δ and θ) as part of the core language, mathematical symbols ($\oplus$ and $\mapsto$) from the standard mathematical toolkit, and the use of a further Greek letter (Φ) introduced by the specifier.

Such sophisticated orthography can be easily written using a pen, pencil or chalk, but poses problems for ASCII-based keyboards. Software tools for Z have resorted to using *mark-up languages*, in which sequences of ASCII characters are used to encode phrases of a Z specification. Mark-up languages have been devised for representing Z specifications embedded in LaTeX [Spivey 1992a, King 1990], troff [Toyn], e-mail [ISO-Z], SGML [Germán *et al.* 1994] and XML [Ciancarini *et al.* 1998] documents, amongst many others. For example, the schema above could be marked-up in LaTeX as

D. Bert et al. (Eds.): ZB 2002, LNCS 2272, pp. 100–119, 2002.

```
\begin{schema}{{\Phi}Update}
    \Delta System
\\  \Delta File
\\  f? : ID
\where
    f? \mapsto \theta~File \in fs
\\  fs' = fs \oplus \{ f? \mapsto \theta File~' \}
\end{schema}
```

and in troff as

```
.ZS \(*FUpdate
\(*DSystem
\(*DFile
f? : ID
.ZM
f? mlet theta File mem fs
fs' = fs fxov { f? mlet theta File ' }
.ZE
```

with accompanying macros defining the rendering of such mark-up.

Z's syntax and semantics have recently been standardised [ISO-Z]. In standardising the syntax of Z, a formalisation of the lexical tokens (such as the keywords, and names of definitions) used in the syntax was needed. Given the complexity of Z's orthography as noted above, and the requirements for internationalisation (the use of non-Latin character sets, such as those of Japanese and Russian) the Z standards panel took the opportunity to go one step further and define the characters from which the lexical tokens are formed. This allows the Z standard to answer such questions as *Can Z names contain superscripts and subscripts?* and *Can Z names contain multiple symbols, and mixes of symbols and letters?*, in the affirmative.

The formalisation of Z's characters has consequences for the mark-ups used in tools. These were described above as using "sequences of ASCII characters [to] encode *phrases* of a Z specification". Now that Z's characters have been formalised, the "phrases" that are encoded are individual characters or sequences of characters.

This paper is a report about the work on standardising these aspects of Z, by two members of the Z standards panel. Sections 2 to 6 provide information for Z authors; sections 7 to 10 provide additional information for Z tool builders.

2   Internationalisation and Character Classes

The Z standard formalises the syntax of the Z notation, including not just the lexical tokens from which phrases are constructed but also the characters that make up these tokens. The character set includes the letters, digits and other symbols of ASCII [ISO-ASCII], and also the mathematical symbols used by Z, and the letters of other alphabets. As few restrictions as possible are imposed on the standard character set, and a mechanism to extend the set is provided.

Other standards bodies have been working specifically on the internationalisation issue. The Universal Multiple-Octet Coded Character Set (UCS) [ISO-UCS, Unicode 2001] includes practically all known alphabets, along with many mathematical and other symbols. The Z standards panel has worked with the STIX project (who provide input on mathematical symbols for Unicode) to ensure that all of Z's standard mathematical symbols are present in UCS. Hence UCS can serve as the definitive representation of essentially any character that might be used in a Z specification.

Each character in UCS has a *code position* and attributes such as *name* and *general property*. The general property distinguishes letters, digits, and other characters, along with further distinctions such as whether a digit is decimal. The characters used in Z are partitioned into four classes: LETTER, DIGIT, SPECIAL and SYMBOL. These are referred to collectively as *Z characters*.[1]

```
ZCHAR   = LETTER | DIGIT | SPECIAL | SYMBOL ;
LETTER  = 'A' | 'B' | 'C' | ... | 'a' | 'b' | 'c' | ...
        | 'Δ' | 'Ξ' | 'θ' | ...
        | 'ℙ' | 'ℕ' | ...
        ;
DIGIT   = '0' | '1' | '2' | ... ;
SPECIAL = '′' | '?' | '!' | '(' | ')' | '[' | ... ;
SYMBOL  = '∧' | '∨' | '⇒' | '.' | ',' | ... ;
```

The characters of the SPECIAL class have certain special roles: some delimit neighbouring tokens, some glue parts of words together, and some encode the boxes around mathematical text. They cannot be used arbitrarily as characters within Z names. Every other character of UCS can be considered to be present in one of the LETTER, DIGIT or SYMBOL classes according to its UCS general property, and can be used as a character within Z names. Some of these characters are explicitly required by the Z standard; any other UCS character can be used in a Z specification in the appropriate class, though particular tools might not support them all. A UCS character's general property can be determined from the Unicode character database [Unicode 2001].

In the unlikely case of a character being needed that is not in UCS, use of that character is permitted, but its class (LETTER, DIGIT or SYMBOL) is not predetermined, and there is less chance of portability between tools supporting it.

Paragraph outlines are encoded as particular UCS characters that are in the SPECIAL class. This allows their syntax to be formalised in the usual way, independent of exactly how those outlines are rendered.

3   Tokenisation

Several requirements were considered by the Z panel in defining the internal structure of Z tokens.

---

[1] The standard dialect of BNF [ISO-BNF] is used in enumerating them.

1. Names should be able to contain multiple symbols (such as $\frown/$ and $::$), and even combinations of letters and symbols (such as $\perp_x$).

2. The user should not need to type white space between every pair of consecutive Z tokens. For example, x+y should be lexed as three tokens x, $+$, and y, not as a single token.

3. Subscripts and superscripts should be permitted. Single digit subscripts, such as in x_0 or $\mathbb{N}_1$ that have conventionally been used as *decorations*, and more sophisticated subscripts and superscripts should be permitted within names, as in, for example, x_+, y_{min}, or even a^{ξ_3} should the specifier so desire.

Lexical tokens are formalised in terms of the classes into which Z characters are partitioned. For example, a token that starts with a decimal digit character is a numeral, and that numeral comprises as many decimal digits as appear consecutively in the input.

```
NUMERAL = DECIMAL , { DECIMAL } ;
```

Names in Z are formed from words with optional strokes, for example *current*, *next'*, *subsequent''*, *input?*, *output!*, *x0* (a two character word with no stroke) and x_0 (the one character word x with the single stroke 0).

```
NAME = WORD , { STROKE } ;
```

Words can be alphanumeric, as in the examples of names above. Alphanumeric word parts are formed from characters of the LETTER and DIGIT classes. The first word part cannot start with a DIGIT character, as that would start a numeral.

```
ALPHASTR = { LETTER | DIGIT } ;
```

Words can be symbolic, as in $\frown/$. Symbolic word parts are formed from characters of the SYMBOL class.

```
SYMBOLSTR = { SYMBOL } ;
```

In addition, words can be formed from several parts "glued" together, allowing a controlled combination of letters and symbols. Each part of a word can be either alphanumeric or symbolic. The glue can be an underscore, for example *one_word* and x_+_y, or subscripting and superscripting motions, for example x_b, y^2, and $z_\oplus$. These motions are encoded as WORDGLUE characters within the SPECIAL class. Making those motion characters visible, the examples can be depicted as $x \searrow b \nwarrow$, $y \nearrow 2 \swarrow$, and $z \searrow \oplus \nwarrow$.

```
WORDGLUE = '_' | '↗' | '↙' | '↘' | '↖' ;
```

To support words that are purely superscripts, the parts being glued can be empty. For example $\sim$ is a word formed from three characters: a superscripting motion glue character, a SYMBOL, and a motion back down again glue character, $\nearrow\sim\swarrow$. Subscripts and superscripts can be nested, for example a^{b^c} is $a \nearrow b \nearrow c \swarrow\swarrow$ and x^{y_z} is $x \nearrow y \searrow z \nwarrow\swarrow$. For historical reasons, however, a subscripted digit at the end of a name, for example p_2, is lexed as a STROKE, not a subscript part of the word.

```
WORDPART = WORDGLUE , ( ALPHASTR | SYMBOLSTR ) ;

WORD     = WORDPART , { WORDPART }
         | ( LETTER | ( DIGIT  — DECIMAL ) ) , ALPHASTR ,
              { WORDPART }
         | SYMBOL , SYMBOLSTR , { WORDPART }
         ;
```

An empty wordpart can be regarded as an ALPHASTR or a SYMBOLSTR: the ambiguity in the formalisation is irrelevant.

The Z character SPACE separates tokens that would otherwise be lexed as a single token.

A reference to a definition whose name is decorated means something different from a decorated reference to a schema definition. Standard Z requires that these two uses of strokes be rendered differently. A STROKE that is part of a NAME is distinguished from a STROKE that decorates the components of a referenced schema by the absence of SPACE before the STROKE. (See section 3.2 in [Toyn 1998] for more detailed motivation of this.) For example, S' is the name comprising the word S and the stroke $'$, whereas $S\ '$ is the decoration expression comprising the (schema) name S and the decorating stroke $'$. A less subtle distinction of decorating strokes is to render them with parentheses—as in $(S)'$—as then the stroke cannot be considered to be part of the name.

This standard lexis for Z is a compromise between flexibility and simplicity. Apart from subscripting and superscripting motions, there is no notation concerned with rendering the characters: nothing is said about typeface, size, orientation or colour, for example. The meaning of a specification is independent of rendering choices. For example, '*dom*', 'dom' and '**dom**' are regarded as representing the same Z token. For historical reasons, however, there are three classes of exception to this rendering rule.

1. The 'doublestruck' letters, such as $\mathbb{N}$ and $\mathbb{F}$, are regarded as distinct from the upper case letters such as N and F respectively.
2. The schema operators $\setminus$, $\upharpoonright$ and $\mathring{9}$ are regarded as distinct from the toolkit operators $\setminus$, $\upharpoonright$, and $\mathring{9}$.
3. The toolkit's unary negation - is regarded as distinct from binary minus $-$.

4   Mark-ups

All of Z notation—its mathematical symbols, any other UCS character, and the paragraph outlines—can be written using a pen, pencil or chalk. But the input devices used to enter Z specifications for processing by tools are usually restricted to ASCII characters. Even those tools that offer virtual keyboards—palettes of symbols—also use ASCII encodings. So there is a need to encode phrases of Z specifications by sequences of ASCII characters. These encodings are called *mark-ups*. For any particular Z specification, there can be many different mark-ups all of which convert to the same sequence of Z characters. For example, the

LATEX mark-ups \dom and dom both convert to the Z character sequence 'dom', but may be rendered differently by LATEX.[2]

The Z standard defines two mark-ups:

1. the e-mail mark-up, suitable for use in ASCII-based plain text messages, and for mnemonic input to Z tools like Formaliser [Stepney];
2. LATEX mark-up, which allows a Z specification to be embedded within a document to be typeset by LATEX [Lamport 1994], and used as the input (in pre-standard versions of the mark-up) to Z tools like CADiZ [Toyn], Z/EVES [Saaltink 1997], *f*uzz [Spivey 1992a], ProofPower [Arthan] and Zeta [Grieskamp].

Use of a standard mark-up allows specifications to be portable between different tools that conform to the standard in this way. The UCS representation could become another basis for portability of raw Z specifications.

Other Z tools may define their own mark-ups (for example, CADiZ also has a troff mark-up). Any particular mark-up is likely to provide the following features:

1. some means of delimiting formal Z paragraphs from informal explanatory text;
2. names for non-ASCII characters;
3. directives for extending the mark-up language to cope with the names of user-defined operators.

Precise details of conversion depend on the particular mark-up.

5   E-mail Mark-up

In the e-mail mark-up, each non-ASCII character is represented as a string visually-suggestive of the symbol, enclosed in percent signs. There are also ways of representing paragraph outlines.

This mark-up is designed primarily to be readable by people, rather than as a purely machine-based interchange format. To this end, it permits the percent signs to be left out where this will not cause confusion to the reader. However, there are no guarantees that such abbreviated text is parsable. If machine processing is required, either all the percent signs should be used, or a mark-up designed for machines, such as the LATEX markup, should be used.

In full e-mail mark-up, the example at the beginning of this paper is

```
+-- %Phi%Update ---
  %Delta%System
  %Delta%File
  f? : ID
```

---

[2] Prior to the Z standard, mark-ups were usually converted directly to lexical tokens, and so these examples could be, and usually were, converted to different tokens. Now that Z characters have been formalised, it is simpler to convert mark-up to sequences of Z characters, but this means the examples will necessarily convert to the same Z token.

```
|--
  f? %|-->% %theta% File %e% fs
  fs' = fs %(+)% { f? %|-->% %theta% File ' }
---
```

In a more readable, abbreviated form, the predicate part might be written as

```
f? |--> %theta File %e fs
fs' = fs (+) { f? |--> %theta File ' }
```

6   LaTeX Mark-up

LaTeX mark-ups were already in widespread use before Z was standardised. The Z standard LaTeX mark-up is closely based on two of the most widely used ones [Spivey 1992a, King 1990] for backwards compatibility. The Z standard requires that the sequence of lexical tokens that is perceived by reading the rendering shall be the same as the sequence of lexical tokens that it defines from the conversion. As the rendering of LaTeX mark-up was already defined, the conversion of LaTeX mark-up has had to be defined carefully. Including changes to Z, this has resulted in the following changes to LaTeX mark-up.

1. A conjecture paragraph is written within a **zed** environment, with the mark-up of its ⊢? keyword being `\vdash?`.
2. A section header is enclosed within a new environment, **zsection**, with the mark-up of its keywords being `\SECTION` and `\parents`. (This capitalisation follows the same pattern as is used with `\IF`, `\THEN`, `\ELSE` and `\LET`, where the corresponding lower-case `\LaTeXcommands` have already been given different definitions in LaTeX.)
3. Mutually recursive free types [Toyn *et al.* 2000] are separated by the & keyword with mark-up `\&`.
4. The set of all numbers $\mathbb{A}$ is marked-up as `\arithmos`.
5. The '`%%Zchar`' mark-up directive defines the conversion of a `\LaTeXcommand` to a Z character,
   `%%Zchar \LaTeXcommand U+nnnn`   or
   `%%Zchar \LaTeXcommand U-nnnnnnnn`
   where **nnnn** is four hexadecimal digits identifying the position of a character in the Basic Multilingual Plane of UCS, and **nnnnnnnn** is eight hexadecimal digits identifying a character anywhere in UCS, as in the following examples.
   `%%Zchar \nat U+2115`
   `%%Zchar \arithmos U-0001D538`
   For such `\LaTeXcommands` that are used as operator words, mark-up directives '`%%Zprechar`', '`%%Zinchar`' and '`%%Zpostchar`' additionally include **SPACE** before and/or after the character in the conversion of `\LaTeXcommand`, as in the following examples
   `%%Zinchar \sqsubseteq U+2291`
   `%%Zprechar \finset U-0001D53D`
   which define conversions from `\sqsubseteq` to ' ⊑ ' and from `\finset` to '𝔽 '.

6. The 'Zword' directive defines the conversion of a `\LaTeXcommand` to a sequence of Z characters, themselves written in the LATEX mark-up,
   `%%Zword \LaTeXcommand Zstring`
   as in the following example.
   `%%Zword \natone \nat_1`
   For such `\LaTeXcommand`s that are used as operator words, mark-up directives 'Zpreword', 'Zinword' and 'Zpostword' additionally include SPACE before and/or after the converted Z characters, as in the following example
   `%%Zinword \dcat \cat/`
   which defines the conversion from `\dcat` to ' $\frown$/ '.

7. The scope of a mark-up directive is the entire section in which it appears,
   excepting earlier directives (so that there can be no recursive application of
   the conversions), plus any sections of which its section is an ancestor.

8. The conversions of commands `\theta`, `\lambda` and `\mu` to Greek letters are
   defined by `%%Zprechar` directives, so that the necessary SPACE in expressions
   such as $\theta\ e$ need not be marked-up explicitly. (This automates a resolution of
   the backwards incompatibility reported in section 3.9 of [Toyn 1998] in the
   case of LATEX mark-up.) The conversions for other Greek letters are defined
   by `%%Zchar` directives. The inclusion of spaces around the conversion of
   a `\LaTeXcommand` can be disabled by enclosing it in braces, allowing the
   remaining conversion to be used as part of a larger word. For example,
   `{\theta}`, `{\lambda}` and `{\mu}` are mark-ups for corresponding letters in
   longer Greek names.

9. Subscripts and superscripts can be single LATEX tokens or be sequences of
   LATEX tokens enclosed in braces, that is, `_\LaTeXtoken`, `_{\LaTeXtokens}`,
   `^\LaTeXtoken` and `^{\LaTeXtokens}`.

10. There is a new symbol $\ominus$ in the toolkit for set symmetric difference, marked-
    up as `\symdiff`.

7   Converting LATEX Mark-up to Z Characters

The Z standard provides an abstract specification of the conversion from LATEX
mark-up to a sequence of Z characters. Some pseudo-code showing how that
specification might be implemented is provided here.

7.1   The Mark-up Function

Much of the conversion of LATEX mark-up to a sequence of Z characters is managed by the *mark-up function*, which maps individual `\LaTeXcommand`s to sequences of Z characters. Mark-up directives provide the information to extend
the mark-up function: the `\LaTeXcommand`, the sequence of Z characters to convert it to, and whether spaces can be converted before and/or after that sequence.

The mark-up for the Z core language can be largely defined by mark-up
directives in the prelude section, as shown in Figure 1.[3] A tool may need to have

---

[3] Each directive is required to be on a line by itself in a specification; in Figure 1 two
columns are used to save space. The prelude section is an ancestor of every Z section,

%%Zchar	\\	U+000A		%%Zprechar	\forall	U+2200
%%Zinchar	\also	U+000A		%%Zprechar	\exists	U+2203
%%Zchar	\znewpage	U+000A		%%Zinchar	\in	U+2208
%%Zchar	\,	U+0020		%%Zinchar	\spot	U+2981
%%Zchar	\;	U+0020		%%Zinchar	\hide	U+29F9
%%Zchar	\:	U+0020		%%Zinchar	\project	U+2A21
%%Zchar	_	U+005F		%%Zinchar	\semi	U+2A1F
%%Zchar	\{	U+007B		%%Zinchar	\pipe	U+2A20
%%Zchar	\}	U+007D		%%Zpreword	\IF	if
%%Zinchar	\where	U+007C		%%Zinword	\THEN	then
%%Zchar	\Delta	U+0394		%%Zinword	\ELSE	else
%%Zchar	\Xi	U+039E		%%Zpreword	\LET	let
%%Zprechar	\theta	U+03B8		%%Zpreword	\SECTION	section
%%Zprechar	\lambda	U+03BB		%%Zinword	\parents	parents
%%Zprechar	\mu	U+03BC		%%Zpreword	\pre	pre
%%Zchar	\ldata	U+300A		%%Zpreword	\function	function
%%Zchar	\rdata	U+300B		%%Zpreword	\generic	generic
%%Zchar	\lblot	U+2989		%%Zpreword	\relation	relation
%%Zchar	\rblot	U+298A		%%Zinword	\leftassoc	leftassoc
%%Zchar	\vdash	U+22A2		%%Zinword	\rightassoc	rightassoc
%%Zinchar	\land	U+2227		%%Zinword	\listarg	{,}{,}
%%Zinchar	\lor	U+2228		%%Zinword	\varg	_
%%Zinchar	\implies	U+21D2		%%Zprechar	\power	U+2119
%%Zinchar	\iff	U+21D4		%%Zinchar	\cross	U+00D7
%%Zprechar	\lnot	U+00AC		%%Zchar	\arithmos	U−0001D538
				%%Zchar	\nat	U+2115

Fig. 1. Definition of the initial mark-up function

the mapping from \SECTION to section built-in for the prelude's section header itself to be recognised. If a tool supports a character set larger than the minimal set required by the Z standard, such as further Greek and 'doublestrike' letters, these also can be introduced by directives in the tool's prelude section.

The mark-up for additional Z notation, such as that of the toolkit, can all be introduced by mark-up directives in the sections where that Z notation is defined. The mark-up function to be used in a particular section is the union of those of its parents extended according to its own mark-up directives.

7.2 The Scanning Algorithm

Converting the LaTeX mark-up of a Z specification to a sequence of Z characters involves more work than merely mapping \LaTeXcommands to sequences of Z characters.

The mark-up directive corresponding to a particular use of a \LaTeXcommand may appear conveniently earlier in the same section, and it may alternatively

---

whether explicitly listed as a parent or not, so the directives of Figure 1 are in scope everywhere.

appear later in the same section or in a parent section. Moreover, the sections may need to be permuted to establish the definition before use order that is necessary prior to further processing. Some preparation needs to be done to ensure that directives are recognised before their \LaTeXcommands are used. This preparation can be done in a separate first pass. The first pass need recognise only section headers and mark-up directives. A specification of it appears in section 8 below.

The second pass is where the LaTeX mark-up is converted to a sequence of Z characters. The conversion takes place in several phases, due to dependencies between them.

The first phase searches out the formal paragraphs, as delimited by \begin and \end commands of particular LaTeX environments (axdef, schema, gendef, zed, and zsection), eliding the intervening informal text. (For some applications it might be more appropriate to retain the informal text, but eliding it simplifies the following description.)

The second phase tokenises the LaTeX mark-up, consuming ASCII characters and producing Z characters. It:

- elides soft space (spaces, tabs and newlines excepting newlines at the ends of directives) and comments (but retains directives);
- converts @ to • and - to U+2212 and ~ to SPACE;
- inserts SPACE around math function characters (+, -, *, •, |) and after math punctuation characters (;, ,) and around sequences of (having removed soft space within) math relation characters (:, <, =, >) if those math characters are not superscripts or subscripts or enclosed in braces, and with any following superscript or subscript appearing before the following SPACE;
- recognises \LaTeXcommands eliding soft space after them and converting \begin{axdef} to AXCHAR, \begin{gendef} to AXCHAR GENCHAR, \begin{schdef} to SCHCHAR, \begin{zed} to ZEDCHAR, \begin{zsection} to ZED, \end{same} to ENDCHAR, \t*digit* to SPACE, *space* to SPACE, and remembering the names of other \LaTeXcommands and whether they were enclosed in braces;
- and retains every other ASCII character xy as UCS character U+00xy.

The third phase converts \LaTeXcommands to their expansions, consuming Z characters and producing Z characters. It:

- inserts SPACE before \LaTeXcommands that require it and were not in braces;
- applies the mark-up function to convert remaining \LaTeXcommands to Z characters, inserting \ before each {, }, ^, _ and \ in the conversion, and processes the result (rejecting any \LaTeXcommand that is not in the domain of the mark-up function);
- inserts SPACE after \LaTeXcommands that require it and were not in braces, postponing that SPACE to after any superscript or subscript;
- replaces ^\LaTeXcommand by ^{*str*} and _\LaTeXcommand by _{*str*}, where *str* is the conversion of \LaTeXcommand with \ inserted before each {, }, ^, _ and \ in the conversion, and processes the result;
- converts ^{*str*} to ↗ *str* ↙ and _{*str*} to ↘ *str* ↖ and ^*c* to ↗ *c* ↙ and _*c* to ↘ *c* ↖;

- inserts GENCHAR after SCHCHAR if appropriate, and SPACE after schema paragraph's NAME;
- elides braces that are not preceded by the \ escape;
- removes all remaining \ escapes (from {, }, ^, _ and \);
- and forwards all other Z characters unchanged.

The fourth phase manages the mark-up function, consuming Z characters and producing Z characters. It:

- recognises section headers, initialising the section's mark-up function using the mark-up functions of its parents, and forwarding its Z characters unchanged;
- recognises mark-up directives, revises the mark-up function of the current section, and elides their Z characters (rejecting any badly-formed mark-up directives and multiple mark-up directives for the same \LaTeXcommand in this same scope);
- and forwards all other Z characters unchanged.

These four phases naturally communicate via queues of characters. Note that the third phase must not run too far ahead of the fourth phase, otherwise it might attempt to convert a \LaTeXcommand before the corresponding directive has caused revision of the mark-up function.

The pseudo-code for the four phases may be somewhat over-specified, that is, some of the individual conversions might work just as well in different phases. It may appear complicated, but—apart from careful placing of spaces around operators—is a fairly direct translation of LaTeX to Z characters, presented in detail to cover all special cases. It is presented in a form that has been abstracted from an implementation for which no mistakes are presently known.

8   Sections

ISO Standard Z specifies the syntax of Z as being a sequence of sections. It specifies the semantics of just those sequences of sections that are written in definition before use order. Sections may be presented to the user in a different order, perhaps for readability reasons. If the sections are presented to a tool in the same order as to the user, the tool needs to permute them into a definition before use order so that their conformance to the standard can be checked. As extensions to the mark-up language can be defined in parent sections and used in subsequent sections, this permutation has to be done on the source mark-up, before the mark-up is converted to a sequence of Z characters.

If there are cycles in the parents relation, it will not be possible to find a definition before use order for the sections; such a specification does not conform to standard Z, and so no further processing of it is necessary.

A further issue regarding sections is where they come from: do they reside in files, and if so how are they arranged? The standard gives no guidance on this, as it inevitably varies between implementations. Nevertheless, there are likely to be considerable similarities between implementations, and so we believe it is useful to present a specification of one particular implementation.

8.1   Specification of Finding and Permuting Sections

This specification takes a filename as input, along with an environment in which that filename is interpreted, retrieves the mark-up of a Z specification from files determined by the filename and environment, and generates mark-up in which the sections are in a declaration before use order. This specification is written with the needs of the CADiZ toolset in mind, but aspects of it are relevant in other contexts. The specification has been typechecked by CADiZ [Toyn].

Introduction. A specification written in standard Z [ISO-Z] comprises a sequence of *sections* [Arthan 1995], each of which has a header giving both its name and a list of the names of its parents. Its meaning includes the paragraphs of its parent sections as well as its own paragraphs.

For backwards compatibility with traditional Z [Spivey 1992b], a bare sequence of paragraphs (an *anonymous section*) is accepted as a specification comprising the sections of the prelude, the mathematical toolkit, and a section containing that sequence of paragraphs. A named section needs to include the toolkit explicitly as a parent if it makes use of those definitions.

The specification below assumes that sections are stored in files, possibly several per file. Each file is viewed as containing a sequence of paragraphs: the specification distinguishes formal paragraphs, informal paragraphs, and section headers. It needs to interpret the content of section headers, but does not need to interpret the mark-up within formal paragraphs.

Any references to parent sections that have not yet been read are presumed to be in files of the same name, and so those files are read. In each file, any formal paragraphs that are not preceded by a section header are treated as if there had been a section header whose name is that of the file and which has *standard_toolkit* as parent. This is similar to the treatment of anonymous sections in the Z standard. A file's name need not be the same as any of the sections it contains, in which case that name is useless from the point of view of finding parent sections, but it is useful as a starting point for a whole specification.

This specification makes use of the standard mathematical toolkit.

section *sortSects* parents *standard_toolkit*

Data types. Strings are encoded as sequences of naturals. These naturals can be viewed as UCS code positions. Here we use CADiZ's non-standard string literal expressions, such as *"ropey example"*, to display such strings.

$$\mid \quad String == \operatorname{seq} \mathbb{N}$$

Names (of both files and sections) are represented by strings. The form of names would be irrelevant to this specification but for literal names such as *"prelude"*.

$$\mid \quad Name == String$$

Only certain kinds of paragraphs need be distinguished. Informal text between formal paragraphs is retained for possible display in the same order between the formal paragraphs. Informal and formal paragraphs are each treated as untranslated strings, but distinguished from each other to enable the detection of anonymous sections. Section headers are treated like paragraphs in this specification.

$$Paragraph ::= Informal \langle\!\langle String \rangle\!\rangle$$
$$| \ Formal \langle\!\langle String \rangle\!\rangle$$
$$| \ SectionHeader \langle\!\langle [name : Name; \ parentSet : \mathbb{F} \ Name] \rangle\!\rangle$$

The file system is modelled as a function from pathnames (formed of directory and file names) to sequences of paragraphs. This avoids having to specify the parsing of mark-up: this specification is independent of any particular mark-up, though implementations of it will be for specific mark-ups. Section headers can have been distinguished by the section keyword, if not distinguished by other mark-up.

$$Directory == Name$$

$$FileSystem == Directory \times Name \nrightarrow seq \ Paragraph$$

Sections are represented as sequences of paragraphs in which an explicit section header begins each section.

$$Section ==$$
$$\{ps : seq_1 \ Paragraph \ |$$
$$head \ ps \in ran \ SectionHeader$$
$$\wedge \ ran(tail \ ps) \cap ran \ SectionHeader = \varnothing\}$$

Environment. This specification operates in an environment comprising: the file system fs; the current working directory name cwd; the name of the directory containing the toolkit sections $toolkitDir$; and an environment variable $SECTIONPATH$ giving the names of other directories from which sections may be read. The environment is modelled as the global state of the specification. Its value is not changed by the specification.

$$fs : FileSystem$$
$$cwd, toolkitDir : Directory$$
$$SECTIONPATH : seq \ Directory$$

Functions. The function $sectionToName$ is given a section and returns the name of that section. The name returned is that in the section header that is the section's first paragraph.

$$sectionToName ==$$
$$\lambda s : Section \bullet ((SectionHeader^{\sim}) \ (head \ s)).name$$

The function *sectionsToParents* is given a set of sections and returns the set containing the names of the parents referenced by those sections.

$$
\begin{array}{l}
sectionsToParents == \\
\quad \lambda ss : \mathbb{F}\ Section \bullet \\
\qquad \bigcup \{s : ss \bullet ((SectionHeader^{\sim})\ (head\ s)).parentSet\}
\end{array}
$$

The function *filenameToParas* is given a search path of directory names and a file name and returns the sequence of paragraphs contained in the first file found with that name in the path of directories to be searched. If no file with that name is found, an empty sequence of paragraphs is returned (and an error should be reported by an implementation).

$$
\begin{array}{l}
filenameToParas : seq\ Directory \times Name \nrightarrow seq\ Paragraph \\
\hline
\forall n : Name \bullet \\
\quad filenameToParas(\langle\rangle, n) = \langle\rangle \\
\forall d : Directory;\ path : seq\ Directory;\ n : Name \bullet \\
\quad filenameToParas(\langle d\rangle \frown path, n) = \\
\qquad if(d, n) \in dom\ fs \\
\qquad then\ fs\ (d, n) \\
\qquad else\ filenameToParas\ (path, n)
\end{array}
$$

The function *filenameToParagraphs* is given a filename and returns the sequence of paragraphs contained in the first file found with that name in the path of directories to be searched. The directory of toolkits is always searched first (so that its sections cannot be overridden), then whatever directories are explicitly listed in the SECTIONPATH environment variable, and finally the current working directory.

$$
\begin{array}{l}
filenameToParagraphs == \\
\quad \lambda n : Name \bullet \\
\qquad filenameToParas\ (\\
\qquad\quad \langle toolkitDir\rangle \frown SECTIONPATH \frown \langle cwd\rangle, n)
\end{array}
$$

The function *addHeader* reads the named file. If the file starts with a section header, but the name of that section differs from that of the file, an empty section is introduced to prevent re-reading of the file. If the file starts with an anonymous section, the sequence of paragraphs is prefixed with a section header. If the anonymous section has any formal paragraphs, it is named after the file, otherwise it is given a different name in case the first named section has that name.

$$
\begin{aligned}
&addHeader == \\
&\quad \lambda n : Name \bullet \\
&\qquad \text{let } ps == filenameToParagraphs \ n \bullet \\
&\qquad\quad (\mu pref, suff : \text{seq } Paragraph \mid pref \frown suff = ps \\
&\qquad\qquad \wedge \text{ ran } pref \cap \text{ran } SectionHeader = \varnothing \\
&\qquad\qquad \wedge \ (suff = \varnothing \vee head\ suff \in \text{ran } SectionHeader) \bullet \\
&\qquad\qquad\quad \text{if } pref = \varnothing \text{ then} \\
&\qquad\qquad\qquad \text{if } sectionToName\ suff = n \text{ then} \langle\,\rangle \\
&\qquad\qquad\qquad \text{else} \langle SectionHeader\ (\!| \\
&\qquad\qquad\qquad\qquad name == n, \\
&\qquad\qquad\qquad\qquad parentSet == \varnothing\ |\!)\rangle \\
&\qquad\qquad\quad \text{else if ran } pref \cap \text{ran } Formal \neq \varnothing \text{ then} \\
&\qquad\qquad\qquad \langle SectionHeader\ (\!| \\
&\qquad\qquad\qquad\qquad name == n, \\
&\qquad\qquad\qquad\qquad parentSet == \{"\,standard\_toolkit"\}\ |\!)\rangle \\
&\qquad\qquad\quad \text{else} \\
&\qquad\qquad\qquad \langle SectionHeader\ (\!| \\
&\qquad\qquad\qquad\qquad name == n \frown "\,informal", \\
&\qquad\qquad\qquad\qquad parentSet == \varnothing\ |\!)\rangle) \\
&\qquad\qquad \frown ps
\end{aligned}
$$

The function *filenameToSections* reads the named file and partitions its sequence of paragraphs into the corresponding sequence of sections.

$$
\begin{aligned}
&filenameToSections == \\
&\quad \lambda n : Name \bullet \\
&\qquad (\mu ss : \text{seq } Section \mid \frown/\ ss = addHeader\ n)
\end{aligned}
$$

The function *readSpec* is given a set of names of files to be read and a set of sections already read from files. It returns the set of sections containing those already read, those read from the named files, and those read from files named as ancestors of other sections in this set. A file is read only if the named parent has not already been found in previous files and is not present anywhere in the current file; the parent section could be defined later in the current file, in which case any file with the name of the parent is not read. The sections should all have different names (otherwise an implementation should report an error); this specification merges sections that are identical.

$$readSpec : \mathbb{F}\,Name \times \mathbb{F}\,Section \to \mathbb{F}\,Section$$

$$\forall ss : \mathbb{F}\,Section \bullet$$
$$\quad readSpec\ (\varnothing\ , ss) = ss$$
$$\forall ns : \mathbb{F}\,Name;\ ss : \mathbb{F}\,Section \bullet$$
$$\quad readSpec\ (ns, ss) =$$
$$\quad\quad \mu ss_2 == \bigcup\{n : ns \bullet \mathrm{ran}(filenameToSections\ n)\}$$
$$\quad\quad\quad |\ \#ss_2 = \#(sectionToName(\!|\ ss_2\ |\!)) \bullet$$
$$\quad\quad\quad\quad readSpec\ ((sectionsToParents\ ss_2\ \backslash$$
$$\quad\quad\quad\quad\quad\quad sectionToName(\!|\ ss\ |\!)) \backslash ns,$$
$$\quad\quad\quad\quad ss \cup ss_2)$$

The function *orderSections* is given a set of sections and returns those sections in a sequence ordered so that every section appears before it is referenced as a parent. The prelude section has to be forced to be first in the sequence, as it might not be explicitly listed as being a parent. The function is partial because of the possibility of cycles in the parents relation, about which an implementation should report errors.

$$orderSections ==$$
$$\quad \{ss : \mathbb{F}\,Section;\ ss_2 : \mathrm{seq}\,Section\ |$$
$$\quad\quad\quad \mathrm{ran}\,ss_2 = ss$$
$$\quad\quad\quad \wedge\ sectionToName(head\ ss_2) = "prelude"$$
$$\quad\quad\quad \wedge\ (\ \forall ss_3 : \mathrm{seq}\,Section\ |\ ss_3 \subseteq ss_2 \bullet$$
$$\quad\quad\quad\quad \{sectionToName(last\ ss_3)\} \cap$$
$$\quad\quad\quad\quad\quad sectionsToParents\ (\mathrm{ran}(front\ ss_3)) = \varnothing\ )\ \bullet$$
$$\quad\quad (ss, ss_2)\}$$

The function *sortSects* specifies the finding and permuting of sections. It takes the name of a file, and returns the ordered sequence of sections from that file and the files of ancestral sections.

$$sortSects ==$$
$$\quad \lambda n : Name \bullet orderSections\ (readSpec\ (\{n\} \cup \{"prelude"\}, \varnothing))$$

Unformalised details. Recognition of mark-up directives, perhaps moving them to the beginnings of their sections, has not been specified above.

The consistency of this specification (that at least one model satisfies it) has not been formally proven.

When this specification was implemented as part of the CADiZ toolset, several additional features were needed, as follows.

CADiZ's input mark-up can contain **quiet** and **reckless** directives, which are intended to disable/enable typesetting and typechecking respectively. These modes are noted as attributes of each paragraph, so that after permutation appropriate **quiet** and **reckless** directives can be inserted in the output.

The toolset can subsequently typeset sections of the Z specification, in the order in which their paragraphs appeared in the original mark-up. This is enabled by the inclusion of filename and line number directives in the output generated by *sortSects*.

The typechecking tool can save the results of typechecking a section and its ancestors in a file, from which a subsequent invocation of the toolset on a larger Z specification can resume. The names of already typechecked sections are given as an additional argument to *sortSects*, which reads all sections as specified here, but then omits the already typechecked ones from its output.

9   Operator Templates

Once the sections have been permuted into definition before use order, and the mark-up has been converted to Z characters, there remains one further pass to be performed before the Z characters can be translated to lexical tokens. This third pass performs some processing of operator templates.

Operator templates are a kind of Z paragraph. For each operator (prefix, infix, postfix or bracketting name) defined in a specification, an operator template is required. This makes explicit the category (relation, function or generic), name, arity, precedence and associativity of the operator. It also indicates whether each operand is expected to be a single expression or a list of expressions, for example, as in the bracketted sequence $\langle A, B, C \rangle$.

Consider how sequence brackets are introduced in the toolkit. The operator template for sequence brackets says that it is a function in which $\langle$ and $\rangle$ are to be used as names in a unary operator whose operand is a list of expressions.

function $(\langle \ ,, \ \rangle)$

The definition of sequence brackets is in terms of seq, which results in the elements of the bracketted sequence being indexed from 1.

$\langle \ ,, \ \rangle [X] == \lambda s : \mathrm{seq}\, X \bullet s$

An introduction to operator templates was given in [Toyn 1998]. Standard Z has evolved since then: the precedences of prefix and postfix operators are no longer explicitly chosen but instead are fixed such that all prefix user-defined operators have higher precedence than all infix user-defined operators, and all postfix user-defined operators have higher precedence than all prefix user-defined operators.[4]

---

[4] This change avoids some awkward cases such as A *infix*3 *prefix*1 B *infix*2 C and A *infix*2 B *postfix*1 *infix*3 C (where the numbers indicate the desired precedence). Should these be parsed as $(A$ *infix*3 $(prefix1\ B))$ *infix*2 C and A *infix*2 $((B\ postfix1)\ infix3\ C)$, or as A *infix*3 $(prefix1\ (B\ infix2\ C))$ and $((A\ infix2\ B)\ postfix1)\ infix3\ C$, or should both be rejected? Existing Z tools gave parses that paired up differently. Other languages offering the same facility (Prolog and Twelf) disagree on which pair of parses is right. Standard Z avoids the problem by requiring high prefix and postfix precedences. This restricted notation is analogous to that of Haskell.

The names within an operator, such as $\langle$ and $\rangle$ in the above example, are assigned to particular token classes to enable parsing of operator names and operator applications. The appropriate token classes are determined from the operator template. There are separate token classes according to whether the name is the first, middle or last name in an operator, whether operands precede and follow it, and whether any preceding operand is a list of expressions. In the example, $\langle$ is assigned to the token class L (standing for leftmost name), and $\rangle$ is assigned to the token class SR (standing for rightmost name preceded by a sequence of expressions).

A name may be used within several different operators, so long as it is assigned to the same token class by all of their templates. For example, here is a specification of sequences indexed from 0.

$$\text{function } (\ (\ \langle^0\ ,,\ \rangle\ )\ )$$

$$\langle^0\ ,,\ \rangle[X] == \lambda s : \text{seq}\, X \bullet \lambda n : 0 .. \# \,\text{dom}\, s - 1 \bullet s(n+1)$$

This reuse of $\rangle$ is permitted because it is assigned to the same SR token class by both operator templates. Reuse of $\rangle$ as an infix, as in $\_ \rangle \_$, would attempt to associate it with a different token class, and so would not be permitted.

Two problems must be solved before the sequence of Z characters can be lexed. First, users may defer presenting operator templates until after the corresponding operator has been defined, yet information from an operator's template is needed to parse the operator's definition. Second, the reuse of names within different operators means that the syntax of operator templates is expressed in terms of the very token classes that operator templates establish.

The solution to the first problem is to permute operator template paragraphs to the beginnings of their sections. The solution to the second problem is to introduce, before an operator template, directives to the lexer to associate its names with token classes. These transformations can be done on the sequence of Z characters, hence the third pass. (In other words, there is no need to have separate implementations of these transformations for each mark-up.)

10   Summary

The processing of mark-up prior to lexing is summarised in Figure 2.

The first pass, section, gathers the mark-up of the specification's sections together, permutes them into a definition before use order, and brings mark-up directives to the starts of their sections.

The second pass, LATEXmark-up, extracts the formal text, tokenises that LATEX mark-up, expands \LaTeXcommands according their mark-up directives, and manages the scopes of those mark-up directives.

The third pass, optemp, brings operator templates to the starts of their sections, and generates lexis directives before them.

The resulting sequence of Z characters is ready to be lexed.

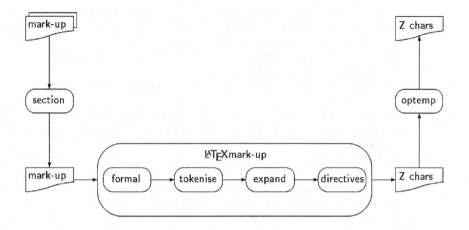

Fig. 2. Processing of mark-up prior to lexing

11   Conclusions

ISO Standard Z has answered the question "What's in a name?" in the Z context. The answer addresses the internationalisation issue. It establishes a new representation for a Z specification: its sequence of Z characters. This representation is intermediate between mark-up and lexical tokens. It simplifies the task of designing a new mark-up for Z, which can be to sequences of Z characters rather than directly to lexical tokens.

The liberal scope rules of sections and operator templates require extra processing of a Z specification prior to lexing. The paper has made explicit an order in which that processing can be done.

Acknowledgements. We thank the members of the Z standards panel for their influence, Sam Valentine for advising on the use of Z in the *sortSects* specification, the STIX committee for arranging the inclusion of Z characters in UCS, and Steve King and anonymous referees for comments on earlier draft versions of this paper.

References

[Arthan] R.D. Arthan. The ProofPower web pages.
    http://www.lemma-one.com/ProofPower/index/index.html.
[Arthan 1995] R.D. Arthan. Modularity for Z.
    http://www.lemma-one.com/zstan_docs/wrk059.ps, September 1995.
[Ciancarini *et al.* 1998] P. Ciancarini, C. Mascolo, and F. Vitali. Visualizing Z notation
    in HTML documents. In *ZUM'98: The Z Formal Specification Notation*, LNCS
    1493, pages 81–95. Springer, 1998.
[Germán *et al.* 1994] D.M. Germán, D. Cowan, and A. Ryman. Comments on the Z
    Interchange Format of the Z Base Standard Version 1.0.
    ftp://ftp.comlab.ox.ac.uk/pub/Zforum/ZSTAN/papers/z-160.ps.gz, 1994.

[Grieskamp] W. Grieskamp. The Zeta web pages. http://uebb.cs.tu-berlin.de/zeta/.

[ISO-ASCII] ISO/IEC 646:1991. *Information Technology—ISO 7-bit Coded Character Set for Information Interchange (3rd edition).*

[ISO-BNF] ISO/IEC   14977:1996(E).   *Information   Technology—Syntactic Metalanguage—Extended BNF.*

[ISO-UCS] ISO/IEC 10646:2000. *Information Technology—Universal Multiple-Octet Coded Character Set (UCS).*

[ISO-Z] ISO/IEC 13568:2001.   *Information   Technology—Z Formal Specification Notation—Syntax, Type System and Semantics: Draft International Standard.* To be published.

[King 1990] P. King. Printing Z and Object-Z LATEX documents. University of Queensland, 1990.

[Lamport 1994] L. Lamport. *LATEX: A Document Preparation System—User's Guide and Reference Manual, 2nd edition.* Addison-Wesley, 1994.

[Saaltink 1997] M. Saaltink. The Z/EVES system. In *ZUM'97: The Z Formal Specification Notation,* LNCS 1212, pages 72–85. Springer, 1997.

[Spivey 1992a] J.M. Spivey. The *f*UZZ manual, 2nd edition. Computer Science Consultancy, 1992.

[Spivey 1992b] J.M. Spivey. *The Z Notation: A Reference Manual, 2nd edition.* Prentice Hall, 1992.

[Stepney] S. Stepney. Formaliser Home Page. http://public.logica.com/~formaliser/.

[Toyn] I. Toyn. CADiZ web pages. http://www-users.cs.york.ac.uk/~ian/cadiz/.

[Toyn *et al.* 2000] I. Toyn, S.H. Valentine, and D.A. Duffy. On mutually recursive free types in Z. In *ZB2000: International Conference of B and Z Users,* LNCS 1878, pages 59–74. Springer, 2000.

[Toyn 1998] I. Toyn. Innovations in the notation of Standard Z. In *ZUM'98: The Z Formal Specification Notation,* LNCS 1493, pages 193–213. Springer, September 1998.

[Unicode 2001] The Unicode Consortium. *The Unicode Standard, Version 3.1,* May 2001. http://www.unicode.org/.

Extraction of Abstraction Invariants for Data Refinement*

Marielle Doche and Andrew Gravell

Department of Electronics and Computer Science
University of Southampton, Highfield
Southampton SO17 1BJ, United-Kingdom
marielle.doche@libertysurf.fr, amg@ecs.soton.ac.uk

Abstract. In this paper, we describe an approach to generating abstraction invariants for data refinement from specifications mixing B and CSP notations. A model-checker can be used to check automatically refinement of the CSP parts. However, we need to translate the CSP into B in order to verify data refinement of the whole specification. The Csp2B tool generates the B specification automatically from the CSP parts. Our proposal is to generate in addition the abstraction invariants, by analysing the labelled transition systems provided by a model-checker. The approach is illustrated with a case study in which a simple distributed information system is specified and two refinements are given, both of which have been fully verified using the proposed combination of model-checking with theorem proving (both automatic and interactive).

Keywords: Formal specification, CSP, failure refinement, B, data refinement, distributed system

1 Introduction

To improve the specification and the validation of complex systems, lots of recent research concerns the integration of two different formal notations to use the advantages of each. For example, in [But99,MS98,FW99,MC99,DS00] the authors combine a behaviour-based notation (CSP) with a model-based one (B, Z or Object-Z). Indeed the B method [Abr96] or the Z notation [Spi92] are suitable for modelling distributed systems, but the sequencing of events is difficult to specify. The CSP language [Hoa85] solves this problem more easily.

The work presented in this paper is based on the Csp2B approach [But99]. In that paper, Butler describes how to combine specifications in CSP and in B, and how to derive automatically B specifications from these combinations. This approach can be applied to derived B machines or B refinements. He also proves that, if they do not share variables, the composition of a CSP and a

---

* We acknowledge the support of the EPRSC (GR/M91013) for the ABCD project
(http://www.dsse.ecs.soton.ac.uk/ABCD/).

D. Bert et al. (Eds.): ZB 2002, LNCS 2272, pp. 120–139, 2002.

B specifications is compositional with respect to the data refinement process, which allows us to refine independently each part. Although it is easy to check refinement of the CSP subpart with a model-checker such as FDR [For97], in a lot of cases, however, we cannot prove refinement of the B subpart on its own, because this in fact depends on the state of the CSP subpart. In such cases, we need to generate from the CSP and B subparts a B machine and its refinement, which can then be verified.

The classical approach to refinement with the B method is data refinement [Abr96] which can be supported by theorem provers such as AtelierB [Ste96] or the B-Toolkit [B-C99]. In data refinement, however, we need to define abstraction invariants that link the variables of the abstract machine with those of the concrete one. This step is often based on the intuition of the specifier and it is difficult to achieve.

In this paper, we propose an approach for reducing the difficulty of this task by generating automatically the abstraction invariants that relate to the CSP subparts. First the FDR tool is used to check refinement of the CSP subparts alone. To do this, it constructs a labelled transition system (LTS). The LTS can be extracted from FDR and used to generate abstraction invariants in B. These can then be conjoined with the abstraction invariants relating to the B subparts.

The following section introduces our example and the Csp2B approach. We discuss some proof issues and when our approach can be applied in section 3. Section 4 describes our approach on a simple case of data refinement. A more complex case, with parallel decomposition, is given in section 5.

2   Csp2B Approach on Our Example

Our work is based on the Csp2B approach proposed by Butler [But99]. The idea of this approach is to increase the descriptive power of B specifications (B machines or B refinements) with the inclusion of CSP processes, which describe the order of events. Moreover, the Csp2B tool automatically translates the CSP processes into B language which can then be checked with a B prover (Atelier B or the B-Toolkit).

2.1   Basic Example

A customer requests some tokens (typically some data, a pension,..) at an office and then collects them at the same office. This is expressed in Csp2B by:

    MACHINE *Tokens*

    SEES *TokensDef*

    ALPHABET
      ReqTokens(off:OFFICE)
      toks ← CollTokens(off:OFFICE)

PROCESS   *Customers =* **Await**
CONSTRAINS   *ReqTokens(off) CollTokens(off)*   **WHERE**

Await $\widehat{=}$ *ReqTokens?off* → **Transact** *(off)*

Transact *(off_ab : OFFICE)* $\widehat{=}$ *CollTokens.off_ab* → **Await**
END

END

Here → is the classical prefixing operator of CSP, the event *ReqTokens* has an input parameter *?off*, the event *CollTokens* has a 'dot' parameter *.off_ab* which means it accepts as input only the value of *off_ab*. The **CONSTRAINS** clause allows us to constrain only a subset of the alphabet or some of the parameters (in this example, the input parameter of the event *CollTokens* is constrained, but not its output). The declaration *Customers =* **Await** defines the initial state of the process *Customers*.This Csp2B description can see the contents of the following B machine, where *home* is a function which associates a home office to each customer (here there are three customers):

MACHINE *TokensDef*

SETS
   $HOME = \{O1, O2, CENTRE\}$

CONSTANTS
   home, OFFICE, initTokens

PROPERTIES
   $OFFICE = HOME$ - $\{CENTRE\}$ $\wedge$
   $initTokens \in \mathbb{N}$ $\wedge$ $initTokens = 6$ $\wedge$
   $home \in \mathbb{N}$ → $HOME \wedge home = \{1 \mapsto O1, 2 \mapsto O2, 3 \mapsto CENTRE\}$

END

Then, the Csp2B tool [But99] translates the constraints on the order of events:

– for each CSP process, a new set and new variables are introduced in the B machine to manage the state of the process;
– each CSP event becomes a B operation, guarded by the state variables (using the B **SELECT** statement).

For our example, we obtain a set *CustomersState* with the values *Await* and *Transact*. Two variables are introduced: *Customers* and *off_ab*. The tool generates the following B machine:

MACHINE *Tokens*

SEES *TokensDef*

SETS
  CustomersState = {*Await, Transact*}

VARIABLES
  Customers, off_ab

INVARIANT
  Customers ∈ *CustomersState* ∧ *off_ab* ∈ *OFFICE*

INITIALISATION
  Customers := *Await* ||
  ANY *new_off_ab* **WHERE**
    new_off_ab ∈ *OFFICE* **THEN** *off_ab* := *new_off_ab* **END**

OPERATIONS

ReqTokens(*off*) =
  PRE *off* ∈ *OFFICE* **THEN**
    SELECT *Customers* = *Await* **THEN**
      Customers := *Transact* || *off_ab* := *off*
    END
  END ;

toks ← **CollTokens**(*off*) =
  PRE *off* ∈ *OFFICE* **THEN**
    SELECT *Customers* = *Transact*   ∧ *off* = *off_ab* **THEN**
      Customers := *Await*
    END
  END

  END

2.2   Conjunction

Moreover, a Csp2B machine M may constrain the order of the operations of an already defined B machine *MActs*. This is defined in the Csp2B description by the clause **CONJOINS** *MActs,* and for each event *Op* of the Csp2B description, the B machine *MActs* contains an operation *Op_Act* with the same interface.

In our example, the B machine *TokensActs* specifies the amount of tokens available for the customer in the system:

MACHINE *TokensActs*

SEES *TokensDef*

VARIABLES
  tokens

INVARIANT
  tokens ∈ ℕ

INITIALISATION
  tokens := initTokens

OPERATIONS

ReqTokens_Act(*off*) = **PRE** *off* ∈ *OFFICE* **THEN** **skip** **END** ;

toks ← **CollTokens_Act**(*off*) =
  PRE *off* ∈ *OFFICE* **THEN**
    IF *tokens* = 0   **THEN** *tokens* := 0 || *toks* := 0
    ELSE
      ANY *tok* **WHERE** *tok* : (1 .. *tokens*) **THEN**
        tokens := *tokens* - *tok* || *toks* := *tok*
      END
    END
  END

END

The Csp2B tool generates a B machine M from the Csp2B description as previously, but the B machine M includes the B machine *MActs*. Now, each operation Op contains a guarded call to the operation Op_Act of the machine *MActs* . Indeed, if Op_Csp is the B statement for the operation Op generated from the Csp2B description we obtain:

Op= Op_Csp || **SELECT** $grd(Op\_Csp)$ **THEN** **Op_Act** **END**

We generate the following B machine (the beginning is the same as previously):

MACHINE *Tokens*

SEES *TokensDef*

INCLUDES *TokensActs*

DEFINITIONS
    grd_Tokens_CollTokens(off)== (*Customers=Transact* ∧ *off=off_ab*);
    grd_Tokens_ReqTokens(off)== (*Customers=Await*)
   ...

OPERATIONS

ReqTokens(*off*) =
  PRE *off* ∈ *OFFICE* **THEN**
    SELECT *grd_Tokens_ReqTokens(off)*
     THEN **ReqTokens_Act**(*off*)
    END
    ||
    SELECT *Customers* = *Await*
     THEN *Customers* := *Transact* || *off_ab* := *off*
    END
  END ;

toks ← **CollTokens**(*off*) =
  PRE *off* ∈ *OFFICE* **THEN**
    SELECT *grd_Tokens_CollTokens(off)*
     THEN *toks* ← **CollTokens_Act**(*off*)
    END
    ||
    SELECT *Customers* = *Transact* ∧ *off* = *off_ab*
     THEN *Customers* := *Await*
    END
  END

END

2.3 Data Refinement

This same approach can be applied to produce a B refinement, which can be verified entirely in B with one of the B provers (Atelier B [Ste96] or the B-Toolkit [B-C99]).

The classical approach to data refinement in B involves introducing concrete variables and extra (hidden) operations. Moreover, to check data refinement, we need to define some abstraction invariants to link the abstract variables with the concrete ones.

The following refinement is defined to refine the previous *TokensActs* machine. Here we give some hints on the internal structure of our system: it is composed of two offices (*O1* and *O2*) and a *Centre*. The tokens about the customer can be held by any of the offices or the centre, thus we introduce the new variables *otokens* and *ctokens*. The abstraction invariant *tokens* = *ctokens* + *otokens(O1)* + *otokens(O2)* means that the global amount of tokens for a customer is the sum of the tokens at the centre and both the offices.

Moreover we introduce new operations to describe the internal communications between the centre and the offices. A customer requests some data at an office (operation **ReqTokens_Act**). If this office holds the data, the customer directly collects them (**CollTokens_Act**), else the office requests the data from the centre (**ReqOff_Act**). If the centre holds the data, it sends them to the office (**SendOff_Act**), else it requests and receives them from the home office of the customer (**QueryHome_Act** and **RecHome_Act**), where the home office of our customer is defined by the *home* function in the machine *TokensDef*. An original description, and some models of this example in different formalisms are given in [HBC+99].

REFINEMENT *TokensRefActs*

REFINES *TokensActs*

SEES *TokensDef*

VARIABLES
   otokens, ctokens

INVARIANT
   $ctokens \in \mathbb{N} \ \wedge \ otokens \in OFFICE \rightarrow \mathbb{N} \ \wedge$
   $tokens = ctokens + otokens(O1) + otokens(O2)$

INITIALISATION
   $ctokens := \text{initTokens} \ || \ otokens := OFFICE \times \{\, 0\}$

OPERATIONS

ReqTokens_Act(*off*) = **PRE** *off* $\in$ *OFFICE* **THEN** skip **END**;

toks $\leftarrow$ **CollTokens_Act**(*off*) =
   PRE *off* $\in$ *OFFICE* **THEN**
       $otokens(off) := 0 \ || \ toks := otokens(off)$
   END;

SendOff_Act(*off*) =
　PRE *off* ∈ *OFFICE* **THEN**
　　ctokens := 0 || *otokens*(*off*) := *otokens*(*off*) + *ctokens*
　END;

RecHome_Act(*off*) =
　PRE *off* ∈ *OFFICE* **THEN**
　　otokens(*off*) := 0 || *ctokens* := *ctokens* + *otokens*(*off*)
　END;

ReqOff_Act(*off*) =
　PRE *off* ∈ *OFFICE* **THEN**
　　SELECT *otokens*(*off*) =0 **THEN skip END**
　END;

QueryHome_Act(*off*) =
　PRE *off* ∈ *OFFICE* **THEN**
　　SELECT *ctokens* = 0 ∧ *home*(1) = *off* **THEN skip END**
　END

END

Unfortunately, it is not possible to prove that *TokensRefActs* refines *TokensActs*: in the case where *tokens>0* and *otokens(off)=0* the concrete operation **CollTokens_Act** does not refine the abstract one. We need more information on the evolution of the variables *ctokens* and *otokens*.

The following Csp2B specification describes the order of the events.[1] Here □ is the external choice of Csp2B.

REFINEMENT *TokensRef*

REFINES *Tokens*

SEES *TokensDef*

CONJOINS *TokensRefActs*

ALPHABET
　ReqTokens(off:OFFICE)
　toks ← CollTokens(off:OFFICE)
　SendOff(off: OFFICE)
　RecHome(off: OFFICE)

---

[1] In practice we cannot include a B refinement in a B machine. Thus in this example, the conjoined B specification *TokensRefActs* is the transcription of the previous B refinement example in a B abstract machine. The abstraction invariant already defined will be introduced in the generated B refinement.

$ReqOff(off: OFFICE)$
$QueryHome(off: OFFICE)$

PROCESS $System = $ **Asleep**
CONSTRAINS $ReqTokens(off)$ $CollTokens(off)$ $SendOff(off)$
$RecHome(off)$ $ReqOff(off)$ $QueryHome(off)$ **WHERE**

Asleep $\widehat{=}$ $ReqTokens?off \rightarrow$ **Request** (off)

Request $(off\_co :OFFICE)$ $\widehat{=}$
    IF $otokens(off\_co) > 0$
    THEN $CollTokens.off\_co \rightarrow$ **Asleep END**
    $\Box$ $ReqOff.off\_co \rightarrow$ **Answer**

Answer $\widehat{=}$
    IF $not(off\_co=home(1))$ **THEN** $(QueryHome.home(1)$
        $\rightarrow RecHome.home(1) \rightarrow SendOff.off\_co \rightarrow$ **Collect**$)$
    ELSE $(SendOff.off\_co \rightarrow$ **Collect**$)$
    END

    Collect $\widehat{=}$ $CollTokens.off\_co \rightarrow$ **Asleep**
END

END

This description can be compiled by the Csp2B tool to produce a B refinement of the machine *Tokens*. The tool produces a new set $SystemState = \{$ *Asleep, Request, Answer, Answer_1, Answer_2, Collect* $\}$ and new variables *off_co* and *System*. In this example, the process **Answer** contains some implicit states (indeed between the events $QueryHome.home(1)$ and $RecHome.home(1)$ and the events $RecHome.home(1)$ and $SendOff.off\_co$). For these implicit states, the Csp2B tool generates then some fresh names (the name of the process followed by an underscore character and a number).

Unfortunately, the tool does not define abstraction invariants to link these new variables with the concrete ones. This must be done manually, which can be difficult in some cases.

3   Discussion

In this section, we are going to discuss the different cases of refinement checking, and when the approach we propose in the sequel can be used.

Figure 1 summarises the Csp2B approach: a Csp description M_CSP is defined which may conjoin a B machine M_Act. The Csp2B tool generates automatically a B machine M. The same approach is applied to define a machine R which is a refinement of M.

To check refinement, three cases are possible:

- there are no conjoined B machines,
- the conjoined B machine M_Act is refined by R_Act,
- we cannot prove that M_Act is refined by R_Act (see our first example)

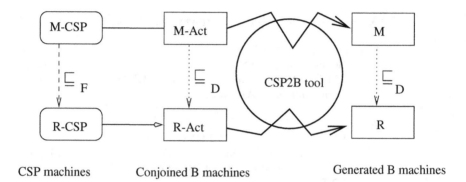

CSP machines   Conjoined B machines   Generated B machines

Fig. 1. Csp2B process

3.1 Without Conjoined Machines

Morgan [Mor90] has defined a correspondence between action systems and CSP. In [WM90] the authors have established a correspondence between failures-divergences refinement and *simulation* for action systems. Butler [But97] has extended this result to the B machine. A machine *Concrete* simulates a machine *Abstract* if *Concrete* is a data refinement of *Abstract* and some progress and non divergence conditions are verified on *Concrete*. Thus in theory, if we have proved failures refinement on a CSP specification, we do not need to prove data refinement on its translation (failures refinement is sufficient because at present data refinement with B does not consider hidden events and divergence).

3.2 Refinement between Conjoined Machine Is Proved

In [But99], Butler has shown that, if the CSP machine doesn't refer the state variables of the conjoined B machine, the parallel operator used to generate the B operations is monotonic with respect to refinement. This allows us to refine independently conjoined B machines and CSP processes. In this case, if $M\_CSP \sqsubseteq_F R\_CSP$ and $M\_Act \sqsubseteq_D R\_Act$ then $M \sqsubseteq_D R$.

If the CSP machine refers to variables of the conjoined B machine, as in our second example, section 5, this result is not valid and the refinement must be proved on the generated B machines. In such a case we can apply our following proposed approach to generate some of the abstraction invariants.

3.3   Refinement between Conjoined Machine Is Not Proved

In this case, we have to prove refinement of the generated B machine (see our first example in sections 2.3 and 4), with some abstraction invariants.

However, when the CSP machines do not refer to variables of the conjoined machines, we can reduce the proof task. In such a case, the parallel operator is monotonic with respect to data refinement, thus the CSP part can be refined independently. Our proposed approach generates abstraction invariants that are independent of the variables of the conjoined B machine, and hence are obviously preserved by this part. Thus in this case, the proof obligations regarding our generated abstraction invariants can easily be discharged.

4   Simple Data Refinement

The aim of this paper is to propose an automatic approach to defining abstraction invariants for the state variables introduced by the Csp2B tool. This approach is based on the analysis of the Labelled Transition Systems (LTS) built from the Csp2B descriptions. Such LTS can easily be obtained with a model-checker like FDR [For97].

4.1   Proposed Approach

Refinement mapping. The FDR tool provides easily and automatically three kinds of checks, increasing in strength:

1. *Trace refinement*: all possible sequences of events for the implementation are possible sequences for the specification.
2. *Failures refinement*: any failure of the implementation (indeed a pair formed by a finite trace of events and the set of refused events after this trace) is a failure of the specification.
3. *Divergence refinement*: any failure of the implementation is a failure of the specification and any divergence of the implementation (when it repeats infinitely often an event) is a divergence of the specification.

Failures refinement, quickly checked with the FDR tool, is a necessary condition for data refinement (indeed B refinement checking does not currently allow to detect divergence of a system).

To make these checks, the tool builds a LTS for the specification and one for the implementation and considers inductively pairs of nodes from the abstract LTS and the concrete one (for more details see [For97,Ros97]). Thus, in case of trace refinement, we can define a relation between abstract states and concrete ones. In [AL91], Abadi and Lamport show that if a concrete transition system refines an abstract one, there is a mapping from the state space of the concrete transition system to the state space of the abstract one, if necessary by adding auxiliary variables.

Formally, we call $M : \Sigma_C \to \Sigma_A$ the refinement mapping, where Σ_C and Σ_A are the sets of nodes respectively of the concrete LTS and of the abstract LTS. $dom(M)$ and $ran(M)$ are respectively the domain and the codomain of M. Given a in $ran(M)$, we denote by $M^{-1}(a)$ the set of nodes of Σ_C which have a as image by M:

$$M^{-1}(a) \hateq \{c | c \in dom(M) \wedge M(c) = a\}$$

In practice, such refinement mappings can be easily obtained with state space reduction algorithms [For97,Ros97]: for any node n of the concrete LTS, we group with n all the nodes reachable from n by an internal event. All these nodes have the same image by M, which can be computed inductively from the initial state.

Acceptance sets. Moreover, the FDR tool provides for each node of an LTS the *acceptance* set, the set of events the node must accept. If trace refinement is verified, the acceptance set of a concrete node is included in the acceptance set of the corresponding abstract node union the set of hidden events (the events or operations present only in the concrete LTS). If failures refinement is verified, we can be sure that for a concrete node if the set is empty, the set of the corresponding abstract node is also empty (since the deadlocks of the implementation are deadlocks of the specification).

The refinement mapping between the LTS provided by FDR and the acceptance sets are our starting point for defining the B abstraction invariants.

For any node n of an LTS, we call $G(n)$ the acceptance set of n. We extend this notation to several nodes : $G(n_1, ..., n_k) = G(n_1) \cup ... \cup G(n_k)$. So, for $a \in ran(M)$ we obtain:

$$G(M^{-1}(a)) = \bigcup_{c \in M^{-1}(a)} G(c)$$

Failures refinement ensures that if $G(M^{-1}(a))$ is empty then $G(a)$ is also empty.

If we call H the set of hidden events, trace refinement ensures:

$$\forall c \in \Sigma_C, G(c) \subseteq G(M(c)) \cup H$$

Finally for any event e, we denote by $grd(e)$ the guard of this event, the condition under which e is enabled, expressed by a predicate.

Abstraction invariant. For each node of Σ_A, we define an invariant.

For the abstract LTS, in a node $a \in \Sigma_A$, at least one guard of an event of $G(a)$ is satisfied. This means that $\bigvee_{e \in G(a)} grd(e)$ is true.

If $a \in ran(M)$, the refinement mapping ensures that at least one of the guard of an event of $G(M^{-1}(a))$ is also satisfied. For each $a \in ran(M)$, we define thus an invariant :

$$(\bigvee_{e \in G(a)} grd(e)) \Rightarrow (\bigvee_{f \in G(M^{-1}(a))} grd(f))$$

If $a \in (\Sigma_A - ran(M))$, this node is not an image of a concrete one, so it can never be reached in the concrete model. So we define an invariant:

$$\neg(\bigvee_{e \in G(a)} grd(e))$$

In practice, the Csp2B tool computes the guards of each CSP event as predicates (see for example $grd\_Tokens\_ReqTokens(off)$ in the B machine $Tokens$ of section 2.1).

4.2   Results on Our Example

Figure 2 shows the concrete LTS directly produced by the FDR model-checker, respectively from the CSP description $Tokens$ and $TokensRef$. The dotted ovals and lines show the refinement mapping. The dashed arrows are internal, or hidden, events.

In the following table we give for each node of the abstract LTS $Tokens$, the acceptance $G(a)$, the set $M^{-1}(a)$, and the set $G(M^{-1}(a))$:

a	$G(a)$	$M^{-1}(a)$	$G(M^{-1}(a))$
0	ReqTokens(O1) ReqTokens(O2)	0	ReqTokens(O1) ReqTokens(O2)
1	CollTokens(O1)	1, 7, 8	CollTokens(O1) ReqOff(O1) SendOff(O1)
2	CollTokens(O2)	2, 3, 4, 5, 6	CollTokens(O2) ReqOff(O2) SendOff(O2) QueryHome(home(1)) RecHome(home(1))

The set of hidden events is:
$H = \{ReqOff(O1), ReqOff(O2), SendOff(O1), SendOff(O2),$
      $QueryHome(O1), QueryHome(O2), RecHome(O1), RecHome(O2)\}$
We can then generate the three following invariants:

$((grd\_Tokens\_ReqTokens(O1) \vee grd\_Tokens\_ReqTokens(O2)) \Rightarrow$
   $(grd\_TokensRef\_ReqTokens(O1) \vee grd\_TokensRef\_ReqTokens(O2)))$

$\wedge$

   $(grd\_Tokens\_CollTokens(O1) \Rightarrow$
    $(grd\_TokensRef\_CollTokens(O1)$ or $grd\_TokensRef\_ReqOff(O1)$
    $\vee grd\_TokensRef\_SendOff(O1)))$

$\wedge$

   $(grd\_Tokens\_CollTokens(O2) \Rightarrow$

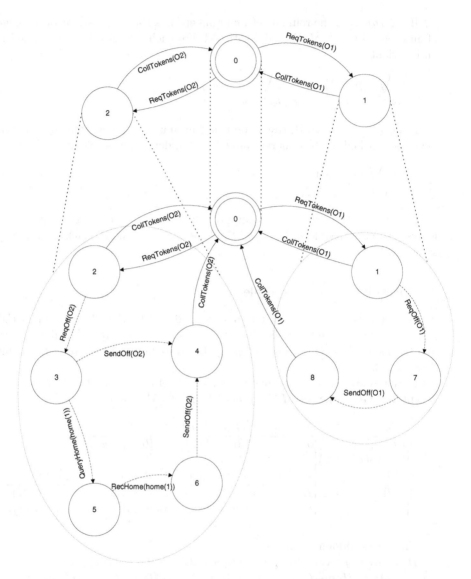

Fig. 2. Abstract and concrete LTS

$(grd\_TokensRef\_CollTokens(O2)$ or $grd\_TokensRef\_ReqOff(O2)$

$\lor$ $grd\_TokensRef\_SendOff(O2)$ or $grd\_TokensRef\_QueryHome(home(1))$

$\lor$ $grd\_TokensRef\_RecHome(home(1))))$

Unfortunately, these invariants and the one defined in the section 2.3 (*tokens = ctokens + otokens(O1) + otokens(O2)*) are not sufficient to prove in B that the B refinement *TokensRef* refines the B machine *Tokens*. We have the same problem with the **CollTokens** operation as in our example of the section 2.3.

Indeed, the previous approach provides only invariants on the variables introduced by Csp2B. We need to add the following invariants, which give conditions on the amount of tokens in different states, for example the first one expresses that when the home office has send the tokens to the centre, there are no more tokens at the home office:

$$((Customers = Answer\_2) \Rightarrow otokens(home(1)) = 0)$$
$$\wedge$$
$$((Customers = Collect \vee Customers = Answer\_1) \Rightarrow ctokens = 0)$$
$$\wedge$$
$$((\neg (Customers = Collect) \vee off\_1=O1) \Rightarrow otokens(O2) = 0)$$
$$\wedge$$
$$((Customers = Collect \wedge off\_1 = O2 \wedge tokens>0) \Rightarrow otokens(O2) >0)$$

With these additional invariants added by hand, it is possible to verify the refinement completely. In this verification, 181 proof obligations were generated with the AtelierB tool, of which 36 were proved manually, the others automatically.

5   Parallel Decomposition

The previous step of refinement has shown different parts in our system:

- a single *Centre*,
- some *Offices*, which can be the home-office of a customer.

A new step of refinement implements this decomposition: in a B refinement we include a machine for each part and we define operations as interactions of the operations of each part. For example the B refinement *TokensRefRefActs* includes the B machine *Centre* and *Offices* (with the renaming in *ce* and *oo* respectively):

REFINEMENT *TokensRefRefActs*

REFINES *TokensRefActs*

SEES *TokensDef*

INCLUDES *ce.Centre, oo.Offices*

INVARIANT
    ctokens = ce.ctokens $\wedge$
    otokens(O1) = oo.otokens(O1) $\wedge$ *otokens(O2) = oo.otokens(O2)*

OPERATIONS

  ReqTokens_Act(*off*) = **oo.ReqTokens**(*off*) ;

  toks ← **CollTokens_Act**(*off*) = *toks* ← **oo.CollTokens**(*off*) ;

  ReqOff_Act(*off*) =
  PRE *off* ∈ *OFFICE* **THEN** **ce.ReqOff**(*off*) ‖ **oo.ReqOff**(*off*) **END**;

  SendOff_Act(*off*) =
  PRE *off* ∈ *OFFICE* **THEN**
    ce.SendOff(*off*) ‖ **oo.SendOff**(*off, ce.ctokens*)
  END;

  ...

END

In a Csp2B notation we can express decomposition as a parallel composition
of several processes:

REFINEMENT *TokensRefRef*

REFINES *TokensRef*

SEES *TokensDef*

CONJOINS *TokensRefRefActs*

ALPHABET
  ReqTokens(off:OFFICE)
  toks ← *CollTokens(off:OFFICE)*
  SendOff(off: OFFICE)
  RecHome(off: OFFICE)
  ReqOff(off: OFFICE)
  QueryHome(off: OFFICE)

PROCESS   *Centre* = **CentreAsleep**
CONSTRAINS   *SendOff(off)*   *RecHome(off)*
    ReqOff(off)   *QueryHome(off)*
WHERE

  CentreAsleep ≙ *ReqOff?off* → **CentreAnswer(off)**

  CentreAnswer(ce_off : OFFICE) ≙
      IF *not(ce_off=home(1))* **THEN** (*QueryHome.home(1)*
        → *RecHome.home(1)* → *SendOff.ce_off* → **CentreAsleep**)
      ELSE (*SendOff.ce_off* → **CentreAsleep**)
      END
END

PROCESS   *Offices =* **OfficeAsleep**

CONSTRAINS   *ReqTokens(off)   CollTokens(off)*

      SendOff(off)   ReqOff(off)

WHERE

OfficeAsleep $\widehat{=}$ *ReqTokens?off* → **OfficeRequest(off)**

OfficeRequest(oo_off : OFFICE) $\widehat{=}$

    IF *otokens(oo_off) > 0*

    THEN *CollTokens.oo_off* → **OfficeAsleep END**

    □ *ReqOff.oo_off* → *SendOff.oo_off* →

        CollTokens.oo_off → **OfficeAsleep**

PROCESS   *Home =* **HomeAsleep**

CONSTRAINS   *RecHome(off)   QueryHome(off)*

WHERE

HomeAsleep $\widehat{=}$ *QueryHome?off* → **HomeRequest(off)**

HomeRequest(ho_off : OFFICE) $\widehat{=}$

    RecHome.(ho_off → **HomeAsleep**

END

END

We can apply exactly the same approach as with simple data refinement: we define a refinement mapping between the concrete and the abstract LTSs, and then we build the abstraction invariant from this mapping and the acceptance sets.

For our example, we obtain for *TokensRefRef* the same LTS as with *TokensRef* (indeed the FDR tool proves they are equivalent).

The following table describes the refinement mapping and the acceptance sets; the set of hidden events is empty:

a	$G(a)$	$M^{-1}(a)$	$G(M^{-1}(a))$
0	ReqTokens(O1) ReqTokens(O2)	0	ReqTokens(O1) ReqTokens(O2)
1	CollTokens(O1) ReqOff(O1)	1	CollTokens(O1) ReqOff(O1)
2	CollTokens(O2) ReqOff(O2)	2	CollTokens(O2) ReqOff(O2)
3	SendOff(O2) QueryHome(home(1))	3	SendOff(O2) QueryHome(home(1))
4	CollTokens(O2)	4	CollTokens(O2)
5	RecHome(home(1))	5	RecHome(home(1))
6	SendOff(O2)	6	SendOff(O2)
7	SendOff(O1)	7	SendOff(O1)
8	CollTokens(O1)	8	CollTokens(O1)

The sets, variables and guards automatically computed by the Csp2B tool are :

REFINEMENT *TokensRefRef*

REFINES *TokensRef*

SEES *POsets*

INCLUDES *TokensRefRefActs*

SETS
$CentreState= \{CentreAsleep, CentreAnswer, CentreAnswer\_1,$
$\quad CentreAnswer\_2\};$
$OfficesState= \{OfficeAsleep, OfficeRequest, OfficeRequest\_1, OfficeRequest\_2\};$
$HomeState = \{HomeAsleep, HomeRequest\}$

VARIABLES
$Centre, \quad ce\_off, \quad Offices, \quad oo\_off, \quad Home, \quad ho\_off$

DEFINITIONS
$grd\_TokensRefRef\_CollTokens(off) ==$
$\quad ((mo.otokens (oo\_off) > 0 \wedge Offices = OfficeRequest \wedge off = oo\_off)$
$\quad \vee (Offices = OfficeRequest\_2 \wedge off = oo\_off)) ;$

$grd\_TokensRefRef\_ReqTokens(off) == (Offices = OfficeAsleep) ;$

$grd\_TokensRefRef\_SendOff ==$
$\quad (((Centre = CentreAnswer\_2 \wedge off = ce\_off)$
$\quad \vee (\neg (\neg (ce\_off = home(1))) \wedge Centre = CentreAnswer$
$\quad\quad \wedge off = ce\_off))$
$\quad \wedge (Offices = OfficeRequest\_1 \wedge off = oo\_off)) ;$

$grd\_TokensRefRef\_RecHome ==$
$\quad ((Centre = CentreAnswer\_1 \wedge off = home (1))$
$\quad \wedge (Home = HomeRequest \wedge off = ho\_off)) ;$

$grd\_TokensRefRef\_ReqOff ==$
$\quad ((Centre = CentreAsleep)$

$$\wedge \ (\ Offices = OfficeRequest \wedge off = oo\_off\ )\ )\ ;$$

$grd\_TokensRefRef\_QueryHome ==$
$(\ (\ \neg\ (\ ce\_off = home\ (\ 1\ )\ )\ \wedge Centre = CentreAnswer$
$\qquad \wedge off = home\ (\ 1\ )\ )$
$\quad \wedge\ (\ Home = HomeAsleep\ )\ )$

...

END

With the abstraction invariants defined following our approach and those defined in the B refinement *TokensRefRefActs*, 398 proof obligations are generated by AtelierB, of which 41 have been proved manually, the others automatically.

6   Conclusion

In this paper, we are interested in an existing approach which combines CSP processes and B descriptions. A tool allows us to generate automatically from this combination B abstract machines or their refinements. However to check refinement we have to define some abstraction invariants to link abstract variables to concrete ones. We have proposed an approach that generates automatically some of these invariants. This approach is based on the labelled transition systems obtained by model-checking the CSP processes. A classical B proof step is then applied to verify refinement of the generated B machines. Moreover, when the conjoined B machines do not depend on the CSP processes, this last refinement step can be discharged. Otherwise, additional invariants must be added manually before verifying refinement.

Acknowledgements. We would like to thank the other members of the ABCD project for fruitful discussions and useful coments on this work. We are especially grateful to Michael Butler for his explaination on the Csp2B approach and his comments on an early version of this paper.

References

[Abr96]   J-R. Abrial. *The B-Book: Assigning Programs to Meanings.* Cambridge University Press, 1996.

[AL91]   M. Abadi and L. Lamport. The existence of refinement mappings. *Theoretical Computer Science*, 82(2), May 1991.

[B-C99]   B-Core (UK) Limited, Oxon, UK. *B-Toolkit, On-line manual.*, 1999. Available at http://www.b-core.com/ONLINEDOC/Contents.html.

[But97]   M. J. Butler. An approach to the design of distributed systems with B AMN. In J. Bowen, M. Hinchey, and D. Till, editors, *Proc. 10th Int. Conf. of Z Users: The Z Formal Specification Notation (ZUM), LNCS 1212*, pages 223–241, Reading, UK, April 1997. Springer-Verlag, Berlin. Available at http://www.dsse.ecs.soton.ac.uk/.

[But99]   M. J. Butler. csp2B: A practical approach to combining CSP and B. In J. M. Wing, J. Woodcock, and J. Davies, editors, *Proc. FM'99: World Congress on Formal Methods*, LNCS 1708, pages 490–508. Springer-Verlag, Berlin, September 1999. Available at http://www.dsse.ecs.soton.ac.uk/.

[DS00]   J. Derrick and G. Smith. Structural refinement in object-z / csp. In *IFM'2000 (Integrated Formal Methods)*, volume 1945 of *LNCS*. Springer-Verlag, 2000. Available at
http://www.cs.ukc.ac.uk/research/tcs/index.html.

[For97]   Formal Systems (Europe) Ltd. *Failures-Divergence Refinement- FDR2 user manual*, Octobre 1997. Available at
www.formal.demon.co.uk/fdr2manual/index.html.

[FW99]   C. Fischer and H. Wehrheim. Model-checking CSP-OZ specifications with FDR. In *First international Conference on Integrated Formal Methods (IFM99)*, pages 315–334. Springer-Verlag, 1999. Available at
http://semantik.Informatik.Uni−Oldenburg.DE/persons/clemens.ficher/.

[HBC+99]   P. Hartel, M. Butler, A. Currie, P. Henderson, M. Leuschel, A. Martin, A. Smith, U. Ultes-Nitsche, and B. Walters. Questions and answers about ten formal methods. In S. Gnesi and D. Latella, editors, *Proc. 4th Int. Workshop on Formal Methods for Industrial Critical Systems*, volume II, pages 179–203, Trento, Italy, July 1999. ERCIM, STAR/CNR, Pisa, Italy. Available at http://www.dsse.ecs.soton.ac.uk/.

[Hoa85]   C.A.R Hoare. *Communicating Sequential Processes*. Prentice Hall, 1985.

[MC99]   I. MacColl and D. Carrington. Specifying interactive systems in Object-Z and CSP. In *First international Conference on Integrated Formal Methods (IFM99)*. Springer-Verlag, 1999. Available at
http://archive.csse.uq.edu.au/ ianm/.

[Mor90]   C.C. Morgan. Of wp and CSP. In W.H.J. Feijen, A.J.M. van Gasteren, D. Gries, and J. Misra, editors, *Beauty is our business: a birthday salute to Edsger W. Dijkstra*. Springer Verlag, 1990.

[MS98]   A. Mota and A. Sampaio. Model-checking CSP-Z. In *Fundamental Approach of Software Engineering (FASE98)*, number 1382 in LNCS, pages 205–220. Springer Verlag, 1998. Available at http://www.di.ufpe.br/ acm/.

[Ros97]   A.W. Roscoe. *The Theory and Practice of Concurrency*. Prentice-Hall, 1997.

[Spi92]   J. M. Spivey. *The Z Notation: A Reference Manual*. Prentice Hall International Series in Computer Science, 2nd edition, 1992.

[Ste96]   Steria, Aix-en-Provence, France. *Atelier B, User and Reference Manuals*, 1996. Available at http://www.atelierb.societe.com/index_uk.html.

[WM90]   J.C.P. Woodcock and C.C. Morgan. Refinement of state-based concurrent systems. In D. Bjorner, C.A.R. Hoare, and H. Langmaack, editors, *VDM'90*, volume 428 of *LNCS*. Springer-Verlag, 1990. Available at http://www.iro.umontreal.ca/labs/teleinfo/PubListIndex.html.

An Approach to Combining B and Alloy

Leonid Mikhailov and Michael Butler

Department of Electronics and Computer Science
University of Southampton
Highfield, Southampton, SO17 1BJ
United Kingdom
mjb@ecs.soton.ac.uk

Abstract. In this paper we propose to combine two software verification approaches, theorem proving and model checking. We focus on the B-method and a theorem proving tool associated with it, and the Alloy specification notation and its model checker "Alloy Constraint Analyser". We consider how software development in B can be assisted using Alloy and how Alloy can be used for verifying refinement of abstract specifications. We demonstrate our approach with an example.

Keywords : B-method, Alloy.

1 Introduction

The approaches to creating verifiably correct systems can be divided in two broad categories: a top down approach when developers start with an abstract specification and gradually refine it to an executable implementation, which is guaranteed to be correct with respect to the specification, and a bottom up approach when developers attempt to implement a specification straight away and later on undertake a verification effort to make sure that their implementation complies to the specification.

The first approach is usually based on some sort of refinement calculus. Showing that a certain refined specification or, in fact, a final implementation complies to the corresponding abstract specification usually involves proving a lot of properties. Theorem proving is a very tedious process involving keeping in mind a multitude of assumptions and transformation rules. To help with this task a number of general purpose theorem provers exist, such as PVS, HOL, etc. [13, 6]. Such theorem provers usually have some automated tactics such as GRIND in PVS which attempt to prove the set goal automatically. As most of the refinement calculi (and formalisations of programming notations) are formulated in undecidable logics (first and higher order logics) proving all goals automatically is impossible. Thus the tool usually produces several subgoals that it didn't manage to resolve automatically and asks user guidance and assistance. The user by applying the set of rules and theorems available in the system attempts to prove the remaining goals.

With the second approach the specifiers usually formulate a number of liveness and safety properties that the implementation is supposed to comply to. It

D. Bert et al. (Eds.): ZB 2002, LNCS 2272, pp. 140–161, 2002.

is, of course, possible to apply general purpose theorem provers for this purpose. However a different verification technique, generally referred to as "model checking" is quite prominent with this approach. The general idea of model checking can be briefly expressed as follows: a program in its abstract representation, and the verification properties to be checked are formulated in some formalism based on logic. Next these formulas are submitted to the tool which tries to find a counter example violating the formulated verification conditions [9,7,10,5].

Both theorem proving and model checking have advantages and disadvantages. The main advantage of theorem proving is that it permits to reason about infinite domains which are the most interesting in practice. A disadvantage is that a significant amount of highly qualified labour is required to verify even a relatively simple program. With theorem proving at times it can be difficult to say whether a property does not prove because the assumptions are not sufficiently strong or just some extra effort and ingenuity is required.

Model checking is much more applicable for finite domains, although there is a lot of ongoing research trying to apply this method to infinite domains. In general, for infinite domains, while model checking can find a counter example demonstrating that the specification is contradictory in one way or another, it may not prove that the specification is correct. In this respect model checking is similar to testing, which also cannot prove the program correct. However what both of these approaches (model checking and testing) can do is to increase our confidence in the system. Another shortcoming of model checking is that it is usually applied for verifying consistency of a rather high level specifications, while ultimately everybody is interesting in the correctness of the software implementing these specifications. Obviously, while a specification can be perfectly correct, the implementation may not be correct. Verifying correctness of the executable programs with respect to their specifications is a topic of ongoing research.

In this paper we propose to combine these two approaches to verification, with the goal being to benefit from the advantages of both theorem proving and model checking. In particular we consider combining the B method and a corresponding tool with the Alloy specification notation and its constraint analyser. The B method is a top down development approach which is supported by industry-strength tools, which integrate a theorem prover for verifying the correctness of the specification and its refinements [1]. The Alloy specification notation is state-based and is supported by the Alloy constraint analyser, which is a finite state model checker [8,9]. We briefly present these specification and verification methods in the following sections. An earlier version of this paper has appeared in proceedings of the 6th International Workshop on Formal Methods for Industrial Critical Systems (FMICS 2001) [12].

The main idea discussed in this paper is as follows. Complete formal proof of all proof obligations generated by the B tool is often practically infeasible. Often a proof obligation cannot be proved for the simple reason that it is not true. That can happen, for example, because a specification of an operation is not logically strong enough. Or, simply, the specification of an operation can be erroneous.

The realisation of impossibility to prove a certain proof obligation usually brings about a realisation that certain amendments can be made to the specification, which would generate additional conjuncts in the hypothesis, permitting proof of the obligation. At times, however, a developer can experience difficulties proving a particularly tricky property, although sufficient hypothesis are present. Distinguishing between these two situations is important, as significant resources can be wasted on trying to show unprovable goals.

Once the B tool has generated proof obligations we try to run an automated theorem prover supplied with the tool. It usually leaves some of the obligations unproved. Our idea is that before actually trying to prove these obligations interactively, we translate them into the Alloy language and run the Alloy constraint analyser on them. Counter examples that the Alloy constraint analyser can generate are usually suggestive. When a developer realizes how a certain instantiation of variables of the counter example invalidates the property under consideration, it becomes clear which amendments can be made to the specification to exclude the counter example. This suggest a certain debugging process, which most certainly has a shorter cycle than when interactive prover is used for finding error. Once the Alloy constraint analyser cannot find a counter example for a sufficiently large instantiation of the domains, it is a good indication that the verified property is probably correct. The developer can then return to the B interactive prover with confidence that this property should be possible to prove. Translation between B specification notation and Alloy is manual at the moment. However, in case certain modification and additions would be made to the Alloy specification notation, such translation could be done automatically in both directions.

We proceed as follows. First, we briefly review specification and verification methods of B and Alloy. We then present a "Student Grades Database" example specification in B and its translation into Alloy. Next, we consider how one can verify consistency and refinement of B machine operations using the Alloy constraint analyser. Finally, we discuss the modifications and changes that have to be introduced into the Alloy specification notation to facilitate automatic translation of B specifications and proof obligations into Alloy. We conclude by reviewing related work and outlining directions of future research.

2   Summary of the Used Formalisms

Let us now briefly present the formalisms of B and Alloy and the development methods associated with them.

2.1   The B Specification and Verification Method

The B method has an associated specification notation, the so-called Abstract Machine Notation (AMN). This specification notation is classified as a state-based notation and is quite similar to such well-known formal notations as Z and VDM [14,16]. Compared to the specification notation of Z, AMN is more appealing to programmers, as it includes such familiar constructs as assignment and **IF   THEN   ELSE**   along with nondeteministic specification statements such as nondeterministic choice **ANY**.   We briefly introduce the necessary subset of AMN as we present an example.

The B method has three development stages: the specification, the refinement, and the implementation. Development in the B method is centred around the concept of *machines*: an abstract machine – **MACHINE**, a refinement machine – **REFINEMENT** , and an implementation machine – **IMPLEMENTATION**. Machines are similar to modules encapsulating their internal state. Machines provide a section for initialising their internal state and operations for accessing and manipulating it.

The developer starts off with translating an informal specification into an abstract **MACHINE**, which is allowed to use only an abstract subset of all available statements in AMN. As a part of the abstract machine specification, the developer has to introduce an invariant, which should be established by initialisation and should hold before and after the execution of all operations of this machine. When the developer submits the produced specification to the tool, it generates a number of theoretically justified verification conditions that are sufficient to establish that the specification is not contradictory, or, in other words, consistent.

Next, the developer defines a **REFINEMENT** machine which, in general, is similar to the abstract specification machine, but is usually more deterministic. The refinement machine must include an invariant which usually consists of two parts, the part restricting the variables introduced in a refinement step, and the so-called "gluing invariant" relating these variables and their counterparts in the corresponding abstract machine. When the refinement is submitted to the tool, the latter generates a number of proof obligations sufficient to establish that the refinement is consistent and that it correctly implements the corresponding specification. The development process can include a chain of **REFINEMENT**s transitively refining the abstract machine with the last refinement in the chain - an **IMPLEMENTATION** machine. The **IMPLEMENTATION** machine maps directly to a programming language such as C or Ada. In this paper, we only focus on the features of abstract machines and refinements which are relevant to our discussion.

As soon as some proof obligations are generated, the developer can try to discharge them using an automated theorem prover incorporated in the tool, which attempts to discharge the generated proof obligations. Typically, there is a number of proof obligations that the automated prover cannot discharge, so the developer can switch the prover to an interactive mode and attempt to prove the remaining proof obligations manually.

The B method is supported by two commercially available tools, B-Toolkit developed and distributed by B-Core company, UK [2], and AtelierB developed and distributed by Steria, France [15]. In general, the tools are quite similar and each of them excels in slightly different aspects of the method. Accordingly, in the following discussion we refer to both of them as "the tool".

2.2   The Alloy Specification and Verification Method

The Alloy specification notation and the Alloy Constraint Analyser are the research products of Daniel Jackson and his colleagues at MIT [8,9]. The Alloy specification language (to which we further refer as Alloy) is also state-based like B. An Alloy specification usually contains several sections. One of the obligatory sections is for variable declaration, where variables can be declared as either atoms, subsets of declared domains, or relations of various kinds connecting these sets and/or domains. Declaration of the variables can be arranged so that the specification would have an implicit invariant restricting the set of possible states in which these variables can be present. In addition, in another section of the specification, the developer can write down an arbitrary number of named explicit invariants that further restrict the state. The developer can also write down a named assertion containing an arbitrary logical formula expressed on the variables of this specification. In yet another section of the specification, the developer can write down named operations modifying variables declared in the specification. Operation specifications describe a relation between pre- and post-states of the variables, similar to operation schemas of Z.

Verification with Alloy typically proceeds in the following manner. After the developer has recorded the variables and all implicit and explicit invariants restricting the set of states the variables can be in, he or she can write down some conjectures about the relation between the declared variables in the form of named assertions. It is then possible to submit such an assertion to the Alloy constraint analyser which tries to find a counter example invalidating the assertion. The Alloy constraint analyser does this by converting the assertion, all related variable declarations, and appropriate invariants to a boolean formula, negating it and submitting it to one of several available general purpose boolean solvers. The chosen solver, in turn, tries to find an instantiation of the variables in the submitted formula making it true. Naturally, to make this process finite, the user of the Alloy constraint analyser is asked to indicate the dimensions of the participating domains.

The developer can also verify the operations defined in the specification against any or all of the invariants. For this, the developer has to mark an operation he or she wants to verify against a particular invariant, and the analyser then tries to find an example instantiation of the variables which satisfies the invariant before an execution of the operation but does not satisfy it after. Internally, the analyser achieves this in a manner similar to verifying assertions.

We briefly introduce the subset of the Alloy specification language necessary for our purposes as we present the example.

It is important to mention that at the moment Alloy does not provide any support for verifying implementations or refined specifications on compliance with the original specification. In this paper we discuss how such features can be introduced to Alloy.

```
MACHINE      DbAbstr
SETS
    STUDENTS ; GRADES
VARIABLES
    abstDb
INVARIANT
    abstDb ∈ STUDENTS ⇸ GRADES
INITIALISATION
    abstDb := ∅
OPERATIONS
    append( st , gr )    ≙
        PRE
            st ∈ STUDENTS ∧ gr ∈ GRADES ∧ st ∉ dom ( abstDb )
        THEN
            abstDb := abstDb ∪ { st ↦ gr }
        END
END
```

Fig. 1. The abstract machine *DbAbstr*

3   Example of Specifications in B and Alloy

In this section we follow the outline of our verification method briefly described in the introduction. Rather than discussing the method on an abstract level, we chose to demonstrate it with an example. Due to numerous restrictions and shortcomings of the Alloy specification notation, we chose a rather simple example of specifying a database of student grades. Yet, verifying this specification arises a multitude of interesting issues that we discuss below.

3.1   Specifying a Student Grades Database in B

Suppose that we would like to create a simple database containing information about students and their grades. On an abstract level, such a database can be modelled as a partial function. The B specification of such a model can be represented as an abstract machine *DbAbstr*, as shown in Fig.1.

This machine introduces two new domains, which are declared in the section **SETS**: *STUDENTS* and *GRADES*. These domains are the fixed sets sometimes referred to as *deferred sets*, as the developer only needs to give them a concrete representation in the implementation.

The next section of the B specification contains declarations of the variables, which hold the state of the machine. In our case, this is the variable *abstDb*.

The **INVARIANT** section holds the invariant of the machine. In general, an invariant is a predicate which is established by the initialisation of state variables and holds before and after execution of all operations declared in the machine. In B, an invariant usually includes predicates that give a type to the state variables declared in the **VARIABLES** section. In our machine, *abstDb* is constrained to be a partial function from the deferred set *STUDENTS* to the deferred set *GRADES*.

In the next section **INITIALISATION**, all variables of the machine must be initialized. Thus, *abstDb* is assigned an empty set.

As follows from the name of the next section, it contains the definitions of all operations defined for this machine. To illustrate our idea, it is sufficient to provide only one operation. Therefore, the machine *DbAbstr* only has an operation **append**, for adding records about students' grades into the database. This operation has a precondition verifying the types of the corresponding parameters and also checking that the submitted student is not already in the database, i.e. in the domain of the partial function *abstDb*. In B, the outcome of an operation is only defined in those states where its precondition evaluates to true.

As soon as the definition of the *DbAbstr* machine is complete, we can run the type checker, the proof obligation generator, and the automated theorem prover on it. Because of the simplicity of *DbAbstr*, the automated theorem prover of the tool can resolve one hundred percent of the generated proof obligations.

Now let us consider a refinement of our student database. In this refinement, shown in Fig.2-3 we implement the student database as a connected list of nodes. The clause **REFINEMENT** declares that the machine is intended to be a refinement of another machine. In the next section of the refinement machine, the developer has to indicate which exactly machine it refines, in our case it is *DbAbstr*. Similarly to abstract machines, refinements can also declare deferred sets. In our case, we declare a new set *LINKS* that will serve as a domain of all links available for building a linked list. Next, the developers can declare some constants original to the refined specification, so we declare a constant *nil* that is used for marking the end of the list. The clause **PROPERTIES** is used for constraining the declared constants, in particular, the developers must indicate the type of the constants: *nil* is an element of the domain *LINKS*.

Next, we declare variables *stDb* ,*grDb* , *next* , and *head* that are used for implementing a linked list. As can be seen from the upper part of the invariant, *stDb* is declared as a partial injective function associating *LINKS* with *STUDENTS*. Note that, as the function is injective, there can be no two different links referring to the same student. On the other hand, *grDb* is declared not as injective function, but simply as a partial function from *LINKS* to *GRADES* – clearly, several students could have received the same grade on an exam. The function *next* represents the linked list itself, and is injective, which helps us later to state that the list is really linked, i.e. all of its nodes can be reached from its *head*.

The invariant in a refinement can, in general, be divided into three parts. The first one describes the types of the variables declared in the refinement. The second one describes the relations between the variables declared in the refinement that are true after the initialisation of these variables and remain true before and after execution of all operations of this machine. In our case, this

REFINEMENT   *DbConcr*
REFINES   *DbAbstr*
SETS
   LINKS
CONSTANTS
   nil
PROPERTIES
   $nil \in LINKS$
VARIABLES
   stDb , *grDb* , *next* , *head*
INVARIANT
   $stDb \in LINKS \rightarrowtail STUDENTS \wedge$
   $grDb \in LINKS \nrightarrow GRADES \wedge$
   $next \in LINKS \rightarrowtail LINKS \wedge$
   $head \in LINKS \wedge$
   $\mathsf{dom} (stDb) = \mathsf{dom} (grDb) \wedge$
   $\mathsf{dom} (grDb) = \mathsf{dom} (next) \wedge$
   $(next = \varnothing \wedge head = nil \vee$
   $(nil \in \mathsf{ran} (next) \wedge nil \notin \mathsf{dom} (next) \wedge head \in \mathsf{dom} (next))) \wedge$
   $(next \neq \varnothing \Rightarrow$
   $\forall\, zz \, . \, (\, zz \in LINKS \wedge zz \in \mathsf{ran} (next) \Rightarrow head \mapsto zz \in next^{*} \,) \,) \wedge$
   $\forall\, link1 \, . \, (\, link1 \in \mathsf{dom} (stDb) \Rightarrow abstDb (stDb (link1)) = grDb (link1) \,) \wedge$
   $\mathsf{dom} (abstDb) = \mathsf{ran} (stDb)$
INITIALISATION
   $stDb , grDb , next , head := \varnothing , \varnothing , \varnothing , nil$

Fig. 2. The refinement machine *DbConcr*

part of the invariant can be subdivided into three conjuncts. The first one states
that the domains of *stDb*, *grDb* , and *next* are equal. This condition guarantees
that students and their grades will be attached to the links connected in the
list. The second one states that either the list is empty and *head* is equal to *nil*
or *head* is in the domain of *next* and *head* is not equal to *nil* and *nil* is not in
the domain but is in the range of *next*. This conjunct describes the structure of
the list, i.e. the list is either empty and the head is pointing to *nil*, or the list
starts from *head* and is terminated by *nil*. The third conjunct states that the list
must always be properly connected, i.e. starting from the head, it should always
be possible to reach the terminating nil. This is expressed by stipulating that
any tuple such that its first element is *head* and its second element is any one
belonging to the range of *next* must belong to the reflexive transitive closure of
the function *next*.

    Finally, the third part of the invariant represents a so-called "gluing invari-
ant" which explains how the state of the abstract machine is represented in terms

OPERATIONS

$append(\, st\, ,\, gr\,)\quad \widehat{=}$

 PRE

  $st \in STUDENTS \wedge gr \in GRADES \wedge st \notin \mathrm{ran}\,(\, stDb\,)$

 THEN

  ANY ll **WHERE** $ll \in LINKS - \mathrm{dom}\,(\, next\,) - \{\, nil\, \}$

  THEN

   IF $next = \varnothing$ **THEN**

    $head := ll$ $\|$

    $next := \{\, ll \mapsto nil\, \}$ $\|$

    $stDb := \{\, ll \mapsto st\, \}$ $\|$

    $grDb := \{\, ll \mapsto gr\, \}$

   ELSE

    $stDb(ll) := st$ $\|$

    $grDb(ll) := gr$ $\|$

    ANY $xx\, ,\, next1$ **WHERE**

     $xx \in \mathrm{dom}\,(\, next\,) \wedge xx \mapsto nil \in next \wedge$

     $next1 \in LINKS \rightarrowtail LINKS \wedge$

     $\forall\, yy\,.\,(\, yy \in LINKS \wedge yy \in \mathrm{dom}\,(\, next\,) - \{\, xx\, \} \Rightarrow$

        $next1\,(\, yy\,) = next\,(\, yy\,)\,) \wedge$

     $next1\,(\, xx\,) = ll \wedge$

     $next1\,(\, ll\,) = nil$

    THEN

     $next := next1$

    END

   END

  END

 END

END

Fig. 3. The refinement machine *DbConcr* (continued)

of the variables of its refinement. In our case it suffices to state that for all links in the domain of *stDb*, the grade recorded in *abstDb* (in the machine *DbAbstr*) for the student associated with a link in *stDb* (in the machine *DbConcr*) is equal to the grade associated with this link in *grDb* (in the machine *DbConcr*). It is also necessary to add that for all records in the abstract database there is a link in the concrete one. We achieve this by stating that the domain of *abstDb* is equal to the range of *stDb*.

As follows from the name of the following section, the variables are initialized in it. All functions are assigned empty sets and the head is assigned *nil* .

On the concrete level, definitions of operations become more elaborate. Preconditions of the operations can only be logically weakened, and they can be expressed on the variables of this refinement machine. Consider the refined **append** operation. First, we create a temporary variable *ll* which represents a new link to be inserted into the list *next*. This variable is assigned a value that is arbitrarily chosen from *LINKS*, is not equal to *nil*, and is a fresh value, i.e. it is not in the domain of *next*.

When appending a new student/grade record to the linked list, there can be two distinct cases, when initially the list is empty and when it is not. In the first case, we assign to *next* a tuple *ll* $\mapsto$ *nil*, thus making *next* represent a list with one element *ll*, terminated by *nil*. We also make *head* to point to *ll* and associate a supplied student and grade with the link *ll*. If the linked list is not empty, we associate the supplied student and grade with the new link *ll*. After this, we create two temporary variables *xx* and *next1*, where *xx* is assigned to refer to the last element in the list before *nil* and *next1* is a copy of *next* in all the links except for the one *xx* is pointed at. In *next1*, *xx* is pointing not to *nil*, but to the new link *ll*, which, in turn, points to *nil*. In fact, *next1* describes a new state of the function *next*. Thus the definition of the operation *append* concludes with the assignment of this new value *next1* to *next*.

For a reader well familiar with the style of B specifications, the specification presented above may appear to be somewhat convoluted, as it is quite easy to significantly shorten the definition of the refined **append** . The style of the specification presented above is motivated by the restrictions of the Alloy specification notation. We discuss these restrictions in the concluding section, as well as the modifications that it would be necessary to make to Alloy in order to permit for more natural specifications in B.

The refinement machine *DbConcr* presented in Fig.2 and Fig.3 appears to be correct, i.e. the definition of the operation *append* is consistent with respect to the invariant of the refinement, and also *append* appears to be a proper refinement of its counterpart in *DbAbcst*. But is it really correct? To be able to verify this conjecture in Alloy, we first need to consider how we can formalise the machine *DbConcr* in Alloy.

3.2   Translating the Student Grades Database to Alloy

Consider the Alloy specification presented in Fig.4 and Fig.5. In the section **domain**, we declare three domain sets with familiar names: STUDENTS, GRADES, and LINKS. The keyword **fixed** is used to indicate that the marked set is unchangeable, remaining invariable before and after all operations. The next section contains the declaration of state variables. Unlike in AMN, the Alloy variable declaration not only lists the variables, but also describes their type, and partially introduces an invariant. For instance, stDb is declared as a partial injective function from LINKS to STUDENTS. The arrow -> is used for constructing general relation types.

```
model DbConcr {
     domain { fixed STUDENTS,  fixed GRADES, fixed LINKS}
     state {
           stDb : LINKS? -> STUDENTS?
           domStDb : LINKS
           ranStDb : STUDENTS
           grDb : LINKS -> GRADES?
           domGrDb : LINKS
           next : LINKS? -> LINKS?
           head : LINKS!
           domNext : LINKS
           ranNext : LINKS
           nil : fixed LINKS!
           next1 : LINKS? -> LINKS?
           domNext1 : LINKS
           ranNext1 : LINKS
     }
     def domStDb { domStDb = {l : LINKS | some l.stDb}}
     def ranStDb {ranStDb = {st : STUDENTS | some st.~stDb}}
     def domGrDb {domGrDb = {l : LINKS | some l.grDb}}
     def domNext {domNext = {l : LINKS | some l.next}}
     def ranNext {ranNext = {l : LINKS | some l.~next}}
     def domNext1 {domNext1 = {l : LINKS | some l.next1}
     def ranNext1 {ranNext1 = {l : LINKS | some l.~next1}}
     cond emptyList {all l : LINKS | no l.next}

     inv StateInv {
           domStDb = domGrDb && domGrDb = domNext
           ( emptyList  && head = nil ||
                ((nil in ranNext) && !(nil in domNext) &&
                                     (head in domNext)) )
           ( !emptyList ->
                (all zz : LINKS | zz in ranNext -> zz in head.*next) )
     }

     op init{
           all l : LINKS | no l.stDb' && no l.grDb' && no l.next'
           head' = nil }
```

Fig. 4. The Alloy representation of *DbConcr*

To constrain a variable to be a relation of a particular kind, such as an injective function, the domain and the range of the relation can be restricted using the so-called multiplicity characters. In the case of stDb, the multiplicity character used is ? which, when attached to the name of the set in the variable declaration, makes it to have zero or one element. As ? is attached to both the

```
   op append(st : STUDENTS!, gr : GRADES!) {
   !(st in ranStDb)
some ll : LINKS - domNext - nil |
       (emptyList  -> head' = ll && ll.next' = nil &&
                         ll.stDb' = st && ll.grDb' = gr &&
                         (all l : LINKS - ll | no l.next' &&
                                     no l.stDb' && no l.grDb' )) &&
       ( !emptyList ->
             ll.stDb' = st &&
             ll.grDb' = gr &&
             some xx : domNext | xx.next = nil &&
                   (all yy : LINKS | yy : (domNext - xx)
                                  -> yy.next1 = yy.next) &&
                   xx.next1 = ll && ll.next1 = nil &&
             (all l : LINKS | l.next' = l.next1) &&
       (all l : LINKS - ll | l.stDb' = l.stDb &&
                             l.grDb' =l.grDb) &&
       head' = head )
   }
}
```

Fig. 5. The Alloy representation of *DbConcr* (Continued)

domain and the range of stDb signifying that for each element in the domain of
stDb there is at most one element in its range and the other way around, i.e.
stDb is injective.

In this specification, we also use the multiplicity character !, which makes
a set to have exactly one element. More information on multiplicity characters
and the Alloy specification notation in general can be found in [8].

In Alloy, domain-valued variables are modelled as subsets of domains rather
than elements of domains, and relational image rather than function application
is used to apply relations to values. Unique values are represented by singleton
sets.

A declaration of the kind domStDb : LINKS declares domStDb to be a subset
of the domain LINKS. The operator : is used in Alloy to indicate a subset relation
while declaring a variable, and the operator in is used for this purpose in other
parts of the specification. The variable domStDb : LINKS serves an auxiliary
purpose only, as the machine *DbConcr* does not have a counterpart for it. This
variable is necessary because Alloy does not have a function *dom* which would
return a domain of a given relation. To circumvent this problem of Alloy, we
have to declare the variable domStDb and constrain it using the definition

```
   def domStDb { domStDb = {l : LINKS | some l.stDb}}
```

which makes domStDb to be equal to the set of such links whose image of stDb
is non-empty. Note the usage of the operator dot (.), which is used for taking
an image of a set through a relation.

An Alloy term `l.stDb` is equivalent to a B term $stDb(l)$.[1] The auxiliary variable `ranStDb` represents the range of the function `stDb` and is defined similarly to `domStDb`. In the definition of `ranStDb` note the usage of the ~ operator, which takes the inverse of the function. As `stDb` is defined as an injective function, its reverse is a function as well. The other variables whose name starts with `dom` or `ran` represent, respectively, domains or ranges of the corresponding functions and are all defined in a similar manner.

The variable `grDb` is represented as a partial function, while `next` is a partial injective function. There is also a declaration of the variable `head`, which is a one element set, and a variable `nil` which is marked with the keyword `fixed` turning it into a constant.

The state of the variables can be further constrained using any number of named invariants. In our case, we have only one invariant `StateInv`, which is, in fact, a translation of the invariant of the machine *DbConcr*, apart from the typing conjuncts. As Alloy prohibits comparisons of structured sets and has no predefined constant for an empty set, we had to introduce a condition `emptyList`, which in B terms is $next = \varnothing$. In Alloy, `*next` represents the reflexive transitive closure of the function `next`. At this point a careful reader could have noticed that the "gluing" part of the *DbConcr* invariant does not have a counterpart in `StateInv`. As Alloy does not support the notion of refinement directly, the invariant of an Alloy model can only refer to the variables defined in this model, while the gluing invariant refers to the variables of *DbAbstr* as well. The gluing invariant is of no significance for verifying consistency of the concrete *append* which is the topic of the next section. However, it is crucial for verifying the correctness of a refinement step. We discuss how to specify a gluing invariant in Alloy in Section 4.2.

The Alloy specification notation does not include a reserved construct for initialisation of variables. However, we can formalise the initialisation section of the *DbConcr* machine as an operation `init` of Alloy. It states that for all elements in the set `LINKS` the image of these elements through `stDb`, `grDb` and `next` is equal to the empty set. In addition, the new value of `head` is equal to the predifined constant `nil`. Clearly this operation establishes `StateInv`.

The definition of the operation `append` in Alloy is, practically, a straightforward translation of its B counterpart. Alloy does not have programming language statements like "**IF   THEN   ELSE** ", neither does it have an assignment statement. Instead, an operation in Alloy must be described as a relation between initial (unprimed) and resulting (primed) states of the variables. A B specification is built on an assumption that only the variables explicitly modified in the specification change, and all the other variables remain unchanged. In Alloy, however, it is necessary to explicitly mention that all the variables that were not modified in the definition of an operation remain in the initial state.

As was already mentioned, it is impossible (at the moment) to compare structured sets in Alloy. Thus, we cannot say $next := \{ll \mapsto nil\}$, but we should

---

[1] Should `stDb` be a general relation, the Alloy term `l.stDb` would translate into $stDb[\{l\}]$ in B.

say that the image of ll through `next` is equal to `nil`, or `ll.next'` = `nil`. The definition of the operation *append* in the refinement machine *DbConcr* is formulated using a temporary variable `next1`. As in Alloy it is impossible to quantify over relations, we had to introduce this temporary variable in the state declaration. As the only invariant binding *next1* is the one making it an injective partial function from LINKS to LINKS, this is the same as stating that there exists some `next1` in the definition of the operation.

It should be noted that not only the style of the B specification presented in this paper was motivated by the need to translate it directly to Alloy, but the style of Alloy specifications was adjusted for this purpose as well. In particular, a specification written in a pure Alloy style instead of using partial functions would rather use total functions. Therefore, domain expressions would not usually be used.

At the moment the translation from B to Alloy is done by hand. However, undoubtedly, the translation between AMN and the Alloy specification notation could be made automatic if Alloy were extended with several features. We will discuss these features in the concluding section of the paper.

4   Verifying Properties in Alloy

Let us now return to the question of whether the specification of the method *append* is correct. First, we take a look at operation consistency, and then consider the correctness of a refinement step.

4.1   Verifying Operation Local Consistency

If we submit `append` along with `StateInv` to the Alloy constraint analyser and indicate that the domains should be instantiated with only three elements, the analyser generates the counter example presented in Fig.6[2]. The counter example clearly violates the invariant, since after execution of the operation, `domNext'` contains `nil`, which contradicts one of the conjuncts in the invariant. Returning to the specification of `append`, it is fairly easy to spot the error. The part of the specification which deals with the case when the list is not empty describes what should be the value of the list `next1` at all the links in the domain of `next` and also at the new link ll we have added. This condition does not exclude, however, that `next1` can have other links. Thus, the Alloy constraint analyser is free to introduce `nil` into the domain of `next1`, which violates `StateInv`.

This problem can be traced back to our B specification. To establish consistency of an operation with respect to the machine invariant the B tool generates a proof obligation stating that the invariant conjoined with the operation's precondition implies the invariant expressed on the new values of the variables. In our case, one of the generated proof obligations would state that the invariant and the precondition of **append** imply that *nil* is not in the domain of

---

[2] We have only left the values of the relevant variables for clarity

```
Analyzing append vs. StateInv ...
Scopes: GRADES(3), LINKS(3), STUDENTS(3)
Conversion time: 10 seconds
Solver time: 13 seconds
Counterexample found:
Domains:
  LINKS = {nil,L0,L1}
Sets:
  domNext = {L0}
  domNext1 = {nil,L0,L1}
  domNext' = {nil,L0,L1}
Relations:
  next = {L0 -> nil}
  next1 = {nil -> L0, L0 -> L1, L1 -> nil}
  next' = {nil -> L0, L0 -> L1, L1 -> nil}
Skolem constants:
  l1 = L1
```

Fig. 6. The counter example for the operation *append*

next1. However, the counter example found by the Alloy constraint analyser demonstrates that these assumptions are not strong enough for resolving this obligation. All attempts to prove such an obligation would be futile.

To fix the problem, we additionally need to state that the list `next1` should only be larger than `next` by one element `l1`:

$$\texttt{domNext1 = domNext + l1}$$

Indeed, this amendment is sufficient to resolve the problem. Now the developer, equipped with the confidence reinforced by the fact that the Alloy constraint analyser cannot find any counter examples, can return to proving the subgoals dealing with the consistency of the operation.

It is also possible to check the consistency of an operation in a different manner. Instead of translating the definition of a B operation into Alloy, it is sufficient to translate the proof obligations generated by the B tool as Alloy assertions and run the Alloy constraint analyser on them similarly to verifying operation refinement as described in the next section.

4.2   Verifying Operation Refinement

The definition of an operation in a refinement machine can be consistent with respect to the local invariant, i.e. the part of the invariant referring only to the variables of the refined machine. However, at the same time the relation between it and its abstract counterpart can be other than refinement. Some of the proof obligations generated by the tool during verification are directed at establishing that abstract and concrete definitions of operations are, in fact, in the refinement relation. We propose to translate such proof obligations into Alloy named assertions in order to check that these proof obligations are indeed provable. Alloy assertions are the logical predicates expressed using the variables of an Alloy specification that are supposed to evaluate to true in any state the variables can be in. Accordingly, the tool attempts to find a state invalidating the predicate in the assertion.

The debugging process that we propose is then as follows. The counter example generated by the analyser can hint at modifications that must be made either to the invariant of the refinement or to the definition of an operation in B. The developer then should make these modifications to the B specification, regenerate the proof obligations, run an automated theorem prover on them, and in case any are left, translate the remaining to Alloy as assertions and repeat the debugging cycle again until the Alloy constraint analyser is unable to generate a counter example in a reasonably large scope. To become one hundred per cent certain that the refinement machine is, in fact, in the refinement relation with its abstract counterpart, the developer can then go on and prove the remaining proof obligations using an interactive theorem prover.

There is, however, a complication. As we have already mentioned, the Alloy specification notation does not provide any support for defining abstract specifications and their refinements separately. In order to express the "gluing" part of the *DbConcr*'s invariant, we have to combine all the definitions of abstract state and the definitions of its concrete implementation in the same model. Therefore, we should extend our model with the definitions for the partial function `abstDb` and its domain `domAbstDb`. The last one is defined similarly to all the other definitions of domains of functions.

```
abstDb : STUDENTS -> GRADES?
domAbstDb : STUDENTS
```

We should also extend the invariant `StateInv` to include the "gluing" conjuncts:

```
all link1 : domStDb | link1.stDb.abstDb = link1.grDb
all st : STUDENTS | some st.~stDb <-> some st.abstDb
```

To demonstrate our approach to verifying refinement, let us now return to our example. To demonstrate our approach to verification, we first need to introduce an error in the definition of *DbConcr*'s *append* that would not invalidate the consistency of the operation with respect to the invariant of the refinement machine, yet would break the refinement relation.

In the B method, the refinement machine can only be proved to be in a refinement relation with its abstract counterpart if all operations of the refinement machine preserve the gluing invariant. In our example, it states that for all links in the domain of *stDb*, the grade recorded in *abstDb* (in the machine *DbAbstr*) for the student associated with a link in *stDb* (in the machine *DbConcr*) is equal to the grade associated with this link in *grDb* (in the machine *DbConcr*). It is also states that the domain of *abstDb* is equal to the range of *stDb*. Obviously,

---

$append(\ st\ ,\ gr\ )\quad\widehat{=}$

    PRE

        $st \in STUDENTS\ \wedge\ gr \in GRADES\ \wedge\ st \notin \mathrm{ran}\ (\ stDb\ )$

    THEN

        ANY ll **WHERE** $ll \in LINKS - \mathrm{dom}\ (\ next\ ) - \{\ nil\ \}$

        THEN

            IF $next = \varnothing$ **THEN**

                $head := ll\ \parallel$

                $next := \{\ ll \mapsto nil\ \}\ \parallel$

                $stDb := \{\ ll \mapsto st\ \}\ \parallel$

                $grDb := \{\ ll \mapsto wrong\ \}$

        ELSE

            ... continuation as in Fig.2

Fig. 7. A fragment of the erroneous definition of the operation *append* invalidating the refinement relation

this invariant would be violated, should we erroneously associate the submitted student not with the submitted grade but with some other *wrong* grade in *append* of *DbConcr* (see Fig.7).

Naturally, we would need to introduce the constant *wrong* in the clause **CONSTANTS** of the machine and give its type in the clause **PROPERTIES**. If we now subject the refinement machine to the standard steps of type checking, proof obligation generation, and automated theorem proving, we will be left with several proof obligations, of which "append.22" is of particular interest (see Fig.8).

The proof obligation "append.22" effectively states that the gluing invariant must hold after the execution of *append*. It must hold under the assumptions that are extracted from the **PROPERTIES** and **INVARIANT** clauses of the *DbAbtr* and *DbConcr* machines and also from the precondition of the *append* operation of this machines and the local information available from the definition of *append* in *DbConcr*.

To verify such a proof obligation in Alloy, we can represent it as a named assertion. When submitted to the constraint analyser, the latter tries to verify whether the predicate in the assertion is true in all states restricted by all invariants of the model. Therefore, while translating a B proof obligation to Alloy, we can omit all those conjuncts on the left hand side of the implication that are repeating the **INVARIANT**s and **PROPERTIES** of the abstract and concrete machines already represented in the state declaration and the invariants of the Alloy model. The obligation "append.22" can be translated as an Alloy assertion, as presented in Fig.9.

go(append.22)

   ”*"Component properties""* ∧

   ...

   ”*'Previous components properties'"* ∧

   ...

   ”*'Previous components invariants'"* ∧

   ...

   ”*'Component Invariant'"* ∧

   ...

   ”*'append preconditions in previous components'"* ∧

   ...

   ”*'append preconditions in this component'"* ∧

   $st \notin$ ran ($stDb$) ∧

   ”*'Local hypotheses'"* ∧

   $ll \in LINKS$ ∧ $ll \notin$ dom ($next$) ∧ $ll \neq nil$ ∧

   $next \neq \varnothing$ ∧ $xx \in$ dom ($next$) ∧ $xx \mapsto nil \in next$ ∧

   $next1 \in LINKS \rightarrow\!\!\!\!\!\rightarrow LINKS$ ∧ $next1^{-1} \in LINKS \rightarrow\!\!\!\!\!\rightarrow LINKS$ ∧

   dom ($next1$) = dom ($next$) ∪ { ll } ∧

   ∀ yy . ($yy \in LINKS$ ∧ $yy \in$ dom ($next$) − { xx } ⇒

             $next1$ (yy) = $next$ (yy)) ∧

   $next1$ (xx) = ll ∧ $next1$ (ll) = nil ∧

   $link1 \in$ dom ($stDb \lhd\!\!\!- \{ ll \mapsto st \}$) ∧

   ”*'Check that the invariant*

   (!link1.(link1: dom(stDb) ⇒ abstDb(stDb(link1)) = grDb(link1)))

   is preserved by the operation - ref 4.4, 5.5'"

   ⇒

   $(abstDb \cup \{st \mapsto gr\})$ (($stDb \lhd\!\!\!- \{ ll \mapsto st \}$) ($link1$))

      = ($grDb \lhd\!\!\!- \{ ll \mapsto wrong \}$) ($link1$)

Fig. 8. The proof obligation "append.22"

Unfortunately, at the moment the Alloy specification notation is not sufficiently rich to always permit a one-to-one translation of B. Alloy does not permit to use set operations such as intersection, union, etc. on structured sets (i.e. relations). Neither it is possible to compare structured sets. In a way, in Alloy it is impossible to state that "a certain relation is such and such", it is only possible to state "a certain relation satisfies these properties", and these "properties" should always be expressed elementwise. Therefore, to express our proof obligation in Alloy, we have to perform a case analysis on the domains of the functions participating in the right hand side of the goal.

```
assert PO22 {
all st : STUDENTS, gr : GRADES, ll : LINKS, xx : LINKS, link1 : LINKS |
  !(st in ranStDb) &&
  !(st in domAbstDb) &&
  !(ll in domNext) &&
  ll != nil &&
  ! emptyList &&
  xx in domNext &&
  xx.next = nil &&
  domNext1 = domNext + ll &&
  (all yy : LINKS | yy : domNext && yy !=xx  -> yy.next1 = yy.next) &&
  xx.next1 = ll &&
  ll.next1 = nil &&
  link1 in domStDb + ll ->
    (link1 in (domStDb - ll) -> (link1.stDb in domAbstDb ->
        (link1 in (domGrDb - ll) -> link1.stDb.abstDb = link1.grDb))) &&
    (link1 in (domStDb - ll) -> (link1.stDb in domAbstDb ->
        (link1 in ll -> link1.stDb.abstDb = wrong))) &&
    (link1 in (domStDb - ll) -> (link1.stDb in st ->
        (link1 in (domGrDb - ll) -> gr = link1.grDb))) &&
    (link1 in (domStDb - ll) -> (link1.stDb in st ->
        (link1 in ll -> gr = wrong))) &&
    (link1 in ll -> (st in domAbstDb ->
        (link1 in (domGrDb - ll) -> st.abstDb = link1.grDb))) &&
    (link1 in ll -> (st in domAbstDb ->
        (link1 in ll -> st.abstDb = wrong))) &&
    (link1 in ll -> (st in st ->
        (link1 in (domGrDb - ll) -> gr = link1.grDb))) &&
    (link1 in ll -> (st in st -> (link1 in ll -> gr = wrong)))
}
```

Fig. 9. The proof obligation "append.22" translated to Alloy

The constraint analyser easily finds a counter example demonstrating that
the assertion PO22 is not always true, i.e. that the submitted grade gr is not
always equal to the constant wrong. If the developer now reverses the definition
of *append* operation to its state before we introduced the "*wrong*" error and goes
through the entire proposed debugging cycle, then the Alloy constraint analyser
will be unable to find a counter example for the corresponding assertion in a
sizable scope.

5   Conclusions and Related Work

As was mentioned above, both B and Alloy are state-based formalisms. Alloy,
rather like Z, describes state changes in terms of pre and post states and is also

formalised in the first order logic. While the B notation was designed to resemble an imperative programming language syntax, the subset of the language that can be used in abstract **MACHINE**s and **REFINEMENT**s is essentially just syntactic sugar for first order logic expressions on the pre and post states of variables. Accordingly, a translation from B to Alloy would be rather straightforward if Alloy did not have certain features.

The ability to work with relations as with sets of tuples appears to be the most important of these features. Alloy should permit all standard operations for manipulating ordinary sets, such as set comparison, set union, set difference, etc. In the absence of this feature, not only specifications are much longer, but also it is impossible to directly express properties of updated relations. This shortcoming of Alloy is apparent in our translation of the proof obligation *append.22*. Also, Alloy should permit quantifying over relations, as under certain conditions the B tool can generate proof obligations having quantification over relational variables. An introduction of the usual functions *dom* and *ran* for taking domain and range of a relation, as well as a constant $\varnothing$ would significantly simplify the resulting Alloy specifications, as it would be possible then, for instance, to describe the domain of a constructed function. Finally, the absence of integers (or even of any finite subset of natural numbers) and arithmetic is a very severe restriction of the current Alloy implementation, making it inapplicable to the majority of practical cases. From our communication with Alloy developers it appears that the next version of Alloy will address most of these issues.

The idea to combine theorem proving with model checking is not new as such. For instance, an objective of Symbolic Analysis Laboratory (SAL) project of SRI International's Computer Science Laboratory [3] is to provide integrated combination of static analysis, model checking and theorem proving techniques for verification of concurrent systems. SAL framework features an intermediate language which serves as a medium for representing the state transition semantics of systems described in Java or Esterel. It also serves as a common representation for driving back-end tools such as PVS theorem prover [13] and SMV model checker [11]. The SAL framework, however, is geared towards verifications of concurrent systems formalisable as transition systems, while in our approach we apply the Alloy constraint analyser for verification of state-based B specifications.

The fact that a significant development effort can be wasted trying to prove false conjectures during formal software development was observed by Juan Bicarregui and Brian Matthews in [4]. They suggest using automatic theorem proving technology in refutation of proof obligations in order to find faulty proof obligations. Although refuting the proof obligation indicates a fault in design, it does not, in itself, helps to identify the source of the problem. The authors propose to use a model generator on the negation of a faulty proof obligation to find a counter example. This, in a way, is similar to the procedure used by Alloy for finding counter examples. Our work on combining B and Alloy, therefore, can be seen as a logical continuation of the direction of research outlined in [4], even though we were not aware of this work at the moment our paper was finished.

Currently, the translation from B to the Alloy specification notation is done by hand. To allow for an automatic translation, the Alloy specification language has to be extended with the features we just mentioned. In principle, we perceive two major ways in which the described approach to verification can be implemented as a tool. The first way is to add Alloy-like features into tools supporting the B method. At the moment, such tools are supplied as integrated sets of utilities for type checking, proof obligation generation, specification animation, and theorem proving. Naturally, a utility permitting for model checking proof obligations would integrate nicely with such tools. In practice, it is often infeasible to adhere to a completely formal development, as theorem proving is a very tedious and lengthy process employing highly qualified personnel. Therefore, the B method is often applied in a so-called "soft" manner, i.e. some of the steps of the method are omitted or validated only informally. For instance, developers might decide to informally review the remaining proof obligations which the automated theorem prover did not manage to resolve. Of course, this approach can compromise the correctness of the resulting system as it is rather easy to overlook an error. In this respect, should a B tool support a model checker similar to Alloy, it would help significantly to avoid errors and, in a way, make such an application of the B method "harder". Obviously, however, verifying proof obligations with a model checker should not discourage the developers from trying to prove the remaining proof obligations interactively. In fact, from the theoretical standpoint, even if a model checker would permit to verify a property on finite subsets of infinite domains, theorem proving must still be used to make certain that the property holds on the entire domain. At the moment we are involved in a project of building a Prolog-based tool for animation and model checking of general B machines in a manner similar to Alloy's.

The second way of implementing the suggested approach to verification as a tool is to add B-like features to the Alloy constraint analyser. In particular, Alloy can be extended to permit for verifying refinement. This would amount to extending the Alloy specification language with special notation for specifying abstract and refined models. The Alloy constraint analyser could be extended with a verification condition generator. Such an extension would open an entirely new scope of potential applications for Alloy.

Acknowledgements. We would like to thank Daniel Jackson for the comments he has provided on an earlier version of this paper. The anonymous referees also helped to improve the paper.

References

1. J.-R. Abrial. *The B-Book: Assigning Programs to Meanings.* Cambridge University Press, 1996.
2. B-Core (UK) Limited, Oxon, UK. *B-Toolkit, On-line manual.*, 1999. Available at http://www.b-core.com/ONLINEDOC/Contents.html.

3. S. Bensalem, C. Muñoz, S. Owre, H. Rueß, J. Rushby, V. Rusu, H. Saïdi, N. Shankar, E. Singerman, and A. Tiwari. An overview of SAL. In C. M. Holloway, editor, *LFM 2000: Fifth NASA Langley Formal Methods Workshop*, pages 187–196, Hampton, VA, June 2000. NASA Langley Research Center.

4. J. C. Bicarregui and B. M. Matthews. Proof and refutation in formal software development. In *3rd Irish Workshop on Formal Software Development*, www.ewic.org.uk, July 1999. British Computer Society, Electornic Workshops in Computing.

5. Formal Systems (Europe) Ltd. *Failures-Divergence Refinement – FDR2 user manual*, October 1997. Available at http://www.formal.demon.co.uk.

6. M. Gordon. Introduction to the HOL system. In M. Archer, J. J. Joyce, K. N. Levitt, and P. J. Windley, editors, *Proceedigns of the International Workshop on the HOL Theorem Proving System and its Applications*, pages 2–3, Los Alamitos, CA, USA, Aug. 1992. IEEE Computer Society Press.

7. G. J. Holzmann. The model checker spin. *IEEE Trans. on Software Engineering*, 23(5):279–295, May 1997.

8. D. Jackson. Alloy: A lightweight object modelling notation. MIT Lab for Computer Science, July 2000.

9. D. Jackson, I. Schechter, and I. Shlyakhter. Alcoa : the alloy constraint analyser. In *Proc. International Conference on Software Engineering*, Limerick, Ireland, June 2000.

10. K. MacMillan. *The SMV Language*. Cadence Berkeley Labs, 1999.

11. K. L. McMillan. *Symbolic Model Checking*. Kluwer Academic Publishers, Norwell Massachusetts, 1993.

12. L. Mikhailov and M. Butler. Combining B and Alloy. In *Proceedings of FMICS'2001*, pages 29–45, Paris, July 2001. INRIA.

13. N. Shankar and J. M. Rushby. *PVS Tutorial*. Computer Science Laboratory, SRI International, Menlo Park, CA, Feb. 1993. Also appears in Tutorial Notes, *Formal Methods Europe '93: Industrial-Strength Formal Methods*, pages 357–406, Odense, Denmark, April 1993.

14. J. M. Spivey. *The Z Notation: A Reference Manual*. Prentice Hall, 1987.

15. Steria, Aix-en-Provence, France. *Atelier B, User and Reference Manuals*, 1996. Available at http://www.atelierb.societe.com/index_uk.html.

16. M. Woodman and B. Heal. *Introduction to VDM*. McGraw-Hill, 1993. ISBN 0-07-707434-3.

Software Construction by Stepwise Feature Introduction

Ralph-Johan Back

Åbo Akademi and Turku Centre for Computer Science
Lemminkainenk. 14, 20520 Turku, Finland
backrj@abo.fi

Abstract. We study a method for software construction that is based on introducing new features to an evolving system one by one. A layered software architecture is proposed to support this method. We show how to describe the software with UML diagrams. We provide an exact semantics for the UML diagrams employed in the software architecture, using refinement calculus as the logical framework and show how to reason about software correctness in terms of these UML diagrams.

1 Introduction

We consider here an approach to software construction that is based on incrementally extending the system with new features, one at a time. We refer to this method as *stepwise feature introduction*. Introducing a new feature may destroy some already existing features, so the method must allow us to check that old features are preserved when adding a new feature to the system. Also, because the software is being built bottom-up, there is a danger that we get into a blind alley: some new feature cannot be introduced in a proper way, because of the way earlier features have been built. Therefore, we need to be able to change the structure of (or *refactor*) the software when needed .

We will here consider stepwise feature introduction from three main points of view: what is an appropriate software architecture that supports this approach, what are the correctness concerns that need to be addressed, and what kind of software process supports this approach. We use the refinement calculus [1,17,9] as our logical basis, but we will not to go in more detail into this theory here, for lack of space.

A software system of any reasonable functionality will have a large number of different, interdependet features, and the overall behavior of the software becomes quite complex. We will therefore need good ways of visualizing the structure of software, in order to identify the features in the system, see how they are put together and how they interact with each other. Let us therefore start by describing stepwise feature introduction by means of an analogue. As all analogues, this one is not perfect, but it gives the proper feeling for the overall structure of software.

A *software system* starts out as a village. We think of the village as a collection of houses that are connected by an infrastructure (roads, telephone lines, etc.).

D. Bert et al. (Eds.): ZB 2002, LNCS 2272, pp. 162–183, 2002.
© Springer-Verlag Berlin Heidelberg 2002

The houses correspond to the *software components* in the system. In the village center, we have *service providers*, like shops, schools, municipal authorities and so on, each providing some special *service* that is needed in the village community. We assume that each service provider is located in a house of its own. The service is used by the residential houses in the village, as well as by other service providers in the village. Thus, a software components can be a service provider, a user or both. The village has a development history: it was once founded with just a few houses, and has grown since then with new houses (providing new services or duplicating existing services). Some houses have been torn down, while others have been modified to accommodate changing service needs.

Each house has a name (e.g., the street address of the building). A person living in a house can communicate with persons living in other houses using the communication infrastructure (by telephone, mail, or by visiting them). They get service by asking a service provider to carry out some task, or to provide some information (so communication between software components is by *message passing*). The person is restricted in his communication by his address book (containing mail addresses and/or telephone numbers) and can communicate directly only with the people that are listed there (or with people whose addresses or telephone numbers he gets from other people who already are in his address book).

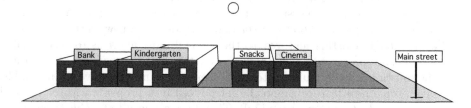

Fig. 1. A village

When the village grows, this simple organization becomes more and more difficult to maintain. The services become more complex, containing more and more specialized services or *service features*, and there are more and more houses being built. The village is turning into a city. The simple shops in the city center are getting bigger, growing into many-store buildings, with one or more departments on each floor, each department providing some specialized service. For instance, what started out as a simple bank with only one counter, for depositing and withdrawing money from accounts, may grow into a big bank building, where the original service is still available on the ground floor, but on the second floor there is a department for trading with stocks, and on the third floor a department for mortgage loans.

The floors of a building correspond to a layering of the software features in a component. At the lower floors are services that have been needed early on, while higher up are specialized services that have been created more recently. We assume that the building technique used for houses in the city is very flexible, so that whenever one needs to extend the service provided by a service provider, one can simply build a new floor on top of the house. It is also easy to tear down a floor, if it is not needed any more. A new service may depend on some older service, provided lower down in the building, but it may also replace some earlier service . We will assume that anybody who contacts a department at some floor will have access to the specialized service provided by that department, but also has access to all the service provided by the departments in the floors below. We assume that a customer that wants a service always contacts the top floor of the building (e.g., takes the elevator to the top floor). This floor provides the most recent service features, as well as all older services.

When a building is extended with a new floor, to provide some additional service, it may require some new service from other service providers in the city. These new service features may either be provided by newly built houses, or by existing houses that are extended with new floors to provide the additional service. Thus, introducing a new service may lead to a small building boom where a number of houses are built or extended at the same time, to provide for related services. The building material used in the houses ages quite rapidly and visibly, so that one can clearly distinguish the houses and extension floors build during one building boom from those built during other booms. In this way, the houses in the city center will get a curious striped appearance, not usually seen in city planning. The related service features in the different buildings show up as a layer in the city scape. (Repairing a house or a floor does not affect the outer appearance of the house, so this pattern is preserved also by renovations).

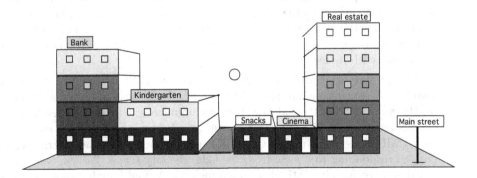

Fig. 2. City scape

The departments are very conservative, in the sense that once they have decided to get service from some other service provider, they will not change the service provider, or request new service from the provider. In particular, this means that a department in a specific layer will not make use of the new service features that are made available by service providers on layers that have been built later. The department can communicate with floors on later layers, because these floors provide all the service features of lower layers, but it does not request any service features that have been introduced after its own layer was built. Of course, a layer higher up in the service provider that the department uses can redefine the contents of an existing service feature, so that the service the department gets is in fact different. The conservatism implies that each layer can function without the layers that have been built later, continuing together with the lower layers to provide the service that was made available when the layer in question was built.

The development of the city is evolutionary, with new layers being built on top of existing ones, without changing the old layers. Sometimes, however, it becomes too difficult to just simply add a new service feature on top of the existing ones. In these cases, one may need to reconstruct part of a building or maybe even part of the city, by changing the layers and the service features they provide so that it becomes easier to add the new service features. This corresponds to *software refactoring* [14]. The city planning authorities insist, however, that the restructuring should be done in such a way that the layered structure of the city is preserved. In particular, each layer should after the reconstruction be able to function without dependence of the layers higher up in the hierarchy, providing all the service features of its own and lower layers.

This picture emphasizes the gradual evolution of a software system towards better and more powerful functionality, punctuated with larger refactorings when the layering scheme runs into trouble. The process is such that the system is always kept functional: at each stage, the system may only have a subset of the desired features implemented, but those that are implemented work ok. Throughout the development, each layer provides a working version of the overall system, with a certain subset of the desired overall service, but which can be tested, experimented with and even used in production. Hence, the system is useful already while it is being built, and the construction and testing of the system can proceed in parallel with its use.

The picture also emphasize a four dimensional view of the city (and of software). Two dimensions are formed by the city buildings, how they are laid out on the map in the city, and how they communicate with each other. It corresponds to a view of the city from above. In software, this view will correspond to a UML class diagrams. A third dimension is added by the layering of the city, corresponding to a view from the side, the city scape. We will capture the layering as regions in UML class diagrams [12]. A final fourth dimension in the city is time, captured by the history of the city, how it has been developed. In software, this dimension is captured by the successive software releases, showing both the slow evolutionary changes and the more drastic software refactorings that have been carried out while constructing the software.

The stepwise feature introduction method has much in common with the original *stepwise refinement* method [13,18]. A large programming problem is broken up into a number of smaller programming problems. Solving these smaller problems one by one will build up the solution to the original problem in an incremental fashion. As in the stepwise refinement method, we hope to make the overall structure of the software system clearer and more easy to understand, make it easier to check its correctness and easier to maintain and adapt the system to changing requirements. Another similarity is that we also emphasize that the software should be checked for correctness while it is being built, rather than in a separate phase after it has been built. A main difference to stepwise refinement is again that we are building the system bottom-up rather than top-down. Also, our emphasis in on object-oriented programming techniques, while stepwise refinement was designed for procedural programs. However, we will show that *refinement calculus*, originally developed as the logical foundation for stepwise refinement, also provides a good foundation for stepwise feature introduction.

In the following sections, we will try to make this picture more precise. We will first describe how to build features into software in layers, and how to argue about the correctness of a system build in this way. We will look at how to describe specifications and implementations in this approach, how to express requirements and how to organize testing of layered software. We will then also describe a diagrammatic approach to reasoning about correctness in layered systems. We will finally give some comments on a software process that supports stepwise feature introduction and also relate some initial experiments from using this software construction method in practice.

2   Feature Extensions, Layers, and Components

A software component provides a *service*. The service consists of a collection of *(service) features*. From a programming language point of view, a feature will be implemented as a class that (usually) extends one or more other classes. This class introduces a new feature while at the same time inheriting all the features introduced in the classes that it extends.

For instance, assume that we have a class that provides a simple text widget in a graphical user interface. This text widget can be placed in a frame in order to be shown on the screen, and it provides service for manipulating the text it contains with the mouse and from the keyboard. The text can be manipulated by inserting new text into it, deleting text, selecting text, and moving the insertion point by clicking the mouse or using the cursor keys. The text widget only works with simple ascii-text.

We can think of a number of features by which we can extend the service provided by this simple text widget, like the following:

Saving. Save the text in the widget to a file, or replace the text in the widget with the contents of a file.

Cut and paste. Cut (or copy) a selected piece of text into the clipboard, and paste the contents of the clipboard into the place where the insertion cursor is.

Styles. Format pieces of text in the widget, e.g. by changing the selected text to boldface, italics, underlined, or colored text.

Extending the simple text widget with a cut-and-paste feature will give us a new text widget that is able to do all the things that the original text widget does, and in addition allows cut, copy and paste in the text. Adding this new feature to the text widget has to be done very carefully: not only do we have to change the original text widget so that cut-copy-and paste is now possible, but at the same time we have to be careful to preserve the old features. Not preserving some old feature, or corrupting it, is considered an error. We refer to this step as a *feature introduction step*.

Consider another extension of the simple text widget, where we add styles to the text. Again, we get a new text widget, with all the features of the original text widget, but which in addition is also capable of showing styled text. Hence, for correctness, we need to show that styles are correctly implemented, and that the old features have been preserved.

Adding cut-and-paste and adding styles are two independent extensions of the simple text widget. We can then construct a new class that inherits from both the cut-and-paste text widget and from the styles text widget, using multiple inheritance. We have to be careful here, however, because inheritance alone is probably not what we want. In this case, the cut-and-paste operations were defined for ascii-text only, so cutting and pasting does not take the styles of the text into account (i.e., only the ascii text is copied, not the styles for the text). What we really want is to copy and paste the styles also. This means that we have a *feature interaction*: the two independent features cannot be combined directly, because they interact in an unwanted way. A *feature combination step* will resolve this unwanted feature interaction. It is a class that extends both the cut-and-paste text class and the style text class, by overriding the behavior of the methods in these classes so that the new features interact in the desired way (here that the styles are also copied and pasted). Sometimes this step can be very simple, at other times feature combination can require a major effort.

The reason for introducing these two features in two separate steps is that we want to separate our concerns. First, we want to work out how to add cut-and-paste to a simple ascii text widget, without the complications of styled text. Then, we want to consider how to add styles to text, without the complications of also having to consider cut-and-paste of the text. Once we have worked out the implementations for these two features, we are ready to tackle the problem of how to combine the two features. This will usually mean that part of the code for each feature has to be modified (by overriding some methods) to take the other feature into considerations. However, this is easier to do once the code for each feature in isolation has been explicitly written out and is open for inspection and analysis. The feature introduction and feature consolidation steps are shown in Figure 3.

Feature combination is thus realized by multiple inheritance, while feature introduction is realized by single inheritance. Feature combination and feature introduction are similar in that in both cases, the features of the superclasses have to be preserved.

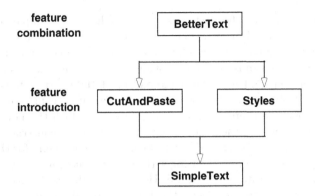

Fig. 3. Feature introduction and combination

An alternative way of building up the better text widget would be to introduce first one feature, and then the other as an extension of the first feature. This means that we need to decide in which order the two features should be introduced. The alternative approaches are illustrated in Figure 4, where extension is indicated by stacking boxes on top of each other. It is not clear which

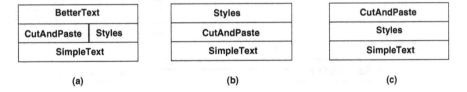

Fig. 4. Alternative extension orders

alternatives is best. Alternative (a) avoids an arbitrary choice between which feature to introduce first, and allows us to consider the two features in separation. On the other hand, it requires a combination step. Alternatives (b) and (c) introduce one feature before the other. This avoids the combination step, so the resulting hierarchy will be simpler, but this can happen at the expense of making the second feature more difficult to introduce.

A component is divided into layers, where one layer can be the combination of a number of independent feature introductions followed by a feature combination step, or just a single feature introduction. A layer will often cut across components, so that the same layer exist on a number of related components. In stepwise feature introduction, the software is constructed layer by layer, with the components in the same layer designed to work together.

As an example, consider again the text widget. Now assume that we are also building an editor that will display this text widget. The editor will contain a pane that shows the text widget, but it will also have a menu for manipulating the text widget. In the first layer, we will have the simple text widget, and an editor which only can display this widget, maybe with a simple menu that only permits quitting the application. This simple editor can in fact also use the feature extensions of the text widget, the cut-and-paste extension and the style extension, and also the combination of these two extensions, the better text widget. However, the user of the simple editor cannot make use of these features, because there are no menu choices by which the new features can be activated. To access these new features, we need to construct an extension of the simple editor (a better editor), which has an edit menu for the cut-and-paste operations, and a style menu for setting and removing styles in the text. The situation is illustrated in Figure 5. Here we use directed arrows to indicate the situation where one class (here the Editor classes) can use another class (here the Text classes) by sending messages to it.

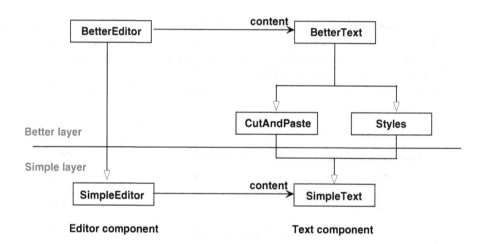

Fig. 5. Interacting components

The new editor is, however, restricted to only work on the better text widget, because the features that it assumes in the menus are only available on this level. Hence, there are two layers in the design, the *Simple* layer and the *Better* layer. The layers are not part of the standard UML notation, but they can be indicated by regions in the diagram. Figure 6 shows the architectural design with components and features, split into layers. Here the layers are defined by horizontal lines. This diagram emphasizes the two primary components, the text widget and the text editor.

3   Correctness Concerns

We will assume that the classes that make up the software are annotated with information about how the classes are intended to be used. In particular, we

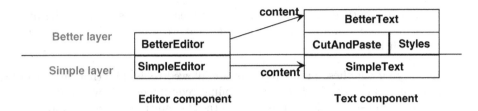

Fig. 6. Layers and components

assume that each class has a *class invariant*, which expresses the conditions on the attributes that must be established when the class is instantiated, and which must be preserved by each operation on the class. Also, we assume that the methods of the class have a *precondition* which states the assumptions that must hold when the method is called, and possibly a *postcondition*, which expresses properties that hold when the method call returns. These are all annotations that are assumed in, e.g., the *design by contract* approach advocated by Meyer and others [15].

A class that implements a feature must satisfy certain conditions in order to be correct, in the sense that it provides a solid step on which other extensions can be built. The main conditions that have to be satisfied are four:

Internal consistency. The class must be *internally consistent*, in the sense that it respects the constraints that are indicated for its part. The most important constraint is that it has to respect its class invariant. Other constraints that may be given in the class are loop invariants and postconditions for methods. In proving internal consistency, it may be assumed that the preconditions for the methods are satisfied upon start of the method. Besides the safety properties expressed as invariants, consistency also requires that each method invocation terminates.

Respect. Besides being consistent with its own constraints, one also has to show that the class respects the constraints of other classes that it uses. This means in particular that the class has to satisfy the preconditions of all the methods that it calls. For dynamically typed programming languages, one also needs to check that the typing is correct, in the sense that no illegal operation is attempted on an object during execution. In practice, this has to be done also for statically typed languages, because the type system provided by the language is usually too crude, and one needs to make finer distinctions in the code (*soft typing*).

Preserving old features. One also needs to show that a feature extension preserve the features of the superclass, i.e., that it is a *superposition refinement* of the superclass. An arbitrary subclass will not automatically be a superposition refinement, it has to be designed and proved to be one.

Satisfies requirements. The feature has been introduced for a specific reason, i.e., we have certain requirements that need to be realized by the feature. We therefore also need to show that the feature does in fact satisfy the requirements given for it. A feature extension may satisfy all the previous correctness criteria, but still not be what we want, so this is also a crucial condition.

In carrying out these proof steps, we are allowed to assume that all features on lower layers do satisfy their corresponding requirements. If we prove that each feature extension satisfies these correctness conditions, then by induction, it will follow that the software system as a whole will satisfy these four correctness conditions. Note that the final software system will be represented by one feature extension class, for which we then have shown the above properties.

We will express the correctness conditions and reason about their satisfaction in the refinement calculus [9]. We will in this paper not go deeper into the formalization of the stepwise feature introduction method in the refinement calculus, as this will be considered in more detail elsewhere.

4   Specifications, Implementations, and Views

The BetterText class described above may not be the only way in which we can construct a class with the required functionality. Maybe there is a class library around with a text widget that already has this functionality. Or maybe the Bet-terText widget is ok as a description of what we want to achieve, but it is not efficient enough. We can then consider the BetterText widget as a *specification* of the service that we want, to be *implemented* by another class (or maybe we want to have different implementations of this service around). This is illustrated in Figure 7. Here we have implemented the BetterText widget with another widget, EfficientText. This text widget in turn makes use of a text widget called LibraryText, from an existing class library. The EfficientText class is essentially

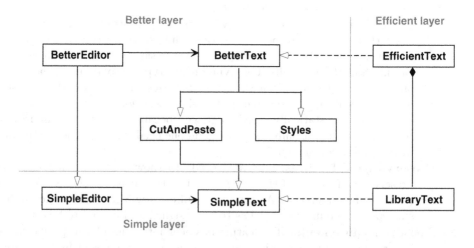

Fig. 7. Implementing the BetterText widget

a wrapper for the LibraryText widget, translating the method calls for the BetterText widget to corresponding calls to the LibraryText widget, to bridge the differences in interfaces. The figure also shows that the SimpleText can be implemented directly by LibraryText, which we assume has all the methods of SimpleText directly available, so no wrapper is needed in this case. Implementations are indicated in the diagrams by dashed open arrows, going from the implementation to the specification.

This figure shows that the specification of the BetterText widget is also built up in layers. Thus, we can use the stepwise feature introduction for constructing the software itself as well as for constructing the specifications for the software components. Many software components are very complex, and provide a number of different, highly interdependent features. The specification of such components can then be simplified by specifying the component one feature at a time, building up the specification in layers, as if it was an executable software component. A more thorough description of layered specifications is given in [7].

We can describe specifications and implementation by our city analogue as well. Consider first SimpleText. It corresponds to a house in the village (or to the bottom floor in a building in the city). If you ask for a service from this service provider, it will pretend to do the service itself, but in reality it is just a front end for another service provider (LibraryText), situated in another city (which has a different layering than our city, or no layering at all). For the BetterText, the same kind of analogue holds: A service request to this department is forwarded to another service provider (in this case EfficientText), and the replies that it gets are given to you as if they had been computed on site. The EfficientText could be in our city, maybe built in a later layer than the BetterText. The user may be under the impression that the service provider is structured as BetterText, but in reality this is just a way of explaining the service that BetterText offers to the user, but is not the way in which the service is actually realized.

For correctness of an implementation, we need to show that the implementation is a *data refinement* of the specification. This can be done using standard data refinement techniques [1,10,16,6], requiring that an abstraction function (or abstraction relation) is given that maps the attributes of the implementing class to the attributes of the specification class. Superposition refinement is a special case of data refinement. A data refinement step may change the collection attributes used in a class to represent the data, adding new attributes and deleting some old attributes. Superposition refinement can only add new attributes, it cannot remove old attributes. Superposition refinement (applied to reactive systems) is described in more detail in [8].

If we prove that BetterText is data refined by EfficientText, then any usage of BetterText in the system can be replaced with using EfficientText, by monotonicity of data refinement. Thus, the diagram in Figure 7 can be simplified to the diagram in Figure 8. Here the BetterEditor is using EfficientText directly, and the SimpleEditor is using LibraryText directly.

A *view* or *interface* of a class can also be seen as a specification of that class. For instance, consider the situation shown in Figure 9. Here the BetterText wid-

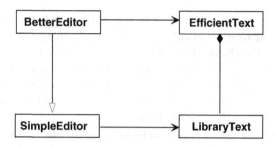

Fig. 8. Using implementation directly

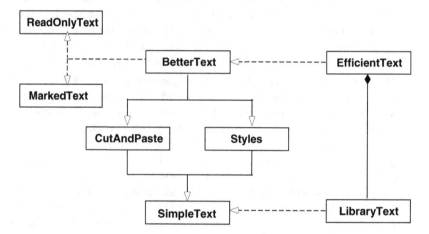

Fig. 9. Different views of the BetterText widget

get is itself seen as an implementation that supports two different views, one that only allows reading of the text but no changes to it (ReadOnlyText) and the other (MarkedText) that allows highlighting of the text using styles, but does not allow other kind of changes to the text. Views are thus just specifications. ReadOnlyText and MarkedText are thus two different specifications that are implemented by the same class, in this case BetterText. Again, correctness concerns require us to show that both specifications are correctly implemented by the BetterText widget, by showing a data refinement between the corresponding entities. Here we need to show that BetterText is a data refinement of, e.g., ReadOnlyText, while in the previous example, we had to show data refinement in the other direction, that BetterText is data refined by EfficientText. The figure also shows that we can cascade implementations: MarkedText is implemented by BetterText which in turn is implemented by EfficientText.

5   Reasoning Based on Diagrams

The diagrams that we have been showing above are more or less standard UML diagrams. However, we can also consider these diagrams as proof diagrams, by adding a simple notational device to the diagrams: an exclamation mark next to a diagram entity means that the correctness condition associated with this entity has been established. More precisely,

- an exclamation mark in a class box means that the class is *internally consistent*,
- an exclamation mark on one end of an association means that the class in the other end is *respecting* the usage of the other class, and an exclamation mark next to an attribute that the class is respecting the usage of the attribute,
- an exclamation mark on an inheritance arrow means that the subclass is a *superposition refinement* of the superclass,
- an exclamation mark on the implementation arrow means that the implementing class is a *data refinement* of the original class.

We can say that a class, association, inheritance arrow or implementation arrow in a diagram is *correct*, if we are allowed to place an exclamation mark next to it.

A question mark next to a diagram entity means that we do not know if the correctness condition associated with this entity is true. No marking for an entity means that we have not yet considered the associated correctness condition, or prefer to ignore this condition.

What it means that the correctness condition has been established can vary from one software project to the next. Sometimes, it just means that we have done a careful check (e.g., a code review) and are convinced that the correctness condition holds here. Sometimes it can mean that we have tested the correctness criterion, and found no violations. Sometimes it can mean that we have carried out a semi-formal or formal proof of the correctness condition.

From a given diagram with markings for the correctness conditions that have been established, we can infer other features of the diagram. An example of this was given in Figure 8, where we inferred that we could replace the use of Better-Text in the BetterEditor with the use of EfficientText. This inference is shown in more detail in Figure 10. In the upper diagram, we indicate that all the classes shown are internally consistent, the association is correct and the implementation is correct. The lower diagram shows that we have changed BetterEditor to use EfficientText rather than BetterText. However, the check marks in the lower diagram are as before, indicating that this change of usage does not invalidate any of the correctness conditions. The fact that the BetterEditor respects the EfficentEditor, as indicated in the lower diagram, follows from the fact that EfficientEditor is a data refinement of the BetterEditor. Data refinement also implies that the BetterEditor component is still internally consistent. As the internal consistency of a component can depend on any semantic properties of the components used, this means that the EfficientEditor satisfies all requirements that the BetterEditor did.

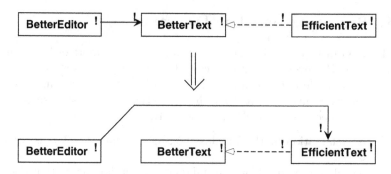

Fig. 10. Reasoning with diagrams

Consider now the case where we have not proved data refinement of BetterText with EfficientText. The situation is shown in the upper diagram in Figure 11. The lower diagram now shows that we do not know whether BetterEditor

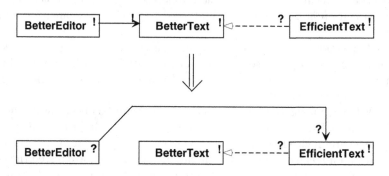

Fig. 11. Reasoning with less information

respects EfficientEditor (even if it did respect BetterText). We also do not know whether BetterEditor itself is internally consistent anymore, because the behavior of the component that it is using may have changed. BetterEditor may be internally consistent, but we have to establish this by a new proof.

The transitivity and monotonicity properties of data refinement and superposition refinement mean that we can infer a number of other correct associations and arrows from the existing ones. As these are implicit in the diagram, we do not need to write out them explicitly. In fact, we do not want to write them out explicitly, as they would only make the diagrams less readable, without adding any information that is not there already. We show in Figure 12 some of the implicit correct arrows that can be inferred from the existing correct arrows in Figure 7. The implicit arrows are thin, the given ones are thick. We omit the exclamation marks in the diagram for simplicity. The diagram shows that SimpleEditor also can use BetterText, CutAndPaste and LibraryText (among

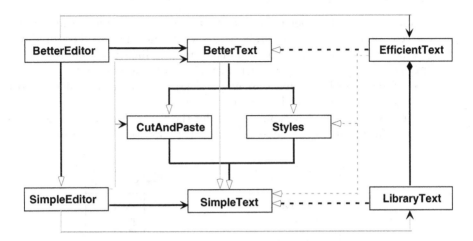

Fig. 12. Implicit correct associations and arrows

others) as content, respecting these classes. This is a consequence of the monotonicity of ordinary program constructs under refinement. It also shows that
BetterText is a superposition refinement of SimpleText directly. In addition, it
shows that EfficienText is a data refinement also of Styles and SimpleText (because superposition refinement is a special case of data refinement). These are
again consequences of the transitivity of data refinement. An earlier attempt to
formalize refinement calculus using diagrams can be found in [3].

6   Requirements and Testing

The previous discussion has concentrated on showing that a software system
is internally consistent, that components are used in a responsible way, that
layering is a superposition refinement in order to preserve features from one
layer to the next, and that implementations are data refinements, so that the
implementation can be used instead of the specification. But we have not said
anything about satisfying requirements yet. The system that we are building
might well be working perfectly but doing something completely different than
what we intended.

We will model requirements by *customer* classes. An customer of a component
uses the service provided by the component in a certain way, checking that the
service it gets is what it expected. We can express a service requirement by a
statement of the form

$$R = [P]; S; \{Q\}$$

Here P and Q are conditions (predicates) on the program variables (object attributes). The *assume statement* $[P]$ will *succeed* directly, if condition P does not
hold, but acts as a skip statement otherwise. The *assertion statement* $\{Q\}$ will

fail directly if Q does not hold, but otherwise it also acts as a skip statement. The requirement R is an executable statement, which may succeed directly (if P does not hold initially), it may terminate in an ordinary way after executing S (if Q holds after executing S), or it may fail, if Q does not hold after executing S. It is also possible to have assume and assert statements inside S, so that execution of S itself may also fail or succeed.

An customer class CheckSimpleText for the SimpleText class could be structured as in Figure 13. The customer class can have one method for checking a

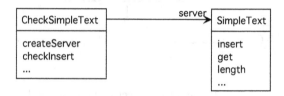

Fig. 13. Checking requirements on SimpleText

number of requirements, or a number of methods, one for each requirement. An example requirement is expressed by the method checkInsert, which expresses that inserting text at a specific place changes the text in the appropriate way. This method is defined as follows (using a Python-like syntax []):

```
def checkInsert(self, t,s,i):
    [0 ≤ i ≤ |t|]
    server:= self.createServer(t)
    server.insert(i,s)
    s1:=server.get()
    {s1 = t[: i] + s + t[i :]}
```

Assume that the method $\mathsf{insert}(i, s)$ inserts string s at position i in the widget, and that $\mathsf{get}()$ returns the text string in the widget. The customer expects (and requires) that the method checkInsert should never fail, no matter how initial text string t, the position index i and the inserted string s are chosen. Here $t[: i]$ denotes the prefix of t up to (but not including) i, and $t[i :]$ denotes the suffix of t starting from i. String concatenation is denoted by $+$.

The method *createServer* creates an instance of the SimpleText class:

```
def createServer(t):
    return SimpleText(t)
```

Requirements as described above are very similar to *unit tests*, except that the methods cannot be executed automatically, they require some parameters to be given during call time. A requirement that does not require any parameters can be seen as a *test*. For instance, the following is a test methods:

```
def testInsert(self):
    self.checkInsert(" abcdefg", "XYZ",2)
```

As no input parameters are needed, this test can be executed automatically, without user interaction.

Consider next the requirements for the CutAndPaste class. We may want to check that cutting a piece of the text and then pasting it to another place has the desired effect. For this purpose, we can define a customer class, CheckCutAndPaste, that has a method that expresses this requirement. This methods could work as follows:

```
def checkCutFollowedByPaste(self,t,i,j,k):
    [ 0 ≤ i ≤ j ≤ |t| ∧ 0 ≤ k ≤ |t| − j + i ]
    server:= self.createServer(t)
    server.cut(i,j)
    server.paste(k)
    s1:= t[:i]+t[j:]
    s2:= s1[:k]+t[i:j]+s1[k:]
    {server.get() = s2}
```

In this class, we redefine the createServer method to instantiate an object of class CutAndPaste:

```
def createServer(t):
    return CutAndPaste(t)
```

We check that class CutAndPaste correctly implements the features that we require, by methods such as checkCutFollowedByPaste . However, when extending the SimpleText with the new cut-and-paste feature, we also need to check that we do not loose any features that we have in the original class. This means that all the requirements that we have expressed for SimpleText should still hold for the new class CutAndPaste. We can check this by defining the class CheckCutAndPaste to be a subclass of CheckSimpleText. In this way, the former class inherits all the requirements expressed as methods in CheckSimpleText. These methods will be applied to an instance of class CutAndPaste, rather than to an instance of class SimpleText, because we have redefined createServer in CheckCutAndPaste.

The way the requirements are checked in these cases is shown in Figure 14. The figure shows the two classes being observed, and the two customer classes. The dashed line shows the instantiation of the text widget when SimpleText requirements are checked from CheckCutAndPaste.

In general, the customer classes will be structured by inheritance in the same way as the classes of the system itself. This is illustrated in Figure 15, where we have indicated the observation classes as shadows for the main classes. The naming of these observation classes can be uniform, the name being derived from the observed class, so we do not need to indicate the names of the classes here. Also, we do not need to indicate the inheritance hierarchy of the customers, as it is the same as that of the observed classes. An exclamation mark on a customer class indicates that the associated requirements are satisfied. In [5] we describe in

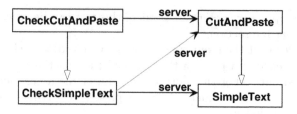

Fig. 14. Layered checking of requirements

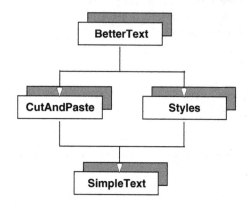

Fig. 15. Service providers (light boxes) and customers (dark boxes)

more detail methods for testing layered software that has been built by stepwise feature introduction.

Our analogue extends to testing and requirements, by considering the requirements, here modelled as customer classes, as the ordinary houses in the village or city. The inhabitants in these houses are the customers of the services provided by the buildings in the village or city center. The customers can have different requirements on the service. We can have two different customers that make requirements on the same service class, or we can have a customer that make requirements on different service classes. Figure 15 shows the special case with one customer for each service class.

7   Software Process

The stepwise feature introduction method does not fit well with the traditional software process, with its strong separation of requirements analysis, design, coding, testing and maintenance. However, it does fit quite well with the *extreme programming* approach that has been advocated by Kent Beck and others [11]. Extreme programming is a collection of techniques and practices that together form an interesting software process for small team system development. Among the practices advocated are short iteration cycles in the development, a concentration on producing code rather than documentation, avoiding to plan for future extensions, and frequent testing and integration of software during development. It also encourages practices like pair-programming and on-site customers, in order to create the right dynamics in the software team.

Extreme programming also emphasizes the incremental construction of software, in a planning game that lets the customers list and rank the features they need in the software, while the programming team estimates the time and other resources needed to implement each feature. The next iteration cycle is then carried out in a fixed time, and includes only those features that the customer has deemed to be most important and which all can be implemented in the time frame provided for the iteration cycle. However, extreme programming literature does not seem to be very precise about how to structure the software that is developed in this way. Stepwise feature introduction can be seen as a complement to the extreme programming process, describing more precisely how to structure the software and build the components that are developed in the incremental fashion that extreme programming proposes.

Extreme programming has also been seen as a new coming of hackerism, as it de-emphasizes documentation and careful design of the software, and emphasizes the coding part of the software process. We have tried above to show that this is not a necessary feature of extreme programming, it is possible to add documentation and correctness concerns as central parts of the incremental software construction process.

The stepwise feature introduction method defines the layering structure of the software, but does not take any position on the order in which the layers are built (except the obvious one, that they should be built from bottom up). In particular, we have not said when requirements and tests should be written and when they should be checked. From the correctness point of view, it does not matter in which order the customer classes and the service classes are written, as long as consistency can be proved for all classes and correctness for all arrows. One well-motivated approach generalizes the testing approach in extreme programming. They propose that a tester (unit test class) for a class should be written before the class itself is written. Also, they propose that all tests are executed automatically after each iteration cycle in the software development cycle.

Let us look at a this approach in more detail. Referring to the example in Figure 14, each iteration will build one layer in the software. In each iteration, we will define the requirements for the layer, design the layer, code the classes that are needed, and check the correctness of the layer. The sequence for the bottom layer in our text editor could be, e.g., as follows:

1. Construct the class CheckSimpleText that expresses the requirements for a simple text widget, and also contains test cases for this text widget, to be executed once the widget has been written.
2. Construct the class SimpleText, using the requirements and tests in CheckSimpleText as design guidelines.
3. Check that the class SimpleText is internally consistent.

4. Check that the CheckSimpleText class is internally consistent and respect the SimpleText class. This will show that the class SimpleText satisfies its requirements.

We can check consistency and correctness by formal or semi-formal proofs, by hand testing the requirements for selected parameter values, or by executing these tests.

To build the next layer, which introduces the cut-and-paste feature, we proceed in a similar way, applying steps 1 - 4 to the classes CheckCutAndPaste and CutAndPaste. In addition, we have to continue with two more steps, that are needed for every extension:

5. Check that the requirements for SimpleText are still satisfied, after the cut-and-paste feature has been introduced. We do this by checking the requirements and running the tests inherited from CheckSimpleText, but using an instance of class CutAndPaste rather than an instance of class SimpleText.
6. Check that CutAndPaste is a superposition refinement of SimpleText.

The above is, of course, a simplification of the actual software construction process. In practice quite a lot of planning is needed to identify the right features to be introduced, plan the layer structure, and solve special technical implementation problems. We have here also not discussed in more detail the need for refactoring of the software, when the layering structure runs into problems. Usually the need for refactoring becomes evident when it becomes increasingly difficult to add new features, with elaborate constructions to get around bad design decisions made in constructing earlier layers. A refactoring will often change the layering structure, sometime more and sometimes less, but the main purpose is to allow the software construction to proceed in a layered fashion after the problematic design decisions have been corrected.

8   Conclusions

We have above described the stepwise feature introduction method in outline. The method in itself is probably not new, a larger piece of software has to be implemented in more or less this way, building one feature at a time. Using inheritance for the layering is also not new, but rather one of the main motivations for having inheritance in the first place. What we have tried here is to make this method more precise, by identifying the correctness conditions that must be satisfied for the final system to be correct. The central idea here is that the layers should be built as superposition refinements on top of each other, i.e., at the same time as new features are introduced one has to check that the old features are preserved. This will distribute the total verification effort over the layers in an even way. A difference between the diagrams that we use and ordinary UML diagrams is that our diagrams have a precise mathematical meaning, being essentially logical statements in the refinement calculus about the classes that make up the software and the relationship between these different classes. We can even describe logical inference steps with the diagrams.

Layering in order to make software easier to construct and understand is also a well-known and important technique, often used in large software systems. However, the layers are usually quite thick, containing a lot of functionality and features, whereas we are here proposing that the software should be built up in very thin layers, each layer only introducing one feature at a time. Also, we insist that each layer, together with the layers lower down in the hierarchy, should be able to function on its own, and that this property should be preserved also after restructuring of the software. This means that the software system is really a collection of software systems, with progressively more functionality. When one is having problems with one layer in the software, it is always possible to regress to a lower layer that works correctly, albeit with less functionality than the higher layer. Also, this approach provides an easy way to branch off the software into different directions, providing different sets of extension features on a common basic set of features. Thus, the approach supports the design of software platforms and software families based on platforms.

We have also considered in more detail the role of specifications, implementation, requirements and testing in this context. It turns out that these notions are all easily expressed in UML and can be explained in refinement calculus terms. They also fit well into the layered software architecture.

We have carried out a few case studies in stepwise feature introduction, and the results have been quite encouraging. In [4], we report on a programming project with 6 undergraduate students and 3 Ph.D. students that combined extreme programming and stepwise feature introduction in a very successful way. The intention now is to continue with a number of larger case studies, some with industrial partners and some done in an academic environment, to check out different aspects of the method. We are also building UML based tools to help in software construction, testing and software project management. In a parallel development, we are exploring the strengths and weaknesses of extreme programming, and in particular the combination of extreme programming with stepwise feature introduction.

References

1. R.J. Back. *Correctness Preserving Program Refinements: Proof Theory and Applications*, volume 131 of *Mathematical Centre Tracts*. Mathematical Centre, Amsterdam, 1980.
2. R.J. Back. A calculus of refinements for program derivations. *Acta Informatica*, 25:593–624, 1988.
3. .R.J. Back. Refinement diagrams. J. M. Morris and R. C. Shaw, editors, *4th Refinement Workshop*, pages 125 - 137, Cambridge, UK, January 1991. Springer-Verlag.
4. R.J. Back, L. Milovanov, I. Porres-Paltor and V. Preoteasa. A Case Study in Extreme Programming and Stepwise Feature Introduction. TUCS Technical Report, in preparation.
5. R.J. Back, L. Milovanov and V. Preoteasa. Testing layered software. TUCS Technical Report, in preparation.

6. R.J. Back, A. Mikhajlova and J. von Wright. Class Refinement as Semantics of Correct Object Substitutability. *Formal Aspects of Computing*, 12: 18-40, 2000.
7. R.J. Back and V. Preoteasa. Layered Specifications: A Case Study. TUCS Technical Report, in preparation.
8. R.J. Back and K. Sere. Superposition Refinement of Reactive Systems. *Formal Aspects of Computing*, 8(3):324–346, 1996.
9. R.J. Back and J. von Wright. *Refinement Calculus: A Systematic Introduction.* Springer-Verlag, 1998.
10. R. J. Back and J. von Wright. Encoding, decoding, and data refinement. *Formal Aspects of Computing*, 2000.
11. K. Beck. *Extreme Programming Explained* Addison-Wesley, the XP Series, 1999.
12. G. Booch, J. Rumbaugh and I. Jacobson. *The Unified Modeling Language User Guide.* Addison-Wesley, 1998.
13. E.W. Dijkstra. Notes on structured programming. In O. Dahl, E.W. Dijkstra, and C.A.R. Hoare, editors, *Structured Programming.* Academic Press, 1971.
14. M. Fowler. *Refactoring : Improving the Design of Existing Code.* Addison-Wesley Object Technology Series 2000.
15. B. Meyer. *Object-Oriented Software Construction*, Prentice Hall, second edition, 1997.
16. W.P. de Roever and K. Engelhardt. *Data Refinement: Model-Oriented Proof Methods and their Comparison.* Cambridge Tracts in Theoretical Computer Science 47, 1998.
17. C.C. Morgan. *Programming from Specifications.* Prentice-Hall, 1990.
18. N. Wirth. Program development by stepwise refinement. *Communications of the ACM*, 14:221–227, 1971.

The Semantics of *Circus*

Jim Woodcock[1] and Ana Cavalcanti[2]

[1] Oxford University Computing Laboratory
Wolfson Building, Parks Road, Oxford, UK
Jim.Woodcock@comlab.ox.ac.uk
[2] Universidade Federal de Pernambuco/Centro de Informática
P.O. Box 7851, 50740-540 Recife PE, Brazil
alcc@cin.ufpe.br

Abstract. *Circus* is a concurrent language for refinement; it is a unification of imperative CSP, Z, and the refinement calculus. We describe the language of *Circus* and the formalisation of its model in Hoare & He's unifying theories of programming.

1 Introduction

An important research agenda in software engineering is the integration of languages and techniques that have proved themselves successful in specific aspects of software development. In particular, there has been interest for many years in combining state-based and behavioural formalisms [36]. There are several proposals for combining Z [4,26] with process algebras [15,19,23]. With *Circus* [32], we also want a calculational method of program development as those of [2,21, 22]; a similar calculus has been proposed for Z [7].

Circus is a unified programming language containing Z and CSP constructs, specification statements, and guarded commands [9]. It includes assignments, conditionals, and loops, and the reactive behaviour of communication, parallelism, and choice. All existing combinations of Z with a process algebra model concurrent programs as communicating abstract data types, but we do not insist on identifying events with operations that change the state. The result is a general programming language suitable for developing concurrent programs.

Imperative refinement calculi are normally given predicate transformers semantics; however, theories of refinement for CSP are based on the failures-divergences model [15,23]. A connection between weakest preconditions and CSP does exist [20], and a sound and complete theory of refinement has been developed based on it [36]. We use the unifying theory of [16], where both state and communication aspects of concurrent systems are integrated in a state-based failures-divergences model. This leads to a simple and elegant definition of refinement and makes a good underpinning for a refinement calculus. Our goals are: ease of use for those familiar with Z and CSP; encapsulation of the model; and the possibility of reusing existing theories, techniques, and tools.

In the next section we give an overview of the structure of *Circus* specifications and define its syntax. Section 3 provides a brief introduction to unifying theories of programming before, in Section 4, we give the formal semantics of *Circus*. Finally, in Section 5 we discuss related and future work.

D. Bert et al. (Eds.): ZB 2002, LNCS 2272, pp. 184–203, 2002.

Program	::=	CircusParagraph*
CircusParagraph	::=	Paragraph
	|	ChannelDefinition | ChanSetDefinition | ProcDefinition
ChannelDefinition	::=	**channel** CDeclaration
CDeclaration	::=	SCDeclaration | SCDeclaration; CDeclaration
SCDeclaration	::=	N^+ | N^+ : Expression | Schema-Exp
ChanSetDefinition	::=	**chanset** N == CSExpression
CSExpression	::=	$\{\!\|\,\|\!\}$ | $\{\!\| N^+ \|\!\}$ | N
	|	CSExpression $\cup$ CSExpression | CSExpression $\cap$ CSExpression
	|	CSExpression $\setminus$ CSExpression
ProcDefinition	::=	**process** N $\widehat{=}$ Process
Process	::=	**begin** PParagraph* • Action **end** | N
	|	Process; Process | Process $\square$ Process | Process $\sqcap$ Process
	|	Process $[\!\|$ CSExpression $\|\!]$ Process | Process $\|\|\|$ Process
	|	Process $\setminus$ CSExpression
	|	Declaration $\odot$ Process | Process$\lfloor$Expression$^+\rfloor$
	|	Process$[N^+ := N^+]$
	|	Declaration • Process | Process(Expression$^+$)
	|	$[N^+]$Process | Process[Expression$^+$]

Fig. 1. *Circus* syntax

2   *Circus*

As with Z specifications, *Circus* programs are sequences of paragraphs: besides the Z paragraphs, we have channel and channel set definitions, and process definitions. In Figures 1 and 2 we present the specification of the *Circus* syntax. We use an extended BNF notation, where SC* and SC$^+$ represent a possibly empty list and a comma-separated list of elements of the syntactic category SC. The syntactic category N is that of the valid Z identifiers. The definition for the Z constructs in the syntactic categories called Paragraph, Schema-Exp, Predicate, and Expression can be found in [26]. In Figure 3, as a simple example, we present a process that generates the Fibonacci sequence; more substantial examples of the use of *Circus* can be found in [32,33].

Processes communicate with each other and the environment by means of channels, whose definitions declare the types of the values that are communicated through them. For the Fibonacci generator, we need to declare the channel *out* through which the Fibonacci sequence is output.

channel *out* : $\mathbb{N}$

Channel declarations may be grouped in schemas. For conciseness, we can define channel sets to be used as arguments to the parallelism and hiding operators.

$$\text{PParagraph} \quad ::= \text{Paragraph} \mid \text{N} \mathrel{\widehat{=}} \text{Action}$$

$$\text{Action} \quad ::= \text{Schema-Exp} \mid \text{CSPActionExp} \mid \text{Command}$$

$$\text{CSPActionExp} ::= Skip \mid Stop \mid Chaos$$
$$\mid \text{Communication} \to \text{Action} \mid \text{Predicate \& Action}$$
$$\mid \text{Action; Action} \mid \text{Action} \mathbin{\square} \text{Action} \mid \text{Action} \mathbin{\sqcap} \text{Action}$$
$$\mid \text{Action} \llbracket \text{CSExpression} \rrbracket \text{Action} \mid \text{Action} \mathbin{\vert\vert\vert} \text{Action}$$
$$\mid \text{Action} \setminus \text{CSExpression} \mid \mu\, \text{N} \bullet \text{Action}$$
$$\mid \text{Declaration} \bullet \text{Action} \mid \text{Action}(\text{Expression}^{+})$$

$$\text{Communication} ::= \text{N CParameter}^{*}$$

$$\text{CParameter} \quad ::= ?\,\text{N} \mid ?\,\text{N : Predicate} \mid !\,\text{Expression} \mid .\,\text{Expression}$$

$$\text{Command} \quad ::= \text{N}^{+} : [\,\text{Predicate}, \text{Predicate}\,] \mid \text{N}^{+} := \text{Expression}^{+}$$
$$\mid \textbf{if } \text{GuardedActions } \textbf{fi}$$
$$\mid \textbf{var } \text{Declaration} \bullet \text{Action} \mid \textbf{con } \text{Declaration} \bullet \text{Action}$$

$$\text{GuardedActions} ::= \text{Predicate} \to \text{Action} \mid \text{Predicate} \to \text{Action} \mathbin{\square} \text{GuardedActions}$$

Fig. 2. *Circus* syntax

process $Fib \mathrel{\widehat{=}} \textbf{begin}$

   $FibState \mathrel{\widehat{=}} [\,x, y : \mathbb{N}\,]$

   $InitFibState \mathrel{\widehat{=}} [\,FibState' \mid x' = y' = 1\,]$
   $InitFib \mathrel{\widehat{=}} out!1 \to out!1 \to InitFibState$

   $OutFibState \mathrel{\widehat{=}} [\,\Delta FibState;\ next! : \mathbb{N} \mid next! = y' = x + y \wedge x' = y\,]$
   $OutFib \mathrel{\widehat{=}} \mu\, X \bullet \textbf{var } next : \mathbb{N} \bullet OutFibState;\ out!next \to X$

   $\bullet\ InitFib;\ OutFib$

end

Fig. 3. Fibonacci generator

The definition of a process gives its name and its specification: state and behaviour. An explicit process specification like that of *Fib* is a Z specification interspersed with action definitions: the state is defined as in Z and the behaviour is defined by a distinguished nameless action at the end. Typically, this action is defined in terms of other actions previously defined in the process specification. In our example, the state contains two components: x and y of type $\mathbb{N}$. A proof-obligation requires us to prove that the invariant is not *false*; in this case and in most cases it is simple. The main action is the CSP sequence *InitFib; OutFib*.

An action can be a schema that specifies an operation over the process state, a command of Dijkstra's language of guarded commands, or formed using CSP operators. We also have specification statements as in Morgan's calculus [21], and logical constants. The action *InitFib* is defined using the prefixing operator: it outputs 1 twice through *out* and then behaves like the initialisation *InitFibState*.

In *OutFib*, firstly, a local variable is declared; secondly, the *next* Fibonacci number is calculated and the state is appropriately updated by *OutFibState*; thirdly, the number is output; and finally, *OutFib* proceeds recursively. The variable *next* is referred to in *OutFibState* as *next!*. Unlike in Z, we interchangeably use dashes and shrieks to decorate after-state variables. Our choice in the example emphasises that *next* is a local variable. The following simpler definition of *OubFib* is possible, though.

$$OutFibState \mathrel{\widehat{=}} [\, \Delta FibState \mid y' = x + y \wedge x' = y \,]$$
$$OutFib \mathrel{\widehat{=}} \mu X \bullet out!(x + y) \rightarrow OutFibState;\ X$$

If, however, an implicit specification of the output is needed, the previous style should be adopted.

As actions, processes can also be combined using CSP operators. The state of the combination encompasses those of the operands, and the behaviour is given by combining the main actions of the operands with the used operator. As an example, consider *Fib* ⦀ *Fib*. Its state comprises two copies of the state components of *Fib*; renaming is used to avoid clashes. As the interleaving operator is used, two Fibonacci sequences are merged arbitrarily through *out*.

The *Circus* indexing operator is novel: the process $i : T \odot P$ behaves like P, but operates on different channels. Each communication over a channel c is turned into a communication of a pair over a channel c_i. The first element of the pair is i, and the second is the value originally communicated. The declaration of c_i is implicit. The index i in $i : T \odot P$ is a parameter; the process $(i : T \odot P)\lfloor e \rfloor$ communicates pairs as explained above, but the first element of the pairs is e: the index.

For example, suppose we want to generate two Fibonacci sequences, but we want to identify the generator of each element. We consider the process $i : \{1, 2\} \odot Fib$. It outputs through channel out_i the pairs $(i, 1)$, $(i, 1)$, $(i, 2)$, $(i, 3)$, $(i, 5)$, and so on, where in each case i is either 1 or 2. The process $(i : \{1, 2\} \odot Fib)\lfloor 1 \rfloor$ produces pairs through out_i whose first elements are 1; similarly for $(i : \{1, 2\} \odot Fib)\lfloor 2 \rfloor$. Finally, $(i : \{1, 2\} \odot Fib)\lfloor 1 \rfloor \mathrel{⦀} (i : \{1, 2\} \odot Fib)\lfloor 2 \rfloor$ produces an arbitrary merge of the two sequences of pairs: the first element of the pair identifies the generator and the second is a Fibonacci number.

In CSP, indexing is achieved by renaming since a communication of the value 2 over c is just an event name $c.2$. In *Circus*, this is not the case and we need to handle channel names and types separately. The reason for this distinction is the need for strong typing of communications in the spirit of Z.

The process $P[oldc := newc]$ is similar to P, but communicates through the channel *newc* where P would communicate through *oldc*. The declaration of *newc*, if necessary, is implicit.

Parametrisation of processes is similar to indexing, but is not related to channels. It declares extra variables, the parameters, which are available in the specification of the process. Instantiation fixes the value of these variables. The Z facility for defining generic paragraphs is extended to processes in *Circus*.

In $[X]P$, the generic parameter X is introduced to be used in P as a type. Instantiation can be explicit as in $P[\mathbb{N}]$ or inferred from the context.

The CSP recursion and prefixing operators are not available for processes. Since there is no notion of a global state, communication is handled by actions. It is also not clear that recursive definitions of processes are useful.

3   Unifying Theories of Programming

In Hoare & He's unifying theory of programming [16], the theory of relations is used as a unifying basis for the science of programming across many different computational paradigms. Programs, designs, and specifications are all interpreted as relations between an initial observation and a subsequent (intermediate or final) observation of the behaviour of a device executing a program.

In their unification, different theories share common ideas: sequence is relational composition; the conditional is a simple Boolean connective; nondeterminism is disjunction; and parallelism is a restricted form of conjunction. The miracle of the refinement calculus is the empty relation and abortion is the universal relation. Making assertions conditional-aborts brings all of assertional reasoning within the scope of the theory. Both correctness and refinement are interpreted as inclusion of relations, and all the laws of a relational calculus are valid for reasoning about correctness in all theories and in all languages.

Particular design calculi and programming languages are differentiated by their alphabet, signature, and healthiness conditions. The *alphabet* of a theory gives names for a range of external observations of program behaviour. By convention, the name of an initial observation is undecorated, but the name of a similar observation taken subsequently is decorated with a dash. This allows a relation to be expressed as in Z by its characteristic predicate. The *signature* provides syntax for denoting the objects of the theory. It names the relations corresponding to primitive operations directly, and provides operators for combining them. The *healthiness conditions* select the objects of a sub-theory from those of a more expressive theory in which it is embedded. Thus programs form a subset of designs, and designs form a subset of specifications.

The alphabet of each theory contains variables to describe all aspects of program behaviour that are considered relevant. In a purely procedural paradigm, these stand for the initial and final values of the global variables accessed and updated by a program block. Some variables are *external*, because they are globally shared with the real world in which the program operates, and so they cannot be declared locally. The first example is the Boolean variable *okay*: it means that the system has been properly started in a stable state; *okay'* means subsequent stabilisation in an observable state. This permits a description of programs that fail due to nonterminating loops or recursion.

In a theory of reactive processes, the variable *tr* records past interactions between a process and its environment; and the Boolean variable *wait* distinguishes the intermediate observations of waiting states from final observations of termination. During a wait, the process can refuse to participate in certain events offered by the environment; these are specified by the variable *ref*.

The signature of a theory varies in accordance with its intended use, whether in specification, in design, or in programming. A specification language has the least restricted signature. Design calculi successively remove unimplementable operators, starting with negation; all operators are then monotonic, and recursion can safely be introduced as a fixed-point operator. In the programming language, only implementable operations are left.

A healthiness condition distinguishes feasible descriptions of reality from infeasible ones. By a suitable restriction of signature, design languages satisfy many healthiness conditions, and programming languages satisfy even more. Hoare & He have shown [16] that all healthiness conditions of interest may be expressed in the form $P = \phi(P)$, where ϕ is an idempotent function mapping all relations to the healthy ones of a particular theory. These idempotents link higher-level designs to lower-level implementations.

The laws of CSP are not true for all predicates over the observational variables; there are eight healthiness conditions for CSP processes: three characterise *reactive* processes in general; two constrain reactive processes to be CSP ones; and a further three are more specific still. The first healthiness condition for a reactive process (**R1**) is that its execution can never undo any event that has already been performed. The second healthiness condition (**R2**) requires that a reactive process's behaviour is oblivious of what has gone before. The third healthiness condition (**R3**) is designed to make sequential composition work as intended. Suppose that we have the sequential composition of two processes P_1 and P_2. When P_1 terminates, the value of $wait'$ is false; therefore, P_2's value of $wait$ is also false, and control passes from P_1 to P_2. On the other hand, if P_1 is still waiting for interaction with its environment, then its value of $wait'$ and P_2's value of $wait$ are both true; in this case, P_2 must leave the state unchanged. Of course, this is sensitive to the previous process's stability: if activated when $okay$ is false, then its only guarantee is to extend the trace.

R is the set of reactive processes, which satisfy these first three healthiness conditions. An interesting subset of **R** satisfies two additional conditions. The first (**CSP1**) states that, if a process has not started, then we can make no prediction about its behaviour. The second (**CSP2**) states that we cannot require a process to abort; this is characterised by the monotonicity of the $okay'$ variable.

CSP is the set of reactive processes that satisfy these two healthiness conditions. Further healthiness conditions are required to capture the standard theory of CSP; each of them is given as a simple unit law. The first (**CSP3**) requires that a CSP process does not depend on the initial value of the *ref* variable when $wait$ is false. Of course, when $wait$ is true, then it must behave as required by **R3**. The second (**CSP4**) requires that the ref' variable is irrelevant after termination. The third (**CSP5**) requires that refusal sets must be subset-closed.

4   The Model of *Circus*

Our model for a *Circus* program is a Z specification that describes processes and actions as relations. The model of a process is itself a Z specification, and the model of an action is a schema.

4.1   Channel Environment

We use the Z mathematical notation as a meta-language. The semantics of a process depends on the channels in scope. These are recorded in an environment:

$$ChanEnv == ChanName \nrightarrow \mathsf{Expression}$$

The given set *ChanName* contains the channel names. The channel environment associates a channel name to its type.

The semantic function $[\![\_]\!]^{\mathcal{CD}}$: ChannelDefinition $\nrightarrow$ *ChanEnv* gives the meaning of channel definitions as channel environments recording their declarations. Untyped channels, used as synchronisation events, and are given the type *Sync*, a given set. For conciseness, we omit the definition of $[\![\_]\!]^{\mathcal{CD}}$, which can be found in [31] along with a few other definitions also omitted from this paper.

A channel is not a value in our model and so we cannot define a set of channels. In a *Circus* program, channel sets are used to abbreviate process expressions like parallelism and hiding. We assume that these process expressions are expanded by replacing references to channel sets with the set of channels it defines. The channel set definitions can then be eliminated.

4.2   Process Environment

A process definition may refer to other processes previously defined and so we also need a process environment that associates process names to their models:

$$ProcEnv == \mathsf{seq}(ProcName \times \mathsf{ZSpecification})$$

The given set *ProcName* contains the valid process names, and ZSpecification is the syntactic category of Z specifications. We use sequences because the order in which processes are declared is relevant: we cannot refer to a process before its definition. This is a restriction inherited from Z that can be lifted by tools, as long as they guarantee that there is an appropriate order to present the specification. The Z specification corresponding to a whole program includes those corresponding to the individual processes in the order that they appear.

A process definition enriches the environment by associating a process name to the Z specification corresponding to the declared process. Moreover, if the process specification involves indexing or channel renaming, new channel names are implicitly declared. The semantic function $[\![\_]\!]^{\mathcal{PD}}$ gives the meaning of a process definition as a process environment that records just the single process it declares, and a channel environment recording the new channels it introduces.

$$[\![\_]\!]^{\mathcal{PD}} : \mathsf{ProcDefinition} \nrightarrow ChanEnv \nrightarrow ProcEnv \nrightarrow (ChanEnv \times ProcEnv)$$

$$[\![\mathbf{process}\ N \mathbin{\widehat{=}} P]\!]^{\mathcal{PD}}\gamma\,\rho = \mathbf{let}\ Ps == [\![P]\!]^{\mathcal{P}}\gamma\,\rho \bullet (Ps.1, \langle(N, Ps.2)\rangle)$$

The semantics *Ps* of *P* is taken in the current channel and process environments; it is a pair containing a channel environment and a Z specification. The semantics of the process definition includes the channel environment and a process environment that associates *N* to the Z specification. The function $[\![\_]\!]^{\mathcal{P}}$ gives the meaning of processes and is defined later on.

4.3   Programs

A program's meaning is given by $[\![\_]\!]^{\mathcal{PROG}}$: Program $\nrightarrow$ ZSpecification. The model of a *Circus* program is its Z paragraphs and the model of its processes. More precisely, the specification starts with the following four paragraphs. First, we define a boolean type *Bool* as a free type with two constants: False and True; we use variables of this type as predicates, for simplicity. The second paragraph declares the given sets *Sync* and *Event*; the former is the type of the synchronisation events, and the latter includes the possible communications of the program.

The third paragraph specifies the components that comprise the state of a process, in addition to the user state components. These are the variables of the unifying theory model and the additional *trace* variable, which records the events that occurred since the last observation. Our processes do not have an alphabet as in [16]; instead we consider the alphabetised parallel operator of [23].

$$ProcessState \,\widehat{=}\, [\, trace, tr : \text{seq } Event; \; ref : \mathbb{P}\, Event; \; okay, wait : Bool\,]$$

Changes to the process state are constrained as specified in the fourth paragraph: valid process observations increase the trace.

$$ProcessStateObs \,\widehat{=}\, [\, \Delta ProcessState \mid tr \text{ prefix } tr' \land trace' = tr' - tr\,]$$

The remaining paragraphs are determined by $[\![prog]\!]^{\mathcal{CPARL}} \; \emptyset \; \emptyset$. This is the semantics of the list of paragraphs that compose the program itself.

4.4   Paragraphs

A *Circus* paragraph can contribute to the semantics of the whole program by extending the Z specification, and the channel and process environments.

$$[\![\_]\!]^{\mathcal{CPAR}} : \text{CircusParagraph} \nrightarrow ChanEnv \nrightarrow ProcEnv \nrightarrow$$
$$(\text{ZSpecification} \times ChanEnv \times ProcEnv)$$

For a Z paragraph Zp, the definition of $[\![\_]\!]^{\mathcal{CPAR}}$ adds Zp to the Z specification as it is, and does not affect the channel or the process environment.

$$[\![Zp]\!]^{\mathcal{CPAR}} \; \gamma \; \rho = (tc\, Zp, \gamma, \rho)$$

Slight changes to the Z paragraph may be needed because of schemas with untyped components, which are assumed to be synchronisation event declarations. The function tc, when applied to a schema that declares such components, yields the schema obtained by declaring the types of these components to be *Sync*.

A channel definition cd gives rise to a few paragraphs in the Z specification and enriches the channel environment.

$$[\![cd]\!]^{\mathcal{CPAR}} \; \gamma \; \rho = \textbf{let } \gamma' == [\![cd]\!]^{\mathcal{CD}} \bullet (events\, \gamma', \gamma \oplus \gamma', \rho)$$

The environment γ' records the channels declared in cd. For each channel c recorded to have type T different from *Sync* in γ', we have that *events* γ' yields an axiomatic description that declares c to be an injective function from T to $Event$. If T is $Sync$, then c is declared to be itself an event. These constants and injective functions are $Event$ constructors.

A process definition pd determines a Z specification of its model, and enriches the process environment and possibly the channel environment as well.

$$[\![pd]\!]^{\mathcal{CPAR}} \ \gamma \ \rho = \mathbf{let} \ pds == [\![pd]\!]^{\mathcal{PD}} \ \gamma \ \rho \bullet ((pds.2\ 1).2, \gamma \oplus pds.1, \rho \oplus pds.2)$$

The semantics of pd is a pair pds containing the channel environment that records the channels pd (implicitly) declares, and a process environment that records the process defined by pd. The Z specification corresponding to the process is the second element of the pair in the first and unique position of the process environment, which is itself the second element of pds.

The function $[\![\_]\!]^{\mathcal{CPARL}}$ maps lists of paragraphs to Z specifications whose paragraphs are obtained from the $Circus$ paragraphs as specified by $[\![\_]\!]^{\mathcal{CPAR}}$.

We eliminate repeated names used across different process definitions by prefixing each name with the name of the process in which it is declared. Also, in the model of a process, where several schemas are defined, their names should be fresh. Below, we leave this assumption implicit.

4.5   Processes

The semantic function $[\![\_]\!]^{\mathcal{P}}$ gives the meaning of a process declaration as a pair containing a channel environment and a Z specification.

$$[\![\_]\!]^{\mathcal{P}} : \mathsf{Process} \nrightarrow ChanEnv \nrightarrow ProcEnv \nrightarrow (ChanEnv \times \mathsf{ZSpecification})$$

For a process name N, the semantics is the current channel environment and $modelOf \ N \ in \ \rho$, which is the Z specification associated to N in ρ.

For an explicit process specification $\mathbf{begin} \ ppl \bullet A \ \mathbf{end}$, the semantics is a Z specification containing the following paragraphs: $ProcObs$, a schema describing the observations that may be made of the process; the Z paragraphs as they are, except for those that are schemas that define operations as these are actions; and, for each action, a schema constraining the process observations. In order to define $ProcObs$ we specify a schema $State$ that defines the process state. It includes the components of the schema $ProcessState$ previously defined and those of the state defined in ppl, which we assume to be named $UserState$.

$$State \mathrel{\widehat{=}} UserState \land ProcessState$$

The state of a $Circus$ process in our model includes the components of the state in its specification, which we refer to as user state, and the observation variables of the unifying theory. A process observation corresponds to a state change.

$$ProcObs \mathrel{\widehat{=}} \Delta UserState \land ProcStateObs$$

As we explain later on, the state can be extended by the declaration of extra variables. Therefore, we actually consider a family of schemas $ProcObs(USt)$; for a schema reference USt that defines the user state, $ProcObs(USt)$ is the schema defined as $ProcObs$ above, except that it includes ΔUSt, instead of $\Delta UserState$.

4.6   Actions

To each action $N \cong A$ corresponds a schema named N. It is determined by the function $[\![\_]\!]^{\mathcal{A}}$ which takes as arguments the current channel environment and the name of the schema that defines the user state.

$$[\![\_]\!]^{\mathcal{A}} : \mathsf{Action} \nrightarrow \mathit{ChanEnv} \nrightarrow \mathsf{N} \nrightarrow \mathsf{Schema\text{-}Exp}$$

The main action is nameless; the schema corresponding to it is given a fresh name. This schema is also determined by the function $[\![\_]\!]^{\mathcal{A}}$.

We distinguish three cases in the definition of the behaviour of an action: the normal case, the case in which the previous operation diverged, and the case in which the previous operation has not terminated.

$$[\![A]\!]^{\mathcal{A}}\gamma \; \mathit{USt} = [\![A]\!]^{\mathcal{A}_{\mathcal{N}}}\gamma \; \mathit{USt} \lor \mathit{Diverge}(\mathit{USt}) \lor \mathit{Wait}(\mathit{USt})$$

The function $[\![\_]\!]^{\mathcal{A}_{\mathcal{N}}}$ characterises the normal behaviour of an action.

$$[\![\_]\!]^{\mathcal{A}_{\mathcal{N}}} : \mathsf{Action} \nrightarrow \mathit{ChanEnv} \nrightarrow \mathsf{N} \nrightarrow \mathsf{Schema\text{-}Exp}$$

It is defined by induction below. The family of schemas $\mathit{Diverge}(\mathit{USt})$ characterise the behaviour of an action in the presence of divergence.

$$\mathit{Diverge}(\mathit{USt}) \cong [\,\mathit{ProcObs} \mid \neg \; \mathit{okay}\,]$$

Divergence is characterised by the fact that *okay* is false; the only guarantee is that *trace* can only be extended: a restriction enforced by $\mathit{ProcessStateObs}$. For $\mathit{Wait}(\mathit{USt})$ we have the following definition.

$$\mathit{Wait}(\mathit{USt}) \cong [\,\Xi \mathit{ProcObs}(\mathit{USt}) \mid \mathit{okay} \land \mathit{wait}\,]$$

The waiting state occurs when both *okay* and *wait* are true: there is no divergence, but the previous action has not terminated; the state does not change. The normal case of the actions behaviour is characterised by the schema below.

$$\mathit{Normal}(\mathit{USt}) \cong [\,\mathit{ProcObs}(\mathit{USt}) \mid \mathit{okay} \land \neg \; \mathit{wait}\,]$$

In this case, there is no divergence and the previous action has terminated.

Schema Expressions. For a schema expression SExp, we have the following.

$$[\![\mathit{SExp}]\!]^{\mathcal{A}_{\mathcal{N}}}\gamma \; \mathit{USt} = \mathit{SExp} \land \mathit{OpNormal} \lor \mathit{OpDiverge}$$

$\mathit{OpNormal}$ describes when SExp is activated in a state that satisfies its precondition; the *trace* is not modified and *okay* and *wait* do not change: the operation terminates successfully.

$$\mathit{OpNormal} \cong [\,\mathit{Normal}(\mathit{USt}) \mid \mathit{trace}' = \langle\rangle \land \mathit{okay}' \land \neg \; \mathit{wait}'\,]$$

If the precondition of SExp is not satisfied, the action diverges.

$$\mathit{OpDiverge} \cong [\,\mathit{Normal}(\mathit{USt}); \; \mathit{SExp} \lor \neg \; \mathit{SExp} \mid \neg \; \mathrm{pre} \; \mathit{SExp} \land \neg \; \mathit{okay}'\,]$$

We include $\mathit{SExp} \lor \neg \; \mathit{SExp}$ to put input and output variables in scope. In the sequel, we define the normal behaviour of the actions.

CSP Expressions. The definition of the normal behaviour of *Skip* is as follows.

$$[\![Skip]\!]^{\mathcal{A}_\mathcal{N}}\gamma \; USt = [Normal(USt) \land \Xi USt \mid trace' = \langle\rangle \land okay' \land \neg \; wait']$$

The user state is not changed, the trace is also not changed, and it terminates. For *Stop*, we have a similar definition, but deadlock is characterised by the fact that *wait'* is true. For *Chaos*, we require $\neg \; okay$, which characterises divergence.

Sequencing is defined in terms of a function *sequence* on *ProcObs(USt)*.

$[\![A; \; B]\!]^{\mathcal{A}_\mathcal{N}}\gamma \; USt$ ————————————————————

$Normal(USt)$

$\theta ProcObs(USt) = \theta([\![A]\!]^{\mathcal{A}}\gamma \; USt) \; sequence \; \theta([\![B]\!]^{\mathcal{A}}\gamma \; USt)$

The function *sequence* takes two process observations and returns the process observation that characterises their sequential composition.

$\_sequence\_ : ProcObs(USt) \times ProcObs(USt) \nrightarrow ProcObs(USt)$

$\forall \; a, b, c : ProcObs(USt) \mid c = a \; sequence \; b \Leftrightarrow$
$\quad before \; c = before \; a \land after \; a = before \; b \land after \; b = after \; c$

The sequential composition is well-defined only if the final state of the first process is equal to the initial state of the second. The functions *before* and *after* project the initial and the final state out of a process observation. If *a* diverges, then we have that $\neg \; a.okay'$ and consequently $\neg \; b.okay$. So, if *b* satisfies the healthiness condition **CSP1**, then the composite *a sequence b* diverges. Similarly, if *a* is waiting, then we have *a.wait'* and so *b.wait*. So, if *b* satisfies the healthiness condition **R3**, then *a sequence b* waits.

We specify a communication as a process observation *Comm(USt)*. Its occurrence is an event characterised by a channel name and a communicated value; in our model, this is an element of *Event*. The process observation *Comm(USt)* takes as input the set of events *accEvents?* that can take place, and outputs the event *e!* that actually takes place. We can make observations at two stages of the communication. The first is when the communication has not yet taken place.

$CommWaiting(USt)$ ————————————————————

$Normal(USt)$
$accEvents? : \mathbb{P} \; Event$
ΞUSt

$trace' = \langle\rangle \land accEvents? \cap ref' = \emptyset \land okay' \land wait'$

The trace is not extended, the acceptable events are not refused, and the user state is not changed. We can also observe a communication after it has occurred.

$$
\begin{array}{|l}
\underline{\ CommDone(\,USt)\ }\rule{4cm}{0.4pt} \\
\ Normal(\,USt) \\
\ accEvents? : \mathbb{P}\ Event \\
\ e! : Event \\
\ \Xi\,USt \\
\hline
\ e! \in accEvents? \wedge trace' = \langle e! \rangle \wedge okay' \wedge \neg\ wait'
\end{array}
$$

The trace is extended with a possible event; the user state is not changed.

$$
\begin{aligned}
Comm(\,USt) \mathrel{\widehat{=}}\ & CommWaiting(\,USt) \vee CommDone(\,USt) \vee \\
& Diverge(\,USt) \vee Wait(\,USt)
\end{aligned}
$$

A communication is actually a more primitive concept than a prefixing, which is the sequential composition of a communication and an action. For a prefixing $c!v \to A$, we have the following semantics.

$$
\begin{array}{|l}
\underline{\ [\![c!v \to A]\!]^{\mathcal{A}_{\mathcal{N}}}\gamma\ USt\ }\rule{3.5cm}{0.4pt} \\
\ Normal(\,USt) \\
\hline
\ \exists\,oc : Comm(\,USt)\ |\ oc.accEvents? = \{c(v)\}\ \bullet \\
\qquad \theta ProcObs(\,USt) = (procObsC\ oc)\ sequence\ \theta([\![A]\!]^{\mathcal{A}}\gamma\ USt)
\end{array}
$$

The only possible communication is $c(v)$. The function $procObsC$ projects out the components of a communication that form a process observation. The semantics of $c.v \to A$ and $c \to A$ can be defined in a similar way. For $c?x \to A$ the definition is different as $c?x$ introduces the variable x in scope for A.

The action $p \mathbin{\&} A$ is enabled only if the p condition holds, in which case, the semantics is that of A; otherwise it behaves as *Stop*.

The behaviour of an external choice $A \mathbin{\square} B$ can be observed in two points. Before the choice is made, the trace is empty, the process is waiting, and the user state has not been changed. The refusal set is characterised by the restrictions of both A and B: an event is refused only if it is refused by both A and B.

$$
\begin{array}{|l}
\underline{\ ExtChoiceWait(\,USt)\ }\rule{4cm}{0.4pt} \\
\ Normal(\,USt) \\
\ \Xi\,USt \\
\hline
\ trace' = \langle\rangle \wedge okay' \wedge wait' \wedge [\![A]\!]^{\mathcal{A}}\gamma\ USt \wedge [\![B]\!]^{\mathcal{A}}\gamma\ USt
\end{array}
$$

If a choice has been made, then either the trace is not empty or the process has diverged or it is not waiting. The behaviour is either that of A or that of B.

$\boxed{\begin{array}{l} \_\,ExtChoiceNotWait(USt)\,\underline{\hspace{4cm}} \\ Normal(USt) \\ \hline (trace' \neq \langle\rangle \vee \neg\ okay' \vee \neg\ wait') \wedge (\llbracket A \rrbracket^{\mathcal{A}}\gamma\ USt \vee \llbracket B \rrbracket^{\mathcal{A}}\gamma\ USt) \end{array}}$

$$\llbracket A \ \Box\ B \rrbracket^{\mathcal{A_N}}\gamma\ USt \mathrel{\widehat{=}} ExtChoiceWait(USt) \vee ExtChoiceNotWait(USt)$$

The internal choice is given simply by disjunction.

We define the semantics of parallelism as shown below.

$\boxed{\begin{array}{l} \_\,\llbracket A \ \llbracket\,C\,\rrbracket\ B \rrbracket^{\mathcal{A_N}}\gamma\ USt\,\underline{\hspace{4cm}} \\ Normal(USt) \\ \hline \exists\ tracea, traceb : \text{seq}\ Event;\ refa, refb : \mathbb{P}\ Event; \\ \quad okaya, okayb, waita, waitb : Bool\ \bullet \\ \quad (\llbracket A \rrbracket^{\mathcal{A}}\gamma\ USt)[tracea, refa, okaya, waita/trace', ref', okay', wait'] \wedge \\ \quad (\llbracket B \rrbracket^{\mathcal{A}}\gamma\ USt)[traceb, refb, okayb, waitb/trace', ref', okay', wait'] \wedge \\ \quad trace' \in tracea \parallel traceb\ sync\ (\llbracket C \rrbracket\gamma) \wedge \\ \quad ref' = (refa \cup refb) \cap \llbracket C \rrbracket\gamma \cup (refa \cap refb) \setminus \llbracket C \rrbracket\gamma \wedge \\ \quad okay' = okaya \wedge okayb \\ \quad wait' = waita \vee waitb \end{array}}$

We include the schemas that define the semantics of A and B. These actions start in the same state, so their restrictions on the initial state are conjoined. We rename the individual final state components to use in the definition of the parallel composition.

An event can be refused by the parallel composition if either A or B can refuse it and they have to synchronise on it, or rather, it is in the synchronisation set C. This set is denoted above by $\llbracket C \rrbracket\gamma$ and includes all events that represent communications over a channel in C according to its type definition in γ. If an event is not in this synchronisation set, then it can only be refused if both A and B can refuse it. The parallel composition diverges if either A or B does, and it terminates when both A and B do. The trace is the combination of the traces of A and B where events in the channel set determined by C are synchronised. The definition of the $\_\parallel\_sync\_$ operator can be found in [23]. The definition of interleaving is similar to that of parallelism.

The semantics of hiding is as follows.

$\boxed{\begin{array}{l} \_\,\llbracket A \setminus C \rrbracket^{\mathcal{A_N}}\gamma\ USt\,\underline{\hspace{4cm}} \\ Normal(USt) \\ \hline \exists\ tracep : \text{seq}\ Event;\ refp : \mathbb{P}\ Event\ \bullet \\ \quad \llbracket A \rrbracket^{\mathcal{A}}\gamma\ USt[tracep, refp/trace', ref'] \wedge \\ \quad trace' = tracep \restriction (Event \setminus \llbracket C \rrbracket\gamma) \\ \quad refp = ref' \cup (\llbracket C \rrbracket\gamma) \end{array}}$

The traces and refusals of $A \setminus C$ are determined in terms of those of A: we have to eliminate all events that represented communications through the channels in C from the trace and from the refusals. In order to deal adequately with the semantics of hiding, we need to have infinite sequences in our model, as suggested in [16]. We leave the complications of this as a future work for now.

The definition of the semantics of a recursive action $\mu X \bullet A$ is standard. We must observe, however, that the action A may use X as an action and, as such, is regarded as a function from actions to actions. For clarity, we refer to this action as $F(X)$.

$$
\begin{array}{|l}
\hline
[\![\, \mu X \bullet F(X)]\!]^{\mathcal{A}_N}\gamma\ USt \\
Normal(USt) \\
\hline
\theta ProcObs(USt) \in \bigcup \{a : SProcess(USt) \mid a \subseteq [\![F(\_)]\!]^{\mathcal{F}}\gamma\ USt(\!| a |\!) \} \\
\hline
\end{array}
$$

In this context, a process is represented as a set of process observations that satisfies the healthiness conditions of CSP. The definition of the set $SProcess(USt)$ of such processes can be found in [31].

We use the semantic function $[\![\_]\!]^{\mathcal{F}}$, which gives the semantics of a function on actions as a function on process observations.

$$
[\![\_]\!]^{\mathcal{F}} : (Action \to Action) \nrightarrow ChanEnv \nrightarrow SchemaName \\
\nrightarrow (ProcObs(USt) \to ProcObs(USt))
$$

The function $[\![F(\_)]\!]^{\mathcal{F}}\gamma\ USt$ can be defined in terms of $[\![\_]\!]^{\mathcal{A}}$ as follows.

$$
\begin{array}{|l}
[\![F(\_)]\!]^{\mathcal{F}}\gamma\ USt : ProcObs(USt) \to ProcObs(USt) \\
\hline
\forall po : ProcObs(USt) \bullet \mathbf{let}\ \theta([\![X]\!]^{\mathcal{A}}\gamma\ USt) == po\ \bullet \\
\quad \exists [\![F(X)]\!]^{\mathcal{A}}\gamma\ USt \bullet f\ po = \theta ProcObs(USt)
\end{array}
$$

According to the definition of $[\![\_]\!]^{\mathcal{A}}$, to determine $[\![F(X)]\!]^{\mathcal{A}}\gamma\ USt$, we need to know $\theta([\![X]\!]^{\mathcal{A}}\gamma\ USt)$. This is specified above as the argument of $[\![F(\_)]\!]^{\mathcal{F}}\gamma\ USt$.

For example, consider $\mu X \bullet out!2 \to X$, in an environment γ in which out is a channel of type integer. The function $[\![out!2 \to \_]\!]^{\mathcal{F}}\ \gamma\ USt$ is as follows.

$$
\begin{array}{|l}
[\![out!2 \to \_]\!]^{\mathcal{F}}\ \gamma\ USt : ProcObs(USt) \to ProcObs(USt) \\
\hline
\forall po : ProcObs(USt) \bullet \exists Output \bullet f\ po = \theta ProcObs(USt)
\end{array}
$$

The schema $Output$ is defined according to the semantics of $out!2 \to X$.

$$
\begin{array}{|l}
\hline
Output \\
Normal(USt) \\
\hline
\exists oc : Comm(USt) \mid oc.accEvents? = \{out(2)\} \bullet \\
\quad \theta ProcObs(USt) = (procObsC\ oc)\ sequence\ po \\
\hline
\end{array}
$$

The semantics of X, which is required in the specification of *Output* is taken to be *po*. The calculation of the semantics of other recursive actions, which apply to X operators whose semantics are given in terms of $[\![X]\!]^{\mathcal{A}}\gamma\ USt$ instead of $\theta([\![X]\!]^{\mathcal{A}}\gamma\ USt)$, requires more effort. We need to manipulate the definition of $[\![F(X)]\!]^{\mathcal{A}}\gamma\ USt$ to express it in terms of $\theta([\![X]\!]^{\mathcal{A}}\gamma\ USt)$.

A parametrised action $D \bullet A$ declares extra variables that can be used in A. We regard them as immutable state components whose values are fixed by initialisation. The semantics of $D \bullet A$ is that of A taken in the extended state. For an instantiation $A(e)$, we define the initial value of the extra state components in the semantics of A to be e and hide them.

Commands. The behaviour of a specification statement $x : [pre, post]$ is highly dependent on whether its precondition holds or not. If it does, then the postcondition must be established and the operation must terminate successfully, but the trace is not affected and only the variables x in the frame can be changed.

$$\begin{array}{|l}
\hline
[\![x : [pre, post]]\!]^{\mathcal{A_N}}\gamma\ USt \\
ProcObs(USt) \\
\hline
Normal(USt) \wedge pre \Rightarrow \\
\quad post \wedge trace' = \langle\rangle \wedge okay' \wedge \neg\ wait' \wedge \alpha USt' \setminus x' = \alpha USt \setminus x \\
\hline
\end{array}$$

We decorate the list of variables x to obtain the list of corresponding dashed variables. In an abuse of notation we use x and x' as sets to define the set of user state components that cannot be changed: all (αUSt) but those in x. The conjunction of equalities $\alpha USt' \setminus x' = \alpha USt \setminus x$ enforces this restriction.

If the precondition holds, but the postcondition cannot be established (under the given circumstances), then we have an infeasible (miraculous) operation. In such a circumstance, the predicate of the above schema is *false*. Therefore, the semantics of $x : [pre, post]$ is a partial relation.

The semantics of the assignment $x := e$ is rather simple. This action does not change the trace and, since we assume the expressions are always well-defined, it does not diverge and terminates. Of course, it sets the final value of x to e. The semantics of the conditional is also standard.

The semantics of variable declaration is given by existential quantification.

$$[\![\mathbf{var}\ x : T \bullet A]\!]^{\mathcal{A_N}}\gamma\ USt = \exists\ x, x' : T \bullet [\![A]\!]^{\mathcal{A}}\gamma\ xUSt(USt)$$

The semantics of the action in the scope of the variable block is taken in the extended user state $xUSt(USt)$ that includes x. The semantics of constant declaration is similar, but given by universal quantification.

4.7   Process Expressions

For the binary operators op, the semantics of P op Q can be given in terms of an explicit process specification.

P op $Q = \textbf{begin}$

> $State \;\hat{=}\; P.State \wedge Q.State$
>
> $P.PPar \uparrow Q.State$
>
> $Q.PPar \uparrow P.State$
>
> $\bullet\; P.Act$ op $Q.Act$

end

The schemas $P.State$ and $Q.State$ are the schemas that define the user state of P and Q. The state of P op Q conjoins the states of P and Q; we assume that name clashes are avoided through renaming.

We also refer to $P.PPar$ and $Q.PPar$, which are the process paragraphs that compose the definitions of P and Q, except $P.State$ and $Q.State$ and the main actions. These are all included in P op Q. We must notice, however, that the schemas in $P.PPar$ that specify an operation on $P.State$ are not by themselves actions in P op Q; we need to lift them to act on the extended state defined by $State$. This is the aim of the operator $\uparrow$, which simply conjoins each such schema with $\Xi Q.State$: actions of P are not supposed to affect the state components that are inherited from Q. Similar comments apply to the actions in $Q.PPar$.

The specification of the action that defines the behaviour of P op Q combines those that specify the behaviour of P and Q, $P.Act$ and $Q.Act$, using op. We observe that $P.Act$ ($Q.Act$) operates on the part of the user state that is due to P (Q) and cannot change the components that are originally part of the Q (P) state. It does not refer to them directly or indirectly, except through schema actions which have been conjoined with $\Xi Q.State$ ($\Xi P.State$).

The semantics of a hiding expression $P \setminus C$ is even simpler. The process paragraphs of P are included as they are; only the main action is modified to include the hiding.

The indexed process $i : T \odot P$ implicitly declares channels c_i, for each channel c used in P. Its semantics, therefore, affects the channel environment.

$$[\![i : T \odot P]\!]^{\mathcal{P}} \gamma \rho = \textbf{let } \gamma' == \{c : used\,P \bullet c\_i \mapsto T \times \gamma\; c\} \bullet$$
$$[\![i : T \bullet (P[c : (used\,P) \bullet c\_i])]\!]^{\mathcal{P}} (\gamma \oplus \gamma')\; \rho$$

The environment γ' records the channels implicitly declared. The set $used\,P$ includes the channels used in P. For each such channel c, γ' records a channel c_i which communicate pairs of values: the index and whatever value was communicated originally. The semantics of $i : T \odot P$ is that of a parametrised process taken in the extended channel environment that includes γ'. The parameter is the index, and the process $P[c : (used\,P) \bullet c\_i]$ is that obtained from P by changing all the references to a used channel c by a reference to the channel c_i. Communications through these channels are also changed so that the index i is also communicated. The generalisation of this and the previous definition for an arbitrary indexed process $D \odot P$ whose indexes are declared by D is lengthy, but straightforward. The semantics of instantiation is given by substitution.

The semantics of renaming is given mainly by substitution on the process definition. In a parametrisation, the parameters are regarded as loose constants. The channel environment in the pair defined by $\llbracket D \bullet P \rrbracket^{\mathcal{P}} \gamma \rho$ is that in $\llbracket P \rrbracket^{\mathcal{P}} \gamma \rho$. The Z specification is also that in $\llbracket P \rrbracket^{\mathcal{P}} \gamma \rho$, preceded by an axiomatic description that introduces D. The instantiation of parametrised processes has the same definition of instantiation of indexed processes: substitution.

The semantics of a generic process $[X]P$ is given by a rewriting of the Z specification denoted by P: each of its paragraphs is turned into a similar generic paragraph with parameter X; the channel environment is the same. Instantiation $P[E]$ of a generic process P is defined by the instantiation of all definitions in the Z specification corresponding to P.

5   Conclusions

This paper presents a unified language for specifying, designing, and programming concurrent systems; it combines Z and CSP in a way suitable for refinement. Fischer reports [11] a survey of related work in combining Z and process algebras: Z and CCS in [14,28]; Z with CSP in [10,24]; and CSP with Object-Z [5] in [10]. The major objective of *Circus* is to provide a theory of refinement and an associated calculus. Nothing in the style of a calculus has been proposed.

Combinations of Object-Z and CSP have been given a failures-divergences model for Object-Z classes [25,12]; data refinement has been briefly explored for such a combination, but no refinement laws have been proposed. Abstract data types have been given behavioural semantics in the failures model [34]. In [6], refinement rules have been proposed to support the development of Java programs, but no semantic model has been provided.

An early paper [17] presents a state-based semantics for a subset of occam in a style similar to ours, using predicates over failures and a stability/termination. The subset includes *Stop*, *Skip*, assignment, communication, conditional, loop, sequence, alternation, and variable declarations. The semantics of parallelism is left as an exercise.

Our semantic definitions are based on those in [16]. We believe, however, that we have provided an accessible presentation of the theory of imperative communicating programs. In doing so, the use of Z as an elegant notation to define relations was very appropriate. It also means that we can make use of tools like Z/Eves [18] to analyse and validate our definitions. Currently, we are encoding our semantics in Z/Eves. It is our plan to prove that all the healthiness conditions hold for our semantic definitions.

We are linking tools for Z and CSP through the unifying theory: we are using FDR [13] and Z/Eves for analysing different aspects of *Circus* specifications. We are also building a tool that calculates the Z specification corresponding to a *Circus* program, producing a specification that is suitable for analysis using Z/Eves.

A new language needs demonstrations of its usefulness; an implementation in the form of tools for analysis and development; and additional theory to underpin and extend. We are considering case studies and examples including the steam boiler control system [1,3,33,30], an IP-packet firewall [35], a smart-card system for electronic finance [27], and a railway signalling application [29].

We are already considering the extension of *Circus* to include the operators of Timed CSP [8]. The resulting language is expected to be adequate to the specification of data, behavioural, and timing aspects of real-time systems. We intend to define its model by extending the unifying theory of programming to cover aspects of time. In order to solve the difficulty with the semantics of the hiding operator, we also plan to extend our model to allow infinite traces [23].

Our main goal, however, is the proposal and proof of refinement laws for *Circus*. We want data refinement rules, and rules that allows the stepwise refinement to code in a calculational way.

Acknowledgments. We would like to thank Augusto Sampaio for his many suggestions on our work. Ana Cavalcanti is partly supported by CNPq, grant 520763/98-0. Jim Woodcock gratefully acknowledges the support of CNPq and the University of Oxford for his visit to the Federal University of Pernambuco.

References

1. J. R. Abrial, E. Borger, and J. Langmaack, editors. *Formal Methods for Industrial Application*, volume 1165 of *Lecture Notes in Computer Science*. Springer Verlag, 1996.

2. R. J. R. Back and J. Wright. *Refinement Calculus: A Systematic Introduction*. Graduate Texts in Computer Science. Springer-Verlag, 1998.

3. J. C. Bauer. Specification for a software program for a boiler water content monitor and control system. Technical report, Institute of Risk Research, University of Waterloo, 1993.

4. S. M. Brien and J. E. Nicholls. Z Base Standard, Version 1.0. Technical Monograph TM-PRG-107, Oxford University Computing Laboratory, Oxford - UK, November 1992.

5. D. Carrington, D. Duke, R. Duke, P. King, G. A. Rose, and G. Smith. Object-Z: An Object-oriented Extension to Z. *Formal Description Techniques, II (FORTE'89)*, pages 281 – 296, 1990.

6. A. L. C. Cavalcanti and A. C. A. Sampaio. From CSP-OZ to Java with Processes (Extended Version). Technical report, Centro de Informática/UFPE, 2001. Available at http://www.cin.ufpe.br/~lmf.

7. A. L. C. Cavalcanti and J. C. P. Woodcock. ZRC - A Refinement Calculus for Z. *Formal Aspects of Computing*, 10(3):267 – 289, 1999.

8. J. Davies. *Specification and Proof in Real-time CSP*. Cambridge University Press, 1993.

9. E. W. Dijkstra. *A Discipline of Programming*. Prentice-Hall, 1976.

10. C. Fischer. CSP-OZ: A combination of Object-Z and CSP. In H. Bowmann and J. Derrick, editors, *Formal Methods for Open Object-Based Distributed Systems (FMOODS'97)*, volume 2, pages 423 – 438. Chapman & Hall, 1997.

11. C. Fischer. How to Combine Z with a Process Algebra. In J. Bowen, A. Fett, and M. Hinchey, editors, *ZUM'98: The Z Formal Specification Notation*. Springer-Verlag, 1998.

12. C. Fischer. *Combination and Implementation of Processes and Data: from CSP-OZ to Java.* PhD thesis, Fachbereich Informatik Universitat Oldenburg, 2000.
13. Formal Systems (Europe) Ltd. *FDR: User Manual and Tutorial, version 2.28,* 1999.
14. A. J. Galloway. *Integrated Formal Methods with Richer Methodological Profiles for the Development of Multi-perspective Systems.* PhD thesis, University of Teeside, School of Computing and Mathematics, 1996.
15. C. A. R. Hoare. *Communicating Sequential Processes.* Prentice-Hall International, 1985.
16. C. A. R. Hoare and He Jifeng. *Unifying Theories of Programming.* Prentice-Hall, 1998.
17. C. A. R. Hoare and A. W. Roscoe. Programs as executable predicates. In *Proceedings of the International Conference on Fifth Generation Computer Systems 1984 (FGCS'84)*, pages 220–228, Tokyo, Japan, November 1984. Institute for New Generation Computer Technology.
18. I. Meisels. *Software Manual for Windows Z/EVES Version 2.1.* ORA Canada, 2000. TR-97-5505-04g.
19. R. Milner. *Communication and Concurrency.* Prentice-Hall, 1989.
20. C. C. Morgan. Of wp and csp. In W. H. J. Feijen, A. J. M. van Gasteren, D. Gries, and J. Misra, editors, *Beauty is our business: a birthday salute to Edsger W. Dijkstra.* Springer, 1990.
21. C. C. Morgan. *Programming from Specifications.* Prentice-Hall, 2nd edition, 1994.
22. J. M. Morris. A Theoretical Basis for Stepwise Refinement and the Programming Calculus. *Science of Computer Programming*, 9(3):287 – 306, 1987.
23. A. W. Roscoe. *The Theory and Practice of Concurrency.* Prentice-Hall Series in Computer Science. Prentice-Hall, 1998.
24. A. W. Roscoe, J. C. P. Woodcock, and L. Wulf. Non-interference through Determinism. In D. Gollmann, editor, *ESORICS 94*, volume 1214 of *Lecture Notes in Computer Science*, pages 33 – 54. Springer-Verlag, 1994.
25. G. Smith. A Semantic Integration of Object-Z and CSP for the Specification of Concurrent Systems Specified in Object-Z and CSP. In C. B. Jones J. Fitzgerald and P. Lucas, editors, *Proceedings of FME'97*, volume 1313 of *Lecture Notes in Computer Science*, pages 62 – 81. Springer-Verlag, 1997.
26. J. M. Spivey. *The Z Notation: A Reference Manual.* Prentice-Hall, 2nd edition, 1992.
27. S. Stepney, D. Cooper, and J. C. P. Woodcock. An Electronic Purse: Specification, Refinement, and Proof. Technical Monograph PRG-126, Oxford University Computing Laboratory, 2000.
28. K. Taguchi and K. Araki. The State-based CCS Semantics for Concurrent Z Specification. In M. Hinchey and Shaoying Liu, editors, *International Conference on Formal Engineering Methods*, pages 283 – 292. IEEE, 1997.
29. J. C. P. Woodcock. Montigel's Dwarf, a treatment of the dwarf-signal problem using CSP/FDR. In *Proceedings of the 5th FMERail Workshop*, Toulouse, France, September 1999.
30. J. C. P. Woodcock and A. L. C. Cavalcanti. A *Circus* steam boiler: using the unifying theory of Z and CSP. Technical report, Oxford University Computing Laboratory, Wolfson Building, Parks Road, Oxford OX1 3QD UK, July 2001.
31. J. C. P. Woodcock and A. L. C. Cavalcanti. *Circus*: a concurrent refinement language. Technical report, Oxford University Computing Laboratory, Wolfson Building, Parks Road, Oxford OX1 3QD UK, July 2001.

32. J. C. P. Woodcock and A. L. C. Cavalcanti. A concurrent language for refinement. In Andrew Butterfield and Claus Pahl, editors, *IWFM'01: 5th Irish Workshop in Formal Methods*. Computer Science Department, Trinity College Dublin, July 2001.

33. J. C. P. Woodcock and A. L. C. Cavalcanti. The steam boiler in a unified theory of Z and CSP. In *8th Asia-Pacific Software Engineering Conference (APSEC 2001)*, 2001.

34. J. C. P. Woodcock, J. Davies, and C. Bolton. Abstract Data Types and Processes. In J. Davies, A. W. Roscoe, and J. C. P. Woodcock, editors, *Millenial Perspectives in Computer Science, Proceedings of the 1999 Oxford-Microsoft Symposium in honour of Sir Tony Hoare*, pages 391 – 405. Palgrave, 2000.

35. J. C. P. Woodcock and Alistair McEwan. Specifying a Handel-C program in the Unifying Theory. In *Proceedings of the Workshop on Parallel Programming*, Las Vegas, November 1999.

36. J. C. P. Woodcock and C. C. Morgan. Refinement of state-based concurrent systems. In D. Bjørner, C. A. R. Hoare, and H. Langmaack, editors, *VDM'90: VDM and Z—Formal Methods in Software Development*, number 428 in LNCS, pages 340–351. Springer, 1990.

Handling Inconsistencies in Z Using Quasi-Classical Logic

Ralph Miarka, John Derrick, and Eerke Boiten

Computing Laboratory, University of Kent, Canterbury, CT2 7NF, UK

{rm17,jd1,eab2}@ukc.ac.uk

Abstract. The aim of this paper is to discuss what formal support can be given to the process of living with inconsistencies in Z, rather than eradicating them. Logicians have developed a range of logics to continue to reason in the presence of inconsistencies. We present one representative of such paraconsistent logics, namely Hunter's quasi-classical logic, and apply it to the analysis of inconsistent Z schemas. In the presence of inconsistency quasi-classical logic allows us to derive less, but more "useful", information. Consequently, inconsistent Z specifications can be analysed in more depth than at present. Part of the analysis of a Z operation is the calculation of the precondition. However, in the presence of an inconsistency, information about the intended application of the operation may be lost. It is our aim to regain this information. We introduce a new classification of precondition areas, based on the notions of definedness, overdefinedness and undefinedness. We then discuss two options to determine these areas both of which are based on restrictions of classical reasoning.

1 Introduction

The purpose of this paper is to discuss how to reason in the presence of inconsistencies in a formal setting. Although this might sound strange, specifications, especially large ones, are often inconsistent at some level. Inconsistencies range from contradictory descriptions of the system at hand to contradictions specified in the operations. A significant proportion of the specification analysis process is then devoted to detecting and eliminating such inconsistencies, because, classically (and intuitively), inconsistencies in specifications are regarded as undesirable.

However, those involved in large scale software engineering in practice treat inconsistencies as a fact of life. They occur frequently in large projects and need to be tolerated (possibly for some time) and managed, rather than eradicated immediately. This has led to a considerable amount of research on the development of tools and techniques for living with inconsistencies (Ghezzi and Nuseibeh, 1998), (Balzer, 1991), (Schwanke and Kaiser, 1988), and handling inconsistencies (Finkelstein et al., 1994), (Hunter and Nuseibeh, 1998). The general aim of such

D. Bert et al. (Eds.): ZB 2002, LNCS 2272, pp. 204–225, 2002.

work is to provide practical support for deciding if, when, and how to remove inconsistencies, and to possibly reason in the presence of inconsistencies.

Although the techniques and tools developed for this approach have had a certain amount of success they have, however, mainly focused on informal and semi-formal specification techniques. There has been recent work on more formal approaches (Hunter and Nuseibeh, 1997), but these have largely concentrated on purely logical issues, not connecting themselves to current specification languages. We are interested in seeing what support we can give for the process of living with inconsistencies in a specification notation, namely Z.

Our purpose here is to explore the issue (rather than offering any definite solutions), discussing how inconsistencies can arise and how they might be handled, especially those present in operations. A number of options are discussed, all of which have the same general aim, namely, in the presence of inconsistency, not to immediately derive falsehood, but rather allow further, intermediate, reasoning on other aspects of the state, operation, or specification. These options include restricting the standard logic used for reasoning about Z specifications, as well as using alternative, so-called paraconsistent logics.

The paper is structured as follows. In Section 2 we present a small supporting example, illustrating some sources of inconsistencies and the problems of analysing such Z specifications. Following this we introduce, in Section 3, one way of supporting the reasoning process in the presence of inconsistencies, by presenting a paraconsistent logic called quasi-classical logic. Further, in Section 4 we use quasi-classical logic to support the process of reasoning in the presence of inconsistency. We exemplify the methods in terms of our example as we go along in this work. We consider the particular problem of deriving preconditions of operations in Section 5. Finally, we give some concluding remarks with links to related and future work in Section 6.

2   Background

In this section we introduce a small example written in the specification language Z. The advantage of Z and other formal methods is the possibility of formally analysing a given specification. We discuss two particular ways of analysing Z specifications. This work is concerned with the notion of inconsistency, hence we present some account of it at the end of this section.

2.1   Example

To motivate our work, we present a simplified example from the life of a motorist. The motorist is the owner of a car. To be allowed to drive the car on public roads, the car needs to pass a safety test, part of which is a tyre inspection. The law (in Germany) says that the car must have the same kind of tyre fitted to both the front and rear wheels.

In the state schema, the Boolean *flat* denotes whether any of the tyres are flat. If not the motorist is permitted to *drive* the car. The *Law* states that the *same* tyres should be used front and back. A single operation is specified, that of changing a tyre. Unfortunately, the spare tyre is of a different type, thus we will break the law as a result of a *Change*.

$[CAR]$

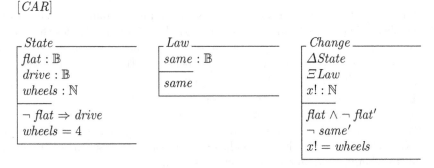

The *Change* operation is clearly inconsistent in an intuitive sense. Once the tyre has been changed, the car is not allowed on the road by the law, because the type of tyre on at least one wheel is now different. However, we might wish to reason about aspects of this specification, for example, that the car is still driveable, since this only depends on the fact that no tyre is flat. Although this example is small and rather artificial, it illustrates the type of reasoning one might wish to perform.

Another operation often performed by a motorist is to refuel their car. We distinguish three kinds of cars: electric cars, cars with diesel engines and cars running on petrol. The electric car needs a power supply to re-charge, whereas the other cars need fuel which can be divided into unleaded, four star and diesel.

$CAR\_TYPE ::= electric \mid diesel \mid petrol$
$FUEL\_TYPE ::= unleaded \mid four\_star \mid diesel\_type$

```
┌─ State2 ──────────────────────┐    ┌─ Choose ────────────────────┐
│ charged : 𝔹                   │    │ ΔState2                     │
│ fuel : FUEL_TYPE              │    │ car? : CAR_TYPE             │
│ amount : FUEL_TYPE → ℕ        │    ├─────────────────────────────┤
└───────────────────────────────┘    │ car? = petrol ⇒ fuel' = unleaded │
                                      └─────────────────────────────┘
```

```
┌─ Refuel ──────────────────────────────────────────────────────┐
│ Choose                                                         │
├────────────────────────────────────────────────────────────────┤
│ (car? = electric ∧ charged') ∨                                 │
│ (car? = petrol ∧ amount'(fuel') = 60 ∧ fuel' = four_star)      │
└────────────────────────────────────────────────────────────────┘
```

This refuel operation is partly inconsistent, because we assign two different types of fuel to be taken when the car requires petrol. It is consistent when applied to electric cars; no refuel operation has been specified for diesel cars. Clearly, this looks like a simple specification error, but in a large specification such errors can be hidden.

2.2   Analysing Z Specifications

One of the benefits of formal methods is the ability to analyse specifications in a mathematical and logical way. In Z, this kind of analysis includes the calculation of the precondition of an operation and the inference of properties of the system before or after the application of an operation. Here, we review the analysis process in the context of inconsistent specifications.

Precondition Calculation. The precondition of an operation is the domain and inputs on which the operation is guaranteed to perform as specified. In Z, the precondition of an operation schema is implicit and therefore has to be extracted. Formally, the precondition of an operation schema Op acting on a state schema $State$ and outputs $outs!$ is defined as:

$$\text{pre } Op = \exists\, State',\, outs! \bullet Op$$

For operations containing inconsistencies, the precondition calculation determines, correctly, that this domain is either empty or partial. For example, the operation $Change$ described above cannot be successfully applied, and indeed we find:

$$\text{pre } Change = [State;\ Law \mid \text{false}]$$

The operation $Refuel$ is only partially applicable because of an inconsistency with the operation $Choose$ that forces unleaded petrol to be used for petrol cars:

$$\text{pre } Refuel = [State2;\ car? : CAR \mid car? = electric]$$

Inferring Properties. In addition to calculating preconditions for operations, it is often desirable to verify that various properties hold for the system being constructed. For example, one might wish for certain system invariants to hold. In our example, one invariant is that there should always be four wheels attached to the car. For example, it can be shown that:

$$State \vdash wheels = 4$$

Using the completeness of the proof system of predicate logic this is equivalent to the statement:

$$State \nvdash wheels \neq 4$$

However, due to the inconsistency in the operation $Change$ it is possible to reason that there are more or less than four wheels present after the operation has been applied, in particular:

$$Change \vdash wheels' = 3$$

2.3   Inconsistencies in Z Specifications

A specification is supposed to be a model of some possible system. If such a specification is inconsistent then it has no models. (Saaltink, 1997) identifies two different types of inconsistency: *global* and *local inconsistency*.

Global Inconsistency. Global inconsistencies are serious, because they make an entire specification unsatisfiable. They occur if some axiom schema, generic schema, or constraint is unsatisfiable. Furthermore, they can arise due to a combination of different paragraphs of a specification, each being consistent. However, set declarations, abbreviations, and schema definitions cannot introduce global inconsistency. Note that no theorem can be trusted that has been proved in a globally inconsistent specification, as its proof is possibly based on a set of impossible assumptions.

Local Inconsistency. A schema can have an inconsistent, i.e. unsatisfiable, predicate. If such a schema is an operation schema, then the operation may not be applicable at all, or only parts of the operation are applicable. This is due to the fact that contradictions in an operation only restrict the precondition of that operation. In the case of the schema describing the state of the system, the entire part of the system governed by that state is not implementable. These kinds of errors are local in the sense that the specification of other components of the system may still be meaningful (although it is usually assumed implicitly, in a state and operation specification that at least one possible (initial) value of the state should exist).

Inconsistency versus Falsity. The distinction between falsity and inconsistency is not made clear within Z. An inconsistent operation behaves like an operation not being specified, i.e. set to false. This in turn makes it much harder to analyse the source of failure of an operation.

Additionally, the way the precondition computation in Z works seems to indicate an ordering of belief, assuming, for example, state schemas to be correct while an operation can be faulty. This leads to operations not being permitted if they are violating the state condition. However, this is not necessarily correct. It could be that the operation is correctly specified but the state specification is flawed.

For example, consider the tyre changing operation described above. The operation *Change* can be considered to be specified correctly. Practically, the specification of the schema *Law* lacks an exception, namely the allowance to drive a car with a replacement tyre of a different type with reduced speed and only to the next garage for the purpose of replacing it.

3   Handling Inconsistency Using Quasi-Classical Logic

The aim of this work is to develop a way to continue to reason in the presence of inconsistency and to be able to infer valid conclusions from inconsistent Z schemas or specifications. In classical predicate logic, on which Z is based, inconsistent information result in triviality, because everything can be inferred from it. This, in turn, renders the information useless, when in fact there may be further valid inferences we wish to make. However, there are several ways of handling inconsistent information. One is to divide the pieces of information into (possibly maximal) consistent subsets (Rescher and Manor, 1970), another is paraconsistent reasoning. The latter allows the derivation of only non-trivial inferences from inconsistent information, i.e. not everything can be inferred.

A paraconsistent logic is a compromise of classical logic, either a weakening of the classical connectives, particularly negation, or of the inference system. The former often results in useful proof rules (like disjunctive syllogism: $\{\alpha, \neg \alpha \lor \beta\} \vdash \beta$) or intuitive equivalences (like $\neg \alpha \lor \beta \equiv \alpha \Rightarrow \beta$) to fail. To preserve the behaviour of the classical connectives, we consider a logic with a weaker inference system. We believe that such a paraconsistent logic is more suitable for our application, because specifiers and analysts will already be familiar with the notation and meaning of the connectives. Here, we present these ideas in a purely logical framework, independent of a specification notation. Subsequently, in Section 4, we will discuss how such a framework could be used within the Z notation.

3.1   The Idea Behind Quasi-Classical Logic

One representative of paraconsistent logics is quasi-classical logic (QCL), developed by (Besnard and Hunter, 1995). It follows the principle of moving away from the view of information being either true or false. We accept that we may have a number of perspectives on information and that these perspectives may contradict each other.

The key to QCL is that it allows only the derivation of information already present in a given theory, even though that theory might be inconsistent. In classical logic, the combination of disjunction introduction, inconsistency and disjunctive syllogism results in the fact that anything is derivable – not so in QCL. However, to cope with this, the proof theory of QCL is more restricted than the proof theory of classical logic.

The restriction imposed is that compositional proof rules (like disjunction introduction) cannot be followed by decompositional proof rules (like conjunction elimination). This results in a logic that is weaker than classical logic. However, an advantage of QCL is that the logical connectives behave classically. The aim of QCL is not so much to reason about the truth in the real world but about handling beliefs. This seems to be compliant with the idea of formal specification where we gather requirements of a system yet to be built.

To give a flavour of QCL we state the following classical derivations which are also derivations in QCL: $\{\alpha, \alpha \Rightarrow \beta\} \vdash_Q \beta$, $\{\neg \beta, \alpha \Rightarrow \beta\} \vdash_Q \neg \alpha$, as well as $\{\alpha \land \beta, \neg \beta \land \gamma\} \vdash_Q \gamma \lor \delta$, where $\vdash_Q$ denotes the QCL consequence relation. Further, we use the symbol $\vdash$ to denote the classical consequence.

In general, the classical properties of reflexivity ($\Delta \cup \{\alpha\} \vdash_Q \alpha$), monotonicity ($\Delta \vdash_Q \alpha$ implies $\Delta \cup \{\beta\} \vdash_Q \alpha$), and-introduction ($\Delta \vdash_Q \alpha$ and $\Delta \vdash_Q \beta$ implies $\Delta \vdash_Q \alpha \wedge \beta$), and or-elimination ($\Delta \cup \{\alpha\} \vdash_Q \gamma$ and $\Delta \cup \{\beta\} \vdash_Q \gamma$ implies $\Delta \cup \{\alpha \vee \beta\} \vdash_Q \gamma$) hold for QCL. Further, the following laws show some of the connections between classical logic and QCL:

- The property of consistency preservation holds:
  $\Delta \vdash_Q \alpha \wedge \neg \alpha$ implies $\Delta \vdash \alpha \wedge \neg \alpha$.
  In particular,
  $\Delta \vdash_Q \alpha$ implies $\Delta \vdash \alpha$.
- The property of supraclassicality fails:
  $\Delta \vdash \alpha$ does not imply $\Delta \vdash_Q \alpha$.
  Consider Δ to be empty, then it is possible to show in classical logic $\vdash \alpha \vee \neg \alpha$ but this does not hold in QCL.

Following the last example, we introduce some classical properties which are not feasible derivations in QCL including the derivations of the counter proofs:

- The property of right modus ponens fails:
  $\Delta \vdash_Q \alpha$ and $\Delta \vdash_Q \alpha \Rightarrow \beta$ does not imply $\Delta \vdash_Q \beta$.
  Consider $\Delta = \{\alpha, \neg \alpha\}$, then $\Delta \vdash_Q \alpha$, and $\Delta \vdash_Q \alpha \Rightarrow \beta$, but $\Delta \nvdash_Q \beta$.
- The property of cut fails:
  $\Delta \cup \{a\} \vdash b$ and $\Gamma \vdash a$ does not imply $\Delta \cup \Gamma \vdash b$.
  Consider that $\{\neg \alpha\} \cup \{\alpha \vee \beta\} \vdash_Q \beta$ and $\{\alpha\} \vdash_Q \alpha \vee \beta$, but $\{\alpha, \neg \alpha\} \nvdash_Q \beta$.
- The property of deduction fails:
  $\Delta \vdash a \Rightarrow b$ does not imply $\Delta \cup \{a\} \vdash b$.
  Consider $\Delta = \{\neg \alpha\}$, then $\Delta \vdash_Q \alpha \Rightarrow \beta$ but $\Delta \cup \{\alpha\} \nvdash_Q \beta$.

More examples of classical properties failing in QCL are given in (Hunter, 2000). In addition, we give a final example of the sort of reasoning from an inconsistent set of information facilitated by QCL: Given $\Delta = \{\alpha \vee \beta, \alpha \vee \neg \beta, \neg \alpha \wedge \delta\}$ possible consequences of Δ include $\alpha \vee \beta, \alpha \vee \neg \beta, \alpha, \beta, \neg \alpha$, and δ but not $\gamma, \neg \delta$, or $\phi \vee \psi$, though $\delta \vee \gamma$ would be possible.

3.2   Proof Theory of Quasi-Classical Logic

The proof theory for QCL's propositional part has been published by (Hunter, 2000). He also shows that QCL is sound and complete with respect to its semantics. Furthermore, a characterisation of the QCL consequence relation is given, separating those classical properties that do hold in QCL (like reflexivity) from those that do not. (Hunter, 2001) extends this work to first order and introduces a proof theory for QCL based on semantic tableaux.

A semantic tableau is a tree-like structure where nodes are labeled with formulae. The idea is that each branch represents the conjunction of the formulae appearing in it and that the tree itself represents the disjunction of its branches. We refer to (Smullyan, 1968) and (Fitting, 1996) who present a thorough overview of the techniques of the semantic tableaux method.

The semantic tableau proof procedure is based on refutation, i.e. to prove X we begin with $\neg X$ and produce a contradiction. This is done by expanding $\neg X$ such that inessential details of its logical structure are removed until a contradiction appears or no expansion rule can be applied. Such expansion results in a tableau tree. For example, to prove the tautology $q \Rightarrow (p \Rightarrow q)$ we construct the following tree:

$$\neg (q \Rightarrow (p \Rightarrow q))$$
$$q, \neg (p \Rightarrow q)$$
$$q, p, \neg q$$

and observe the contradiction between q and $\neg q$. The tableau is closed and thus the tautology is proven.

However, this approach does not work directly for QCL since the truth and falsehood of a predicate are decoupled. Therefore, q being satisfiable does not mean that $\neg q$ is not satisfiable, i.e. it is not possible to construct a contradiction in the same way as above. To overcome this obstacle Hunter introduces signed formulae denoted by $*$, representing that a formula is unsatisfiable. Then showing q and q^* yields a refutation, as well as $\neg q$ and $(\neg q)^*$, because a formula cannot be satisfiable and unsatisfiable at the same time.

In the definition of the quasi-classical (QC) semantic tableau, there are two types of tableau expansion rules, the S-rules and the U-rules. All the S-rules assume the formula above the line to be satisfiable and the U-rules assume it to be unsatisfiable. Basically, the S-Rules correspond to the decompositional rules of QCL, and the U-rules are a variant of the compositional rules. Both types of expansion rules are defined in Appendix A, as well as further details, like the definitions of $\sim$ and $\otimes$.

A QC semantic tableau for a database Δ and a query α is a tree such that: (1) the formulae in $\Delta \cup \{\alpha^*\}$ are at the root of the tree; (2) each node of the tree has a set of signed formulae; and (3) the formulae at each node are generated by an application of one of the decomposition rules on a signed formula at ancestors of that node.

A QC tableau is closed if and only if every branch is closed. A branch is closed if and only if there is a formula β for which β and β^* belong to that branch. A branch is open if there are no more rules that can be applied, and it is not closed. A tableau is open if there is at least one open branch.

Hunter showed that a set of assumptions Δ implies a query α by QCL denoted $\Delta \vdash_Q \alpha$, if and only if a QC tableau for Δ and query α is closed.

Example. To show that the disjunctive syllogism holds, i.e. $\{\alpha, (\neg \alpha \vee \beta)\} \vdash_Q \beta$ we construct the following closed tableau:

$$\alpha, \neg \alpha \vee \beta, \beta^*$$

$$(\sim \neg \alpha)^* \qquad \otimes(\neg \alpha \vee \beta, \neg \alpha)$$
$$(\neg \neg \alpha)^* \qquad\qquad \beta$$
$$\alpha^*$$

3.3   Non-derivable in QCL

Classical logicians may find that there is one major drawback to QCL, namely that it is not possible to show the classical tautologies from an empty set of assumptions. For example, the tautology $q \Rightarrow (p \Rightarrow q)$ as given above cannot be verified using QCL, e.g. the following tableau is not closed:

$$(q \Rightarrow (p \Rightarrow q))^*$$
$$(\neg q \vee (p \Rightarrow q))^*$$
$$(\neg q)^*, (p \Rightarrow q)^*$$
$$(\neg q)^*, (\neg p \vee q)^*$$
$$(\neg q)^*, (\neg p)^*, q^*$$

It is not possible to construct a refutation, because an unsatisfiable formula can only be decomposed into unsatisfiable formulae, hence, no contradiction with a satisfiable formula can be derived. Therefore, no tautology can be shown from the empty set of assumptions. However, it is not clear that it is a drawback for the application of QCL in the context of formal specification, because any derivation is based on a non-empty set of assumptions. Furthermore, when trying to prove a tautology the attempt of performing the proof will indicate a set of necessary assumptions. For example, to close the above tableau, we would need either q, $\neg q$ or $\neg p$ in the set of assumptions. In particular, the formula $q \vee \neg q$ is a good assumption to state, because the truth of q does certainly not influence the tautology.

In the future, Hunter's work will have to be further extended to incorporate equality theory to be a truly alternative logic for Z. This can be done following (Fitting, 1996) or (Beckert, 1997). This is, however, a matter of further research and not the focus of the main issue we discuss here.

4   Reasoning in the Presence of Inconsistencies

The aim of this section is to demonstrate the use of quasi-classical logic to help us analyse possibly inconsistent Z specifications. QCL allows fewer properties to be derived from inconsistent specifications, in return for an increase in "usefulness" of the properties that can be derived. We illustrate this with the derivation of some properties from the example given in Section 2. We believe that those properties are intuitively valid while others should not hold in any case, not even due to inconsistencies. This approach will enable the specifier to validate the specification in more depth without being forced to remove inconsistencies immediately.

A schema in Z consists of a declaration part and a predicate. We assume the predicate to be type correct, i.e. it conforms to the declaration. We reason using QCL in a similar way to reasoning with the usual Z logic. For example, derivations using QCL include those of the form $Schema \vdash_Q p$ which has the same meaning as $Schema \vdash p$, except that QCL has been used in the derivation of the predicate p. (See (Woodcock and Davies, 1996) for a description of the formal meaning of $Schema \vdash p$.)

In our example, the operation *Change* conflicts with the schema *Law*. Despite this, we can show that after changing the tyre four wheels are connected to the car.

$$Change \vdash_Q wheels' = 4$$

This follows directly from the root of the tableau (omitting the type definitions):

$$\neg\, flat \Rightarrow drive,\, wheels = 4,\, \neg\, flat' \Rightarrow drive',\, wheels' = 4,\, same,$$
$$same',\, flat \wedge \neg\, flat',\, \neg\, same',\, x! = wheels,\, (wheels' = 4)^*$$

However, we cannot derive anymore that:

$$Change \vdash_Q wheels' = 3$$

or that there are any other number of wheels apart from four. Therefore we know that there will be exactly four wheels on our car after changing one tyre.

Similarly, it follows from the root of the tableau that no tyre is flat after applying the operation *Change*:

$$Change \vdash_Q \neg\, flat'$$

Furthermore, the tyres are not flat and, hence, it is possible to drive the car:

$$Change \vdash_Q drive'$$

The proof of this follows from the tree below, where we simplify the tree by removing unnecessary detail from the root of the tree (this simplification, however, does not affect the proof itself).

$$same',\, \neg\, same',\, \neg\, flat' \Rightarrow drive',\, flat \wedge \neg\, flat',\, (drive')^*$$
$$|$$
$$flat,\, \neg\, flat'$$
$$|$$
$$\neg\,\neg\, flat' \vee drive'$$
$$|$$
$$flat' \vee drive'$$

$$(\sim flat')^* \qquad \otimes(flat' \vee drive',\, flat')$$
$$|\qquad\qquad\qquad |$$
$$(\neg\, flat')^* \qquad\qquad drive'$$

This tree is closed and the proof is, therefore, complete. Observe that the contradiction of *same'* could not contribute to the proof.

We also introduced the operations *Refuel* and *Choose* which contradicted each other in the choice of petrol for a petrol car. However, we can still infer that the car will be full with petrol, i.e. $amount'(fuel') = 60$ at the end of the *Refuel* operation, or it will be *charged* if it is an electric car:

$$Refuel \vdash_Q amount'(fuel') = 60 \vee charged'$$

The QCL tableau to show this is:

$$car? = petrol \Rightarrow fuel' = unleaded,$$
$$(car? = electric \wedge charged') \vee$$
$$(car? = petrol \wedge amount(fuel') = 60 \wedge fuel' = four\_star),$$
$$(amount'(fuel') = 60 \vee charged')^*$$

$$(car? = electric \wedge charged') \vee car? = petrol,$$
$$(car? = electric \wedge charged') \vee (amount'(fuel') = 60 \wedge fuel' = four\_star)$$

$$(car? = electric \wedge charged') \vee amount'(fuel') = 60,$$
$$(car? = electric \wedge charged') \vee fuel' = four\_star$$

$$car? = electric \vee amount'(fuel') = 60,$$
$$charged' \vee amount'(fuel') = 60$$

$charged'$	$amount'(fuel') = 60$
$(amount'(fuel') = 60)^*, (charged')^*$	$(amount'(fuel') = 60)^*, (charged')^*$

Quasi-classical logic allows one to derive non-trivial conclusions from inconsistent information. We demonstrated the application of QCL by analysing an inconsistent Z specification in terms of derivations of properties we considered important to verify.

On the one hand, quasi-classical inferences are a subset of those possible by classical logic. Therefore, using QCL as an underlying logic for Z will not introduce undesired logical consequences. On the other hand, QCL cannot deal with classical tautologies from the empty set of assumptions. Otherwise, QCL yields the same consequences as classical reasoning in the context of consistent specifications. This means that the specifier has to make all assumptions explicit, a task anyway enforced by formal specification development. In further research we will investigate whether QCL's consequence relation can replace the classical consequence relation in Z.

5   Preconditions in the Presence of Inconsistencies

The precondition of an operation is the predicate that has to be fulfilled to apply the operation successfully. In Z specifications such preconditions are implicit, i.e. they have to be calculated from the operation. The calculation leads to two situations: either the operation can be applied successfully or not. However, allowing for inconsistencies in specifications we will distinguish three possibilities:

(1) the operation can be applied consistently, i.e. it is defined,
(2) the operation can be applied but is inconsistent, i.e. it is overdefined,
(3) the operation cannot be applied, i.e. it is undefined.

This distinction breaks the situation in Z that an inconsistently specified operation is, in terms of the results of calculating the precondition, the same as an operation that has not been specified. We wish to explore whether this division is more useful when analysing specifications and whether we can use it to support further development (like refinement) without resolving inconsistencies immediately.

For example, the precondition of the *Refuel* operation introduced in Section 2 can be divided into the following three categories: (1) $type(car?) = electric$, (2) $type(car?) = petrol$, and (3) any other situation, which amounts to $type(car?) = diesel$. Observe that the normal precondition *pre* covers (1), so $\neg$ *pre* is the combination of (2) and (3). The problem we investigate here is how to calculate the regions (2), and consequently (3).

Actually, we concentrate on a slight modification of the above problem. We calculate the combination of the defined and overdefined region. Both approaches presented below use the known definition of the precondition, i.e.

$$\text{pre } Op = \exists \, State', outs! \bullet Op$$

but apply a different set of simplification rules to this abstract formula. To obtain the three precondition regions we proceed the following way: Use standard logic to determine the defined region ($\text{pre}_d \, Op$). Use the classical precondition definition but apply a restricted set of simplification rules to determine the combination of the defined and overdefined region, i.e. $\text{pre}_{comb} \, Op = \text{pre}_d \, Op \vee \text{pre}_{od} \, Op$ and construct the intersection of both to obtain only the overdefined region. The undefined area is then the complement of the combined region, i.e. $\neg \, \text{pre}_{comb} \, Op$.

5.1   The Two-Point Rule

One possible way to determine the alternative precondition regions is to use the standard Z logic, but to restrict the inference of false when the operation defines inconsistent after states, i.e. the contradiction law will not be applied to after states. The reasoning process is then continued by distributing the inconsistency. The aim is to determine where the operation was intended to work, rather than where it would be applicable classically.

In standard Z logic, one of the fundamental rules to simplify the precondition of an operation is the One-Point-Rule (Woodcock and Davies, 1996, p.48):

$$\frac{\exists\, x : S \bullet (p(x) \wedge x = t)}{t \in S \wedge p(t)} \quad [\; \begin{array}{l} \text{One-Point Rule (OPR),} \\ x \ \text{not free in } t \end{array} \;]$$

However, when applied to an inconsistent predicate it always results in *false*. For example:

$\qquad$ pre *Refuel*

$\equiv \{$Definition of pre + Schema Expansion$\}$

$\qquad \exists\, charged', amount', fuel' \bullet$

$\qquad car? = petrol \Rightarrow fuel' = unleaded \wedge ((car? = electric \wedge charged') \vee$

$\qquad\qquad (car? = petrol \wedge amount'(fuel') = 60 \wedge fuel' = four\_star))$

$\equiv \{\exists\text{-Distribution}\}$

$\qquad \exists\, amount', fuel' \bullet car? = petrol \Rightarrow fuel' = unleaded \wedge$

$\qquad\qquad ((\exists\, charged' \bullet (car? = electric \wedge charged')) \vee$

$\qquad\qquad (car? = petrol \wedge amount'(fuel') = 60 \wedge fuel' = four\_star))$

$\equiv \{$OPR on *charged'* + Rewrite$\}$

$\qquad car? = electric \wedge \exists\, fuel' \bullet (\neg\, car? = petrol \vee fuel' = unleaded) \vee$

$\qquad\qquad \exists\, amount', fuel' \bullet ((\neg\, car? = petrol \vee fuel' = unleaded) \wedge$

$\qquad\qquad\qquad car? = petrol \wedge amount'(fuel') = 60 \wedge fuel' = four\_star)$

$\equiv \{$OPR on *fuel'* twice$\}$

$\qquad (car? = electric \wedge \neg\, car? = petrol) \vee$

$\qquad\qquad \exists\, amount' \bullet (\neg\, car? = petrol \vee four\_star = unleaded) \wedge$

$\qquad\qquad\qquad\qquad car? = petrol \wedge amount'(four\_star) = 60$

$\equiv \{$Contradiction + Domain Knowledge$\}$

$\qquad (car? = electric \wedge (car? = diesel \vee car? = electric))$

$\equiv \{$Absorption$\}$

$\qquad car? = electric$

In this derivation, the information that the postconditions of *Refuel* and *Choose* are contradictory for $car? = petrol$ is lost. For this reason we introduce a variant of the OPR that preserves this information and allows one to continue to reason despite the presence of inconsistencies.

Looking at the forward direction of the OPR we actually want to distribute substitution through the predicate p. Therefore, we require:

$$\frac{\exists\, x : S \bullet (p(x) \wedge x = t_1 \wedge x = t_2)}{t_1 \in S \wedge t_2 \in S \wedge p(t_1) \wedge p(t_2)} \quad [\; \begin{array}{l} \text{Two-Point Rule (2PR),} \\ x \ \text{not free in } t_1 \ \text{or } t_2 \end{array} \;]$$

Applying the 2PR can result in a consistent predicate even though it was inconsistent before, i.e. $p(t_1) \wedge p(t_2)$ is satisfiable though $t_1 = t_2$ is not. For example, given $p(x) = x \leq 5$, $t_1 = 3$, and $t_2 = 4$ then $3 = 4$ is not satisfiable, but $3 \leq 5 \wedge 4 \leq 5$ is.

Applying the 2PR to our example above we reason:

$$\vdots$$

$$car? = electric \wedge \exists\, fuel' \bullet (\neg\, car? = petrol \vee fuel' = unleaded) \vee$$
$$\exists\, amount', fuel' \bullet ((\neg\, car? = petrol \vee fuel' = unleaded) \wedge$$
$$car? = petrol \wedge amount'(fuel') = 60 \wedge fuel' = four\_star)$$
$$\equiv \{\text{OPR on } fuel' \text{ once} + \text{Rewrite}\}$$
$$(car? = electric \wedge \neg\, car? = petrol) \vee \exists\, amount', fuel' \bullet$$
$$(\neg\, car? = petrol \wedge car? = petrol \wedge$$
$$amount'(fuel') = 60 \wedge fuel' = four\_star) \vee$$
$$(fuel' = unleaded \wedge car? = petrol \wedge$$
$$amount'(fuel') = 60 \wedge fuel' = four\_star)$$
$$\equiv \{\text{Contradiction}\}$$
$$(car? = electric \wedge \neg\, car? = petrol) \vee \exists\, amount', fuel' \bullet car? = petrol \wedge$$
$$fuel' = unleaded \wedge amount'(fuel') = 60 \wedge fuel' = four\_star$$
$$\Rightarrow \{\text{2PR on } fuel'\}$$
$$(car? = electric \wedge \neg\, car? = petrol) \vee \exists\, amount' \bullet$$
$$car? = petrol \wedge amount'(unleaded) = 60 \wedge amount'(four\_star) = 60$$
$$\equiv \{\text{OPR on } amount'\}$$
$$(car? = electric \wedge \neg\, car? = petrol) \vee car? = petrol$$
$$\equiv \{\text{Rewrite} + \text{Contradiction}\}$$
$$car? = electric \vee car? = petrol$$

Informally, we are using the 2PR as follows. If t_1 and t_2 are equivalent in 2PR, this results in the forward direction of the OPR. Otherwise, we have an inconsistent situation, because x cannot take more than one value. In such cases, we split the predicate p into two and substitute t_1 in one and t_2 in the other occurrence of p. This might distribute inconsistency to the several instances of p, but it may also remove inconsistency.

As with the One-Point rule, the 2PR is applied to remove the existential quantification from a predicate. In the case of a consistent predicate, we preserve information by applying the One-Point rule, which is indeed an equivalence operation. The Two-Point rule is applied when we have an inconsistency, i.e. "too much information". In this situation, we are interested in reducing the amount of information, possibly removing the inconsistency, and thus the 2PR is applied only in one direction. Using the 2PR in the reverse direction would introduce information, which would be inappropriate. For example, consider the predicate $3 \leq 5 \wedge 4 \leq 5$, we do not infer $\exists x \bullet x \leq 5 \wedge x = 3 \wedge x = 4$ because this introduces an inconsistency.

In the case where the after state of an operation is functionally determined by the before state, the use of the 2PR enables us to determine where the operation was intended to be applied, even though the actual definition may be inconsistent. For example, the defined region of *Refuel*, pre_d *Refuel*, has the predicate $car? = electric$, and the combination of the defined and overdefined region, pre_{comb} *Refuel*, has the predicate $car? = electric \vee car? = petrol$. The overdefined region, pre_{od} *Refuel*, is, therefore, given by the predicate $car? = petrol$.

5.2   Using Quasi-Classical Logic's Equivalences

In the previous subsection we considered the use of classical logic with restrictions to support precondition calculation. Quasi-classical logic has already been successfully applied in Section 4 to reason in the presence of inconsistency. Therefore, it seems natural to ask how QCL can be used to simplify the precondition of an inconsistent operation. In Appendix B we state some laws of quasi-classical logic, mainly equivalence laws, which will be applied subsequently.

Example. Again, we calculate the precondition of the operation *Refuel*, this time using the QCL equivalences. We abbreviate the declared names by their first letter, with the exception of *four_star* being denoted 4^*. The derivation then proceeds as follows:

$$\exists\, c', a', f' \bullet c? = p \Rightarrow f' = u \wedge$$
$$((c? = e \wedge c') \vee (c? = p \wedge a'(f') = 60 \wedge f' = 4^*))$$
$$\equiv_Q \{\exists\text{-Distribution}\}$$
$$\exists\, a', f' \bullet (c? = p \Rightarrow f' = u) \wedge$$
$$((\exists\, c' \bullet c? = e \wedge c') \vee (c? = p \wedge a'(f') = 60 \wedge f' = 4^*))$$
$$\equiv_Q \{\text{Idempotency of } c', \text{OPR on } c', \text{One Law for Equality}\}$$
$$\exists\, a', f' \bullet (c? = p \Rightarrow f' = u) \wedge$$
$$(c? = e \vee (c? = p \wedge a'(f') = 60 \wedge f' = 4^*))$$
$$\equiv_Q \{\wedge\text{-Distribution}, \exists\text{-Distribution}\}$$
$$(\exists\, f' \bullet (c? = p \Rightarrow f' = u) \wedge c? = e) \vee$$
$$(\exists\, a', f' \bullet (c? = p \Rightarrow f' = u) \wedge (c? = p \wedge a'(f') = 60 \wedge f' = 4^*))$$
$$\equiv_Q \{\wedge\text{-Distribution}, \exists\text{-Distribution}\}$$
$$(c? = e \wedge \neg\, c? = p) \vee (\exists\, f' \bullet c? = e \wedge f' = u) \vee$$
$$(\exists\, a', f' \bullet (c? = p \Rightarrow f' = u) \wedge (c? = p \wedge a'(f') = 60 \wedge f' = 4^*))$$
$$\equiv_Q \{\text{Idempotency of } f', \text{OPR on } f', \text{One Law for Equality}\}$$
$$(c? = e \wedge \neg\, c? = p) \vee c? = e \vee$$
$$(\exists\, a', f' \bullet (c? = p \Rightarrow f' = u) \wedge (c? = p \wedge a'(f') = 60 \wedge f' = 4^*))$$
$$\equiv_Q \{\text{OPR on } f', \text{One Law for Equality}\}$$
$$(c? = e \wedge \neg\, c? = p) \vee c? = e \vee$$

$$(\exists\, a' \bullet ((c? = p \Rightarrow 4^* = u) \wedge c? = p \wedge a'(4^*) = 60)$$
$$\equiv_Q \{\text{OPR on } a', \text{ One Law for Equality}\}$$
$$(c? = e \wedge \neg\, c? = p) \vee c? = e \vee ((c? = p \Rightarrow 4^* = u) \wedge c? = p)$$
$$\equiv_Q \{\text{Absorption, Implication Law}\}$$
$$c? = e \vee (c? = p \wedge 4^* = u)$$

i.e.

$$\text{pre}_{comb}\ \textit{Refuel} =_Q (\textit{car?} = \textit{electric} \vee (\textit{car?} = \textit{petrol} \wedge \textit{four\_star} = \textit{unleaded})).$$

We interpret this result as follows. The operation *Refuel* is applicable if the given car is an electric car, or it is a petrol car but four star and unleaded are the same. Note, this calculation used laws of equality which have not been incorporated into QCL yet. We therefore emphasise that the final form of this proof will depend on the exact shape of a modified logic QCL$_=$, i.e. quasi-classical logic with equality.

We separate the defined area, i.e. pre$_d$ *Refuel*, from the result of the above derivation and determine the overdefined area:

$$\text{pre}_{od}\ \textit{Refuel} =_Q (\textit{car?} = \textit{petrol} \wedge \textit{four\_star} = \textit{unleaded}).$$

The overdefined area derived using quasi-classical equivalences is different to the one derived using the Two-Point rule. It provides more information on the source of inconsistency, e.g. that unleaded should be the same as four star.

6   Conclusion and Future Work

6.1   Conclusion

The aim of this work was to discuss how to reason in the presence of inconsistent Z schemas and still to be able to derive useful information from them. We introduced quasi-classical logic to support the process of non-trivial reasoning despite inconsistencies. We replaced the classical proof system by QCL's version and derived properties from our example specification. We also indicated that certain undesired inferences are not possible anymore and, therefore, demonstrated the usefulness of this approach. However, we only considered local inconsistencies, although this method can be extended to deal with global inconsistency as well.

We also decided to split the precondition of an operation into three areas: the defined, overdefined and undefined one. This enhanced separation of the precondition allows the analyst to investigate where an operation has been overdefined more clearly. We assume that this distinction will be beneficial in a theory of refinement to be developed later. We introduced two possible ways to determine these regions. However, using either way resulted in a combination of the defined and overdefined regions. This is different to the classical approach which does not distinguish between the overdefined and undefined area. Combining classical precondition calculation with our results enabled us to separate all three regions.

Conclusion drawn from inconsistent specifications using QCL are useful in the sense that they only represent information present in the specification rather than information introduced due to triviality. It is clearly understood that inconsistencies need to be removed at some stage of the development process but it may well be beneficial to delay such step. QCL itself is very similar to classical logic apart from its treatment of inconsistencies. Classical consistent specifications can be analysed using QCL almost without any changes. However, inconsistent specifications need a more subtle treatment due to the failure of several inference rules in the presence of inconsistency. In particular, to represent *false* the specifier cannot use a contradiction anymore.

6.2   Related Work

There are three categories of related work we will discuss: previous publications on handling inconsistency in Z, the problems of managing inconsistency in other specification languages, and work on other paraconsistent logics to handle inconsistencies in a more general setting.

Present literature on handling inconsistency in Z is largely concerned with preventing and eliminating inconsistencies, rather than managing them. For example, (Valentine, 1998) presents "sufficient conditions and stylistic guidelines for achieving [consistency]" and proofs for the success of this approach. This work is based on the assumption that "consistency is essential for a Z specification to have any useful meaning". This is correct in the standard approach to Z. However, considering non-standard interpretations, this is not necessarily the case anymore. Our work shows that it is possible to derive useful conclusions from inconsistent Z specifications.

(Arthan, 1991) describes work in progress on a high integrity proof tool for Z specifications. One concern is the inconsistency of sets of axiomatic and generic Z schemas, i.e. global inconsistency. He proposes a rule of *conservative extensions* allowing new objects to be defined only if an appropriate consistency condition has been proved. If not, the new objects will be redefined such that no inconsistency occurs. The approach of (Arthan, 1991) could be adapted to local inconsistent schemas by weakening their schema property to *true* so that no conclusions can be drawn from a locally inconsistent predicate. Although this avoids the problem of triviality, it comes at a loss of information, whereas our approach is based on the information present in a specification.

The work by (Saaltink, 1997) is concerned with the analysis of Z specifications using the automated theorem prover Eves. Consistency checking is one possible analysis that can be performed. Further, preconditions calculated using Z/Eves remove all information based on inconsistencies. This work is related to finding inconsistencies, rather than inconsistency management.

In the introduction to this paper we presented links to the current work on managing inconsistencies in the software development process in general terms. However, it seems impossible to find specific work on inconsistency management for any particular specification language, like B (Abrial, 1996), VDM, LOTOS, CSP, or others. All formal approaches on living with inconsistencies deal with the use of (non-standard) predicate logic as specification language.

Work on developing paraconsistent logics is also relevant. We refer to (Hunter, 1998) and (Batens et al., 2000) for a brief overview. Batens' inconsistency-adaptive logics (Batens, 1999; Batens, 2000) are a set of paraconsistent logics that aim at handling consistent theories exactly like classical logic but to be adaptive to inconsistency and not to infer everything from it. Its dynamic proof theory, however, is not as close to common logical reasoning as the proof theory of QCL. Further, work on applying paraconsistent logics to mathematics (Mortensen, 1995; da Costa, 2000) may be of value to the work on inconsistency management in Z because of Z's foundation in set theory.

6.3   Future Work

One major motivation for this work is the belief in a theory that allows continued development of specifications despite the presence of inconsistencies. Refinement is one of the processes of specification development from an abstract form to a more concrete representation. Refinement is also the process of adding information. This can, however, lead to the introduction of inconsistencies. The idea behind the alternative precondition regions is to support refinement in the presence of overdefinedness. Current investigations suggest that a combination of classical and quasi-classical refinement rules can support detection and controlled removal of inconsistencies. However, this relation requires further investigation.

A theory of refinement in the presence of inconsistency will then contribute to work on viewpoint specifications (Boiten et al., 1999), where the unification of two or more viewpoints is defined as the least common refinement of the viewpoints. So far, the verification of this property also contains a consistency check between the viewpoint specifications. However, this forces removal of the inconsistency to unify the viewpoints. Our aim is to support viewpoint unification and the analysis of the resulting specification without necessarily removing the inconsistency.

In Section 3 we mentioned that we need to enhance quasi-classical logic to be a true contender for an alternative paraconsistent logic for Z. An essential extension to QCL is the incorporation of a theory of equality.

In connection to our previous work on un(der)definedness in Z (Miarka et al., 2000), it seems worthwhile to investigate further the duality of un(der)definedness and overdefinedness, i.e. inconsistency. Also, we identified in that piece of work that inconsistency issues can arise between the schema components. It will be interesting to see whether our current work can be beneficial to our previous work and whether both can be combined.

Though the work presented here is mainly concerned with local inconsistencies, we also mentioned the problem of global inconsistencies. It will be interesting to investigate the application of paraconsistent logics, like QCL, to develop a schema calculus which is more robust in the presence of inconsistencies.

Acknowledgement. We like to thank Anthony Hunter for the discussions about QCL. Further, we acknowledge all the anonymous referees for their corrections and helpful comments to improve this work.

References

Abrial, J.-R. (1996). *The B-Book: Assigning Programs to Meanings.* Cambridge University Press.

Arthan, R. D. (1991). Formal Specification of a Proof Tool. In Prehn, S. and Toetenel, H., editors, *Proceedings of Formal Software Development Methods (VDM '91)*, Lecture Notes in Computer Science 552, pages 356–370, Berlin, Germany. Springer.

Balzer, R. (1991). Tolerating Inconsistency. In *Proceedings of the 13th International Conference on Software Engineering*, pages 158–165. IEEE Computer Society Press / ACM Press.

Batens, D. (1999). Inconsistency-Adaptive Logics. In Orlowska, E., editor, *Logic at Work. Essays Dedicated to the Memory of Helena Rasiowa*, Studies in fuzziness and soft computing, Volume 24, pages 445–472, Heidelberg, New York. Physica-Verlag.

Batens, D. (2000). A Survey of Inconsistency-Adaptive Logics. In (Batens et al., 2000), pages 49–73.

Batens, D., Mortensen, C., Priest, G., and Bendegem, J.-P. V., editors (2000). *Frontiers of Paraconsistent Logic.* Research Studies Press Ltd., Baldock, Hertfordshire, England.

Beckert, B. (1997). Semantic Tableaux with Equality. *Journal of Logic and Computation*, 7(1):39–58.

Besnard, P. and Hunter, A. (1995). Quasi-Classical Logic: Non-Trivializable Classical Reasoning from Inconsistent Information. In Froidevaux, C. and Kohlas, J., editors, *Proceedings of the ECSQARU European Conference on Symbolic and Quantitive Approaches to Reasoning and Uncertainty*, Lecture Notes in Artificial Intelligence 946, pages 44–51, Berlin. Springer Verlag.

Boiten, E. A., Derrick, J., Bowman, H., and Steen, M. W. A. (1999). Constructive consistency checking for partial specification in Z. *Science of Computer Programming*, 35(1):29–75.

da Costa, N. C. (2000). Paraconsistent Mathematics. In (Batens et al., 2000), pages 165–179.

Finkelstein, A., Gabbay, D., Hunter, A., Kramer, J., and Nuseibeh, B. (1994). Inconsistency Handling in Multi-Perspective Specifications. *IEEE Transactions on Software Engineering*, 20(8):569–578.

Fitting, M. C. (1996). *First-Order Logic and Automated Theorem Proving.* Graduate Texts in Computer Science. Springer-Verlag, Berlin, 2nd edition.

Ghezzi, C. and Nuseibeh, B. (1998). Guest Editorial: Introduction to the Special Section: Managing Inconsistency in Software Development. *IEEE Transactions on Software Engineering*, 24(11):906–907.

Hunter, A. (1998). Paraconsistent Logic. In Besnard, P. and Hunter, A., editors, *Reasoning with Actual and Potential Contradictions*, Volume II of Handbook of Defeasible Reasoning and Uncertain Information (Gabbay, D. and Smets, Ph., editors), pages 13–44. Kluwer Academic Publishers, Dortrecht, The Netherlands.

Hunter, A. (2000). Reasoning with Contradictory Information Using Quasi-Classical Logic. *Journal of Logic and Computation*, 10(5):677–703.

Hunter, A. (2001). A Semantic Tableau Version of First-Order Quasi-Classical Logic. In Benferhat, S. and Besnard, P., editors, *Symbolic and Quantitative Approaches to Reasoning with Uncertainty, Proceedings of the 6th European Conference, EC-SQARU 2001, Toulouse, France*, Lecture Notes in Artificial Intelligence 2143, pages 544–555. Springer Verlag.

Hunter, A. and Nuseibeh, B. (1997). Analysing Inconsistent Specifications. In *Proceedings of the 3rd International Symposium on Requirements Engineering (RE'97)*, pages 78–86. Annapolis, USA, IEEE Computer Society Press.

Hunter, A. and Nuseibeh, B. (1998). Managing Inconsistent Specifications: Reasoning, Analysis, and Action. *ACM Transactions on Software Engineering and Methodology*, 7(4):335–367.

Miarka, R., Boiten, E., and Derrick, J. (2000). Guards, Preconditions, and Refinement in Z. In Bowen, J. P., Dunne, S., Galloway, A., and King, S., editors, *ZB2000: Formal Specification and Development in Z and B / First International Conference of B and Z Users*, Lecture Notes in Computer Science 1878, pages 286–303, Berlin Heidelberg New York. Springer-Verlag Berlin.

Mortensen, C. (1995). *Inconsistent Mathematics*. Kluwer Academic Publishers Group, Dordrecht, The Netherlands.

Rescher, N. and Manor, R. (1970). On Inference from Inconsistent Premises. *Theory and Decision*, 1:179–217.

Saaltink, M. (1997). The Z/EVES User's Guide. Technical Report TR-97-5493-06, ORA Canada, 267 Richmond Road, Suite 100, Ottawa, Canada.

Schwanke, R. W. and Kaiser, G. E. (1988). Living with Inconsistency in Large Systems. In *Proceedings of the International Workshop on Software Version and Configuration Control*, pages 98–118, Grassau, Germany.

Smullyan, R. M. (1968). *First-Order Logic*, Volume 43 of *Ergebnisse der Mathematik und ihrer Grenzgebiete*. Springer-Verlag, New York.

Valentine, S. H. (1998). Inconsistency and Undefinedness in Z – A Practical Guide. In Bowen, J. P., Fett, A., and Hinchey, M. G., editors, *ZUM'98: The Formal Specification Notation*, Lecture Notes in Computer Science 1493, pages 233–249, Berlin Heidelberg New York. Springer Verlag.

Woodcock, J. and Davies, J. (1996). *Using Z - Specification, Refinement, and Proof*. Prentice Hall International Series in Computer Science. Prentice Hall Europe. Online: http://softeng.comlab.ox.ac.uk/usingz/ (last access 18/10/2001).

Appendix A: The QCL Expansion Rules

In this appendix we introduce the tableaux expansion rules for quasi-classical logic (see Section 3). The rules are divided into S-rules and U-rules. The S-rules consider a formula above the line as satisfiable, whereas the U-rules consider a formula above the line as unsatisfiable. The presentation of these rules is preceded by the definition of necessary terminology.

Given a language $\mathcal{L}$, the set of tableaux of all formula over $\mathcal{L}$ is denoted $\mathcal{T}(\mathcal{L})$. Let α be an atom and $\sim$ a complementation operation such that $\sim \alpha$ is $\neg\,\alpha$ and $\sim (\neg\,\alpha)$ is α. The abbreviation $\otimes(\alpha_1 \vee \ldots \vee \alpha_n, \alpha_i)$ is defined as the clause obtained by removing α_i from $\alpha_1 \vee \ldots \vee \alpha_n$.

The following are the S-rules for QC semantic tableaux, where t is in $\mathcal{T}(\mathcal{L})$ and t' is in $\mathcal{T}(\mathcal{L})$ but not occurring in the tableau constructed so far. The $|$ symbol denotes the introduction of a branch point in the QC semantic tableau.

The disjunction S-rules:

$$\frac{\alpha_1 \vee \ldots \vee \alpha_n}{(\sim \alpha_i)^* \mid \otimes(\alpha_1 \vee \ldots \vee \alpha_n, \alpha_i)} \quad [\text{where } \alpha_1, \ldots, \alpha_n \text{ are literals}]$$

$$\frac{\alpha_1 \vee \ldots \vee \alpha_n}{\alpha_1 \mid \ldots \mid \alpha_n} \quad [\text{where } \alpha_1, \ldots, \alpha_n \text{ are literals}]$$

The rewrite S-rules:

$$\frac{\alpha \wedge \beta}{\alpha, \beta} \qquad \frac{\neg\,\neg\,\alpha \vee \gamma}{\alpha \vee \gamma} \qquad \frac{\neg\,(\alpha \wedge \beta) \vee \gamma}{\neg\,\alpha \vee \neg\,\beta \vee \gamma}$$

$$\frac{\neg\,(\alpha \vee \beta) \vee \gamma}{(\neg\,\alpha \wedge \neg\,\beta) \vee \gamma} \qquad \frac{\alpha \vee (\beta \wedge \gamma)}{(\alpha \vee \beta) \wedge (\alpha \vee \gamma)} \qquad \frac{\alpha \wedge (\beta \vee \gamma)}{(\alpha \wedge \beta) \vee (\alpha \wedge \gamma)}$$

The quantification S-rules:

$$\frac{(\forall X \bullet \alpha(X)) \vee \gamma}{\alpha(t) \vee \gamma} \qquad \frac{(\neg\,\exists X \bullet \alpha(X)) \vee \gamma}{\neg\,\alpha(t) \vee \gamma}$$

$$\frac{(\exists X \bullet \alpha(X)) \vee \gamma}{\alpha(t') \vee \gamma} \qquad \frac{(\neg\,\forall X \bullet \alpha(X)) \vee \gamma}{\neg\,\alpha(t') \vee \gamma}$$

The first disjunction S-rule links to the following U-rules for the QC semantic tableaux, where t is in $\mathcal{T}(\mathcal{L})$ and t' is in $\mathcal{T}(\mathcal{L})$ but not occurring in the tableau constructed so far. The $|$ symbol denotes the introduction of a branch point in the QC semantic tableau.

The disjunction U-rule: $\quad \dfrac{(\alpha \vee \beta)^*}{\alpha^*, \beta^*}$

The conjunction U-rule: $\quad \dfrac{(\alpha \wedge \beta)^*}{\alpha^* \mid \beta^*}$

The rewrite U-rules:

$$\frac{(\neg\,\neg\,\alpha \vee \gamma)^*}{(\alpha \vee \gamma)^*} \qquad \frac{(\neg\,(\alpha \wedge \beta) \vee \gamma)^*}{(\neg\,\alpha \vee \neg\,\beta \vee \gamma)^*} \qquad \frac{(\neg\,(\alpha \vee \beta) \vee \gamma)^*}{((\neg\,\alpha \wedge \neg\,\beta) \vee \gamma)^*}$$

The quantification U-rules:

$$\frac{((\forall X \bullet \alpha(X)) \vee \gamma)^*}{(\alpha(t') \vee \gamma)^*} \qquad \frac{((\neg\,\exists X \bullet \alpha(X)) \vee \gamma)^*}{(\neg\,\alpha(t') \vee \gamma)^*}$$

$$\frac{((\exists X \bullet \alpha(X)) \vee \gamma)^*}{(\alpha(t) \vee \gamma)^*} \qquad \frac{((\neg\,\forall X \bullet \alpha(X)) \vee \gamma)^*}{(\neg\,\alpha(t) \vee \gamma)^*}$$

Please note that γ can be "empty". Furthermore, we will add the following two rewrite rules: the S-rule for implication, and the U-rule for implication:

$$\frac{(\alpha \Rightarrow \beta) \vee \gamma}{(\neg\, \alpha \vee \beta) \vee \gamma} \qquad \frac{((\alpha \Rightarrow \beta) \vee \gamma)^*}{((\neg\, \alpha \vee \beta) \vee \gamma)^*}$$

Rules for equality: $(x = x)^*$ closes a branch

$$\frac{\begin{array}{c} x = y \\ \alpha(x) \end{array}}{\alpha(y)} \qquad \frac{\begin{array}{c} x = y \\ \alpha(y) \end{array}}{\alpha(x)} \qquad \frac{(x = y \wedge \alpha(x))^*}{(\alpha(y))^*} \qquad \frac{(x = y \wedge \alpha(y))^*}{(\alpha(x))^*}$$

Appendix B: Quasi-Classical Laws

Here we present a list of laws which are valid in QCL. We use the notation $a \dashv\vdash_Q b$ to denote that $\{a\} \vdash_Q b$ and $\{b\} \vdash_Q a$. These equivalences can be proved using the tableaux rules from Appendix A.

Commutativity:

$a \vee b \dashv\vdash_Q b \vee a$

$a \wedge b \dashv\vdash_Q b \wedge a$

De Morgan Laws:

$\neg\,(a \wedge b) \dashv\vdash_Q (\neg\, a) \vee (\neg\, b)$

$\neg\,(a \vee b) \dashv\vdash_Q (\neg\, a) \wedge (\neg\, b)$

Idempotent Laws:

$a \vee a \dashv\vdash_Q a$

$a \wedge a \dashv\vdash_Q a$

Double Negation Law:

$a \dashv\vdash_Q \neg\,\neg\, a$

One Law for Equality:

$a \wedge x = x \dashv\vdash_Q a$

One-Point Rule:

$(\exists x \bullet p(x) \wedge x = t) \dashv\vdash_Q p(t)$

Some Quantification Laws:

$\exists x \bullet (a(x) \vee b(x)) \dashv\vdash_Q (\exists x \bullet a(x)) \vee (\exists x \bullet b(x))$

$\exists x \bullet (a(x) \wedge b) \dashv\vdash_Q (\exists x \bullet a(x)) \wedge b$

Associativity:

$(a \vee b) \vee c \dashv\vdash_Q a \vee (b \vee c)$

$(a \wedge b) \wedge c \dashv\vdash_Q a \wedge (b \wedge c)$

Distributivity:

$a \vee (b \wedge c) \dashv\vdash_Q (a \vee b) \wedge (a \vee c)$

$a \wedge (b \vee c) \dashv\vdash_Q (a \wedge b) \vee (a \wedge c)$

Absorption Laws:

$a \vee (a \wedge b) \dashv\vdash_Q a$

$a \wedge (a \vee b) \dashv\vdash_Q a$

Implication Laws:

$(a \Rightarrow b) \wedge (a \Rightarrow c) \dashv\vdash_Q a \Rightarrow (b \wedge c)$

$(a \Rightarrow b) \vee (a \Rightarrow c) \dashv\vdash_Q a \Rightarrow (b \wedge c)$

$(b \Rightarrow a) \vee (c \Rightarrow a) \dashv\vdash_Q (b \wedge c) \Rightarrow a$

$(b \Rightarrow a) \wedge (c \Rightarrow a) \dashv\vdash_Q (b \vee c) \Rightarrow a$

$a \wedge (a \Rightarrow b) \dashv\vdash_Q a \wedge b$

Loose Specification and Refinement in Z

Eerke Boiten

Computing Laboratory, University of Kent at Canterbury
Canterbury Kent, CT2 7NF, UK
Phone: +44-1227-769184, Fax: +44-1227-762811
E.A.Boiten@ukc.ac.uk
http://www.cs.ukc.ac.uk/people/staff/eab2/

Abstract. Z is a specification notation with a model-based semantics, but in contrast to most such languages, its normal refinement relation is not defined in terms of model containment. This paper investigates this phenomenon, leading to a variety of observations concerning the relation between Z semantics and refinement.

Keywords: Z, refinement, semantics, model containment, states-and-operations, loose specification.

1 Introduction

For many specification languages, there is a tight link between their semantics and refinement relations – often one is defined in terms of the other. Not so, however, for Z: its semantics is fully formalised, but the notion of refinement is not directly related, and often described as "informal".

This paper investigates the interplay between Z semantics and Z refinement, suggesting a partial reconciliation between the two. Section 2 describes the problem in detail, based on two formal definitions of refinement: MC-refinement (Model Containment) based on semantics, and SO-refinement (State and Operations) formalising traditional Z refinement. In Section 3 we investigate an alternative interpretation of states-and-operations specifications that allows operation refinement to be subsumed by MC-refinement. Then we consider the interplay between axiomatic definitions and given types and SO-refinement in Section 4, leading to a partial formalisation of combined SO- and MC-refinement using the notions of *existential* and *universal* parameters of a specification. Section 5 discusses some further issues concerning refinement of type parameters. The final section summarizes our observations.

2 Refinement versus Semantics

A common criticism of Z is that it is a specification notation without a formal development method [6]. This is partly due to the fact that the use of Z for system development (the "states-and-operations" style) is not formalised within the

D. Bert et al. (Eds.): ZB 2002, LNCS 2272, pp. 226–241, 2002.
© Springer-Verlag Berlin Heidelberg 2002

notation. In particular, the use of decorations (primes, question marks, exclamation marks) and the Δ and Ξ notations are only conventions with no semantic meaning attached. Thus, there is a development method which aims to address "formal development", but it is not formally embedded within the language.

This is probably caused by a reluctance on the Z community's part to pin down the notation to a specific use. Part of the success of Z must be due to its flexibility, and it appears to be much more commonly used as a "data dictionary" definition language than in the states-and-operations style. Thus, it would be counterproductive to assign to Z a formal semantics which goes beyond the intuitive standard set theoretic one. On the other hand, the usual conventions have been recorded in an "informative" appendix to the standard [12], and various efforts have been made to formalise the use of Z for system development, as discussed below.

Closely related to this is the issue of a logic for Z. For development and verification, this is clearly a necessity. However, the standard only contains a semantics, and states that any logic that respects the semantics is acceptable. Henson and Reeves [9,10,11] have defined variants of Z, taking logic rather than semantics as the starting point. In [9] they give some indication of how an approach like this may be used for formalising states-and-operations.

Woodcock, Davies and Bolton [5,16] as well as Derrick and Boiten [7] give relational interpretations of the states-and-operations style. Effectively these use a subset of Z as a meta-language, in which the state-based specification language and its properties are encoded. Using this encoding, "standard" downward and upward simulation rules for Z are verified [16], also for the "guarded" interpretation[1] of pre-conditions [5]. In addition, refinement relations that change inputs and outputs, and refinements concerned with internal actions and change of granularity are verified in this manner in [7].

The refinement rules for this style in Z are very much centered on *schemas* representing states and operations. The implementation freedom represented by these rules is mostly constrained to reducing "non-determinism": a relation between states represents relations that are included in it[2], in particular the *functions* from before- to after-states. Even for data refinement this is the case, although the relations considered there are between sequences of inputs and outputs. In the rest of this paper, we will refer to this kind of refinement as "SO-refinement" (for States-and-Operations), with operation refinement, data refinement, etc., as special cases. The full formalisation given in [7] makes this a well-defined notion; related notions of refinement for Z described there are based on generalisations of the underlying relational characterisation of refinement.

This interpretation of refinement contrasts with its definition in many other specification languages. Z is a model-based language, i.e., the semantics of Z

---

[1] In the "guarded" interpretation, an operation is refused outside its pre-condition; thus, in this interpretation pre-conditions may not be widened in refinement.

[2] This is strictly true only in the "guarded" interpretation, or once we have "totalised" the relations to account for the particular interpretation of the domains of these relations.

links any Z specification to a collection of models that satisfy the specification's properties. The natural way of defining refinement for such languages is by *model containment*: *A* is a refinement of *B* if all the models of *A* are also models of *B*.[3] The main advantage of having such a definition of refinement is obvious: it is determined already by the semantics, so it does not require an extra layer of definition and interpretation of the language. In addition, compositional properties of refinement follow from compositionality of models: where models are constructed independently, refinements can take place independently, too. In the rest of this paper, we will refer to this kind of refinement as "MC-refinement" (for Model Containment).

There are some specification constructs in Z for which MC-refinement would be intuitively meaningful. These are constructs that allow multiple models, and are commonly referred to as constructs for "loose specification". These include *axiomatic definitions* (often called *constant* definitions) and *given types*. In general, a model of a Z specification is a collection of bindings, which assign to each top-level name in the specification a value from the universe of possible Z values.

Example 1. The specification containing only

$$\begin{array}{|l} x : \mathbb{N} \\ \hline x \geq 17 \end{array}$$

has as its models all the possible bindings of the name x to natural numbers ≥ 17. A specification containing only the declaration

$$[PEOPLE]$$

has as its models all sets of Z values bound to the name *PEOPLE*. As these declarations are independent, the models of their combination are obtained by taking all possible combinations of their models.

An MC-refinement of the combined specification is obtained by adding the constraint

$$\begin{array}{|l} \#PEOPLE = x \end{array}$$

allowing only those models where x is bound to the number of elements of the set that *PEOPLE* is bound to, for example excluding all models where *PEOPLE* is infinite.   □

A strengthening like this may or may not be a meaningful extension of the SO-refinement rules. This depends on the way in which the names *PEOPLE* and x are used in a states-and-operations specification, see Section 4.

Conversely, SO-refinements are not normally MC-refinements.

---

[3] A common extra restriction is that *A* may only have an empty set of models when *B* is also unsatisfiable.

Example 2. Consider the state schema *ABool* and the operations *AnOp* and *NotherOp* below.

```
┌─ ABool ──────────────────────────────────────────
│ x : 𝔹
│
└──────────────────────────────────────────────────
```

```
┌─ AnOp ─────────────────       ┌─ NotherOp ──────────────
│ ΔABool                        │ ΔABool
│                               │
└─────────────────────          ├─────────────────────────
                                │ x′
                                │
                                └─────────────────────────
```

Clearly *NotherOp* is an SO-refinement of *AnOp*.[4]

However, the specification containing *ABool* and *AnOp* has a single model, which assigns to the name *ABool* the set of bindings

$$\{\!\langle\!| \ x == true \ |\!\rangle, \langle\!| \ x == false \ |\!\rangle\}$$

and to *AnOp* a single set of bindings

$$\{\!\langle\!| \ x == true, x' == true \ |\!\rangle, \langle\!| \ x == true, x' == false \ |\!\rangle,$$
$$\langle\!| \ x == false, x' == true \ |\!\rangle, \langle\!| \ x == false, x' == false \ |\!\rangle\}$$

For another operation to be an MC-refinement of this, it should have the same model, or if we allow that, no models at all. Replacing *AnOp* by *NotherOp*, however, also leads to a *single* model, with the same value for *ABool* and for *NotherOp* the set

$$\{\!\langle\!| \ x == true, x' == true \ |\!\rangle, \langle\!| \ x == false, x' == true \ |\!\rangle\}$$

Thus, there is an issue with naming, but this could be removed by insisting that refinement preserves the names of operations, as is the case in B [1]. More seriously, the *set of* models of *NotherOp* is not at all included in the *set of* models of *AnOp*: they both have a singleton set of models, and these models are different. MC-refinement is concerned with the models of a specification; operation refinement here is evidenced by inclusions between particular *sets* that are parts of single models.   □

A more positive observation is that the relational interpretation as defined in [7,16] commutes with model containment. That is, it makes no difference whether we (MC-)constrain a SO-specification first, and then take its relational interpretation, or whether we constrain the relational interpretation of the SO-specification. This is obvious once we realise that the only sources of loose specification in an SO-specification are predefined names (constants and given types) used in defining the states and operations – which will occur in the same rôle in the relational interpretation.

---

[4] This is trivially true as both operations are universally applicable and there are no inputs or outputs; feel free to imagine both having an output *x*! equal to *x'* to make it less trivial.

3 Models of States-and-Operations

As the example above shows, the models of an operation are not related in the most obvious way to the models of possible SO-refinements of that operation. In fact, these models, do not even represent relations, because schema bindings are unordered labeled tuples. However, it is actually possible to change the interpretation of an SO-specification in such a way that *operation refinement* coincides with MC-refinement.

For that purpose, we need to translate non-determinism, and also the possibility to widen pre-conditions, into loose specification. Thus, in an operation schema, the predicate will not be taken at face value, but as a loosely specified predicate, whose models are all its operation refinements.

This leads to the following. A schema

```
┌─ Op ──────────────────────────────────────
│ ΔState
│ ──────────────────────────────────────────
│ pred
└──────────────────────────────────────────
```

is interpreted as a combination of a schema with a fixed predicate p, and an axiomatic definition which constrains p to all allowed refinements of *pred*:

```
┌─ Op ──────────────────────────────────────
│ ΔState
│ ──────────────────────────────────────────
│ p
└──────────────────────────────────────────
```

```
┌──────────────────────────────────────────
│ p : ℙ ΔState
│ ──────────────────────────────────────────
│ ∀ State; State' •
│     (∃ State' • pred) ⇒ ∃ State' • p
│     (∃ State' • pred) ∧ p ⇒ pred
└──────────────────────────────────────────
```

The first condition in the loose specification of p is that for any given *State*, p should be true for some *State'* whenever *pred* is – this is the "applicability" condition of operation refinement. The second condition is that if p links a *State* and a *State'* within the "pre-condition" of the operation, the states should also be linked by *pred*. This is the "correctness" condition of operation refinement.

This interpretation represents an operation schema by all its operation refinements. For example, under this interpretation, the specification containing *AnOp* in Example 1 has 9 models[5]. One of these assigns to the name *AnOp* the set of bindings

$$\{ \langle\!| \ x == true, x' == true \ |\!\rangle, \langle\!| \ x == false, x' == true \ |\!\rangle \}$$

which is also the set assigned to *NotherOp* in the *single* model of its specification.

---

[5] One containing 4 pairs, 4 containing 3 pairs, and 2*2 containing 2 pairs of a before- and after-state.

We would obtain an alternative but (in terms of possible refinements) equivalent representation if we insisted on p being a (partial) "function" from *State* to *State'*, representing *pred* by all its "implementations". Henson and Reeves [9] present essentially this construction for operations as well. To avoid needing an extra layer of interpretation like we have here, they define operations as a separate syntactic category in their variant of Z.

Although this approach appears to work for operation refinement, it is dubious whether it would be practicable for full SO-refinement, i.e. including data refinement. This is because the particular state schema used for an abstract data type is not part of its relational semantics. Thus, the loose specification representing it would need to quantify over all possible concrete and abstract representations of the state. A further complication would be the need to have both upwards and downwards simulation, or powersimulations, to ensure completeness. Finally, for data refinement it is not sufficient to consider operations individually – the initial state, and the other operations need to be known as well. (For detailed explanation of all these issues and extensive references to the literature see [7].) All in all, encoding data refinement as MC-refinement seems possible in principle but undesirable in practice.

4   The Rôle of Constants and Given Types

In the Z literature, different views on the relation between states-and-operations and loose specification are presented. Spivey [14] refers to non-deterministic operation schemas as "loose specification", thereby suggesting that the distinction between SO-refinement and MC-refinement is accidental. However, he does suggest different circumstances in which constants and given types may be instantiated. He also observes that loose specification may aid in specifying a *family* of abstract data types, where sometimes all members of the family must be implemented, and sometimes only a single one.

Woodcock and Davies [16, Section 23.1] explicitly distinguish between loose specification and non-deterministic specification, and between SO-refinement and MC-refinement. They also indicate that constants may be fixed at different stages in development. However, they give no verdict on when derived restrictions on constants should imply the failure of SO-refinement.

Barden, Stepney and Cooper [2, Section 22.5] distinguish three different kinds of "under-specification": non-determinism, loose specification, and undefinedness. They observe that *only* using loose specification it is possible to define specification elements that take on a fixed but unknown value from a specified range of values; using non-determinism they might also take on varying values within that range. Otherwise they do not claim any fundamental difference between the three kinds of under-specification.

To illustrate the various rôles that constants and given types may take, let us consider a number of examples. In each of these, we will present an SO-refinement that turns out to be valid *only* in conjunction with some MC-refinement – by constraining the possible values of a given type or constant. Intuitively, this appears to be acceptable in some cases and not in others. By formalising these intuitions, we provide a basis for integrating SO-refinement and MC-refinement.

4.1   Examples

Example 3. Consider the following specification. There is a given set of *PEOPLE*, some of whom are members of the set *pp*.

$[PEOPLE]$

__ *People* _____
$pp : \mathbb{P}\,PEOPLE$

Two attempts at specifying the operation of joining *pp* are:

__ *Join$_1$* _____	__ *Join$_2$* _____
$\Delta People$	$\Delta People$
$p? : PEOPLE$	$p? : PEOPLE$
$p? \notin pp$	$p? \in pp$
$pp' = pp \cup \{p?\}$	$pp' = pp \cup \{p?\}$

An attempt to prove that *Join$_1$* is refined by *Join$_2$* should intuitively fail – one is applicable when *p?* is not yet a member, the other when *p?* is already a member. However, these pre-conditions are both equally false when $PEOPLE = \varnothing$, and a proof of refinement will reduce to that condition.   □

Example 4. It is always difficult to decide on the representation of an error situation like "result not found". Consider, for example, the following specification

$[PEOPLE, DIGIT]$

$phonebook : PEOPLE \nrightarrow \text{seq}\,DIGIT$

$\#\{n : \text{ran}\,phonebook \bullet \#n\} \le 1$

(all phone numbers are equally long).

__ *GiveNo* _____
$x? : PEOPLE$
$n! : \text{seq}\,DIGIT$

$x? \in \text{dom}\,phonebook \Rightarrow n! = phonebook(x?)$
$x? \notin \text{dom}\,phonebook \Rightarrow n! \notin \text{ran}\,phonebook$

The second predicate in the *GiveNo* schema is to ensure that the value returned for a person not in the phone book can never be a valid phone number. Now one might decide to implement this second predicate by

$$x? \notin \text{dom}\, phonebook \Rightarrow \text{ran}\, n! = \{0\}$$

using the "obvious" error code of a sequence of zeroes. This, however, is only correct if nobody has zero as their phone number. It remains to be seen whether this is an acceptable constraint on our "parameter" *phonebook*. The implementation might be rejected for being too restrictive, or the extra constraint might be validated by the customer, thereby strengthening the initial specification. □

Example 5. In [3] we considered a refinement of a transmission protocol. The abstract specification considered an end-to-end transmission; the concrete specification characterised a transmission along a predefined route, where the number of intermediate stations was given by a loosely specified constant. The concrete specification of an over-eager receiver turned out to be only correct if the number of intermediate stations was exactly 0, and was therefore dismissed.   □

4.2   Analysis of Examples

The emerging conclusion from these examples appears to be: it all depends on the *context* whether MC-refinement is acceptable to verify an SO-refinement. However, below we formalise this rôle of the context.

In our review of the literature above we omitted a crucial term from the discussion by Spivey: loosely specified specification elements may be *parameters*. This is to circumvent the lack of an explicit parametrisation mechanism in Z – note that "sections" in the Z standard [12] do not have parameters. Thus, where families of abstract data types need to be described, the parameters tend to be included as top-level constants.

Below, we will distinguish *universal* and *existential* parameters. Universal parameters are parameters in the traditional sense – universality implies that refinement will need to work for *all* possible values of the parameters to be acceptable. An existential parameter is one that is only required to lead to a correct SO-refinement for *some* of its possible values.

4.3   Universal Parameters

Parameters *can* actually be represented explicitly in Z when they are *types*, using *generic schemas*. However, as there is no encapsulation in an SO-specification, this parameter has to be listed explicitly in every schema.

Example 6. An alternative but semantically non-equivalent representation of Example 3 would be

```
┌─ People [X] ─────────────────────
│  pp : ℙ X
└──────────────────────────────────
```

$$
\begin{array}{|l}
\hline
Join_1\,[X]\;\rule{4cm}{0.4pt} \\
\Delta People \\
p? : X \\
\hline
p? \notin pp \\
pp' = pp \cup \{p?\} \\
\hline
\end{array}
\qquad
\begin{array}{|l}
\hline
Join_2\,[X]\;\rule{4cm}{0.4pt} \\
\Delta People \\
p? : X \\
\hline
p? \in pp \\
pp' = pp \cup \{p?\} \\
\hline
\end{array}
$$

requiring no declaration of X.

The semantics of this specification would not be a set of models for every different possible value of X as in Example 3, but a *single* model that assigns *functions* to the names $People$, $Join_1$ and $Join_2$. These functions take a value for the parameter X, and produce from it a value for the instantiated generic schema $People[X]$ (etc.)

For $Join_1[X]$ to (SO-)refine $Join_2[X]$, clearly once again we would obtain the condition that $X = \varnothing$. Intuitively, the refinement should be deemed to fail for that reason. In any other theory of parameterised data types, a parameterised data type is refined by another parameterised data type only if the refinement holds for *every* instantiation. Generic definitions in Z that constrain their parameters are possible, but may lead to inconsistency of the specification as a whole when instantiated.   □

This intuitive line of reasoning can be formalised by calling such a name a *universal* parameter. We define the term "universal parameter" formally by stating what the proof obligation for universal parameters is in the context of an SO-refinement proof. This extends the SO-specification method by explicitly allowing a context of definitions to be taken into account.

Definition 1. *If a top level name in a Z specification is declared to be a* universal parameter, *then SO-refinements of that specification are correct only if their models assign the same collection of possible values to that name as the models of the original specification.*

When there are multiple universal parameters, the above definition should be applied to the tuple of all universal parameters, i.e. all *combinations* of their values should be preserved by SO-refinement.

The constants in Examples 4 and 5 can clearly be interpreted as universal parameters. The refinement for Example 4 would be rejected, unless the initial specification could be strengthened to guarantee that 0 is not a valid phone number. Similarly, the refinement which introduced no more than 0 intermediate stations into a point-to-point protocol in Example 5 [3] was rejected as unsuitable.

Designating $PEOPLE$ a universal parameter makes the specifications of Examples 3 and 6 equivalent as far as SO-refinement is concerned: they both require SO-refinement to hold for every possible instantiation of the parameter. The use of generic schemas makes the rôle of the parameter explicit and unambiguous. For that reason, generic schemas should be preferred to given types as universal parameters. The practical counter-argument is that this requires an explicit instantiation with *every* occurrence of every schema name involved.

4.4   Existential Parameters

Other loosely specified constants have a dual rôle to universal parameters: they represent implementation parameters, which may be made more precise in the SO-refinement process. An example could be an error-detecting or error-correcting encoding of bytes, for which it is known that extra bits will be necessary, but not in advance how many. Maybe less obviously, given types may also play this rôle. Consider a design of a database, where the database record is initially taken from a given type. Functions are introduced from the record to its fields, and properties of these functions are proved (for example, which fields are candidate keys). Eventually, the abstract record will be replaced by a concrete one consisting of particular fields, or combinations of such tuples.

Parameters of this kind will be declared as *existential* parameters. The following definition also extends the (informal) states-and-operations method by taking a context of loosely specified definitions into account.

Definition 2. *If a top level name in a Z specification is declared to be an* existential parameter, *then SO-refinements of that specification are correct* only if *they preserve consistency of the specification.*

The reference to consistency is because unsatisfiable axiomatic definitions are a cause of inconsistency in Z specifications [15]. If the abstract specification was inconsistent already, any SO-refinement is probably acceptable; otherwise, the SO-refinement may only induce such constraints on the loose definitions in the concrete specification that it remains consistent.

A very important requirement when using existential parameters in practice is that the derived additional constraints are recorded in the refined specification.

Consider the following.

Example 7. We have a SO-specification containing an under-specified boolean constant

$$\boxed{\; b : \mathbb{B} \;}$$

Thus, models will have b bound to either *true* or *false*. The rest of the specification is

$State$	Op_1
$x : \mathbb{Z}$	$\Delta State$
	$x! : \mathbb{Z}$
$Init$	
$State'$	$b \Rightarrow x' = x + 1$
	$\neg b \Rightarrow x' = x - 1$
$x' = 0$	$x! = x'$

This specification has two possible models: one which outputs the numbers $0, 1, 2, 3, \ldots$ and one which outputs the numbers $0, -1, -2, -3, \ldots$.

We declare b to be an *existential* parameter. Thus, we would be happy with either of the output sequences.

Now, we perform an operation refinement, replacing Op_1 by

$$
\begin{array}{|l}
\underline{\ Op_2\ } \\
\Delta State \\
x! : \mathbb{Z} \\
\hline
b \Rightarrow x' = x + 1 \\
\neg b \Rightarrow x' = x - 2 \\
x! = x' \\
\end{array}
$$

For this SO-refinement to be correct, b needs to be *true*. Because b is an existential parameter, we only need to ensure that the concrete specification allows *some* value for b, inducing some MC-refinement if necessary. In the concrete specification, we do need to constrain b by replacing its declaration by

$b == true$

This is evident once we consider a possible operation refinement of Op_2, namely

$$
\begin{array}{|l}
\underline{\ Op_3\ } \\
\Delta State \\
x! : \mathbb{Z} \\
\hline
b \Rightarrow x' = x + 2 \\
\neg b \Rightarrow x' = x - 2 \\
x! = x' \\
\end{array}
$$

This operation refines Op_2, provided we can constrain b to be *false*. Now imagine we had not recorded that b needs to be *true* in the second specification. Then we could refine Op_1 by Op_2, and subsequently Op_2 by Op_3 – however, transitivity does not hold as there is no satisfiable constraint on b that allows Op_1 to be refined by Op_3.   □

Thus, not recording the (MC-)constraints induced by the (SO-)refinement obligations leads to a non-transitive, or unsound, refinement relation.

A situation in which MC-refinements may not be sound relates to Z's use of maximal sets as types in combination with schema negation. Consider the following MC-refinement.

Example 8. The specification

$[TOONE]$

$$
\begin{array}{|l}
zero, one : TOONE \\
\hline
one \neq zero \\
\forall x : TOONE \bullet x = one \lor x = zero \\
\end{array}
$$

which is also known as the expansion of the free type definition

$TOONE ::= zero \mid one$

has as possible models for $TOONE$ any two element set. Any further constraint would either lead to the same set of models, or make the specification inconsistent. An MC-refinement which changes the syntactic category of $TOONE$ could be

$TOONE == \{0, 1\}$

$$
\begin{array}{|l}
\hline
zero, one : TOONE \\
\hline
zero = 0 \wedge one = 1 \\
\end{array}
$$

However, now consider the schema

$$
\begin{array}{|l}
\hline
\_B \\
\hline
x : TOONE \\
\hline
\end{array}
$$

The schemas $\neg B$ and $B \vee \neg B$ using the two definitions of $TOONE$ (repeated for convenience) are given side by side below:

$TOONE ::= zero \mid one$ $\qquad\qquad$ $TOONE == \{0, 1\}$

$$
\begin{array}{|l}
\hline
\_\neg B \\
\hline
x : TOONE \\
\hline
false \\
\end{array}
\qquad
\begin{array}{|l}
\hline
\_\neg B \\
\hline
x : \mathbb{Z} \\
\hline
x \notin \{0, 1\} \\
\end{array}
$$

$$
\begin{array}{|l}
\hline
\_B \vee \neg B \\
\hline
x : TOONE \\
\hline
true \\
\end{array}
\qquad
\begin{array}{|l}
\hline
\_B \vee \neg B \\
\hline
x : \mathbb{Z} \\
\hline
true \\
\end{array}
$$

Now consider the schema expression denoting the "complement" of B:

$\exists (B \vee \neg B) \bullet \neg B$

In the original specification, this denotes the empty schema "$false$", but in the MC-refinement it denotes the empty schema "$true$". Note that the mathematical toolkit of Z provides no way to denote the complement of a set (with respect to the maximal type of its elements) – the above example shows how this *is* possible using schema negation.

Thus, this kind of MC-refinement may not be sound in the context of schema negation. Although this appears a rather undesirable property of the Z semantics, one might argue that referring to the complement of a given set could never be a good idea.

4.5   Multiple Parameters

The classification of parameters is not complete once we have identified universal and existential parameters. One can also imagine situations where, for each possible value of a universal parameter, a particular value of a (dependent existential) parameter should be found which allows refinement to proceed. Indeed, in general we should expect multiple parameters not only to be quantified by $\forall$ and $\exists$, but also by $\forall \ldots \exists$.

5   Further Refinement for Type Parameters

Example 8 described an MC-refinement. However, if we change the names in the second specification (to avoid clashes), and document the relation between the names as part of a *retrieve relation*, this kind of transition can be made in *data refinement* as well. When members of the given type only occur as state components, this can eliminate the given type using traditional tools.

Example 9. Using the definitions of Example 8, and

$$\begin{array}{|l}
\hline
succ : TOONE \nrightarrow TOONE \\
\hline
succ = \{zero \mapsto one\} \\
\end{array}$$

consider the SO-specification

$$\begin{array}{|l}
\hline
\_State \underline{\hspace{3cm}} \\
count : TOONE \\
mem : \mathbb{P}\,\mathbb{N} \\
\hline
\end{array}
\qquad
\begin{array}{|l}
\hline
\_Init \underline{\hspace{2cm}} \\
State' \\
\hline
count' = zero \\
mem' = \varnothing \\
\hline
\end{array}$$

$$\begin{array}{|l}
\hline
\_Add \underline{\hspace{2.5cm}} \\
\Delta State \\
x? : \mathbb{N} \\
\hline
mem' = mem \cup \{x?\} \\
(count, count') \in succ \\
\hline
\end{array}
\qquad
\begin{array}{|l}
\hline
\_Remove \underline{\hspace{2cm}} \\
\Delta State \\
x! : \mathbb{N} \\
\hline
\{x!\} = mem \setminus mem' \\
(count', count) \in succ \\
\hline
\end{array}$$

The data refinement that replaces the state components of type $TOONE$ by numbers or booleans is left to the reader.

At this stage, it does not matter whether $TOONE$ is a universal or existential parameter, because no MC-refinement is involved in the data refinement step. Technically the name $TOONE$ remains part of the specification, but it no longer affects the rest of the specification.   □

Occurrences of the parameters of *generic* schemas could also be removed using data refinement in a similar way – consider, for example, the analogue of the above specification where *TOONE* is a parameter of every schema. Removing an unused schema parameter is formally not a correct MC-refinement, as the Z semantics distinguishes (at the level of models) constant functions versus constants.

However, one situation where given types and type parameters cannot be factored out through data refinement is when they are a part of the observable behaviour of the specification. The relational interpretation of SO-specifications [7,16] makes only the situations in which an operation is applicable and the inputs and outputs of operations observable. Thus, when values of given types occur as inputs or outputs of an operation, a generalisation of data refinement is necessary to implement the given type. One such generalisation is IO-refinement [4,7]. Hayes and Sanders [8] make a case for separating out the implementation details of IO from the functionality of operations. Clearly an input that is taken from a given type represents such an abstraction of the actual input. IO-refinement formalises the introduction and modification of such a separation as refinement steps, provided certain restrictions hold on the abstraction relations for IO. In particular, they should be total on abstract inputs and outputs, and injective for abstract outputs. This means that every concrete output should uniquely identify a concrete one.

Example 10. The following system describes a bank with a static clientele, all of whom hold a single account only. For that reason, the link between customers and account numbers is described axiomatically: a universal parameter of the specification. The given type *CUSTOMER* is an existential parameter.

$[CUSTOMER]$

$$\mid accountno : CUSTOMER \rightarrowtail \mathbb{N}$$

___ *BankState* _____
$accts : \mathbb{N} \nrightarrow \mathbb{Z}$
_____
dom $accts$ = ran $accountno$

___ *Init* _____
BankState'
_____
ran $accts \subseteq \{0\}$

___ *Deposit* _____
$\Delta BankState$
$c? : CUSTOMER$
$n? : \mathbb{N}$
_____
let $i == accountno(c?) \bullet accts' = accts \oplus \{i \mapsto accts(i) + n?\}$

The models of *CUSTOMER* are already tightly constrained by totality and injectivity of *accountno*: there should be exactly as many customers as there are account numbers. Thus, a possible MC-refinement replaces *CUSTOMER* by ran *accountno*. The corresponding IO-refinement performs the same replacement in *Deposit*, giving

$$
\begin{array}{l}
\underline{\quad Deposit_C \quad\quad\quad\quad\quad\quad\quad\quad\quad\quad\quad\quad\quad\quad\quad\quad\quad\quad\quad\quad\quad\quad} \\
\Delta BankState \\
c? : \mathbb{N} \\
n? : \mathbb{N} \\
\underline{\quad\quad\quad\quad\quad\quad\quad\quad\quad} \\
c? \in \mathrm{ran}\ accountno \\
accts' = accts \oplus \{c? \mapsto accts(c?) + n?\}
\end{array}
$$

Another way of looking at this refinement step is as an MC-refinement of the SO-specification *and its models*, which then justifies the replacement of *CUSTOMER* without an appeal to IO-refinement.   □

The above example also relates to the discussion about given types as abstract database records in Section 4.2. If we view *CUSTOMER* above as representing the full customer record, the account number is indeed a candidate key for it under the given constraints on *accountno*.

6   Concluding Comments

This paper contributes to the formalisation of the states-and-operations development method in Z by recognising the rôles of loosely specified names as parameters, and by providing a vocabulary that allows different uses of such parameters to be declared and formalised. In particular, *existential* parameters indicate values that may be constrained in refinement steps, whereas refinements should hold for *all* values of *universal* parameters. Where such parameters are *types*, generic schemas provide an adequate and unambiguous mechanism for universal parameters. Thus, it makes sense to reserve the use of given types for *existential* parameters. Axiomatic definitions should be annotated to indicate whether they are existential, universal, or some combination of these.

In Object-Z [13], similar issues arise. Constants in a class definition may actually get different values in different instances of the class. As in Z, there is no mechanism for parametrising classes by anything other than a type.

The story for B is different, as SO-refinement is built into the language, and there is no independent notion of semantics. It has a more extensive and more explicit parametrisation mechanism than Z, machines may have parameters that are scalars or sets. Deferred sets and (concrete) constants play a rôle similar to existential parameters. However, they may only be given a value (e.g. in the implementation); deferred sets may not be constrained through the PROPERTIES clause. Abstract constants [1, Page 261] may be refined, and can be used for loose specification. All in all, B offers a full treatment of universal parameters, and support for existential parameters, but not as much for their gradual refinement.

Acknowledgements. To John Derrick and Ralph Miarka for comments on drafts and discussions; to Martin Henson and Ian Toyn for correspondence on various issues; to the referees for corrections and advice, in particular for pointing out the soundness problem now illustrated in Example 8.

References

1. J.-R. Abrial. *The B-Book: Assigning programs to meanings.* Cambridge University Press, 1996.
2. R. Barden, S. Stepney, and D. Cooper. *Z in Practice.* Prentice Hall, 1994.
3. E.A. Boiten, H. Bowman, J. Derrick, and M. Steen. Viewpoint consistency in Z and LOTOS: A case study. In J. Fitzgerald, C.B. Jones, and P. Lucas, editors, *FME'97: Industrial Application and Strengthened Foundations of Formal Methods*, volume 1313 of *Lecture Notes in Computer Science*, pages 644–664. Springer-Verlag, September 1997.
4. E.A. Boiten and J. Derrick. IO-refinement in Z. In A. Evans, D. Duke, and T. Clark, editors, *3rd BCS-FACS Northern Formal Methods Workshop*. Springer-Verlag, September 1998. http://www.ewic.org.uk/.
5. C. Bolton, J. Davies, and J.C.P. Woodcock. On the refinement and simulation of data types and processes. In K. Araki, A. Galloway, and K. Taguchi, editors, *International conference on Integrated Formal Methods 1999 (IFM'99)*, pages 273–292, York, July 1999. Springer-Verlag.
6. A.L.C. Cavalcanti and J.C.P. Woodcock. ZRC – a Refinement Calculus for Z. *Formal Aspects of Computing*, 10(3):267–289, 1998.
7. J. Derrick and E.A. Boiten. *Refinement in Z and Object-Z: Foundations and Advanced Applications.* Springer-Verlag, 2001.
8. I.J. Hayes and J.W. Sanders. Specification by interface separation. *Formal Aspects of Computing*, 7(4):430–439, 1995.
9. M.C. Henson and S. Reeves. New foundations for Z. In J. Grundy, M. Schwenke, and T. Vickers, editors, *International Refinement Workshop & Formal Methods Pacific '98*, Discrete Mathematics and Theoretical Computer Science, pages 165–179, Canberra, September 1998. Springer-Verlag.
10. M.C. Henson and S. Reeves. Revising Z: Part I - Logic and semantics. *Formal Aspects of Computing*, 11(4):359–380, 1999.
11. M.C. Henson and S. Reeves. Revising Z: Part II - Logical development. *Formal Aspects of Computing*, 11(4):381–401, 1999.
12. ISO/IEC. Formal Specification – Z Notation – Syntax, Type and Semantics: 2nd Final Committee Draft. International Standard CD 13568.2, International Standards Organization, 2000.
13. G. Smith. *The Object-Z Specification Language.* Kluwer Academic Publishers, 2000.
14. J. M. Spivey. *The Z notation: A reference manual.* Prentice Hall, 2nd edition, 1992.
15. S.H. Valentine. Inconsistency and undefinedness in Z – a practical guide. In J. P. Bowen, A. Fett, and M. G. Hinchey, editors, *ZUM'98: The Z Formal Specification Notation*, volume 1493 of *Lecture Notes in Computer Science*, pages 233–249. Springer-Verlag, September 1998.
16. J.C.P. Woodcock and J. Davies. *Using Z: Specification, Refinement, and Proof.* Prentice Hall, 1996.

On Using Conditional Definitions in Formal Theories

Jean-Raymond Abrial[1] and Louis Mussat[2]

[1] Consultant, Marseille France `jr@abrial.org`
[2] DCSSI, Paris France `louis.mussat@sgdn.pm.gouv.fr`

Abstract. In this paper, our intention is to explore the notion of definition in formal theories and, in particular, that of *conditional definitions*.
We are also interested in analyzing the consequences of the latter on the
structure of corresponding *proof systems*. Finally, we shall investigate the
various ways such proof systems can be *simplified*.

1 Introduction

In formal texts, conditional definitions lead to such oddities as division by zero,
the minimum of an empty set, or, more generally, the application of a function
to an argument lying outside its domain. In the presence of such *ill-defined
expressions*, people usually divide in two groups. A first group considers that such
pathologies are totally uninteresting, that professional mathematicians never
write things of that kind, and that it would correspond, in every day life, to
people not mastering their mother tongue (so that the problem really is that of
having people first learn how to correctly express themselves). A second group,
especially among the formal developers, considers that this is a serious question,
that it may lead to crashes, and that it is thus a problem that one must face and
somehow "solve" (we shall see below that a quite a number of "solutions" have
been proposed). In this paper, we again study the problem and try to give our
view and contribution to it. Our "solution" is certainly not entirely novel but
it opens, we think, a number of interesting ways to be explored in the domain
of *mechanized proof strategy*. As our work is done in relation with B, we shall
make our investigation on a specific formal theory, namely set theory as it has
been re-constructed in the B-Book [1].

Before developing our main subject, it is certainly worthwhile recalling what
we understand by a proper concept of *definition* in a formal theory. For this, we
shall follow the way this notion was presented in an introductory book of logic
written by P. Suppes [9] in the fifties, where an entire chapter is devoted to that
question. The present paper is to be read *with that very understanding of the
concept of definition in mind*.

1.1 On Definitions

Suppes first introduces two criteria characterizing the definition of new symbols
in a formal theory: (1) the *Criterion of Eliminability*, and (2) the *Criterion*

D. Bert et al. (Eds.): ZB 2002, LNCS 2272, pp. 242–269, 2002.

of Non-creativity. The former requires that "any definition introducing a new symbol may be used to eliminate all subsequent meaningful occurrences (of that new symbol)". The latter requires that any definition of a new symbol "(does not) make possible the derivation of some previously unprovable theorem stated wholly in terms of primitive and previously defined symbols". In other words, a proper definition must not add any extra "power" to a theory, it is just a useful, but not indispensable, *extension* of it.

Then Suppes proposes some precise rules by which new symbols can be introduced while guaranteeing both previous criteria. He makes the distinction between rules defining either new *predicate* symbols or new *term* symbols. In both cases, this is done by means of equivalences. We shall continue to follow Suppes in the presentation of these rules.

1.2   Defining a New Predicate Symbol by an Equivalence

A new predicate symbol P involving a single argument $\mathbf{E}$, thus forming the new predicate $\mathsf{P}(\mathbf{E})$, is introduced by an axiomatic equivalence of the following form:

$$\vdash\ \ \mathsf{P}(\mathbf{E})\ \Leftrightarrow\ \mathbf{D_E}$$

where (1) $\mathbf{D_E}$ is a formula only depending on the variable $\mathbf{E}$, and (2) $\mathbf{D_E}$ is a formula only containing previously defined symbols of the theory. Traditionally, the new construct, called the *definiendum*, is situated on the left hand side of the equivalence sign while the other side contains the proposed definition, called the *definiens*. For example, in a theory of natural numbers, the binary infix operator "$<$" is defined as follows in terms of "$\leq$":

$$\vdash\ \ \mathbf{E_1}<\mathbf{E_2}\ \Leftrightarrow\ \mathbf{E_1}\leq\mathbf{E_2}\ \wedge\ \mathbf{E_1}\neq\mathbf{E_2}$$

1.3   Defining a New Term Symbol by an Equivalence

A new term symbol T involving a single argument $\mathbf{E}$ is introduced by an axiomatic equivalence of the form:

$$\vdash\ \ \mathbf{F}=\mathsf{T}(\mathbf{E})\ \Leftrightarrow\ \mathbf{D_{E,F}}$$

where (1) $\mathbf{D_{E,F}}$ is a formula only depending on the variables $\mathbf{E}$ and $\mathbf{F}$, (2) $\mathbf{D_{E,F}}$ is a formula only containing previously defined symbols of the theory, and finally (3) the following *justifying theorems*

$$\vdash\ \ \forall\,(\mathbf{F},\mathbf{G})\cdot(\mathbf{D_{E,F}}\wedge\mathbf{D_{E,G}}\Rightarrow\mathbf{F}=\mathbf{G}\,)$$
$$\vdash\ \ \exists\,\mathbf{F}\cdot\mathbf{D_{E,F}}$$

must be both provable under the axioms and the previously introduced definitions of the theory. In other words the new term denotes a *unique* entity, which indeed *exists*. We notice that the justifying theorem concerning uniqueness guarantees the criterion of non-creativity since otherwise one could derive an obvious contradiction, and the justifying theorem concerning the existence guarantees that the definition is not void. For example, in a theory of integers, the binary infixed symbol of subtraction can be defined as follows:

$$\vdash \quad \mathbf{F} = \mathbf{E_1} - \mathbf{E_2} \quad \Leftrightarrow \quad \mathbf{E_1} = \mathbf{F} + \mathbf{E_2}$$

And the two justifying theorems, which are valid, are then the following:

$$\vdash \quad \forall\,(\mathbf{F}, \mathbf{G}) \cdot (\,\mathbf{E_1} = \mathbf{F} + \mathbf{E_2} \;\wedge\; \mathbf{E_1} = \mathbf{G} + \mathbf{E_2} \;\Rightarrow\; \mathbf{F} = \mathbf{G}\,)$$
$$\vdash \quad \exists\,\mathbf{F} \cdot (\,\mathbf{E_1} = \mathbf{F} + \mathbf{E_2}\,)$$

1.4   Defining a New Term Symbol by an Equality

There is another way of introducing a new term symbol T with an argument $\mathbf{E}$: this consists in using a simple axiomatic *equality* of the form

$$\vdash \quad \mathsf{T}(\mathbf{E}) = \mathbf{D_E}$$

where (1) $\mathbf{D_E}$ is a formula only depending on the variable $\mathbf{E}$ and (2) $\mathbf{D_E}$ is a formula only containing previously defined symbols of the theory. The advantage of using that kind of definition is that it does not require proving the two justifying theorems since such a definition is the same as the following:

$$\vdash \quad \mathbf{F} = \mathsf{T}(\mathbf{E}) \quad \Leftrightarrow \quad \mathbf{F} = \mathbf{D_E}$$

and thus the justifying theorems become the following, which are logically valid:

$$\vdash \quad \forall\,(\mathbf{F}, \mathbf{G}) \cdot (\,\mathbf{F} = \mathbf{D_E} \;\wedge\; \mathbf{G} = \mathbf{D_E} \;\Rightarrow\; \mathbf{F} = \mathbf{G}\,)$$
$$\vdash \quad \exists\,\mathbf{F} \cdot (\,\mathbf{F} = \mathbf{D_E}\,)$$

One might wonder then why we are not using that second form systematically in order to introduce new term symbols. The answer is that it is not always possible simply because there are sometimes no explicit term, which can be made easily equal to the new introduced construct (as in the case of subtraction above). But clearly, we shall try to use that second form as often as we can.

1.5   About Recursive "Definitions"

At this point, it is worth departing a little from Suppes presentation of definitions and consider the, so-called, *recursive definitions*, in order to see whether they agree with the concept of definitions we have presented. The typical (and over-used) example of a recursive "definition" is that of the symbol factorial introducing the factorial function in a theory of natural numbers:

$$\vdash \quad \mathsf{factorial}\,(0) = 1 \;\wedge\; \mathsf{factorial}\,(\mathbf{n} + 1) = (\mathbf{n} + 1) \times \mathsf{factorial}\,(\mathbf{n})$$

Clearly this formulation is not a proper definition (according to the concept of definition presented here) of the unary construct factorial $(\mathbf{n})$ as it does not comply with the above way of defining new term symbols. Nor does it represent a definition of the constant symbol factorial for the same reasons. One has however the feeling that it is a correct definition. It is indeed, but not on that form. We have to use an equivalence and thus define the *constant* symbol factorial as follows

$$\vdash \quad \mathbf{F} = \mathsf{factorial} \quad \Leftrightarrow \quad \begin{pmatrix} \mathbf{F} \in \mathbb{N} \to \mathbb{N} \quad \wedge \\ \mathbf{F}(0) = 1 \quad \wedge \\ \forall \mathbf{n} \cdot (\, \mathbf{n} \in \mathbb{N} \; \Rightarrow \; \mathbf{F}(\mathbf{n}+1) = (\mathbf{n}+1) \times \mathbf{F}(\mathbf{n}) \,) \end{pmatrix}$$

This is now a correct definition from a "syntactic" point of view. To make it a genuine definition, one has, of course, to prove both justifying theorems asserting that the definiens introduces a unique function $\mathbf{F}$. As one knows, it is possible to do so by using an adequate mathematical theory (uniqueness requiring however some more precision).

Because definitions of that form are very useful but rather heavy to formulate as such in practice, people *abbreviate* them by just introducing the essence of them, corresponding to some parts only of the above definiens, hence the "definition" we have given initially. So far so good because, in this example, the proper definition can always be (virtually) reconstructed and the justifying theorems proved. But notice that it is only possible because of the very *shape* of the "definition". Should this shape be different (not following the inductive construction of the set of natural numbers) then the justification is not possible. The moral of the story is that recursive "definitions" are perfectly correct abbreviations of genuine definitions *provided they obey certain specific rules*. When it is not the case, then, again, such abbreviations are not definitions according to the way this concept is presented here.

1.6   Conditional Definition

Coming back to our main development, we follow again Suppes for the concept of conditional definitions. Sometimes it is necessary to define new symbols that could only be used *provided certain conditions are met*. In the case of a new term symbol introduced by an equivalence (the only case we consider here, the other ones being very close to it), the corresponding rule is as follows: a new term symbol T involving an argument $\mathbf{E}$ is *conditionally defined* when it is introduced by an axiomatic equivalence of the form:

$$\mathbf{C_E} \quad \vdash \quad \mathbf{F} = \mathsf{T}(\mathbf{E}) \; \Leftrightarrow \; \mathbf{D_{E,F}}$$

where (1) $\mathbf{F}$ is not free in the condition $\mathbf{C_E}$, (2) $\mathbf{D_{E,F}}$ is a formula only depending on the variables $\mathbf{E}$ and $\mathbf{F}$, (2) $\mathbf{C_E}$ and $\mathbf{D_{E,F}}$ are formulae only containing previously defined symbols of the theory, and finally (3) the two justifying theorems

$$\mathbf{C_E} \quad \vdash \quad \forall (\mathbf{F}, \mathbf{G}) \cdot (\, \mathbf{D_{E,F}} \wedge \mathbf{D_{E,G}} \; \Rightarrow \; \mathbf{F} = \mathbf{G} \,)$$
$$\mathbf{C_E} \quad \vdash \quad \exists \mathbf{F} \cdot \mathbf{D_{E,F}}$$

must be both provable under the axioms and the previously introduced definitions of the theory. As can be seen, the usage of conditional definitions is rather delicate because one must always be careful that they are indeed used under the required conditions. Moreover, they clearly *limit the criterion of eliminability* to the situations where the condition holds. For example, in a theory of real numbers, the binary infix symbol of division is defined conditionally as follows in terms of multiplication:

$$E_2 \neq 0 \ \vdash \ F = E_1/E_2 \ \Leftrightarrow \ E_1 = F \times E_2$$

And the two justifying theorems, which are valid, are then the following:

$$E_2 \neq 0 \ \vdash \ \forall (F, G) \cdot (E_1 = F \times E_2 \ \wedge \ E_1 = G \times E_2 \ \Rightarrow \ F = G)$$
$$E_2 \neq 0 \ \vdash \ \exists F \cdot (E_1 = F \times E_2)$$

One could perhaps argue that conditional definitions are not needed at all, as it would suffice to insert the conditions in the definiens in order to get rid of it. In the case of the division of the reals, this yields

$$\vdash \ F = E_1/E_2 \ \Leftrightarrow \ E_2 \neq 0 \wedge E_1 = F \times E_2$$

In other words, the very fact of writing E_1/E_2 would "automatically" imply that E_2 is different from 0. Let us see however what are now the two justifying theorems. After some obvious transformations, we obtain the following

$$\vdash \ E_2 \neq 0 \ \Rightarrow \ \forall (F, G) \cdot (E_1 = F \times E_2 \ \wedge \ E_1 = G \times E_2 \ \Rightarrow \ F = G)$$
$$\vdash \ E_2 \neq 0 \ \wedge \ \exists F \cdot (E_1 = F \times E_2)$$

The uniqueness theorem clearly holds (it is in fact the same as the previous one). But unfortunately, the existence theorem does not, precisely when E_2 is equal to 0. An implication is clearly needed here, *not a conjunction*.

The example of the conditional definition of the division of the reals we have just recalled, shows that one must be extremely careful in not asserting things too quickly concerning E_1/E_2. This could be the case, should we blindly apply the definition of the division without being aware that E_2 might be equal to 0. As a matter of fact, every pupil quickly learns (his teacher is particularly attentive to that) that "simplification by E_2" in the expression $(E_1/E_2) \times E_2$ can only be performed *provided $E_2 \neq 0$ holds*.

What is particularly irritating about such matters is that it pollutes our mind while we are trying to solve some "interesting" problem. In an ideal world, would not it be nice to have the possibility to eliminate *once and for all* such special cases and consider only a problem that has thus been simplified? That is, one where we have the *guarantee* that such pathologies can never occur *by the very construction of the statement of the problem*. Notice that this approach is that taken implicitly by the working mathematician: before engaging in a problem, he is very careful of totally eliminating such odd things as division by zero, divergent series, and the like by means of certain *preliminary* treatments.

This problem raised by formal texts containing potentially "ill-defined" expressions such as $3/0$ is, of course, not new. People have proposed many different solution to that question. Reviews of such solutions can be found in [6] and [8]. In what follows, we shall quickly overview some of them before introducing ours.

1.7   Various Approaches

(1) One of the most famous approaches is that advocated by C.B. Jones and his colleagues [2], [5], [7]. It consists in considering a, so-called, three-valued logic,

within which the *valuation* of predicates can be true, false or else undefined. It is a solution aiming at completely integrating the problem of ill-definedness within the formal logic, in contrast to that proposed by others, which tends, on the contrary, to eliminate the problem as much as possible using various artifacts. It has received a considerable interest in various community and in particular in that of VDM. For instance, within that approach, the following predicate is said to be undefined (as well, by the way, as the equality $3/0 = 3/0$, which thus necessitates to redefine equality):

$$2 + 3/0 = 2$$

(2) Another solution [9] consists in totally eliminating conditional definitions by *forcing* them to be always unconditional. Within that approach, one would, for example, systematically give the "value" 0 to a term of the form $\mathbf{E}/0$. In this context, the equality $2 + 3/0 = 2$ is *perfectly valid*, although rather hard to swallow. With that approach however, classical equality is saved, since clearly $3/0 = 3/0$ holds.

(3) Yet another solution [6] consists in claiming that a term of the form $\mathbf{E}/0$ denotes a genuine real number, but that this number is *unknown*. This approach is called *under-specification*. Here the definition is unconditional, as in the previous approach, but also *incomplete*, in contrast with the previous one. What is hard to accept, however, in this approach is the very fact that $3/0$ *denotes a real number*. This claim is not supported by any classical mathematical construction of such numbers (Dedekind, Cauchy). Notice that there is no means of proving nor refuting the equality $2 + 3/0 = 2$, but, as in the previous one, equality is saved (a real number is indeed equal to itself, thus $3/0=3/0$ holds).

(4) A more drastic approach [9] is one by which the definition of new term symbols is banished. The introduction of new predicate symbols is the only kind of allowed definition. With that approach, an "equality" is redefined for each implicit new term. For instance, division is defined by the predicate symbol Div, where $\mathsf{Div}(\mathbf{F}, \mathbf{E_1}, \mathbf{E_2})$ stands for $\mathbf{F} = \mathbf{E_1}/\mathbf{E_2}$. It is thus defined as follows:

$$\vdash \; \mathsf{Div}(\mathbf{F}, \mathbf{E_1}, \mathbf{E_2}) \; \Leftrightarrow \; \neg\,\mathsf{Zero}(\mathbf{E2}) \,\wedge\, \mathsf{Mult}(\mathbf{E_1}, \mathbf{F}, \mathbf{E_2})$$

where $\mathsf{Zero}(\mathbf{E_2})$ stands for $\mathbf{E_2} = 0$ and $\mathsf{Mult}(\mathbf{E_1}, \mathbf{F}, \mathbf{E_2})$ stands for $\mathbf{E_1} = \mathbf{F} \times \mathbf{E_2}$. The equality $2 + 3/0 = 2$ becomes the following monster:

$$\mathsf{Zero}(x) \,\wedge\, \mathsf{Two}(y) \,\wedge\, \mathsf{Three}(z) \,\wedge\, \mathsf{Div}(t, z, x) \,\wedge\, \mathsf{Plus}(u, y, t) \Rightarrow u = y$$

reducing to the following, which is thus true (since the antecedent of the implication is false):

$$\mathsf{Zero}(x) \,\wedge\, \mathsf{Two}(y) \,\wedge\, \mathsf{Three}(z) \,\wedge\, \neg\,\mathsf{Zero}(x) \,\wedge\, \mathsf{Mult}(z, t, x) \,\wedge\, \mathsf{Plus}(u, y, t) \Rightarrow$$
$$u = y$$

As can be seen, the problem of ill-definedness is now completely evacuated (since there are no terms except variables). Are we, however, ready to accept the rather heavy price?

(5) For the sake of completeness, we must mention the approach [10] taken in IMPS where an undefined predicate is false. It is advocated that this approach is that "commonly used by mathematicians" and "taught to American students in high school and college".

(6) A "final" solution [3], [4] is based on the idea that a formal language containing potentially ill-defined expressions can be given a semantic interpretation within a three-valued logical domain. So far, it is thus very close to the approach of C.B. Jones. Besides a number of technicalities, the main difference, however, lies in the practical treatment of terms and predicates that are interpreted by an undefined value. Rather than being fully integrated within the logic as in C.B. Jones work, they are only "marginally" integrated as is explained in what follows. In this approach, all terms and predicates are associated with a two-valued logical pseudo-operator, $\mathcal{D}$, which is given the following interpretation: $\mathcal{D}(\mathbf{F})$ is given the value false when the formula (predicate or term) $\mathbf{F}$ is interpreted as undefined, and the value true otherwise. Notice that $\mathcal{D}(\mathbf{F})$ can never be interpreted as undefined. Given a predicate $\mathbf{P}$, it is argued that $\mathcal{D}(\mathbf{P})$ can always be constructed. If $\mathcal{D}(\mathbf{P})$ is then proved successfully then $\mathbf{P}$ is guaranteed to only contain sub-predicates and sub-terms that are not undefined. $\mathcal{D}(\mathbf{P})$ has acted as a *filter*. As a consequence, the interpretation (and proof) of $\mathbf{P}$ can proceed from there within a *pseudo-two-valued logic*: in fact, a three-valued one where undefined is now guaranteed to be never encountered. Notice that, in this approach, the equality $2 + 3/0 = 2$ does not pass the filter and is thus simply not considered at all as a formal text to be proved. Interestingly enough, the status of the equality $2 + 3/0 = 2$ as well as that of its negation $2 + 3/0 \neq 2$ can be summarized in the following table for the various approaches we have considered:

	$2 + 3/0 = 2$	$2 + 3/0 \neq 2$
1	undefined	undefined
2	true	false
3	unprovable	unprovable
4	true	true
5	false	false
6	rejected	rejected

1.8   Our Approach

Our approach is very much in the spirit of the last one. It is very close to it, all our results are the same. Only the construction (justification) is *completely different*. We do not make any detour through a three-valued logic. In fact, we do not make any detour through any semantical interpretation at all. We entirely remain within the *syntactic manipulation of proofs*. Our approach just reduces, in a sense, to the development of a mere *proof strategy*.

Our work has been motivated by the very observation, made by the second author of this article, that in the previous approach, when given the task of proving a predicate $\mathbf{P}$, one proves in fact *more* than just it, one rather proves $\mathcal{D}(\mathbf{P}) \wedge \mathbf{P}$. Let's call this predicate $\mathcal{N}(\mathbf{P})$. It was then found that the syntactic transformation $\mathcal{N}(\mathbf{P})$ can be given quite a simple (inductive) definition for the different kind of syntactic constructs of our formal language. It was then easy to reconstruct the filter $\mathcal{D}(\mathbf{P})$ and prove that a successful proof of that filter implies that the proof of $\mathbf{P}$ can then be done safely with our "standard" prover, only slightly modified (simplified) by just *unprotecting* all the conditional definitions it may contain so as to transform them into mere unconditional definitions. This allows one then to possibly fully *eliminate* all the symbols introduced by any kind of definition: we have indeed fullfiled again the criterion of eliminability.

The purpose of this paper is thus to study under which circumstances it is possible to introduce new term symbols by means of conditional definitions in a formal theory as if the definitions in question were unconditional, thus recovering completely the right to subsequently *eliminate these symbols* without bothering about the validity of such an elimination. Taking again the example of $(\mathbf{E_1}/\mathbf{E_2}) \times \mathbf{E_2}$, we want to be in situations where the possibility to simplify by $\mathbf{E_2}$ is *always* guaranteed.

Although our approach seems to be close to the previous ones, one must be aware that it is, in essence, *quite distinct*. This is so because we do not think in terms of any *valuation*. We do not reason in terms of any semantics. For that reason, a predicate, in our framework, can never said to be *undefined*, in exactly the same way as it cannot said to be *true* or *false*. A predicate, at best, is proved or refuted (its negation being proved), or else not (yet) proved nor refuted. What is fundamental then is to determine what the rules of the "proving" game are. In our view, the predicate $0/\mathbf{E} = 0$ is not stamped with any "bad mark": it is something that we can write. However, it is something that we cannot prove nor refute because we do not know whether $\mathbf{E}$ is equal to 0 or not. What our method tells us is exactly this: as $0/\mathbf{E} = 0$ does not pass the filter (meaning for us that no proof can successfully be conducted), you better give more information (i.e. $\mathbf{E} \neq 0$) so that a proof can be successfully performed. By doing this, we do not modify "classical" logic, we do not introduce any special logic at all. We do not modify equality in any way: $\vdash \mathbf{E} = \mathbf{E}$ is still an axiom, and the Leibnithz Law still a rule of inference. The only price we pay is to sometimes reject perfectly provable formulae such as $\mathbf{E}/\mathbf{F} = \mathbf{E}/\mathbf{F}$. Rejecting means, again, that we are asking for more hypotheses.

2   Application to Set Theory

In what follows, we shall illustrate our approach on a specific (although general enough) formal theory, namely Set Theory. In this short section, we recall the way it was introduced in [1] as an *extension* of Predicate Calculus with Equality. We shall also present the proving system we shall use in order to discharge theorems within Set Theory.

2.1   First Order Predicate Calculus with Equality

The language of the classical First Order Predicate Calculus with Equality (and pairs) is defined by the following syntax where the syntactic category *prd* stands for predicates, vrb for variables, trm for terms, and idt for identifiers. This syntax can be disambiguated by means of the usual ingredients, namely the usage of parentheses and the assignment of precedences in decreasing orders to $\neg$, $\wedge$, and $\Rightarrow$.

$$
\begin{array}{rcl}
prd & ::= & prd \wedge prd \\
    &   & prd \Rightarrow prd \\
    &   & \neg\, prd \\
    &   & \forall\, vrb \cdot prd \\
    &   & trm = trm
\end{array}
\qquad
\begin{array}{rcl}
trm & ::= & vrb \\
    &   & trm , trm \\[1ex]
vrb & ::= & idt \\
    &   & vrb , vrb
\end{array}
$$

We suppose that we have at our disposal a Proof System for Predicate Calculus with Equality. Let's call it **PSPC**. This might be, for example, the classical corresponding Sequent Calculus. We shall not present here the axioms and inference rules of this calculus.

2.2   Extending Predicate Calculus to Handle Set Theory

The basic syntax of Set Theory is an extension of the previous one. Now appear the new predicate construct of *set membership* and the new syntactic category of *set*. The three set constructs are the classical cartesian product, comprehension and power set. Notice that, in this syntax, sets are terms whereas some terms are not sets (for instance, pairs).

$$
\begin{array}{rcl}
prd & ::= & \ldots \\
    &   & trm \in set \\[1ex]
trm & ::= & \ldots \\
    &   & set
\end{array}
\qquad
\begin{array}{rcl}
set & ::= & set \times set \\
    &   & \{\, vrb \mid vrb \in set \wedge prd \,\} \\
    &   & \mathbb{P}\,(set) \\
    &   & idt
\end{array}
$$

2.3   A Simplified Axiomatization of Set Theory

The following axioms are mere *linguistic* axioms expressing the way set membership can be defined in a straightforward way for the three basic constructs we have introduced (note that in axiom **A2**, **x** must not be free in **S**, and in axiom **A3**, **x** must not be free in **S** and **T**):

$$
\begin{array}{rlll}
\vdash & \mathbf{E, F} \in \mathbf{S} \times \mathbf{T} & \Leftrightarrow \mathbf{E} \in \mathbf{S} \wedge \mathbf{E} \in \mathbf{T} & \mathbf{A1} \\
\vdash & \mathbf{E} \in \{\, \mathbf{x} \mid \mathbf{x} \in \mathbf{S} \wedge \mathbf{P_x} \,\} & \Leftrightarrow \mathbf{E} \in \mathbf{S} \wedge \mathbf{P_E} & \mathbf{A2} \\
\vdash & \mathbf{S} \in \mathbb{P}\,(\mathbf{T}) & \Leftrightarrow \forall \mathbf{x} \cdot (\mathbf{x} \in \mathbf{S} \Rightarrow \mathbf{x} \in \mathbf{T}) & \mathbf{A3}
\end{array}
$$

In this axiomatization we have used boldface characters to indicate that the corresponding identifiers denote some *meta-variables*. More precisely, **E** and **F** are meta-variables standing for *trm*, **S** and **T** are meta-variables standing for *set*, **P** is a meta-variable standing for *prd*, and finally **x** is a meta-variable standing for *vrb*.

Notice that the above axioms, although they may appear to, do *not* constitute proper definitions of the symbol $\in$. Simply because that symbol does appear on both sides of the equivalence sign. They constitute however three genuine definitions of some specializations of set membership, namely "belonging to a cartesian product", "belonging to a set comprehension", and "belonging to a power set". Such definitions clearly obey our requirements for unconditional definitions and as such *can be eliminated*. Another axiom, which is slightly different from the previous ones, is that of *extensionality* relating set equality and set membership

$$\vdash \quad \mathbf{S} \in \mathbb{P}(\mathbf{T}) \wedge \mathbf{T} \in \mathbb{P}(\mathbf{S}) \;\Rightarrow\; \mathbf{S} = \mathbf{T} \qquad \mathbf{A4}$$

2.4   Extending the Formal Language by Means of Pure Definitions

Next is a sample of *pure* definitions allowing us to extend our language by introducing respectively disjunction, existential quantification, and set inclusion.

$$
\begin{array}{lll}
\vdash & \mathbf{P} \vee \mathbf{Q} \;\Leftrightarrow\; \neg \mathbf{P} \Rightarrow \mathbf{Q} & \mathbf{D1} \\
\vdash & \exists \mathbf{x} \cdot \mathbf{P} \;\Leftrightarrow\; \neg \forall \mathbf{x} \cdot \neg \mathbf{P} & \mathbf{D2} \\
\vdash & \mathbf{S} \subseteq \mathbf{T} \;\Leftrightarrow\; \mathbf{S} \in \mathbb{P}(\mathbf{T}) & \mathbf{D3}
\end{array}
$$

These are indeed pure definitions because: (1) the introduced symbols or constructs are *new*, (2) they appear in the definiendum only, and (3) they are defined in terms of *previous symbols*. As a consequence, they just represent some (very) *useful abbreviations* that can be eliminated *in all circumstances*. From definition **D3** and axiom **A3**, we immediately obtain the following *derived* axiom:

$$\vdash \quad \mathbf{S} \subseteq \mathbf{T} \;\Leftrightarrow\; \forall \mathbf{x} \cdot (\mathbf{x} \in \mathbf{S} \;\Rightarrow\; \mathbf{x} \in \mathbf{T}) \qquad \mathbf{A5}$$

2.5   A Proof System for Set Theory

Our proof system for Predicate Calculus extended with Set Theory is a *straight-forward extension* of **PSPC**. Since any predicate **P** involving sets can obviously be reduced to set memberships, we can use the four axioms **A1** to **A5** to gradually equivalently *eliminate* the various set constructs until there only remains some irreducible "meaningless" set memberships. At this point, we can use our standard **PSPC** on the resulting predicate **Q**, which, again, is equivalent to the original predicate **P**. An example of such a procedure is given by the proof of the following predicate **P**: $S \subseteq T \,\wedge\, T \subseteq U \,\Rightarrow\, S \subseteq U$ leading to the proof of the following equivalent predicate **Q**:

$$\forall x \cdot (\, x \in S \,\Rightarrow\, x \in T \,) \,\wedge\, \forall x \cdot (\, x \in T \,\Rightarrow\, x \in U \,) \,\Rightarrow\, \forall x \cdot (\, x \in S \,\Rightarrow\, x \in U \,)$$

As can be seen, the three predicate $x \in S$, $x \in T$, and $x \in U$ are now "meaning-less". They could be replaced by the three predicates U_x, V_x, and W_x without changing the resulting proof, which, in this case, is completely trivial.

The previous proof technique, which can be used for proving all set-theoretic statements, is not claimed to be "the" proof technique to be used in such a case. We shall use it, however, in this paper because it is very simple. On the other hand, our thesis is that other techniques, using ad-hoc proof rules acting directly at the set-theoretic level, can certainly be deduced from that one (in that the proof rules in question can certainly be proved using it).

What should be clear is that this technique of proof (using a preliminary translation of a set-theoretic predicate into a "pure" predicate) requires that the set-theoretic constructs can *always be eliminated*. This is where the usage of *conditional* definitions might be very problematic since, as we know, such definitions limit the criterion of eliminability. Our main problem thus will be to somehow succeed in using conditional definitions *as if they were not conditional*.

We now enter in the heart of our subject. In the next section, we shall see how "simple" conditional definitions can be eliminated from a formal text to be proved, provided it passes the filter of the standard decision procedure of *type-checking*. In the section to follow after that one, we shall consider more elaborate extensions necessitating a less primitive filter, one that will involve making some *genuine proofs*, which will not be trivially discharged by a decision procedure (although quite simple in general).

3   Filtering the Formal Text with Type-Checking

In this section, we shall extend our set-theoretic language by means of a number of definitions aimed at introducing all the classical constructs, namely union, intersection, difference and the like. In doing so, we shall see that such definitions need to be conditional. Fortunately, the conditions are not very "strong", so that it will be possible to eliminate them after filtering successfully the formal text by means of a simple *decision procedure*: type-checking.

3.1   Extending the Formal Language by Means of Simple Conditional Definitions

Next is another series of extensions of our formal language, extensions corre-sponding to the classical set-theoretic symbols of union, intersection and differ-ence. These symbols are introduced by means of some *conditional* definitions.

$$\mathbf{A} \subseteq \mathbf{S} \; ; \; \mathbf{B} \subseteq \mathbf{S} \; \vdash \; \mathbf{A} \cup \mathbf{B} = \{\mathbf{x} \,|\, \mathbf{x} \in \mathbf{S} \,\wedge\, (\mathbf{x} \in \mathbf{A} \,\vee\, \mathbf{x} \in \mathbf{B})\} \qquad \mathbf{D6}$$
$$\mathbf{A} \subseteq \mathbf{S} \; ; \; \mathbf{B} \subseteq \mathbf{S} \; \vdash \; \mathbf{A} \cap \mathbf{B} = \{\mathbf{x} \,|\, \mathbf{x} \in \mathbf{S} \,\wedge\, (\mathbf{x} \in \mathbf{A} \,\wedge\, \mathbf{x} \in \mathbf{B})\} \qquad \mathbf{D7}$$
$$\mathbf{A} \subseteq \mathbf{S} \; ; \; \mathbf{B} \subseteq \mathbf{S} \; \vdash \; \mathbf{A} - \mathbf{B} = \{\mathbf{x} \,|\, \mathbf{x} \in \mathbf{S} \,\wedge\, (\mathbf{x} \in \mathbf{A} \,\wedge\, \mathbf{x} \notin \mathbf{B})\} \qquad \mathbf{D8}$$

As previously, the symbols are new, they appear in one side only of the equality sign, and they are all defined in terms of previous symbols. The essential difference with the previous definitions lies in the hypotheses that can be seen on the left hand side of the $\vdash$ symbol. The rôle of such hypotheses is to exhibit a set **S** within which **A** and **B** are included so that union, intersection and difference can be defined by means of *set comprehension* (i.e. as subsets of **S**) in an obvious manner. Notice finally that since the introduced symbols indirectly involve the extra variable **S**, it has to be proved that these definitions lead to the same results whatever the set in question (unicity).

A consequence of the presence of such hypotheses is that the introduced symbols *cannot always be eliminated*: this can arise only when the hypotheses in question are present. Next is an example where this is the case: this is a formal proof of $A \subseteq S \,;\, B \subseteq S \;\vdash A \subseteq A \cup B$

$$
\begin{array}{lllll}
(1) & A \subseteq S \; ; \; B \subseteq S & & \vdash & A \subseteq A \cup B & \\
(2) & A \subseteq S \; ; \; B \subseteq S \; ; \; x \in A & \vdash & x \in A \cup B & \mathbf{A5} \\
(3) & A \subseteq S \; ; \; B \subseteq S \; ; \; x \in A & \vdash & x \in \{x \,|\, x \in S \,\wedge\, (x \in A \,\vee\, x \in B)\} & \mathbf{D6} \\
(4) & A \subseteq S \; ; \; B \subseteq S \; ; \; x \in A & \vdash & x \in S \,\wedge\, (x \in A \,\vee\, x \in B) & \mathbf{A2} \\
(5) & A \subseteq S \; ; \; B \subseteq S \; ; \; x \in A & \vdash & x \in A \,\vee\, x \in B & \mathbf{A5} \\
\end{array}
$$

As can be seen, definition **D6** can be applied on line **(2)** since the proper hypotheses (i.e. $A \subseteq S$ and $B \subseteq S$) are present. By looking more carefully at this proof, we can observe that the transformation of $x \in A \cup B$ at line **(2)** into $x \in A \vee x \in B$ at line **(5)** takes three steps, which seems rather tedious for such a triviality. We also notice that, on line **(4)**, we have to discharge the little predicate $x \in S$, which seems rather irrelevant considering what our main problem is. In fact, it is possible to drastically shorten this (and similar) proof by means of the following three rules, which are easily *derivable* from the previous definitions:

$$\mathbf{A} \subseteq \mathbf{S} \; ; \; \mathbf{B} \subseteq \mathbf{S} \; \vdash \; \mathbf{E} \in \mathbf{A} \cup \mathbf{B} \;\Leftrightarrow\; \mathbf{E} \in \mathbf{A} \,\vee\, \mathbf{E} \in \mathbf{B} \qquad \mathbf{R1}$$
$$\mathbf{A} \subseteq \mathbf{S} \; ; \; \mathbf{B} \subseteq \mathbf{S} \; \vdash \; \mathbf{E} \in \mathbf{A} \cap \mathbf{B} \;\Leftrightarrow\; \mathbf{E} \in \mathbf{A} \,\wedge\, \mathbf{E} \in \mathbf{B} \qquad \mathbf{R2}$$
$$\mathbf{A} \subseteq \mathbf{S} \; ; \; \mathbf{B} \subseteq \mathbf{S} \; \vdash \; \mathbf{E} \in \mathbf{A} - \mathbf{B} \;\Leftrightarrow\; \mathbf{E} \in \mathbf{A} \,\wedge\, \mathbf{E} \notin \mathbf{B} \qquad \mathbf{R3}$$

In these rules, the union, intersection and difference of two sets are not defined directly as sets, they are rather defined indirectly by giving an equivalence to the corresponding memberships. What is very interesting about these rules is that the predicate **E** $\in$ **S** has completely disappeared. In fact, **S** is now only mentioned in the hypotheses, where the predicates $\mathbf{A} \subseteq \mathbf{S}$ and $\mathbf{B} \subseteq \mathbf{S}$ just appear as *witnesses* showing that $\mathbf{A}$ and $\mathbf{B}$ are both included in the same set $\mathbf{S}$. These rules lead to the following shorter proof, where it can now be observed that $\mathbf{S}$ does not appear on the right hand side of $\vdash$ as was the case in previous proof. As can be seen, the proof is now reduced almost to the essential.

$$
\begin{array}{llll}
(1) & A \subseteq S \;;\; B \subseteq S & \vdash\; A \subseteq A \cup B & \\
(2) & A \subseteq S \;;\; B \subseteq S \;;\; x \in A \;\vdash\; x \in A \cup B & & \mathbf{A5} \\
(3) & A \subseteq S \;;\; B \subseteq S \;;\; x \in A \;\vdash\; x \in A \lor x \in B & & \mathbf{R1}
\end{array}
$$

On this proof, we can observe that Rule $\mathbf{R1}$ can easily be applied at line (2) because we exactly have the proper hypotheses at hand. Unfortunately, the situation is not always that convenient. For example, suppose we want to prove part of the distributivity of intersection over union, namely $A \subseteq S \;;\; B \subseteq S \;;\; C \subseteq S \;\vdash\; (A \cup B) \cap C \subseteq (A \cap C) \cup (B \cap C)$. Since we shall certainly apply rules $\mathbf{R1}$ and $\mathbf{R2}$ in order to decompose union and intersection, we shall need the missing hypotheses $A \cup B \subseteq S$, $A \cap C \subseteq S$ and $B \cap C \subseteq S$. In the absence of specific rules, the generation of such hypotheses might be tedious. Fortunately, this can be shortened thanks to the following derived *closure* rules:

$$
\begin{array}{lll}
\mathbf{A} \subseteq \mathbf{S} \;;\; \mathbf{B} \subseteq \mathbf{S} \;\vdash\; \mathbf{A} \cup \mathbf{B} \subseteq \mathbf{S} & \mathbf{R4} \\
\mathbf{A} \subseteq \mathbf{S} \;;\; \mathbf{B} \subseteq \mathbf{S} \;\vdash\; \mathbf{A} \cap \mathbf{B} \subseteq \mathbf{S} & \mathbf{R5} \\
\mathbf{A} \subseteq \mathbf{S} \;;\; \mathbf{B} \subseteq \mathbf{S} \;\vdash\; \mathbf{A} - \mathbf{B} \subseteq \mathbf{S} & \mathbf{R6}
\end{array}
$$

On this occasion, we also notice that similar closure rules can be proved for the basic operators (pairing, cartesian product, set comprehension, and power sets):

$$
\begin{array}{lll}
\mathbf{E} \in \mathbf{S} \;;\; \mathbf{F} \in \mathbf{T} & \vdash\; \mathbf{E}, \mathbf{F} \in \mathbf{S} \times \mathbf{T} & \mathbf{R7} \\
\mathbf{A} \subseteq \mathbf{S} \;;\; \mathbf{B} \subseteq \mathbf{T} & \vdash\; \mathbf{A} \times \mathbf{B} \subseteq \mathbf{S} \times \mathbf{T} & \mathbf{R8} \\
\mathbf{A} \subseteq \mathbf{S} & \vdash\; \{\mathbf{x} \mid \mathbf{x} \in \mathbf{A} \land \mathbf{P}\} \subseteq \mathbf{S} & \mathbf{R9} \\
\mathbf{A} \subseteq \mathbf{S} & \vdash\; \mathbb{P}(\mathbf{A}) \subseteq \mathbb{P}(\mathbf{S}) & \mathbf{R10}
\end{array}
$$

The proof now proceeds as follows :

$$
\begin{array}{llll}
(1) & A \subseteq S \;;\; B \subseteq S \;;\; C \subseteq S \;\vdash\; (A \cup B) \cap C \subseteq (A \cap C) \cup (B \cap C) & \\
(2) & \cdots \;;\; A \cup B \subseteq S \quad\vdash\; (A \cup B) \cap C \subseteq (A \cap C) \cup (B \cap C) & \mathbf{R4} \\
(3) & \cdots \;;\; A \cap C \subseteq S \quad\vdash\; (A \cup B) \cap C \subseteq (A \cap C) \cup (B \cap C) & \mathbf{R5} \\
(4) & \cdots \;;\; B \cap C \subseteq S \quad\vdash\; (A \cup B) \cap C \subseteq (A \cap C) \cup (B \cap C) & \mathbf{R5} \\
(5) & \cdots \;;\; x \in (A \cup B) \cap C \quad\vdash\; x \in (A \cap C) \cup (B \cap C) & \mathbf{A5} \\
& \cdots
\end{array}
$$

We have organized the proof in such a way that all the needed hypotheses are generated at the beginning, so that the proof can then proceed with everything at hand.

3.2   Simple Type-Checking

In the previous proof, the steps **(2)**, **(3)**, and **(4)** seem to be extremely general. They consist in generating the proper hypotheses corresponding to some of the sub-formulae of the main statement to prove. We notice that the *goal*, situated on the right hand side of $\vdash$, remains the same during these steps. In the case where more complicated statements have to be proved, it seems that the number of such initial steps might be rather large. On the other hand, these steps, which are clearly very mechanical and systematic, are the consequence of applying the closure rules **R4**, **R5**, and **R6**. Would it be possible to have, in a certain way, these steps generated automatically ?

The answer is yes and it is given by *type-checking*. It is outside the scope of this paper to re-formulate the theory of type-checking as applied to set theory [1]. What will be said here is only the following: provided the statement to prove *does type-check* then each of its sub-terms has a *type*, which, *for the moment*, is a certain *set* to which it belongs. In other words, provided type-checking is successful, then the required hypotheses (and more) are all guaranteed even *without being generated at all*.

3.3   Simplifying the Proof System

The last statement of previous section has an extremely important consequence. Provided, again, type-checking is performed systematically and successfully (we remind the reader that it is performed by an automatic decision procedure), then it is possible to *remove the conditions* in rules **R1**, **R2** and **R3** since we know that the corresponding hypotheses are always (virtually) there. We can thus rephrase and simplify our proof system for set theory (as considered so far) as follows:

$\vdash$	$\mathbf{E, F \in S \times T}$	$\Leftrightarrow$	$\mathbf{E \in S \ \wedge \ E \in T}$	**A1**
$\vdash$	$\mathbf{E \in \{x \mid x \in S \wedge P_x\}}$	$\Leftrightarrow$	$\mathbf{E \in S \ \wedge \ P_E}$	**A2**
$\vdash$	$\mathbf{S \in \mathbb{P}(T)}$	$\Leftrightarrow$	$\forall x \cdot (x \in S \Rightarrow x \in T)$	**A3**
$\vdash$	$\mathbf{S \subseteq T}$	$\Leftrightarrow$	$\forall x \cdot (x \in S \Rightarrow x \in T)$	**A5**
$\vdash$	$\mathbf{S \subseteq T \ \wedge \ T \subseteq S}$	$\Leftrightarrow$	$\mathbf{S = T}$	**A4**
$\vdash$	$\mathbf{E \in A \cup B}$	$\Leftrightarrow$	$\mathbf{E \in A \ \vee \ E \in B}$	**R1**
$\vdash$	$\mathbf{E \in A \cap B}$	$\Leftrightarrow$	$\mathbf{E \in A \ \wedge \ E \in B}$	**R2**
$\vdash$	$\mathbf{E \in A - B}$	$\Leftrightarrow$	$\mathbf{E \in A \ \wedge \ E \notin B}$	**R3**

Given a set-theoretic predicate to prove, what should be completely clear from these rules is that the set-theoretic constructs contained in it (at least those envisaged so far) can always be eliminated by a straightforward translation process. The net result is an equivalent predicate containing basic set membership only. The proof can then proceed from there *within pure predicate calculus* (the remaining set membership being "uninterpreted").

3.4   Constructing the Typing System

Similarly, we can put together the derivable closure rules as follows:

$$
\begin{array}{llll}
\mathbf{E} \in \mathbf{S} \ ; \ \mathbf{F} \in \mathbf{T} & \vdash & \mathbf{E}, \mathbf{F} \in \mathbf{S} \times \mathbf{T} & \text{R7} \\
\mathbf{A} \subseteq \mathbf{S} \ ; \ \mathbf{B} \subseteq \mathbf{T} & \vdash & \mathbf{A} \times \mathbf{B} \subseteq \mathbf{S} \times \mathbf{T} & \text{R8} \\
\mathbf{A} \subseteq \mathbf{S} & \vdash & \{\, \mathbf{x} \mid \mathbf{x} \in \mathbf{A} \wedge \mathbf{P} \,\} \subseteq \mathbf{S} & \text{R9} \\
\mathbf{A} \subseteq \mathbf{S} & \vdash & \mathbb{P}(\mathbf{A}) \subseteq \mathbb{P}(\mathbf{S}) & \text{R10} \\
\mathbf{A} \subseteq \mathbf{S} \ ; \ \mathbf{B} \subseteq \mathbf{S} & \vdash & \mathbf{A} \cup \mathbf{B} \subseteq \mathbf{S} & \text{R4} \\
\mathbf{A} \subseteq \mathbf{S} \ ; \ \mathbf{B} \subseteq \mathbf{S} & \vdash & \mathbf{A} \cap \mathbf{B} \subseteq \mathbf{S} & \text{R5} \\
\mathbf{A} \subseteq \mathbf{S} \ ; \ \mathbf{B} \subseteq \mathbf{S} & \vdash & \mathbf{A} - \mathbf{B} \subseteq \mathbf{S} & \text{R6}
\end{array}
$$

We notice that our closure rules all have the same shape. On the right hand-side of $\vdash$, we have a predicate expressing that a certain term $\mathbf{T}$, depending on a number of meta-variables (i.e. $\mathbf{E}$, $\mathbf{F}$, $\mathbf{A}$, $\mathbf{B}$), belongs (since set inclusion can always be transformed in power set membership) to a certain set S made of *other* meta-variables (i.e. $\mathbf{S}$, $\mathbf{T}$). The set S is either a power set or a cartesian product. On the lefthand side of $\vdash$, we have a number of hypotheses where each meta-variable of T is supposed to belong to a certain set involving only the meta-variables of S, the set in question being made of a simple meta-variable, a cartesian product, or a power set.

From this very shape, these closure rules can be read as follows: if each of the components of a certain construct has a well-defined type then the construct in question also has a well-defined type. As can be seen, a type is either simple, or a cartesian product of types or else a power set of types. Typing gives the *dimensionality* of terms. As in physics where you do not think of adding a distance to a mass, here we do not envisage taking the union of a simple set with, say, the cartesian product of two sets.

3.5   Inference Type-Checking

The first proof which we used when starting this discussion was that of $A \subseteq S \ ; \ B \subseteq S \ \vdash \ A \subseteq A \cup B$. It is now quite clear that, taking account of the simplified proof system we have presented in the previous section, the presence of the hypotheses $A \subseteq S$ and $B \subseteq S$ does not seem to be necessary any more since from now on the rules have no conditions. What are these hypotheses needed for thus ? Well, they are *just* there to ensure the possibility of a correct type-checking of the predicate $A \subseteq A \cup B$.

This seems to be a bit superfluous. We wonder whether we could type-check *without* them. The answer is *inference type-checking*. It gives us exactly what we need, that is, the possibility, when successful, to *infer* from the very shape of a formula that a type *does exist* for each of its sub-terms. This is, in fact, the only thing we need, since in our proofs (as already noticed), we never use the types in question. The ultimate simplification of the proof is thus the following:

$$
\begin{array}{llll}
\textbf{(1)} & & \vdash & A \subseteq A \cup B \\
\textbf{(2)} & x \in A & \vdash & x \in A \cup B & \textbf{A5} \\
\textbf{(3)} & x \in A & \vdash & x \in A \lor x \in B & \textbf{R1}
\end{array}
$$

3.6 Practical Conclusion

At this point, we can formulate our proof stategy as follows for proving a set-theoretic statement **P**:

(1) Type-check **P**
(2) Eliminate set membership predicates as much as possible in **P**
(3) Prove the resulting predicate using **PSPC**

4 Filtering the Formal Text with Extended Type-Checking

In this section, we now extend our set-theoretic language by means of a number of more elaborate definitions aimed at introducing constructs such as $\mathbf{f(E)}$ (the application of a partial function $\mathbf{f}$ to an argument $\mathbf{E}$). Other similar extensions could be introduced at this point, we shall not do so however in order to simplify the presentation since the development concerning these other extensions is of the same nature as that presented in what follows.

The corresponding definitions will appear to be more complicated than the ones we have considered so far. And we shall discover that a simple decision procedure such as type-checking *is not sufficient any more*.

4.1 More Extensions

Encouraged by the previous results, we can envisage more extensions such as those corresponding to the set of *binary relations* from one set to another, the *domain* and *range* of a relation, and the set of *partial functions* from one set to another. For this, we shall use our new approach, consisting in defining these symbols by means of pure definitions and then deriving the corresponding closure (typing) rules.

$$
\begin{array}{llll}
\vdash & S \leftrightarrow T & = & \mathbb{P}(S \times T) & \text{D9} \\
\vdash & E \in \mathrm{dom}\,(r) & \Leftrightarrow & \exists y \cdot (\, E, y \in r\,) & \text{D10} \\
\vdash & F \in \mathrm{ran}\,(r) & \Leftrightarrow & \exists x \cdot (\, x, F \in r\,) & \text{D11} \\
\vdash & \mathrm{fnc}\,(f) & \Leftrightarrow & \forall (x, y, z) \cdot (\, x, y \in f \wedge x, z \in f \Rightarrow y = z\,) & \text{D12} \\
\vdash & f \in S \nrightarrow T & \Leftrightarrow & f \in S \leftrightarrow T \wedge \mathrm{fnc}\,(f) & \text{D13}
\end{array}
$$

The corresponding closure rules are as follows

$$
\begin{array}{llll}
A \subseteq S \;;\; B \subseteq T & \vdash & A \leftrightarrow B \subseteq S \times T & \text{R11} \\
r \subseteq S \times T & \vdash & \mathrm{dom}\,(r) \subseteq S & \text{R12} \\
r \subseteq S \times T & \vdash & \mathrm{ran}\,(r) \subseteq T & \text{R13} \\
A \subseteq S \;;\; B \subseteq T & \vdash & A \nrightarrow B \subseteq S \times T & \text{R14}
\end{array}
$$

4.2   Where Things Are Getting Odd Again

Our next extension consists in introducing the classical construct $f(E)$ denoting the application of a function f to its argument E. Notice that the new symbol that is introduced here is certainly not f, this is in fact a hidden binary symbol, let's call it apply, which together with its arguments yields $\mathrm{apply}\,(f, E)$. As we have done it before, this notation will be introduced by means of a definition. But as the direct definition of $f(E)$ by means of an equality is not convenient, we shall use an equivalence, namely that of the equality predicate $F = f(E)$. The first idea that comes immediately to mind corresponds to the following equivalence:

$$\vdash \quad F = f(E) \quad \Leftrightarrow \quad E, F \in f$$

But, for this to constitute a proper *definition*, we must prove the following *justifying theorems*

$$
\begin{aligned}
&\vdash \quad \forall (F, G) \cdot (E, F \in f \wedge E, G \in f \Rightarrow F = G\,) \\
&\vdash \quad \exists F \cdot (\, E, F \in f\,)
\end{aligned}
$$

By generalizing the first one to all E and using definitions **D12** and **D10**, we obtain the following:

$$
\begin{aligned}
&\vdash \quad \mathrm{fnc}\,(f) \\
&\vdash \quad E \in \mathrm{dom}\,(f)
\end{aligned}
$$

In order to evacuate the justifying theorems (but not the main problem, unfortunately !), our definition has thus to be made conditional:

$$\mathrm{fnc}\,(f) \;;\; E \in \mathrm{dom}\,(f) \;\vdash\; F = f(E) \;\Leftrightarrow\; E, F \in f \qquad \text{D14}$$

The corresponding closure rule can easily be derived from this:

$$\mathbf{f} \subseteq \mathbf{S} \times \mathbf{T} \ ; \ \mathbf{E} \in \mathbf{S} \ ; \ \mathsf{fnc}\,(\mathbf{f}) \ ; \ \mathbf{E} \in \mathsf{dom}\,(\mathbf{f}) \ \vdash \ \mathbf{f}(\mathbf{E}) \in \mathbf{T} \qquad \mathbf{R15}$$

It seems then that all our previous efforts are now vain. Thanks to type-checking, supposedly performed initially and successfuly on a predicate to prove, we had been able to transform all our previous conditional definitions into pure definitions. And, here again, comes a definition with some conditions that are of quite a different nature in comparison to those we had previously. As a result, the present conditions cannot be eliminated by assuming simple type-checking. Moreover, the typing system itself is now *polluted* by some conditions, which do not correspond at all to the typing assumptions we had in all our previous closure rules.

Nevertheless, we still believe that the complete elimination of these conditions in the definition and, consequently, in the closure rules is fundamental in order to drastically simplify matters in the proofs. But it seems that we are now *asking for the impossible*.

4.3   Saving the Typing System

Let us start by the apparently simpler task, namely that of de-polluting the typing system. For this let's rewrite the previous closure rule without the last two assumptions.

$$\mathbf{f} \subseteq \mathbf{S} \times \mathbf{T} \ ; \ \mathbf{E} \in \mathbf{S} \ \vdash \ \mathbf{f}(\mathbf{E}) \in \mathbf{T}$$

This is clearly mathematically wrong. Nevertheless, we have the feeling that there is a certain "truth" in this rule. In fact, under these limited assumptions, it certainly cannot be said that $\mathbf{f}(\mathbf{E})$ belongs to $\mathbf{T}$. But something can be said, namely that $\mathbf{f}(\mathbf{E})$ has the *dimension* of $\mathbf{T}$. We shall thus say that $\mathbf{f}(\mathbf{E})$ *is of type* $\mathbf{T}$, which from now on is *not the same* any more as saying that $\mathbf{f}(\mathbf{E})$ belongs to $\mathbf{T}$. In order to make the closure rules correct again, we have to introduce two new symbols: $\twoheadleftarrow$ to mean *is of type* (this correponds to $\in$) and $\preceq$ to mean *is of super-type* (this corresponds to $\subseteq$). All our closure rules now have to be rewritten as follows:

$$
\begin{array}{lll}
\mathbf{E} \twoheadleftarrow \mathbf{S} \ ; \ \mathbf{F} \twoheadleftarrow \mathbf{T} & \vdash & \mathbf{E}, \mathbf{F} \twoheadleftarrow \mathbf{S} \times \mathbf{T} & \mathbf{R7} \\
\mathbf{A} \preceq \mathbf{S} \ ; \ \mathbf{B} \preceq \mathbf{T} & \vdash & \mathbf{A} \times \mathbf{B} \preceq \mathbf{S} \times \mathbf{T} & \mathbf{R8} \\
\mathbf{A} \preceq \mathbf{S} & \vdash & \{\,\mathbf{x}\,|\,\mathbf{x} \in \mathbf{A} \wedge \mathbf{P}\,\} \preceq \mathbf{S} & \mathbf{R9} \\
\mathbf{A} \preceq \mathbf{S} & \vdash & \mathbb{P}(\mathbf{A}) \preceq \mathbb{P}(\mathbf{S}) & \mathbf{R10} \\
\mathbf{A} \preceq \mathbf{S} \ ; \ \mathbf{B} \preceq \mathbf{S} & \vdash & \mathbf{A} \cup \mathbf{B} \preceq \mathbf{S} & \mathbf{R4}
\end{array}
$$

$$
\begin{array}{llll}
A \preceq S \;;\; B \preceq S & \vdash & A \cap B \preceq S & R5 \\
A \preceq S \;;\; B \preceq S & \vdash & A - B \preceq S & R6 \\
A \preceq S \;;\; B \preceq T & \vdash & A \leftrightarrow B \preceq S \times T & R11 \\
r \preceq S \times T & \vdash & \mathrm{dom}\,(r) \preceq S & R12 \\
r \preceq S \times T & \vdash & \mathrm{ran}\,(r) \preceq T & R13 \\
A \preceq S \;;\; B \preceq T & \vdash & A \nrightarrow B \preceq S \times T & R14 \\
f \preceq S \times T \;;\; E \leftprec S & \vdash & f(E) \leftprec T & R15
\end{array}
$$

What is important about this new typing is the following: when the typing of a predicate is successful, then as soon as the proper conditions concerning all occurrences of the construct $f(E)$ in that predicate are met then it can be said that each of its sub-terms *belongs to its type*.

But, clearly, typing *alone* cannot guarantee this. As a consequence, what we had done previously, namely making all our definitions pure, is not valid any more since it was based on the very fact that each sub-term could be made members of certain sets. Things are not yet saved.

Our problem is thus now the following: how could we guarentee *in advance* (by some process to invent) that all these conditions are met on a predicate so that (1) the definition of $f(E)$ can be made unconditional, (2) the sub-terms of the predicate can then be guaranteed to belong to their types, and (3) by extension, all other definitions can again be made unconditional? As can be seen, things are very intricately mixed here.

4.4   Facing the Wall

Let's start gently then. Suppose we have an "atomic" predicate to prove: one involving only equality, set membeship or set inclusion as its main operator. Let's denote this predicate by $A_{f(E)}$, thus indicating that we have a single occurrence of $f(E)$ in it: this is an obvious simplification, which is presently used to make things more easily tractable, it will be subsequently generalized. This predicate is to be proved without assumptions (this is again a simplification).

$$\vdash A_{f(E)}$$

Suppose that $A_{f(E)}$ does type-check (in the new sense). In order to proceed further (that is, use our pure definitions to perform the proof), we must have the guarantee that the conditions $\mathrm{fnc}\,(f)$ and $E \in \mathrm{dom}\,(f)$ are met. So, we have no choice but to *prove* them, that is:

$$\vdash \mathrm{fnc}\,(f) \;\wedge\; E \in \mathrm{dom}\,(f)$$

Provided the previous proof is successful, then we could in principle put the two conditions as extra hypotheses (by applying the, so-called, *Cut Rule*) and proceed from there under this umbrella. We shall not do so however because we now feel free to use our definition of $f(E)$ without the extra conditions. These extra hypotheses are not needed : this is a technique we have learned from the simpler conditional definitions envisaged earlier.

And then we can obviously proceed as before, that is use *all our pure definitions* in order to gradually eliminate all the set-theoretic constructs. But, to begin with, can we eliminate $\mathbf{f(E)}$? The answer is positive thanks to the so-called "one point rule" allowing us to replace equivalently $\mathbf{A_{f(E)}}$ as follows (where y is supposed to be a "fresh" variable):

$$\vdash\ \forall y \cdot (y = \mathbf{f(E)} \ \Rightarrow\ \mathbf{A}_y)$$

This leads to the following by applying the (now pure) definition of $y = \mathbf{f(E)}$:

$$\vdash\ \forall y \cdot ((\mathbf{E}, y) \in \mathbf{f} \ \Rightarrow\ \mathbf{A}_y)$$

yielding eventually

$$(\mathbf{E}, y) \in \mathbf{f} \ \vdash\ \mathbf{A}_y$$

As can be seen, the term $\mathbf{f(E)}$ has now completely disappeared.

4.5   A Calculus of Syntactic Transformation

To summarize at this point, what we have done is to prove *a little more* than just $\mathbf{A_{f(E)}}$. Putting the extra together with the main, we have indeed proved

$$\vdash\ \mathsf{fnc}\,(\mathbf{f}) \ \wedge\ \mathbf{E} \in \mathsf{dom}\,(\mathbf{f}) \ \wedge\ \mathbf{A_{f(E)}}$$

In other words, we have *tranformed* our original predicate $\mathbf{A}$ into a slightly more complicated one. Let's give a name, $\mathcal{N}(\mathbf{A})$, to this transformation and also a name, $\mathcal{D}(\mathbf{A})$ to the extension. We have thus

$$\mathcal{N}(\mathbf{A}) \ = \ \mathcal{D}(\mathbf{A}) \wedge \mathbf{A}$$

It is now easy to generalize this transformation to predicates involving the logical operators. The syntactic tranformation $\mathcal{N}$ clearly distributes over conjunction, so that we have $\mathcal{N}(\mathbf{P} \wedge \mathbf{Q}) = \mathcal{N}(\mathbf{P}) \wedge \mathcal{N}(\mathbf{Q})$. The case of the negation is interesting. Suppose we have to prove $\neg \mathbf{A}$. The condition $\mathcal{D}(\mathbf{A})$ for applying the conditional definition remaining obviously the same, we have to prove $\vdash \mathcal{D}(\mathbf{A}) \wedge \neg \mathbf{A}$, that is $\vdash \neg (\mathcal{D}(\mathbf{A}) \Rightarrow \mathbf{A})$. By defining the following dual transformation: $\mathcal{T}(\mathbf{A}) = \mathcal{D}(\mathbf{A}) \Rightarrow \mathbf{A}$. We obtain $\mathcal{N}(\neg \mathbf{A}) = \neg \mathcal{T}(\mathbf{A})$, which can be subsequently generalized to any predicate $\mathbf{P}$, yielding $\mathcal{N}(\neg \mathbf{P}) = \neg \mathcal{T}(\mathbf{P})$. All this leads to the following syntactic calculus where $\mathbf{P}$ and $\mathbf{Q}$ are predicates and where $\mathbf{A}$ is an atomic predicate, which does not contain any logical operators:

$\mathcal{N}(\mathbf{A})$	$= \mathcal{D}(\mathbf{A}) \wedge \mathbf{A}$		$\mathcal{T}(\mathbf{A})$	$=$	$\mathcal{D}(\mathbf{A}) \Rightarrow \mathbf{A}$
$\mathcal{N}(\mathbf{P} \wedge \mathbf{Q})$	$= \mathcal{N}(\mathbf{P}) \wedge \mathcal{N}(\mathbf{Q})$		$\mathcal{T}(\mathbf{P} \wedge \mathbf{Q})$	$=$	$\mathcal{T}(\mathbf{P}) \wedge \mathcal{T}(\mathbf{Q})$
$\mathcal{N}(\neg \mathbf{P})$	$= \neg \mathcal{T}(\mathbf{P})$		$\mathcal{T}(\neg \mathbf{P})$	$=$	$\neg \mathcal{N}(\mathbf{P})$
$\mathcal{N}(\mathbf{P} \Rightarrow \mathbf{Q})$	$= \mathcal{T}(\mathbf{P}) \Rightarrow \mathcal{N}(\mathbf{Q})$		$\mathcal{T}(\mathbf{P} \Rightarrow \mathbf{Q})$	$=$	$\mathcal{N}(\mathbf{P}) \Rightarrow \mathcal{T}(\mathbf{Q})$
$\mathcal{N}(\forall \mathbf{x} \cdot \mathbf{P})$	$= \forall \mathbf{x} \cdot \mathcal{N}(\mathbf{P})$		$\mathcal{T}(\forall \mathbf{x} \cdot \mathbf{P})$	$=$	$\forall \mathbf{x} \cdot \mathcal{T}(\mathbf{P})$

4.6   Extended Typing

Unfortunately, although very simple and appealing, the previous calculus is not exactly what we need. In fact, and as already explained, what we are aiming at is to exhibit a preliminary proof (a filter) allowing us to then use the pure definitions within the main proof. This is exactly what we have done on the atomic predicate $\mathbf{A}$ where we first proved $\mathcal{D}(\mathbf{A})$, then $\mathbf{A}$. This is really that process that has to be generalized. The $\mathcal{D}$ operator, only used so far for atomic predicates, has thus to be generalized to any predicate. Generalizing first the relationship between $\mathcal{D}(\mathbf{A})$, $\mathcal{N}(\mathbf{A})$, and $\mathcal{T}(\mathbf{A})$, we obtain:

$$
\begin{aligned}
\mathcal{N}(\mathbf{P}) &= \mathcal{D}(\mathbf{P}) \wedge \mathbf{P} \\
\mathcal{T}(\mathbf{P}) &= \mathcal{D}(\mathbf{P}) \Rightarrow \mathbf{P}
\end{aligned}
$$

By rewriting $\neg\,\mathcal{T}(\mathbf{P})$ as $\mathcal{D}(\mathbf{P}) \wedge \neg\mathbf{P}$, we can see that $\neg\,\mathcal{T}(\mathbf{P}) \vee \mathcal{N}(\mathbf{P})$ is equivalent to $\mathcal{D}(\mathbf{P})$. As a consequence, we have

$$
\mathcal{D}(\mathbf{P}) = \mathcal{T}(\mathbf{P}) \Rightarrow \mathcal{N}(\mathbf{P})
$$

Notice that $\mathcal{D}(\mathbf{P})$ is thus the same as $\mathcal{N}(\neg\mathbf{P}) \vee \mathcal{N}(\mathbf{P})$, which is probably more intuitive than $\mathcal{T}(\mathbf{P}) \Rightarrow \mathcal{N}(\mathbf{P})$, since we have the strong feeling that $\mathcal{D}(\neg\mathbf{P})$ is the same as $\mathcal{D}(\mathbf{P})$. By applying this and the previous calculus, we obtain:

$$
\begin{aligned}
& \mathcal{D}(\mathbf{P} \wedge \mathbf{Q}) \\
={} & \mathcal{T}(\mathbf{P} \wedge \mathbf{Q}) \Rightarrow \mathcal{N}(\mathbf{P} \wedge \mathbf{Q}) \\
={} & \mathcal{T}(\mathbf{P}) \wedge \mathcal{T}(\mathbf{Q}) \Rightarrow \mathcal{N}(\mathbf{P}) \wedge \mathcal{N}(\mathbf{Q}) \\
={} & \neg\,\mathcal{T}(\mathbf{P}) \vee \neg\,\mathcal{T}(\mathbf{Q}) \vee (\mathcal{N}(\mathbf{P}) \wedge \mathcal{N}(\mathbf{Q})) \\
={} & (\mathcal{D}(\mathbf{P}) \wedge \neg\mathbf{P}) \vee (\mathcal{D}(\mathbf{Q}) \wedge \neg\mathbf{Q}) \vee (\mathcal{D}(\mathbf{P}) \wedge \mathcal{D}(\mathbf{Q}) \wedge \mathbf{P} \wedge \mathbf{Q}) \\
={} & (\mathcal{D}(\mathbf{P}) \wedge \neg\mathbf{P}) \vee (\mathcal{D}(\mathbf{Q}) \wedge \neg\mathbf{Q}) \vee (\mathcal{D}(\mathbf{P}) \wedge \mathcal{D}(\mathbf{Q}));
\end{aligned}
$$

Similar results can be obtained in the other cases leading to the following table:

$$
\begin{aligned}
\mathcal{D}(\mathbf{P} \wedge \mathbf{Q}) &= (\mathcal{D}(\mathbf{P}) \wedge \mathcal{D}(\mathbf{Q})) \vee (\mathcal{D}(\mathbf{P}) \wedge \neg\mathbf{P}) \vee (\mathcal{D}(\mathbf{Q}) \wedge \neg\mathbf{Q}) \\
\mathcal{D}(\neg\mathbf{P}) &= \mathcal{D}(\mathbf{P}) \\
\mathcal{D}(\mathbf{P} \Rightarrow \mathbf{Q}) &= (\mathcal{D}(\mathbf{P}) \wedge \mathcal{D}(\mathbf{Q})) \vee (\mathcal{D}(\mathbf{P}) \wedge \neg\mathbf{P}) \vee (\mathcal{D}(\mathbf{Q}) \wedge \mathbf{Q}) \\
\mathcal{D}(\forall\mathbf{x} \cdot \mathbf{P_x}) &= \forall\mathbf{x} \cdot \mathcal{D}(\mathbf{P_x}) \vee \exists x \cdot (\mathcal{D}(\mathbf{P_x}) \wedge \neg\mathbf{P_x})
\end{aligned}
$$

When the predicate $\mathbf{A}$ is atomic, we have to consider several cases:

1. $\mathbf{A}$ does not contain any occurrences of terms of the shape $\mathbf{f}(\mathbf{E})$ or of the shape $\{\mathbf{x}\,|\,\mathbf{x} \in \mathbf{S} \wedge \mathbf{P_x}\}$,

2. $\mathbf{A}_{\mathbf{f(E)}}$ contains occurrences of terms of the shape $\mathbf{f(E)}$ but $\mathbf{f}$ and $\mathbf{E}$ do not contain occurrences of the shape $\mathbf{g(F)}$ or of the shape $\{\mathbf{x} \,|\, \mathbf{x} \in \mathbf{S} \,\wedge\, \mathbf{P_x}\}$,
3. $\mathbf{A}_{\{\mathbf{x}\,|\,\mathbf{x}\in\mathbf{S}\,\wedge\,\mathbf{P_x}\}}$ contains occurrences of terms of the shape $\{\mathbf{x} \,|\, \mathbf{x} \in \mathbf{S} \,\wedge\, \mathbf{P_x}\}$ but $\mathbf{S}$ does not contain occurrences of the shape $\mathbf{f(E)}$ or of the shape $\{\mathbf{x} \,|\, \mathbf{x} \in \mathbf{T} \,\wedge\, \mathbf{Q_x}\}$

Under these restrictions, which *dictates a certain order* for performing the syntactic transformation $\mathcal{D}$ (it involves starting the transformation $\mathcal{D}$ on the deepest sub-term first), the following table shows the proposed values when the predicate $\mathbf{A}$ is atomic:

$$
\begin{aligned}
\mathcal{D}(\mathbf{A}) &= \text{true} \\
\mathcal{D}(\mathbf{A}_{\mathbf{f(E)}}) &= \mathsf{fnc}\,(\mathbf{f}) \,\wedge\, \mathbf{E} \in \mathsf{dom}\,(\mathbf{f}) \,\wedge\, \forall \mathbf{y} \cdot (\, \mathbf{y} = \mathbf{f(E)} \,\Rightarrow\, \mathcal{D}(\mathbf{A_y})\,) \\
\mathcal{D}(\mathbf{A}_{\{\mathbf{x}\,|\,\mathbf{x}\in\mathbf{S}\,\wedge\,\mathbf{P_x}\}}) &= \forall \mathbf{x} \cdot (\, \mathbf{x} \in \mathbf{S} \,\Rightarrow\, \mathcal{D}(\mathbf{P_x})\,) \,\wedge\, \\
&\quad\ \ \forall \mathbf{y} \cdot (\, \mathbf{y} = \{\mathbf{x} \,|\, \mathbf{x} \in \mathbf{S} \,\wedge\, \mathbf{P_x}\} \,\Rightarrow\, \mathcal{D}(\mathbf{A_y})\,)
\end{aligned}
$$

Conversely, given the above definitions of $\mathcal{D}(\mathbf{P})$, it would have been possible to prove that $\mathcal{N}(\mathbf{P})$ is equal to $\mathcal{D}(\mathbf{P}) \,\wedge\, \mathbf{P}$, and also that $\mathcal{T}(\mathbf{P})$ is equal to $\mathcal{D}(\mathbf{P}) \,\Rightarrow\, \mathbf{P}$.

4.7   Filtering with $\mathcal{D}(\mathbf{P})$

Given a predicate $\mathbf{P}$, *nothing guarantees* for the moment that $\mathcal{D}(\mathbf{P})$, as we have proposed it in the previous tables, can act as the *filter* we are aiming at: this has to be proved rigorously. The problem can be stated as follows. Suppose we have proved $\mathcal{D}(\mathbf{P})$, then it must be shown that occurrences of terms of the shape $\mathbf{f(E)}$ in $\mathbf{P}$ can be safely *eliminated* using definition $\mathbf{D14}$, where the condition $\mathsf{fnc}\,(\mathbf{f}) \,\wedge\, \mathbf{E} \in \mathsf{dom}\,(\mathbf{f})$ *has been dropped*. For this, we consider the following syntactic transformation $\mathcal{E}(\mathbf{P})$ aiming at transforming a predicate $\mathbf{P}$ by *unconditionally eliminating* all occurrences of terms of the shape $\mathbf{f(E)}$.

$$
\begin{aligned}
\mathcal{E}(\mathbf{P} \wedge \mathbf{Q}) &= \mathcal{E}(\mathbf{P}) \wedge \mathcal{E}(\mathbf{Q}) \\
\mathcal{E}(\neg \mathbf{P}) &= \neg \mathcal{E}(\mathbf{P}) \\
\mathcal{E}(\mathbf{P} \Rightarrow \mathbf{Q}) &= \mathcal{E}(\mathbf{P}) \Rightarrow \mathcal{E}(\mathbf{Q}) \\
\mathcal{E}(\forall \mathbf{x} \cdot \mathbf{P_x}) &= \forall \mathbf{x} \cdot \mathcal{E}(\mathbf{P_x})
\end{aligned}
$$

When the predicate $\mathbf{A}$ is atomic, we have to consider the same cases as above for $\mathcal{D}$, which, again, dictate to perform the elimination $\mathcal{E}$ in a certain order (this involves making the elimination from the inside up):

$$
\begin{aligned}
\mathcal{E}(\mathbf{A}) &= \mathbf{A} \\
\mathcal{E}(\mathbf{A}_{\mathbf{f(E)}}) &= \forall \mathbf{y} \cdot (\, (\mathbf{y}, \mathbf{E}) \in \mathbf{f} \,\Rightarrow\, \mathcal{E}(\mathbf{A_y})\,) \\
\mathcal{E}(\mathbf{A}_{\{\mathbf{x}\,|\,\mathbf{x}\in\mathbf{S}\,\wedge\,\mathbf{P_x}\}}) &= \forall \mathbf{y} \cdot (\, \mathbf{y} = \{\mathbf{x} \,|\, \mathbf{x} \in \mathbf{S} \,\wedge\, \mathcal{E}(\mathbf{P_x})\} \,\Rightarrow\, \mathcal{E}(\mathbf{A_y})\,)
\end{aligned}
$$

Notice that the elimination of $\mathbf{f(E)}$ in $\mathbf{A_{f(E)}}$ has been done *without taking care of the condition* $\mathsf{fnc}\,(\mathbf{f}) \;\wedge\; \mathbf{E} \in \mathsf{dom}\,(\mathbf{f})$. Now the *main result* is this

$$\mathcal{D}(\mathbf{P}) \;\Rightarrow\; (\mathbf{P} \;\Leftrightarrow\; \mathcal{E}(\mathbf{P}))$$

In other words, we state exactly what we need: provided we have a proof of $\mathcal{D}(\mathbf{P})$, then the elimination of terms of the shape $\mathbf{f(E)}$ in $\mathbf{P}$ can be done with our definition **D14** *where the required conditions have been dropped,* thus resulting in an equivalent predicate. The proof is done by structural induction in the appendix.

4.8   Strengthening the Condition $\mathcal{D}(\mathbf{P})$

The calculation of $\mathcal{D}(\mathbf{P})$ we have done in previous section is not very encouraging (far too complicated). In particular the disjunctive forms we have obtained are rather repulsive. The idea then is to have a stronger but simpler filter called $\mathcal{L}(\mathbf{P})$. For this, we only select the two first disjuncts of $\mathcal{D}(\mathbf{P} \wedge \mathbf{Q})$ and $\mathcal{D}(\mathbf{P} \Rightarrow \mathbf{Q})$, and the first disjunct of $\mathcal{D}(\forall \mathbf{x} \cdot \mathbf{P})$. This leads to the following where we now have some conjunctive forms leading to an easy *decomposition*:

$$
\begin{aligned}
\mathcal{L}(\mathbf{P} \wedge \mathbf{Q}) &= \mathcal{L}(\mathbf{P}) \wedge (\mathbf{P} \Rightarrow \mathcal{L}(\mathbf{Q})) \\
\mathcal{L}(\neg\mathbf{P}) &= \mathcal{L}(\mathbf{P}) \\
\mathcal{L}(\mathbf{P} \Rightarrow \mathbf{Q}) &= \mathcal{L}(\mathbf{P}) \wedge (\mathbf{P} \Rightarrow \mathcal{L}(\mathbf{Q})) \\
\mathcal{L}(\forall \mathbf{x} \cdot \mathbf{P}) &= \forall \mathbf{x} \cdot \mathcal{L}(\mathbf{P})
\end{aligned}
$$

In the atomic cases, we have the usual restrictions and the following straightforward transformation:

$$
\begin{aligned}
\mathcal{L}(\mathbf{A}) &= \text{true} \\
\mathcal{L}(\mathbf{A_{f(E)}}) &= \mathsf{fnc}\,(\mathbf{f}) \wedge \mathbf{E} \in \mathsf{dom}\,(\mathbf{f}) \wedge \forall \mathbf{y} \cdot (\mathbf{y} = \mathbf{f(E)} \Rightarrow \mathcal{L}(\mathbf{A_y})) \\
\mathcal{L}(\mathbf{A_{\{x\,|\,x\in S\,\wedge\,P_x\}}}) &= \forall \mathbf{x} \cdot (\mathbf{x} \in \mathbf{S} \Rightarrow \mathcal{L}(\mathbf{P_x})) \wedge \\
&\quad\ \forall \mathbf{y} \cdot (\mathbf{y} = \{\mathbf{x}\,|\,\mathbf{x} \in \mathbf{S} \wedge \mathbf{P_x}\} \Rightarrow \mathcal{L}(\mathbf{A_y}))
\end{aligned}
$$

The following can then easily be proved by structural induction:

$$\mathcal{L}(\mathbf{P}) \;\Rightarrow\; \mathcal{D}(\mathbf{P})$$

4.9   Proving $\mathcal{L}(\mathbf{P})$

The previous result allows us to use $\mathcal{L}(\mathbf{P})$ rather than $\mathcal{D}(\mathbf{P})$ as a filter. The former is stronger but more tractable than the latter. An interesting question is now that concerning the proof of $\mathcal{L}(\mathbf{P})$. Can we prove it by using the simplified system or are we required to be careful in expanding the potential conditional definitions it contains? The question is essentially that of proving

$$\boxed{\mathcal{L}(\mathcal{L}(\mathbf{P}))}$$

since, should it be the case, then we have $\mathcal{D}(\mathcal{L}(\mathbf{P}))$ and we can thus prove $\mathcal{L}(\mathbf{P})$ with the simplified system. The proof is, again, by structural induction.

4.10   Practical Conclusion

At this point, we can re-formulate our proof strategy as follows for proving a set-theoretic statement $\mathbf{P}$:

> **(1)** Type-check $\mathbf{P}$
> **(2)** Calculate $\mathcal{L}(\mathbf{P})$
> **(3)** Eliminate set-theoretic formulae as much as possible in $\mathcal{L}(\mathbf{P})$
> **(4)** Prove the resulting predicate using **PSPC**
> **(5)** Eliminate set-theoretic formulae as much as possible in $\mathbf{P}$
> **(6)** Prove the resulting predicate using **PSPC**

5   An Example

Suppose we want to prove the following predicate $\mathbf{P}$:

$$f \in S \twoheadrightarrow T \,\wedge\, g \in T \twoheadrightarrow U \,\wedge\, x \in \mathsf{dom}\,(f) \;\Rightarrow\; g(f(x)) \in U$$

The type checking of $\mathbf{P}$ is clearly successful. Let's first compute $\mathcal{L}(\mathbf{P})$:

$$\begin{pmatrix} f \in S \twoheadrightarrow T \,\wedge \\ g \in T \twoheadrightarrow U \,\wedge \\ x \in \mathsf{dom}\,(f) \end{pmatrix} \;\Rightarrow\; \begin{pmatrix} \mathsf{fnc}\,(f) \,\wedge \\ x \in \mathsf{dom}\,(f) \,\wedge \\ \mathsf{fnc}\,(g) \,\wedge \\ f(x) \in \mathsf{dom}\,(g) \end{pmatrix}$$

Part of the consequent is easily discharged. It just remains $f(x) \in \mathsf{dom}\,(g)$, which we can expand using the "one point" rule, and the purified definitions of $f(x)$ and $\mathsf{dom}\,(g)$. After some simplifications, this yields the following, which we obviously *cannot prove*:

$$\exists y \cdot (x, y \in f) \Rightarrow \forall y \cdot (x, y \in f \Rightarrow \exists z \cdot (y, z \in g))$$

In other words, our predicate **P** does not pass the filter $\mathcal{L}(\mathbf{P})$. At this point, we figure out that we had forgotten an assumption concerning the relationship between the range of f and the domain of g. Here is the modification of our original predicate **P** that becomes the new predicate **Q** (we have added the antecedent $\mathsf{ran}\,(f) \subseteq \mathsf{dom}\,(g)$):

$$\begin{pmatrix} f \in S \nrightarrow T \ \wedge \\ g \in T \nrightarrow U \ \wedge \\ \mathsf{ran}\,(f) \subseteq \mathsf{dom}\,(g) \ \wedge \\ x \in \mathsf{dom}\,(f) \end{pmatrix} \Rightarrow g(f(x)) \in U$$

The calculation of the corresponding new $\mathcal{L}(\mathbf{Q})$ leads to the proof of the following, which now holds trivially:

$$\forall y \cdot (\exists x \cdot (x, y \in f) \Rightarrow \exists z \cdot (y, z \in g)) \ \wedge$$
$$\exists y \cdot (x, y \in f)$$
$$\Rightarrow$$
$$\forall y \cdot (x, y \in f \Rightarrow \exists z \cdot (y, z \in g))$$

At this point, we proceed with the proof of **Q**. And for doing this, we use the simplified system. After some simplifications, this leads to proving the following, which holds trivially:

$$\forall (x, y) \cdot (x, y \in f \Rightarrow x \in S \ \wedge \ y \in T) \ \wedge$$
$$\forall (y, z) \cdot (y, z \in g \Rightarrow y \in T \ \wedge \ z \in U)$$
$$\Rightarrow$$
$$\forall (y, z) \cdot (x, y \in f \ \wedge \ y, z \in g \Rightarrow z \in U)$$

Acknowledgements. We thank B. Stoddart for numerous very interesting electronic discussions.

References

1. J.R. Abrial. *The B-Book:Assigning Programs to Meanings*. Cambridge University Press (1996).
2. H. Barringer, J.H. Cheng, C.B. Jones. *A Logic Covering Undefinedness in Program Proofs*. Acta Informatica 21: 251-269 (1984).
3. P. Behm, L. Burdy, J.M. Meynadier. *Well defined B*. Second B International Conference. (Bert editor) LNCS 1393 Springer (1998).
4. L. Burdy *Traitement des expressions dépourvues de sens de la théorie des ensembles*. Thèse de Doctorat (2000).
5. J.H. Cheng, C.B. Jones *On the Usability of Logics which Handle Partial Functions*. Proceedings of Third Refinement Workshop. (1990).
6. D. Gries. *Foundations for Calculational Logic* in *Mathematical Methods in Program Development* (M. Broy and B. Schieder Editors). Springer (1996).

7. C.B. Jones *Partial Functions and Logics: a warning.* Information Processing Letter 54 (1995).
8. B. Stoddart, S. Dunne, A. Galloway. *Undefined Expressions and Logic in Z and B.* Formal Methods in System Design, 15 (1999).
9. P. Suppes. *Introduction to Logic.* Wadsworth International Group (1957).
10. W.M. Farmer and J.D. Guttman. *A Set Theory with Support for Partial Functions* in *Partiality and Modality.* (E. Thijsse, F. Lepage, and H. Wansing Editors). Special issue of *Logica Studia* (2000).

PROOF of $\mathcal{D}(\mathbf{P}) \Rightarrow (\mathbf{P} \Leftrightarrow \mathcal{E}(\mathbf{P}))$. The proof is by structural induction.

(1) *Base Case*: $\mathbf{A}$ does not contain any occurrences of terms of the shape $\mathbf{f(E)}$ or of the shape $\{\mathbf{x} \,|\, \mathbf{x} \in \mathbf{S} \wedge \mathbf{P_x}\}$. We have to prove $\mathcal{D}(\mathbf{A}) \Rightarrow (\mathbf{A} \Leftrightarrow \mathcal{E}(\mathbf{A}))$, which is obvious.

(2) *Inductive Case* : $\mathbf{A_{f(E)}}$. We assume the inductive hypothesis

$$\forall \mathbf{y} \cdot (\mathcal{D}(\mathbf{A_y}) \Rightarrow (\mathbf{A_y} \Leftrightarrow \mathcal{E}(\mathbf{A_y})))$$

and we have to prove $\mathcal{D}(\mathbf{A_{f(E)}}) \Rightarrow (\mathbf{A_{f(E)}} \Leftrightarrow \mathcal{E}(\mathbf{A_{f(E)}}))$. For this, we assume $\mathcal{D}(\mathbf{A_{f(E)}})$, that is

$$\mathsf{fnc}\,(\mathbf{f}) \;\wedge\; \mathbf{E} \in \mathsf{dom}\,(\mathbf{f}) \;\wedge\; \forall \mathbf{y} \cdot (\mathbf{y} = \mathbf{f(E)} \Rightarrow \mathcal{D}(\mathbf{A_y}))$$

and we have to prove $\mathbf{A_{f(E)}} \Leftrightarrow \mathcal{E}(\mathbf{A_{f(E)}})$, that is

$$\mathbf{A_{f(E)}} \Leftrightarrow \forall \mathbf{y} \cdot ((\mathbf{y}, \mathbf{E}) \in \mathbf{f} \Rightarrow \mathcal{E}(\mathbf{A_y}))$$

By applying the *conditional* definition **D14** (which we can do since the proper conditions, namely $\mathsf{fnc}\,(\mathbf{f}) \wedge \mathbf{E} \in \mathsf{dom}\,(\mathbf{f})$, are assumed), we obtain equivalently:

$$\mathbf{A_{f(E)}} \Leftrightarrow \forall \mathbf{y} \cdot (\mathbf{y} = \mathbf{f(E)} \Rightarrow \mathcal{E}(\mathbf{A_y}))$$

But, according to the assumption $\forall \mathbf{y} \cdot (\mathbf{y} = \mathbf{f(E)} \Rightarrow \mathcal{D}(\mathbf{A_y}))$ and the induction hypothesis $\forall \mathbf{y} \cdot (\mathcal{D}(\mathbf{A_y}) \Rightarrow (\mathbf{A_y} \Leftrightarrow \mathcal{E}(\mathbf{A_y})))$ we have $\forall \mathbf{y} \cdot (\mathbf{y} = \mathbf{f(E)} \Rightarrow (\mathbf{A_y} \Leftrightarrow \mathcal{E}(\mathbf{A_y})))$, we can thus replace $\mathcal{E}(\mathbf{A_y})$ by $\mathbf{A_y}$ in $\forall \mathbf{y} \cdot (\mathbf{y} = \mathbf{f(E)} \Rightarrow \mathcal{E}(\mathbf{A_y}))$, leading to the following, which is obvious (one point rule):

$$\mathbf{A_{f(E)}} \Leftrightarrow \forall \mathbf{y} \cdot (\mathbf{y} = \mathbf{f(E)} \Rightarrow \mathbf{A_y})$$

(3) *Inductive Case* : $\mathbf{A_{\{x\,|\,x \in S \wedge P_x\}}}$. We assume both inductive hypotheses

$$\forall \mathbf{y} \cdot (\mathcal{D}(\mathbf{A_y}) \Rightarrow (\mathbf{A_y} \Leftrightarrow \mathcal{E}(\mathbf{A_y})))$$
$$\forall \mathbf{x} \cdot (\mathcal{D}(\mathbf{P_x}) \Rightarrow (\mathbf{P_x} \Leftrightarrow \mathcal{E}(\mathbf{P_x})))$$

and we have to prove $\mathcal{D}(\mathbf{A_{\{x\,|\,x \in S \wedge P_x\}}}) \Rightarrow (\mathbf{A_{\{x\,|\,x \in S \wedge P_x\}}} \Leftrightarrow \mathcal{E}(\mathbf{A_{\{x\,|\,x \in S \wedge P_x\}}}))$. For this, we assume $\mathcal{D}(\mathbf{A_{\{x\,|\,x \in S \wedge P_x\}}})$, that is

$$\forall \mathbf{x} \cdot (\mathbf{x} \in \mathbf{S} \Rightarrow \mathcal{D}(\mathbf{P_x})) \;\wedge\; \forall \mathbf{y} \cdot (\mathbf{y} = \{\mathbf{x} \,|\, \mathbf{x} \in \mathbf{S} \wedge \mathbf{P_x}\} \Rightarrow \mathcal{D}(\mathbf{A_y}))$$

and we have to prove $\mathbf{A_{\{x\,|\,x \in S \wedge P_x\}}} \Leftrightarrow \mathcal{E}(\mathbf{A_{\{x\,|\,x \in S \wedge P_x\}}})$, that is

$$\mathbf{A}_{\{x\,|\,x\in S\,\wedge\,P_x\}} \;\Leftrightarrow\; \forall y \cdot (\,y = \{x\,|\,x\in S\,\wedge\,\mathcal{E}(P_x)\} \;\Rightarrow\; \mathcal{E}(A_y)\,)$$

According to the first assumption $\forall x \cdot (\,x \in S \;\Rightarrow\; \mathcal{D}(P_x)\,)$ and the second inductive hypothesis $\forall x \cdot (\,\mathcal{D}(P_x) \;\Rightarrow\; (P_x \Leftrightarrow \mathcal{E}(P_x))\,)$, we have $\forall x \cdot (\,x \in S \;\Rightarrow\; (P_x \Leftrightarrow \mathcal{E}(P_x))\,)$, we can thus replace $\mathcal{E}(P_x)$ by P_x in $\{x\,|\,x\in S\,\wedge\,\mathcal{E}(P_x)\}$ and obtain equivalently:

$$\mathbf{A}_{\{x\,|\,x\in S\,\wedge\,P_x\}} \;\Leftrightarrow\; \forall y \cdot (\,y = \{x\,|\,x\in S\,\wedge\,P_x\} \;\Rightarrow\; \mathcal{E}(A_y)\,)$$

But, according to the second assumption $\forall y \cdot (\,y = \{x\,|\,x\in S\,\wedge\,P_x\} \;\Rightarrow\; \mathcal{D}(A_y)\,)$ and the first induction hypothesis $\forall y \cdot (\,\mathcal{D}(A_y) \;\Rightarrow\; (A_y \Leftrightarrow \mathcal{E}(A_y))\,)$ we have $\forall y \cdot (\,y = \{x\,|\,x\in S\,\wedge\,P_x\} \;\Rightarrow\; (A_y \Leftrightarrow \mathcal{E}(A_y))\,)$, we can thus replace $\mathcal{E}(A_y)$ by A_y in $\forall y \cdot (\,y = \{x\,|\,x\in S\,\wedge\,P_x\} \;\Rightarrow\; \mathcal{E}(A_y)\,)$, leading to the following, which is obvious (one point rule)

$$\mathbf{A}_{\{x\,|\,x\in S\,\wedge\,P_x\}} \;\Leftrightarrow\; \forall y \cdot (\,y = \{x\,|\,x\in S\,\wedge\,P_x\} \;\Rightarrow\; A_y\,)$$

(4) *Inductive Case :* $P \wedge Q$. We assume both inductive hypotheses

$$\mathcal{D}(P) \;\Rightarrow\; (P \Leftrightarrow \mathcal{E}(P))$$
$$\mathcal{D}(Q) \;\Rightarrow\; (Q \Leftrightarrow \mathcal{E}(Q))$$

and we have to prove

$$\mathcal{D}(P \wedge Q) \;\Rightarrow\; (P \wedge Q \Leftrightarrow \mathcal{E}(P \wedge Q))$$

We thus now assume $\mathcal{D}(P \wedge Q)$, that is

$$(\mathcal{D}(P) \wedge \mathcal{D}(Q)) \;\vee\; (\mathcal{D}(P) \wedge \neg P) \;\vee\; (\mathcal{D}(Q) \wedge \neg Q)$$

and we have to prove $P \wedge Q \Leftrightarrow \mathcal{E}(P \wedge Q)$, that is

$$P \wedge Q \;\Leftrightarrow\; \mathcal{E}(P) \wedge \mathcal{E}(Q)$$

The disjunctive shape of the assumption suggests a proof by cases. The first case, $\mathcal{D}(P) \wedge \mathcal{D}(Q)$, is trivial according to the two inductive hypotheses. We assume then the second case:

$$\mathcal{D}(P) \wedge \neg P$$

From $\mathcal{D}(P)$ and the first induction hypothesis, we deduce

$$P \wedge Q \;\Leftrightarrow\; P \wedge \mathcal{E}(Q)$$

which holds since we have $\neg P$. The last case, $\mathcal{D}(Q) \wedge \neg Q$, is proved in a similar fashion.

(5) *Inductive Case:* $\neg P$. Proof omitted.

(6) *Inductive Case:* $P \Rightarrow Q$. Proof omitted.

(7) *Inductive Case:* $\forall x \cdot P_x$. We assume the inductive hypothesis

$$\forall \mathbf{x} \cdot (\mathcal{D}(\mathbf{P_x}) \; \Rightarrow \; (\mathbf{P_x} \Leftrightarrow \mathcal{E}(\mathbf{P_x})))$$

and we have to prove

$$\mathcal{D}(\forall \mathbf{x} \cdot \mathbf{P_x}) \; \Rightarrow \; (\forall \mathbf{x} \cdot \mathbf{P_x} \Leftrightarrow \mathcal{E}(\forall \mathbf{x} \cdot \mathbf{P_x}))$$

We thus now assume $\mathcal{D}(\forall \mathbf{x} \cdot \mathbf{P_x})$, that is

$$\forall \mathbf{x} \cdot \mathcal{D}(\mathbf{P_x}) \; \vee \; \exists \mathbf{x} \cdot (\mathcal{D}(\mathbf{P_x}) \wedge \neg \mathbf{P_x})$$

and we have to prove $\forall \mathbf{x} \cdot \mathbf{P_x} \Leftrightarrow \mathcal{E}(\forall \mathbf{x} \cdot \mathbf{P_x})$, that is

$$\forall \mathbf{x} \cdot \mathbf{P_x} \quad \Leftrightarrow \quad \forall \mathbf{x} \cdot \mathcal{E}(\mathbf{P_x})$$

The disjunctive shape of the assumption suggests a proof by cases. The first case, $\forall \mathbf{x} \cdot \mathcal{D}(\mathbf{P_x})$ is trivial. We assume the second case

$$\exists \mathbf{x} \cdot (\mathcal{D}(\mathbf{P_x}) \wedge \neg \mathbf{P_x})$$

which allows us to prove easily the following, which is equivalent to $\forall \mathbf{x} \cdot \mathbf{P_x} \Leftrightarrow \forall \mathbf{x} \cdot \mathcal{E}(\mathbf{P_x})$

$$\exists \mathbf{x} \cdot \neg \mathbf{P_x} \quad \Leftrightarrow \quad \exists \mathbf{x} \cdot \neg \mathcal{E}(\mathbf{P_x})$$

QED

A Theory of Generalised Substitutions

Steve Dunne

School of Computing and Mathematics, University of Teesside
Middlesbrough, TS1 3BA, UK
s.e.dunne@tees.ac.uk

Abstract. We augment the usual wp semantics of substitutions with an explicit notion of frame, which allows us to develop a simple self-contained theory of generalised substitutions outside their usual context of the B Method. We formulate three fundamental healthiness conditions which semantically characterise all substitutions, and from which we are able to derive directly, without need of any explicit further appeal to syntax, a number of familiar properties of substitutions, as well as several new ones specifically concerning frames. In doing so we gain some useful insights about the nature of substitutions, which enables us to resolve some hitherto problematic issues concerning substitutions within the B Method.

1 Introduction

In the 1970s Dijkstra[8] invented predicate-transformer semantics, interpreting each command of his guarded command language as a weakest-precondition (wp) predicate transformer. The standard semantics of generalised substitutions[1] follows Dijkstra by interpreting each substitution as a wp predicate-transformer. So then, are *skip* and $x := x$ equivalent as generalised substitutions? Certainly, if we interpret them purely as wp predicate-transformers, the answer must be yes. But such an answer is more problematic than might first appear. If two substitutions S and T are genuinely equal then, thanks to Leibniz, we should be able to interchange them at will in any context without affecting the overall meaning of our formal text. This clearly isn't the case with *skip* and $x := x$. Consider the multiple composition

$$skip \parallel x := x + 1$$

If we attempt to replace the *skip* by $x := x$ in this we obtain

$$x := x \parallel x := x + 1$$

This isn't even considered to be a well-formed substitution expression in conventional B, since both components of the would-be composition are acting on the same variable x, which orthodox B à la B-Book[1] of course forbids.

A second question concerns a substitution's characteristic predicate prd which Abrial[1] defines for substitutions in the context of a machine with variable x. But we sometimes wish to consider substitutions even in an abstract

D. Bert et al. (Eds.): ZB 2002, LNCS 2272, pp. 270–290, 2002.

machine with no state variables. How should we define the prd of a substitution in such a machine?

Again, in section 7.2.3 of [1] Abrial warns that at most one operation of an included machine can be called from within an operation of the including machine, otherwise we could break the invariant of the included machine. He illustrates the point with a pathological example. But Abrial's stricture seems based on the precautionary principle alone. We have no underlying theory to explain precisely why every such multiple set of operation calls should be proscribed.

We believe generalised substitutions admit an interesting theory in their own right, independent of the context of any B abstract machine, and that such a theory can usefully inform our understanding of their role both within the B method and elsewhere. Our purpose in this paper is to develop such a theory, and to demonstrate its application in resolving satisfactorily issues like those above when we examine them in its light.

2   Preliminaries

Before we proceed to define the elements of the generalised substitution language we need to establish some necessary concepts and notations which we will employ in those definitions.

2.1   Concerning Predicates

We will deal extensively with predicates over a given alphabet of variables. It is therefore useful to describe here some of the basic concepts and properties of such predicates, upon which later we will frequently rely.

We define an implication ordering $\ggg$ for predicates over a given alphabet. Let Q and R be predicates over an alphabet comprising a list of variables α. Then we have

Definition 1   (Implication ordering)

$$Q \ggg R \quad =_{df} \quad \forall \alpha \bullet Q \Rightarrow R$$

The constant predicates false and true are respectively bottom and top of this ordering. For any predicate P it follows that false $\ggg P \ggg$ true .

We can now define our fundamental notion of equality between predicates: let Q and R be predicates over the same alphabet; then we have

Definition 2   (Equality for predicates)

$$Q = R \quad =_{df} \quad (Q \ggg R) \wedge (R \ggg Q)$$

We introduce two further useful notations:

Syntactic substitution. Let x be a variable, Q a predicate and E an appropriately typed expression; then $Q\langle E/x \rangle$ denotes the syntactic substitution of every free occurrence of x in Q by E. This notation generalises in the obvious way so that, if x is a list of variables and E is tuple-valued expression of corresponding type, $Q\langle E/x\rangle$ denotes the syntactic substitution of every free occurrence of each variable of x in Q by the corresponding component of E. In the degenerate case where x is the empty list —and hence necessarily E is the (unique) empty tuple value— then $Q\langle E/x\rangle = Q$.

Non-freeness. Let x be a variable and Q a predicate. Then $x \setminus Q$ means "x does not appear free in Q". Again this notation generalises so that, if x is a list of variables, $x \setminus Q$ means "no variable of x appears free in Q."

2.2   Concerning Lists and Frames

Lists. A *list* is a finite sequence of basic variables without repetition. Variables can also denote lists, so a variable is either a basic variable or a list. In fact, if a list comprises one basic variable then the list and its single constituent variable are synonymous, so even a basic variable can be regarded as a (singleton) list.

List decoration. We use systematic decoration for lists so that, for example, if x is a list then x' represents the corresponding list obtained by dashing each constituent variable of x, while x_0 represents that obtained by zero-subscripting each constituent variable in x.

Equality for lists. If x and y are lists the predicate $x = y$ holds exactly if x and y are the same length and corresponding pairs of constituents denote the same value. Thus, for example, if x is $[x_1, x_2]$ and y is $[y_1, y_2]$, then $x = y$ holds if $x_1 = y_1 \wedge x_2 = y_2$.

Frames. We call a collection of basic variables a *frame*. A frame differs from a list in that there is no significance in the order of its constituents. If x is a list then $|x|$ denotes the frame derived from x. If u and v are frames, then $u \cup v$ denotes the new frame obtained by aggregating u and v, and $u - v$ denotes the residual frame obtained by removing from u any variables it shares with v. We say the frame v *extends* the frame u if u is contained in v, written $u \subseteq v$.

The empty list and empty frame. We denote both the empty list and its associated empty frame by $\emptyset$. The empty list is associated with the unique empty-tuple value corresponding, for example, to the empty binding of the schema type with no components in Z, or the origin vector which is the sole member of a degenerate zero-dimensional vector space. If x and y are empty lists they both necessarily denote this empty-tuple value, so $x = y$ is necessarily true and $x \neq y$ is necessarily false.

The alphabet. A generalised substitution is always interpreted in a context provided by a set of variables which forms the reference frame or *alphabet* of all variables whose names are in scope. If we ever have to make explicit reference to it as a frame, we denote the alphabet by α. We assume an implicit lexical ordering over the alphabet.

Characteristic list of a frame. We associate with each frame a characteristic list obtained by placing its constituent variables in lexical order. So where the context demands it, a frame can be interpreted as a list. For example, given a frame u we may sometimes write a predicate such as $u = u'$ or a substitution such as $u := u'$. In such cases we are slightly overloading our notation since the u that appears in such a predicate or substitution must be interpreted as the characteristic list of frame u, and correspondingly u' as the dash-decorated version of that list.

2.3 Active Frame of a Substitution

The use of a frame in the specification of an operation or statement to circumscribe which variables of the state it may update, is well-established in formalisms such as Z[20], VDM[12] and the refinement calculus[2,4,15,13,14]. And in B itself Bicarregui and Ritchie[6] remind us that an abstract machine's (explicit and included) state variables provide a frame for the operations of that machine. In [9] we refined such a notion of frame to apply at the level of an individual generalised substitution, by defining what we called the *active frame* of a substitution. Our operational intuition is that these are the variables to which the substitution may assign values.

We denote the active frame of a generalised substitution S by $frame(S)$. Rouzaud[17] has a similar notion. There he denotes by $M(S)$ the set of modified variables of a substitution S. A generalised substitution may make passive reference to variables in the alphabet outside its active frame. For example, the active frame of $x := y$ is just $|x|$ although it makes passive reference to y too. The active frame may be empty as in *skip*, or as in $x < 7 \mid skip$ which makes only passive reference to x. We distinguish between *skip* and $x := x$ since they have different active frames.

3 The Generalised Substitution Language

We maintain that the meaning of a generalised substitution S comprises two distinct elements: its active frame $frame(S)$, and its wp predicate-transformer $[S]$ which acts on postconditions: that is, predicates over the current alphabet. A corollary is that two generalised substitutions are equal exactly if they have the same frame and wp. Conversely, two substitutions differ even when, always exhibiting the same wp effect, they differ only in their frames.

3.1 Basic Substitutions and Constructors

The basic substitutions and constructors of our generalised substitution language are defined in Table 1. We make the following observations on some of the constructs appearing there:

Table 1. Basic Generalised Substitutions

name	syntax	frame	$[\_]Q$
skip	$skip$	$\emptyset$	Q
assignment	$x := E$	$\lvert x \rvert$	$Q\langle E/x\rangle$
preconditioned substitution	$P \mid S$	$frame(S)$	$P \wedge [S]Q$
guarded substitution	$P \Longrightarrow S$	$frame(S)$	$P \Rightarrow [S]Q$
frame extension	S_y	$frame(S) \cup \lvert y \rvert$	$[S]Q$
bounded choice	$S[\,]T$	$frame(S) \cup frame(T)$	$[S]Q \wedge [T]Q$
unbounded choice	$@\,z\,.\,S$	$frame(S) - \lvert z \rvert$	$\forall z \bullet [S]Q$
sequential composition	$S \,;\, T$	$frame(S) \cup frame(T)$	$[S][T]Q$

Assignment.   In the assignment $x := E$ we allow x to be a basic variable or list, where E is a well-defined expression of correspondingly appropriate type. In particular, if x is a list E must be a tuple whose dimension corresponds with that of x. Recall when x is empty $Q\langle E/x\rangle = Q$, so in the degenerate case where x is the empty list (and E therefore the empty tuple-value) $x := E$ has an empty frame and acts as the identity wp predicate transformer, making it indistinguishable from $skip$. Thus $skip$ is really this special case of assignment.

Frame extension.   This takes the form S_y, where S is a substitution and y is a list of variables. Clearly, if $\lvert y \rvert$ is a subframe of $frame(S)$ then $S_y = S$. We also observe that $skip_x = x := x$. Our operational intuition of S_y is that it behaves like S but in addition explicitly conserves the existing values of the variables of the residual frame $\lvert y \rvert - frame(S)$. The distinction between S_y and S is only significant in the context of a parallel composition (see Section 6) of either of these substitutions with another substitution which affects y.

Bounded choice.   The form $S \,[\,] \, T$ represents a nondeterministic choice between the substitutions S and T. Our operational intuition is that on any occasion when we execute $S \,[\,]\, T$ either S or T will actually execute. The choice is entirely outside our control, being vested as we imagine in a demon who inhabits the machine on which our program is being executed.

Unbounded choice.   This takes the form $@\,z\,.\,S$ where S is a generalised substitution, and z is a list of variables fresh with respect to the current alphabet. The alphabet of S is understood as the current alphabet augmented by $\lvert z \rvert$. The unbounded choice $@\,z\,.\,S$ represents S attenuated by the nondeterministic choice of any feasible value for variable z. The frame of $@\,z\,.\,S$ is obtained by removing each variable of $\lvert z \rvert$ from S's frame, should it be there. In the degenerate case where z is the empty list, $@\,z\,.\,S$ is equivalent to S.

3.2   Calling an Operation of an Included Machine

Our frame extension construct can immediately shed some light on the meaning in B of calling an operation of an included machine. Let the variables of such an included machine M be represented by the frame y and let $Op1$ be an operation of M whose body is characterised by a substitution S. Then if we call $Op1$ the meaning of that invocation is precisely expressed by the substitution S_y.

So we can justify Abrial's stricture against the parallel invocation of two operations from the same included machine. Let $Op2$ be a second operation of M whose body is characterised by the substitution T. If we call $Op2$ the meaning of that invocation is correspondingly expressed by T_y. Thus the meaning of a parallel invocation of $Op1$ and $Op2$ is $S_y \parallel T_y$, which conventional B regards as an ill-formed multiple composition since both its components act on the same frame y.

3.3   Some Extreme Substitutions

It is illuminating to consider all the substitutions we can describe on the empty frame with false as our only precondition or guard. We define and informally name these in Table 2. We have already encountered *skip*. The substitution *magic*

Table 2. Extreme Substitutions on the Empty Frame

informal name	definition
skip	*skip*
magic	false $\Longrightarrow$ *skip*
abort	false | *skip*

is the everywhere infeasible substitution: it always refuses (fails) to execute. The substitution *abort* always behaves abortively. It can never guarantee to establish any postcondition.

3.4   Assertions and Assumptions

Floyd[10] articulated in the 1960s the notions of assertion and assumption which have been associated with his name ever since. The form $P \Longrightarrow skip$ is the manifestation in our generalised substitution language of the *Floyd assumption* which Back and von Wright[4] denote by [P]: it has no effect providing P holds, but otherwise the program is deemed to have succeeded miraculously without executing any further. The form $P \mid skip$, on the other hand, expresses the *Floyd assertion* which would be denoted by $\{P\}$ in [4]: this has no effect providing P holds, but otherwise causes the program to abort.

4   Of Demons and Miracles

The interplay between guarded substitutions and demonic choice is interesting. It exposes what we might call the demon's innate abhorrence of miracles, demonic choice being, so to speak, angelic with respect to feasibility. For example, our everywhere-infeasible substitution *magic* acts as a unit for nondeterministic choice: that is, for any generalised substitution S we have $S \; [] \; magic \; = \; S$. (The proof is trivial.) Thus is the demon's freedom of choice is constrained by feasibility: when offered a choice between a miraculous and a feasible alternative, he must always take the latter.

Indeed, the demon is more constrained than might appear from the immediate choice confronting him: offered a choice between two apparently feasible immediate alternatives, it may be the case that one of those choices leads to an infeasible substitution later, perhaps much later, in the program. The demon must recoil from even such deferred infeasibility. For example, it is easy to show that the program

$$(x := 1 \; [] \; x := 2) \; ; x = 1 \Longrightarrow skip$$

is equivalent to $x := 1$. Nelson[16] suggests we adopt either of two equivalent operational intuitions about the demon's behaviour: either we attribute to him a clairvoyant ability to look into the future course of execution of the program to foresee any infeasibility that would result from taking a particular choice, so as to avoid that choice; or, alternatively, we can imagine that he makes his choice blindly but that execution will backtrack to let him reconsider if infeasibility is subsequently encountered. We observe that in the little program above the $x = 1 \Longrightarrow skip$ is a *post hoc* assumption. It acts as a retrospective "choice filter" to ensure the demon makes the choice we want in the preceding substitution.

We should point out that although we have described the demon's behaviour as "angelic with respect to feasibility", this doesn't mean that such *post hoc* assumptions endow our generalised substitution language with a genuine angelic nondeterminism. For example, consider the program $(abort \; [] \; skip) \; ; S$. There is no assumption, or indeed any other substitution S, that will oblige the demon to eschew the choice of *abort*.

5   Characteristic Predicates

In this section we describe several predicates that can be derived from a generalised substitution, which characterise various important aspects of that substitution's behaviour. The names trm, fis and prd were coined by Abrial.

trm(S)   the termination predicate of S characterises from where execution of S is guaranteed to terminate, and is therefore safe from the risk of non-termination. A substitution must terminate from where it is at least guaranteed establish true, which means it must produce a result of some sort. Thus we define it by

$$trm(S) \quad =_{df} \quad [S] \; true$$

We note that for any substitution S, we have $S = trm(S) \mid S$. Indeed we can think of $trm(S)$ as S's intrinsic precondition, since it is the strongest precondition we can apply to S without changing it.

fis(S)   the feasibility predicate $fis(S)$ characterises from where execution of S is feasible (non-miraculous). A program behaves miraculously where it can guarantee to establish false; conversely it behaves feasibly where it isn't guaranteed to establish false. Thus we define it as

$$fis(S) \quad =_{df} \quad \neg\, [S]\ false$$

We note that for any substitution S, we have $S = fis(S) \Longrightarrow S$. Indeed we can think of $fis(S)$ as the intrinsic guard of a substitution S, since it is the strongest guard we can apply to S without changing it.

prd(S)   the before-after predicate $prd(S)$ reconciles the predicate-transformer interpretation of a generalised substitution with an alternative relational interpretation. It relates before-values of the variables in S's frame to their potential after-values following execution of S. Let $frame(S)$ be s; then we define $prd(S)$ by

$$prd(S) \quad =_{df} \quad \neg\, [S]\ s \neq s'$$

Here s' is assumed to be fresh with respect to the alphabet and represents the after-value of s. Because every substitution S possesses a specified frame which we use in our definition of $prd(S)$, our notion of prd is an absolute one. In contrast, Abrial's notion of prd in [1] is only relative to a given (list of) variable(s) indicated by a subscript, as for example in $prd_x(S)$ or $prd_{x,y}(S)$. If the frame of S is empty the predicate $s \neq s'$ degenerates to false. Hence, for example,

$$prd(skip) \quad = \quad \neg\, [skip]\ false \quad = \quad true$$

prd₀(S)   this derivative of $prd(S)$ is obtained by replacing s and s' repectively by s_0 and s, where s is $frame(S)$. Formally, we have

$$prd_0(S) \quad =_{df} \quad prd(S)\langle s_0, s/s, s'\rangle \qquad\qquad s = frame(S)$$

6   A Liberalised Parallel Composition

We define a parallel composition of substitutions S and T. It takes the form $S \parallel T$, and is defined as follows, where $frame(S) = s$ and $frame(T) = t$:

$$S \parallel T \quad =_{df} \quad trm(S) \wedge trm(T) \mid @s', t'.\ prd(S) \wedge prd(T) \Longrightarrow s, t := s', t'$$

We infer from this definition that

$$frame(S \parallel T) \quad = \quad s \cup t$$
$$trm(S \parallel T) \quad = \quad trm(S) \wedge trm(T)$$
$$prd(S \parallel T) \quad = \quad (trm(T) \Rightarrow prd(S)) \wedge (trm(S) \Rightarrow prd(T))$$

Our parallel composition is more liberal than that defined in [1], with which it coincides if the frames of S and T are disjoint, in that it doesn't require those frames to be disjoint. Similar liberal parallel compositions have been defined by Bert *et al* in [5] and Chartier in [7]. When the frames of S and T exactly coincide, our $\|$ corresponds to Back and Butler[3]'s *fusion*. Our parallel operator is a particular case of Hoare and He[11]'s *parallel by merge* where

> Each process is first executed on its *private* version of the shared variables independently. When all have terminated, their updates on the shared variables are *merged* and written back to the *global* version of the shared variables.

The merge in our case attempts to reconcile exactly the respective two updates of each shared variable. Where this proves impossible the composition is deemed infeasible.

6.1   Rewrite Rules for $\|$

Because of our liberalisation of Abrial's original definition of $\|$ to permit overlapping frames, we must supplement the rewrite rules for $\|$ given by Abrial in [1] with the following new one covering two parallel assignments to the same variable:

$$x := E \;\|\; x := F \quad = \quad E = F \Longrightarrow x := E$$

We also have to modify Abrial's rewrite rule for distributing parallel composition across demonic choice, giving the following rule for any three generalised substitutions S, T and U, where $frame(S) = s$ and $frame(T) = t$:

$$(S \,[\!]\, T) \;\|\; U \quad = \quad (S_t \,\|\, U) \,[\!]\, (T_s \,\|\, U)$$

7   Refinement of Generalised Substitutions

A comprehensive notion of refinement over all generalised substitutions on an alphabet must take frames into account. For example, can we refine *skip* by $x := x$, or indeed $x := x$ by *skip*? To deal with frames we have an extra requirement for refinement concerning frame inclusion. Thus, if S and T are generalised substitutions over the same alphabet, then S is refined by T, written $S \sqsubseteq T$, if and only if, for every postcondition Q over the alphabet,

$$frame(S) \subseteq frame(T) \qquad \text{and} \qquad [S]\,Q \;\Rrightarrow\; [T]\,Q \tag{GSref}$$

Under our generalised substitution refinement *skip* is indeed refined by $x := x$ but the converse is not so. The subclass of generalised substitutions on a given frame u form a complete sublattice under $\sqsubseteq$. Its bottom is $abort_u$ and its top is $magic_u$. If S and T are two such substitutions, their glb is $S \,[\!]\, T$ and their lub is given by $trm(S) \vee trm(T) \mid @u'.\,prd(S) \wedge prd(T) \Longrightarrow u := u'$.

The bottom of the whole class of substitutions over an alphabet α is *abort*. The top of the whole class is $magic_\alpha$.

7.1 Monotonicity

In a refinement calculus, to quote Back and Butler[3],

> ...the required behaviour of the program is specified as an abstract, possibly non-executable, program which is then refined by a series of correctness-preserving transformations into an efficient, executable program. The notion of correctness-preserving transformation is modelled by a refinement relation between programs which is transitive, thus supporting stepwise refinement, and [with respect to which the program constructors are] monotonic, thus supporting piecewise refinement.

Our generalised substitution constructors can indeed all be shown quite easily to be monotonic with respect to our generalised-substitution refinement ordering $\sqsubseteq$. In particular, incorporating frame-inclusion as one of the conditions in our definition of $\sqsubseteq$ helps to ensure that our parallel composition $\parallel$ is monotonic with respect to $\sqsubseteq$.

8 Recursion

We adhere to a standard treatment of recursion in total correctness for our generalised substitutions. A recursive program expression is interpreted as a least fixed-point in our refinement ordering.

8.1 Least Fixed-Points and Transfinite Induction

Nelson[16] describes a generalisation of Tarski[19]'s famous fixed-point theorem sufficient to guarantee that if F is a monotonic function on a complete partial order such as $\sqsubseteq$ then F will have a least fixed-point. Furthermore the least fixed-point will be expressible as the limit of an ordinal-indexed series of approximations starting from the bottom of the ordering. We have to use ordinals rather than just natural numbers because unless F is continuous we might have to traverse beyond ω, the ordinal limit of the natural numbers, to reach our fixed point. Smith[18] gives a concise appreciation of the issue. The relevant fixed-point theory is presented in depth by Back and von Wright[4]. Where $F(X)$ is a $\sqsubseteq$-monotonic function on generalised substitutions, we denote the generalised substitution which is its least fixed-point by $\mu X \bullet F(X)$.

8.2 A Fundamental Iteration Construct

We can now define the fundamental iteration construct of our generalised substitution language, which Abrial calls the *opening* of a substitution. Later we utilise this construct in defining a conventional while loop. It takes the form $S\hat{\ }$ where S is a generalised substitution. It is defined as the least fixed-point X in the refinement ordering of $(S \, ; X) \, [] \, skip$. That is

$$S\hat{\ } \quad =_{df} \quad \mu X \bullet (S \, ; X) \, [] \, skip$$

In their refinement calculus Back and von Wright[4] call the corresponding fixed-point construction on their programs *strong iteration.* Our operational intuition of $S\hat{}$ is that it represents the demon's free choice of an arbitrary number of executions of S, a choice ostensibly ranging from no executions, being equivalent to $skip_s$, up to everlasting repetition of S, which is equivalent to $abort_s$, where as usual s denotes $frame(S)$. Of course the demon is constrained by feasibility. He can only choose a particular number of executions of S if that number is feasible, and this will depend on S.

For example, consider $magic\hat{}$. Since $magic$ is always infeasible, the only number of executions of $magic$ which is feasible is none. The demon's hand is forced: he has to choose zero executions, so $magic\hat{}$ is $skip$.

We also note that $skip\hat{}$ is $abort$. This can be demonstrated by the method of successive approximations, taking $abort$ as our initial and also final approximation, since

$$(skip \; ; \; abort) \; [] \; skip \quad = \quad abort \; [] \; skip \quad = \quad abort$$

making it a fixed point of $(skip \; ; \; X) \; [] \; skip$.

9   Some Derived Substitutions

We now augment our repertoire of basic substitutions with the useful derived substitutions listed in Table 3.

Table 3. Some Derived Generalised Substitutions

substitution name	syntax
indeterminate assignment	$x : P$
conditional	if G then S else T end
short conditional	if G then S end
while loop	while G do S end

Indeterminate assignment: takes the form $x : P$ where x is a (list of) variable(s) from the current alphabet α and P is a predicate over $\alpha \cup x_0$. It represents the assignment to x of any value satisfying P. If x_0 appears in P it denotes the before-value of x. For example, the indeterminate assignment $x : x < x_0$ reduces x by an indeterminate amount. If x is a list of variables, then x_0 is the corresponding list of zero-subscripted variables, any of which may appear in P to denote the before-value of the corresponding variable of x. Formally, we have

$$x : P \quad =_{df} \quad @ \; x' \; . \; P\langle x, x'/x_0, x \rangle \Longrightarrow x := x'$$

In particular, the completely unconstrained assignment α : true, which sets every variable of the global frame quite indeterminately, corresponds to Nelson[16]'s command *havoc*. We note that the Morgan[13] specification statement $w :$ [*Pre*, *Post*] is equivalent to the substitution *Pre* | $w : Post$.

Conditional: takes the familiar form   if G then S else T end   where G is a predicate and S and T are substitutions. We define it by

$$\text{if } G \text{ then } S \text{ else } T \text{ end} \quad =_{df} \quad (G \Longrightarrow S) \;[\!]\; (\neg\, G \Longrightarrow T)$$

This definition exploits the demon's aversion to miracles to ensure he makes the appropriate choice. It is in interesting to note that what has long been regarded by programmers as a fundamental programming construct is in fact merely a compounding of the two more fundamental constructs of guarding and nondeterministic choice.

Short conditional: takes the familiar form   if G then S end   where G is a predicate and S is a substitution. It is defined by

$$\text{if } G \text{ then } A \text{ end} \quad =_{df} \quad (G \Longrightarrow S) \;[\!]\; (\neg\, G \Longrightarrow skip)$$

While loop: takes the familiar form   while G do A end   where G is a predicate and S is a substitution. It is defined by

$$\text{while } G \text{ do } S \text{ end} \quad =_{df} \quad (G \Longrightarrow S)^{\char94} \; ; \neg\, G \Longrightarrow skip$$

This definition is given by Abrial in [1] and echoed by Back and von Wright's essentially similar construction in Lemma 21.8 of [4]. Although our while-loop definition looks inscrutable, it is possible to gain a level of operational insight about how it works. Recall our operational intuition of an opening like $(G \Longrightarrow S)^{\char94}$ as a demonic choice of an arbitrary number of executions of $G \Longrightarrow S$. The $\neg\, G \Longrightarrow skip$ following it is a *post hoc* assumption: it constrains the demon to choose exactly whichever finite repetition of $G \Longrightarrow S$ (there can only be one, if it exists at all) is feasible but makes G false. Conversely, if everlasting repetition of $G \Longrightarrow S$ is feasible then the demon is obliged to choose it by default, so the iteration doesn't terminate.

10   Semantics of Generalised Substitutions

Here we establish three fundamental healthiness conditions which semantically characterise any generalised substitution. They provide the basis by which we can subsequently assess the soundness of the definitions of the constructs which comprise our generalised substitution language. They can also be used to assess the soundness of any further construct which might be proposed in the future to extend the language. We will then use these three conditions as axioms from which to derive the propositions which embody our theory of generalised substitutions.

10.1   Healthiness Conditions

We address the following important question about any attempt to define a given new generalised substitution S by specifying its frame $frame(S)$ and wp predicate transformer $[S]$: do we have to respect any constraints at all, or are we free to define each of $frame(S)$ and $[S]$ arbitrarily?

Reflecting on this question leads us to three conclusions: first, any syntax-oriented specification of $[S]$ in terms of a syntactic transformation of any predicate formula must yield a well-defined predicate-transformer function at the semantic level. That is to say, logically equivalent postcondition formulas must be transformed by $[S]$ into logically equivalent precondition formulas. In other words, as a predicate-transformer $[S]$ must satisfy Leibniz's Equality Law.

Secondly, $[S]$ must be positively conjunctive: that is, it must distribute over any non-empty conjunction of postconditions.

Thirdly, there is a mutual constraint between $frame(S)$ and $[S]$ which ensures that in normal behaviour S's effect is circumscribed by its frame: if x is a variable outside $frame(S)$ then any feasible non-aborting execution of S will leave x unchanged; the only way S can affect x is by acting miraculously or aborting. So a sound definition of a generalised substitution S must respect the following three conditions:

Leibniz   (GS1)

$$Q \; = \; R \quad \vdash \quad [S]Q \; = \; [S]R$$

positive conjunctivity   (GS2)

$[S]$ distributes through all non-empty conjunctions

frame circumscription   (GS3)

for any postcondition Q such that $frame(S) \setminus Q$ then

$$[S] \, Q \quad = \quad trm(S) \; \wedge \; (fis(S) \Rightarrow Q)$$

It is straightforward to show that all our substitution constructs conform to the above conditions. For example, we prove $x := E$ upholds GS1:

$Q = R$

$\equiv$ {defn of = for predicates}

$\forall \alpha \bullet Q \Leftrightarrow R$

$\equiv$ {let β, x be α}

$\forall \beta \bullet \forall x \bullet Q \Leftrightarrow R$

$\equiv$ {change name of bound variable x to x'}

$\forall \beta \bullet \forall x' \bullet (Q \Leftrightarrow R)\langle x'/x \rangle$

$\equiv$ {spurious quantification over x, ok since $x \setminus (Q \Leftrightarrow R)\langle x'/x \rangle$}

$\forall x \bullet \forall \beta \bullet \forall x' \bullet (Q \Leftrightarrow R)\langle x'/x \rangle$

$$\equiv \qquad \{\alpha \text{ is } \beta, x\}$$

$$\forall \alpha \bullet \forall x' \bullet (Q \Leftrightarrow R)\langle x'/x \rangle$$

$$\Rrightarrow \qquad \{\text{specialise } x' \text{ to } E\}$$

$$\forall \alpha \bullet (Q \Leftrightarrow R)\langle E/x \rangle$$

$$\equiv \qquad \{\text{distribute substitution through subformulas}\}$$

$$\forall \alpha \bullet Q\langle E/x \rangle \Leftrightarrow R\langle E/x \rangle$$

$$\equiv \qquad \{\text{defn of } [x:=E]\}$$

$$\forall \alpha \bullet [x := E]Q \Leftrightarrow [x := E]R$$

$$\equiv \qquad \{\text{defn of } = \text{ for predicates}\}$$

$$[x := E]Q \;=\; [x := E]R$$

$\square$

10.2 Some Properties of Substitutions

It is well known that conjunctivity entails the weaker property of monotonicity, which in turn entails what is sometimes called weak disjunctivity. Thus from GS2 we infer the following:

Proposition 1   (monotonicity and weak disjunctivity)

Let S be any generalised substitution and Q and R be predicates. Then

$$Q \Rrightarrow R \qquad \vdash \quad [S]Q \Rrightarrow [S]R \qquad\qquad \text{monotonicity} \qquad (1.1)$$

and

$$[S]Q \vee [S]R \quad \vdash \quad [S](Q \vee R) \qquad\qquad \text{weak disjunctivity} \quad (1.2)$$

Proof of 1.1 :

	$Q \Rrightarrow R$	{hypothesis}
$\equiv$	$Q \wedge R \;=\; Q$	{logic}
$\Rrightarrow$	$[S](Q \wedge R) \;=\; [S]Q$	{Leibniz GS1}
$\equiv$	$[S]Q \wedge [S]R \;=\; [S]Q$	{conjunctivity GS2}
$\equiv$	$[S]Q \Rrightarrow [S]R$	{logic}

Proof of 1.2 :

	$Q \Rightarrow Q \vee R$	{tautology}	
$\Rrightarrow$	$[S]Q \Rightarrow [S](Q \vee R)$	{monotonicity}	1
	$R \Rightarrow Q \vee R$	{tautology}	
$\Rrightarrow$	$[S]R \Rightarrow [S](Q \vee R)$	{monotonicity}	2
	$[S]Q \vee [S]R \Rightarrow [S](Q \vee R)$	{logic 1, 2}	

$\square$

Using monotonicity we can immediately establish a number of relationships between the characteristic predicates we defined in section 5 :

Proposition 2   (fis, trm and prd)

Let S be any generalised substitution and Q be any postcondition. Then the following tautologies hold:

$$\vdash \quad trm(S) \ \lor \ prd(S) \tag{2.1}$$

$$\vdash \quad prd(S) \ \Rightarrow \ fis(S) \tag{2.2}$$

$$\vdash \quad fis(S) \ \lor \ [S]Q \tag{2.3}$$

$$\vdash \quad [S]Q \ \Rightarrow \ trm(S) \tag{2.4}$$

Proof of 2.1 :

Let $frame(S) = s$

$$
\begin{array}{lll}
& s \neq s' \ \Rightarrow \ \text{true} & \{\text{tautology}\} \\
\Rightarrow & [S]\,s \neq s' \ \Rightarrow \ [S]\,\text{true} & \{\text{monotonicity}\} \\
\equiv & \neg\,[S]\,s \neq s' \ \lor \ [S]\,\text{true} & \{\text{logic}\} \\
\equiv & prd(S) \ \lor \ trm(S) & \{\text{defns of prd and trm}\}
\end{array}
$$

Proof of 2.2 :

Let $frame(S) = s$

$$
\begin{array}{lll}
& \text{false} \ \Rightarrow \ s \neq s' & \{\text{tautology}\} \\
\Rightarrow & [S]\,\text{false} \ \Rightarrow \ [S]\,s \neq s' & \{\text{monotonicity}\} \\
\equiv & \neg\,[S]\,s \neq s' \ \Rightarrow \ \neg\,[S]\,\text{false} & \{\text{logic}\} \\
\equiv & prd(S) \ \Rightarrow \ fis(S) & \{\text{defns of prd and fis}\}
\end{array}
$$

Proof of 2.3 :

$$
\begin{array}{lll}
& \text{false} \ \Rightarrow \ Q & \{\text{tautology}\} \\
\Rightarrow & [S]\,\text{false} \ \Rightarrow \ [S]\,Q & \{\text{monotonicity}\} \\
\equiv & \neg\,[S]\,\text{false} \ \lor \ [S]\,Q & \{\text{logic}\} \\
\equiv & fis(S) \ \lor \ [S]\,Q & \{\text{defn of fis}\}
\end{array}
$$

Proof of 2.4 :

$$
\begin{array}{lll}
& Q \ \Rightarrow \ \text{true} & \{\text{tautology}\} \\
\Rightarrow & [S]\,Q \ \Rightarrow \ [S]\,\text{true} & \{\text{monotonicity}\} \\
\equiv & [S]\,Q \ \Rightarrow \ trm(S) & \{\text{defn of trm}\}
\end{array}
$$

$\square$

The next proposition describes a generalised substitution's wp predicate-transformer effect on a disjunction when one of the disjuncts is independent of the substitution's frame. Note that its proof depends on GS3 as well as GS1 and GS2.

Proposition 3   (Disjunctivity with a frame-independent disjunct)

Let S be a generalised substitution and Q and R be predicates where R is independent of $frame(S)$: that is, $frame(S) \setminus R$. Then

$$[S]\,(Q \vee R) \quad = \quad [S]\,Q \vee (trm(S) \wedge R)$$

Proof of $\Rightarrow$:

$$
\begin{array}{llll}
& (Q \vee R) \wedge \neg R & \Rightarrow & Q & \{\text{tautology}\} \\
\Rightarrow & [S]\,((Q \vee R) \wedge \neg R) & \Rightarrow & [S]\,Q & \{\text{monotonicity}\} \\
\equiv & [S]\,(Q \vee R) \wedge [S]\,\neg R & \Rightarrow & [S]\,Q & \{\text{conjunctivity}\} \\
\equiv & [S]\,(Q \vee R) \wedge trm(S) \wedge (fis(S) \Rightarrow \neg R) & \Rightarrow & [S]\,Q & \{\text{GS3, } frame(S) \setminus \neg R\} \\
\equiv & [S]\,(Q \vee R) \wedge (fis(S) \Rightarrow \neg R) & \Rightarrow & [S]\,Q & \{\text{Prop 2.4, logic}\} \\
\equiv & [S]\,(Q \vee R) & \Rightarrow & (fis(S) \wedge R) \vee [S]\,Q & \{\text{logic}\} \\
\equiv & [S]\,(Q \vee R) & \Rightarrow & (fis(S) \vee [S]\,Q) \wedge (R \vee [S]\,Q) & \{\text{logic}\} \\
\equiv & [S]\,(Q \vee R) & \Rightarrow & \text{true} \wedge (R \vee [S]\,Q) & \{\text{Prop 2.3}\} \\
\equiv & [S]\,(Q \vee R) & \Rightarrow & R \vee [S]\,Q & \{\text{logic}\} \\
\equiv & [S]\,(Q \vee R) & \Rightarrow & (trm(S) \wedge R) \vee [S]\,Q & \{\text{Prop 2.4, logic}\}
\end{array}
$$

Proof of $\Leftarrow$:

$$
\begin{array}{llll}
& [S]Q \vee [S]R & \Rightarrow & [S](Q \vee R) & \{\text{weak disjunctivity}\} \\
\equiv & [S]Q \vee (trm(S) \wedge (fis(S) \Rightarrow R)) & \Rightarrow & [S](Q \vee R) & \{\text{GS3, } frame(S) \setminus R\} \\
\equiv & trm(S) \wedge ([S]Q \vee (fis(S) \Rightarrow R)) & \Rightarrow & [S](Q \vee R) & \{\text{Prop 2.4, logic}\} \\
\equiv & trm(S) \wedge (R \vee (fis(S) \Rightarrow [S]Q)) & \Rightarrow & [S](Q \vee R) & \{\text{logic}\} \\
\equiv & trm(S) \wedge (R \vee [S]Q) & \Rightarrow & [S](Q \vee R) & \{\text{Prop 2.3}\} \\
\equiv & (trm(S) \wedge R) \vee [S]Q & \Rightarrow & [S](Q \vee R) & \{\text{Prop 2.4, logic}\}
\end{array}
$$

$\square$

11   A Normal Form for Substitutions

We can now establish the following important result. In the B-Book[1] Abrial proves the same result by structural induction over the syntax of the generalised substitution language, but our proof here is more general since it depends ultimately only on our three healthiness axioms GS1, GS2 and GS3.

Proposition 4   (Normal form of a generalised substitution)

Let S be a generalised substitution and let $frame(S)$ be s. Then

$$S \;\; = \;\; trm(S) \mid @\, s' .\; prd(S) \Longrightarrow s := s'$$

Proof :

To prove two substitutions are equal we must show they have the same frame and the same wp effect on any postcondition. In this case trivially the frame of the right-hand substitution is s too. We now prove that if Q is a postcondition, so that necessarily $s' \setminus Q$, then

$$[S]Q \;\; = \;\; [trm(S) \mid @\, s' .\; prd(S) \Longrightarrow s := s']\, Q$$

We have the tautology

$$
\begin{array}{llll}
 & Q & = & \forall s' \bullet \; s' = s \;\Rightarrow\; Q\langle s'/s\rangle & \{\text{1-point rule, } s' \setminus Q\} \\
\Rightarrow & [S]Q & = & [S]\,(\forall s' \bullet \; s' = s \;\Rightarrow\; Q\langle s'/s\rangle) & \{\text{GS1 Leibniz}\} \\
 & & = & \forall s' \bullet \; [S](s' = s \;\Rightarrow\; Q\langle s'/s\rangle) & \{\text{conjunctivity}\} \\
 & & = & \forall s' \bullet \; [S](s' \neq s \;\vee\; Q\langle s'/s\rangle) & \{\text{logic}\} \\
 & & = & \forall s' \bullet \; [S]\, s' \neq s \;\vee\; (trm(S) \wedge Q\langle s'/s\rangle) & \{\text{Prop 3, } s \setminus Q\langle s'/s\rangle\} \\
 & & = & \forall s' \bullet \; trm(S) \;\wedge\; ([S]\, s' \neq s \;\vee\; Q\langle s'/s\rangle) & \{\text{Prop 2.4, logic}\} \\
 & & = & trm(S) \;\wedge\; \forall s' \bullet \; [S]\, s' \neq s \;\vee\; Q\langle s'/s\rangle & \{s' \setminus trm(S)\} \\
 & & = & trm(S) \;\wedge\; \forall s' \bullet \; \neg\, [S]\, s' \neq s \;\Rightarrow\; Q\langle s'/s\rangle & \{\text{logic}\} \\
 & & = & trm(S) \;\wedge\; \forall s' \bullet \; prd(S) \;\Rightarrow\; Q\langle s'/s\rangle & \{\text{defn of prd}\} \\
 & & = & [trm(S) \mid @\, s' .\; prd(S) \Longrightarrow s := s']\, Q & \{\text{defns in GSL}\} \\
\end{array}
$$

$\square$

An immediate corollary is that for any substitution S with frame s
$S \;=\; trm(S) \mid s : prd_0(S)$. So a substitution S with frame s is equivalent to the Morgan[13] specification statement $s : [trm(S)\, ,\, prd_0(S)]$.

12   More about Extended Substitutions

The following proposition tells us about the prd of an extended substitution:

Proposition 5   (prd of an extended substitution)

Given a generalised substitution S and a list of variables y, let z be the residual frame of $|y|$ outside $frame(S)$: that is, $z \;=\; |y| - frame(S)$. Then

$$prd(S_y) \;\;\; = \;\;\; prd(S) \;\wedge\; (trm(S) \Rightarrow z = z')$$

Proof:

Let $frame(S) \;=\; s$, so $frame(S_y) \;=\; |s, z|$.

Then $prd(S_y)$

$$
\begin{array}{llll}
= & \neg\,[S_y]\,s, z \neq s', z' & \{\text{defn of prd}\} \\
= & \neg\,[S]\,s, z \neq s', z' & \{\text{defn of } S_y\} \\
= & \neg\,[S](s \neq s' \vee z \neq z') & \{\text{expand } s, z \neq s', z'\} \\
= & \neg\,([S]\,s \neq s' \ \vee \ (trm(S) \wedge z \neq z')) & \{\text{Prop 3, } s \setminus z \neq z'\} \\
= & \neg\,[S]\,s \neq s' \ \wedge \ (trm(S) \Rightarrow z = z') & \{\text{logic}\} \\
= & prd(S) \ \wedge \ (trm(S) \Rightarrow z = z') & \{\text{defn of prd}\}
\end{array}
$$

$\square$

The next proposition gives an alternative interpretation of an extended substitution, which vindicates our intuition about such an extension:

Proposition 6   (Extending a substitution by $\|$)

Given a generalised substitution S and a list of variables y, let z be the residual frame of $|y|$ outside $frame(S)$: that is, $z = |y| - frame(S)$. Then

$$
S_y \ = \ S \,\|\, z := z
$$

Proof:

Trivially, the frames and trms the lhs and rhs agree. By Proposition 5 the prd of the lhs is $prd(S) \ \wedge \ (trm(S) \Rightarrow z = z')$, which coincides with that of the rhs, given that $trm(z := z)$ is true and $prd(z := z)$ is $z = z'$.

$\square$

13   More about Refinement

Proposition 7   (Refinement by trm and prd)

Let S and T be generalised substitutions over an alphabet, and let $frame(S) = s$ and $frame(T) = t$. Then

$$
S \sqsubseteq T \ \ = \ \ s \subseteq t \ \wedge \ (trm(S) \Rrightarrow trm(T)) \ \wedge \ (prd(T) \Rrightarrow prd(S_t))
$$

Proof of $\Rightarrow$

$$
\begin{array}{lll}
& s \subseteq t & \{\text{GSref}\} \hspace{3cm} 1 \\[4pt]
& frame(S_t) \ = \ s \cup t & \{\text{defn of } S_t\} \hspace{2.1cm} 2 \\[4pt]
& frame(S_t) = t & \{1, 2\} \hspace{2.9cm} 3 \\[4pt]
& [S]\,Q \Rightarrow [T]\,Q & \{\text{GSref}\} \hspace{3cm} 4 \\[4pt]
& [S]\,\text{true} \Rightarrow [T]\,\text{true} & \{4, \text{let } Q \ = \ \text{true}\} \\[4pt]
\equiv & trm(S) \Rightarrow trm(T) & \{\text{defn of trm}\}
\end{array}
$$

$$[S]\, t \neq t' \quad \Rightarrow \quad [T]\, t \neq t' \qquad\qquad \{4,\ \text{let } Q \;=\; t \neq t'\}$$

$$\equiv \quad \neg\,[T]\, t \neq t' \quad \Rightarrow \quad \neg\,[S]\, t \neq t' \qquad \{\text{contra-implication}\}$$

$$\equiv \quad prd(T) \Rightarrow prd(S_t) \qquad\qquad\qquad \{3,\ \text{defn of prd}\}$$

Proof of $\Leftarrow$

Let z be $t - s$, so t is $|s, z|$. Then $[S]\, Q$

$=$ $\qquad$ {Prop 4}

$trm(S) \;\wedge\; (\forall\, s' \bullet prd(S) \Rightarrow Q\langle s'/s\rangle$

$=$ $\qquad$ {1-point rule, $z' \setminus prd(S)$}

$trm(S) \;\wedge\; (\forall\, s', z' \bullet prd(S) \wedge z = z' \Rightarrow Q\langle s', z'/s, z\rangle)$

$=$ $\qquad$ {logic, $s', z' \setminus trm(S)$}

$trm(S) \;\wedge\; (\forall\, s', z' \bullet prd(S) \wedge (trm(S) \Rightarrow z = z') \Rightarrow Q\langle s', z'/s, z\rangle)$

$=$ $\qquad$ {Prop 5}

$trm(S) \;\wedge\; (\forall\, s', z' \bullet prd(S_t) \Rightarrow Q\langle s', z'/s, z\rangle)$

$=$ $\qquad$ $\{t' \;=\; s', z'\}$

$trm(S) \;\wedge\; (\forall\, t' \bullet prd(S_t) \Rightarrow Q\langle t'/t\rangle)$

$\Rightarrow$ $\qquad$ {by hypothesis $trm(S) \Rightarrow trm(T)$ and $prd(T) \Rightarrow prd(S_t)$}

$trm(T) \;\wedge\; (\forall\, t' \bullet prd(T) \Rightarrow Q\langle t'/t\rangle)$

$=$ $\qquad$ {Prop 4}

$[T]\, Q$

$\square$

A slightly disquieting consequence of our definition of substitution refinement is that a demonic choice like $S\ [\,]\ T$ is no longer in general refined by each of its components S and T. It will of course be so refined if the frames of S and T coincide. However, by means of frame extension we can recover a more general formulation about the refinement of such a choice, as expressed in the following proposition:

Proposition 7 (Refinement of demonic choice)

If S and T are substitutions such that $frame(S) = s$ and $frame(T) = t$ then

$$S\ [\,]\ T \sqsubseteq S_t \qquad \text{and} \qquad S\ [\,]\ T \sqsubseteq T_s$$

Proof: follows immediately from the definition of frame extension, which gives us that $frame(S_t) \;=\; frame(T_s) \;=\; s \cup t$, and that the wp effects of S_t and T_s are respectively the same as those of S and T.

$\square$

14   Conclusion

We contend our simple innovation of recognising the frame of a substitution as part of its formal meaning has been effective in unravelling some knotty issues in B concerning the nature of substitutions. Our framed substitutions have a rich theory in their own right which we have sought to illustrate with the propositions we have presented in this paper. In demonstrating that substitutions can be contemplated outside their familiar context of B abstract machines we believe we have achieved a separation of concerns which can be of value not only to B users and B theorists but possibly to other potential users of the generalised substitution language too.

One of the anonymous reviewers commented that there is still work to be done in giving a semantics to B which adequately explains the syntactic constraints imposed on the various forms of composition of machines. We certainly concur with such a view, and would like to hope our theory of generalised substitutions may provide a useful foundation for such an endeavour.

Acknowlegements. I am indebted to the two anonymous reviewers and Louis Mussat for pointing out shortcomings in the original draft of this paper. As ever, I am grateful to my colleague Bill Stoddart for his vital interest and encouragement when the ideas presented here were in embryonic form.

References

1. J.-R. Abrial. *The B-Book: Assigning Programs to Meanings.* Cambridge University Press, 1996.
2. R.-J. Back. A calculus of refinements for program derivations. *Acta Informatica,* 25:593–624, 1988.
3. R.-J. Back and M.J. Butler. Fusion and simultaneous execution in the refinement calculus. *Acta Informatica,* 35(11):921–940, 1998.
4. R.-J. Back and J. von Wright. *Refinement Calculus: A Systematic Introduction.* Springer-Verlag New York, 1998.
5. D. Bert, M.-L. Potet, and Y. Rouzaud. A study on components and assembly primitives in B. In H. Habrias, editor, *Proceedings of the First B Conference,* pages 47–62. IRIN, Nantes, 1996.
6. J.C. Bicarregui and B. Ritchie. Invariants, frames and postconditions: a comparison of the VDM and B notations. In J. Woodcock and P.G. Larsen, editors, *Proceedings of FME'93,* number 670 in Lecture Notes in Computer Science. Springer-Verlag, 1993.
7. P Chartier. Formalisation of B in Isabelle/HOL. In D. Bert, editor, *B'98: Recent Advances in the Development and Use of the B Method; Proceedings of the Second International B Conference, Montpellier, France,* number 1393 in Lecture Notes in Computer Science, pages 66–82. Springer-Verlag, 1998.
8. E.W. Dijkstra. *A Discipline of Programming.* Prentice-Hall International, 1976.
9. S.E. Dunne. The Safe Machine: a new specification construct for B. In J.M. Wing, J. Woodcock, and J. Davies, editors, *FM'99 - Formal Methods,* number 1708 in Lecture Notes in Computer Science, pages 472–489. Springer-Verlag, 1999.

10. R.W. Floyd. Assigning meanings to programs. *Proceedings of Symposia in Applied Mathematics*, 19:19–32, 1967.
11. C.A.R. Hoare and He Jifeng. *Unifying Theories of Programming*. Prentice Hall, 1998.
12. C.B. Jones. *Systematic Software Development Using VDM (2nd edn)*. Prentice-Hall, 1990.
13. C.C. Morgan. The specification statement. *ACM Transactions on Programming Languages and Systems*, 10(3), 1988.
14. C.C. Morgan. *Programming from Specifications (2nd edn)*. Prentice Hall International, 1994.
15. J.M. Morris. A theoretical basis for stepwise refinement and the programming calculus. *Science of Computer programming*, 9:287–306, 1987.
16. G. Nelson. A generalisation of Dijkstra's calculus. *ACM Transactions on Programmg Languages and Systems*, 11(4), 1989.
17. Y. Rouzaud. Interpreting the B-Method in the Refinement Calculus. In J.M. Wing, J. Woodcock, and J. Davies, editors, *FM'99 - Formal Methods*, number 1708 in Lecture Notes in Computer Science, pages 411–430. Springer-Verlag, 1999.
18. G. Smith. Recursive schema definitions in Object-Z. In J.P. Bowen, S.E. Dunne, A. Galloway, and S. King, editors, *ZB2000: Formal Specification and Development in B and Z*, number 1878 in Lecture Notes in Computer Science, pages 42–58. Springer-Verlag, 2000.
19. A. Tarski. A lattice theoretical fixed point theorem and its applications. *Pacific Journal of Mathematics*, 5:285–309, 1955.
20. J. Woodcock and J. Davies. *Using Z: Specification, Refinement and Proof*. Prentice Hall, 1996.

Reinforced Condition/Decision Coverage (RC/DC): A New Criterion for Software Testing

Sergiy A. Vilkomir and Jonathan P. Bowen

South Bank University, Centre for Applied Formal Methods
School of Computing, Information Systems and Mathematics
103 Borough Road, London SE1 0AA, UK
{vilkoms, bowenjp}@sbu.ac.uk
http://www.cafm.sbu.ac.uk/

Abstract. A new Reinforced Condition/Decision Coverage (RC/DC) criterion for software testing is proposed. This criterion provides further development of the well-known Modified Condition/Decision Coverage (MC/DC) criterion and is more suitable for testing of safety-critical software. Formal definitions in the Z notation for RC/DC, as well as MC/DC, are presented. Specific examples of using of these criteria are considered and some features are formally proved.

1 Introduction

Software testing criteria determine requirements for the scope and the volume of software testing. The requirements for testing logical structure of programs are specified using so-called control-flow criteria. The aim of these criteria is testing *decisions* (the program points at which the control flow can divide into various paths) and *conditions* (atomic predicates which form component parts of decisions) in a program.

One of the simplest control-flow criteria is the decision coverage criterion which states that every decision in the program has taken all possible outcomes at least once [13]. This criterion requires only two test cases for each binary decision. The greatest number of test cases is required by the multiple condition coverage criterion which states that all possible combinations of condition outcomes in each decision have been invoked a minimum of once [13].

The decision coverage criterion is weak and not sufficient, especially for the testing of safety-critical software. The multiple condition coverage criterion requires 2^n test cases for a decision made up of n conditions and this is often not possible in practice. So it is necessary to use an intermediate criterion that combines sufficient scope of testing with a relatively small number of test cases. One of such criteria is the Modified Condition/Decision Coverage (MC/DC), which requires testing of the independent influence of every condition on the decision. This criterion has been introduced in the RTCA/DO-178B standard [16], which provides regulatory requirements for avionics software.

We considered the formal definitions of the main control-flow criteria in [17,19]. For those requiring more background information, a review of control-flow criteria with a fuller list of relevant references is available [19]. In this paper we propose the further

D. Bert et al. (Eds.): ZB 2002, LNCS 2272, pp. 291–308, 2002.
© Springer-Verlag Berlin Heidelberg 2002

development of this approach. The paper is structured as follows. Section 2 presents a detailed analysis of MC/DC. A new version of the definition in the Z notation [1] is proposed and the explanation how this formal approach can eliminate the ambiguity of informal definitions is given. A specific example using MC/DC is considered, illustrating the interdependence of the conditions and decisions. We analyze a major shortcoming of the MC/DC criterion, namely the deficiency of requirements for the testing of the "false operation" type of failures. Examples of situations when failures of this type are present are considered. These have especially vital importance for safety-critical applications in particular.

To eliminate the shortcoming of MC/DC, we propose a new Reinforced Condition/Decision Coverage (RC/DC) criterion, which is considered in Section 3. Z schemas for the formal definition of RC/DC and examples of its application are provided.

2   MC/DC

2.1   General Definition

The definition of the MC/DC criterion, according to [16], is the following:

> *Every point of entry and exit in the program has been invoked at least once, every condition in a decision in the program has taken on all possible outcomes at least once, every decision in the program has taken all possible outcomes at least once, and each condition in a decision has been shown to independently affect the decision's outcome. A condition is shown to independently affect a decision's outcome by varying just that condition while holding fixed all other possible conditions.*

The maximum number of required tests for a decision with n conditions is $2n$. The place of MC/DC in the hierarchy of control-flow criteria is given in Figure 1. The definitions and analysis of these criteria are considered in [13,14,15,19,22].

The first part of the MC/DC definition (*every point of entry and exit in the program has been invoked at least once*) is just the standard statement coverage criterion. This part is traditionally added to all control-flow criteria and is not directly connect with the main point of MC/DC.

The second and the third parts of the definition are just the condition and decision coverage criteria. The inclusion of these parts in the definition of MC/DC could be considered excessive because satisfiability of the condition and decision coverage results from the main part of the MC/DC definition: *each condition has been shown to independently affect the decision's outcome.*

The key word in this definition is "independently"; i.e., the aim of MC/DC is the elimination during testing of the mutual influence of the individual conditions and the testing of the correctness of each condition separately.

Investigation of MC/DC has initially been considered in [4,5,10]. Detailed consideration of the different aspects of this criterion was carried out more recently (1999–2001) in [2,6,7,9]. The successful practical application of MC/DC for satellite control software

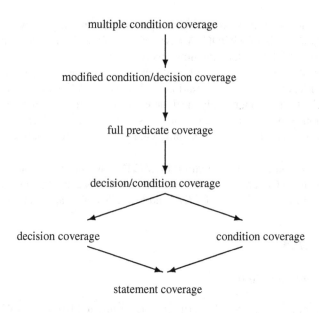

Fig. 1. The hierarchy of control-flow criteria.

has been evaluated [8] though the difficulties during the analysis of this type of coverage (e.g., it is extremely expensive to carry out and can affect staff morale and time) were also addressed [3]. The application of this criterion in the testing of digital circuits was considered in [12]. A number of software tools (LDRA Testbed, McCabe, ATTOL, CodeTEST, Cantata++, etc.) support the MD/DC criterion.

However, it should be noted that the original definition of the MC/DC criterion allows different interpretations and understanding during the application of the criterion. The informal definition gives no precise answer to some practical questions. For example:

- How to handle the situation when it is impossible to vary a condition and a decision while holding fixed all other conditions (see Section 2.3 for an example of such a situation). To assume that such a condition does not satisfy the MC/DC criterion is probably not the best way of dealing with this situation. In any case, such conditions should be checked during testing.
- How to understand multiple occurrences of a condition in a decision. For example, for a decision of the form $(A \wedge B) \vee (\neg A \wedge C)$, should we assume three conditions $(A, B,$ and $C)$ or four (the first $A, B, C,$ and the second A) conditions? Both approaches have been used [4,9] but the last one seems unnatural for many situations.
- How to treat degenerate conditions and decisions, which are either all 0 or all 1. Of course, the appearance of such conditions should attract the attention of the tester and be justified. But if it is valid for some reason, does it mean that in this case MC/DC is not satisfied because such conditions cannot be varied?

– How to consider coupled conditions, i.e., conditions that cannot be varied independently. According to [4], two or more conditions are strongly coupled if varying one always varies the other, and weakly coupled if varying one sometimes, but not always, varies the others. However, it is questionable whether strongly coupled conditions really exist.

For a more precise definition of the MC/DC criterion, eliminating inaccuracies and answering the above questions, we propose a formal definition of MC/DC using the Z notation.

2.2   Formal Definition in the Z Notation

We considered the formal definitions in the Z notation for all main control-flow criteria, including MC/DC, in [19]. In this paper, a slightly different approach is proposed. Within the framework of this approach, we distinguish the difference between input variables for a whole program and input variables for an individual decision. It allows us to consider possible changes of values of input variables and the multiple execution a decision in a loop. A more precise definition of the notion of a decision is introduced gradually and used in the next section for the formulation of a formal definition for RC/DC.

For defining the criteria, the given sets $STATEMENT$ and $INPUT$ are used:

$$[STATEMENT, INPUT]$$

The $STATEMENT$ set contains all possible program statements. The second set contains all possible values of all input, output and internal variables during the running of the program. In practice, the elements of $INPUT$ are values associated with variables under consideration at any particular part of the program, but this level of detail is not required at this level of abstraction. We use the name $INPUT$ to emphasize the fact that the members of this set are the input data relative to decisions and conditions in a program. The set $startinput$ is a set of the values of the input variables only and is a subset of $INPUT$. Some of the values of the input variables (namely, the $starttest$ set) are used as testing data.

$$
\begin{array}{|l}
startinput, starttest : \mathbb{P}\, INPUT \\
statementinput : STATEMENT \times INPUT \nrightarrow \mathbb{P}\, INPUT \\
\hline
starttest \subseteq startinput \\
\mathrm{dom}\; statementinput = STATEMENT \times startinput
\end{array}
$$

The $statementinput$ function returns (for each program statement and each fixed values of input variables) all values of all program variables which are possible when this statement is executed. It can be several different sets of values (several different elements of $INPUT$) for one decision because of potential multiple execution of this decision in a loop.

Now we can create a Z schema[1] for the statement coverage criterion, which is a component of all other control-flow criteria including MC/DC.

---

[1] All schemas in this paper have been checked using the ZTC type-checker package [11].

StatementCoverage _____

$st : STATEMENT$

_____

$\bigcup\{i : starttest \bullet statementinput(st, i)\} \neq \varnothing$

If for some statement st the value of $statementinput$ is the empty set for all testing data, this means that this statement is never executed during test runs of a program; i.e., the testing data does not satisfy the statement coverage criterion.

We now introduce some definitions. The $Bool$ set contains values for logical variables: 1 ($TRUE$) and 0 ($FALSE$):

$$Bool == \{0, 1\}$$

We encode this as numbers since this is a standard nomenclature by many testers. We use the elements of this set as values of $cond$, a set of partial "logical" functions on $INPUT$:

$$cond == INPUT \nrightarrow Bool$$

The following schema describes a decision. We consider a decision as a program statement ($decst$) and an associated logical function ($value$).

Dec _____

$decst : STATEMENT$
$value : cond$
$decinput, decinput0, decinput1, testset : \mathbb{P}_1 INPUT$
$argdec : \mathbb{P}_1 cond$

_____

$decinput = \mathrm{dom}\ value = \bigcup\{i : startinput \bullet statementinput(decst, i)\}$
$decinput0 = \{i : decinput \mid value\ i = 0\}$
$decinput1 = \{i : decinput \mid value\ i = 1\}$
$\langle decinput0, decinput1\rangle$ partitions $decinput$
$testset = \bigcup\{i : starttest \bullet statementinput(decst, i)\}$
$argdec \subseteq \{c : cond \mid \mathrm{dom}\ c = decinput \wedge \mathrm{ran}\ c = Bool\}$

The $decinput$ set contains data (values of input, output, and internal variables), which activates the given decision. The $decinput0$ and $decinput1$ sets contain data, for which the decision equals 0 and 1 respectively. These sets partition $decinput$; i.e., $decinput0 \cup decinput1 = decinput$ and $decinput0 \cap decinput1 = \varnothing$.

The $argdec$ set contains all conditions (atomic predicates), which make a decision. For example, if a decision is of the form $A_1 \vee B_1$, then the A_1 and B_1 conditions form part of the decision and $argdec$ is $\{A_1, B_1\}$ (see also an example in section 2.3). These conditions are the arguments of the logical formula, which determines the decision $value$ function uniquely.

The $testset$ set contains data from $INPUT$, which activating the given decision during test runs of a program, i.e., when the $starttest$ set is used as input data. The members of $testset$ are testing data for a given decision. If a decision statement is executed multiple times inside a loop, one test case for the whole program from *starttest* (one test run) generates several test cases for the decision from *testset*.

The relation *InputPairs* describes pairs of data from *INPUT*. It is convenient to use this type because we always apply a pair of test sets for testing when varying each condition or decision.

$$InputPairs == INPUT \leftrightarrow INPUT$$

Consider *DecModified*, a modified version of the *Dec* schema. Two sets of pairs of data are considered for each condition.

DecModified

Dec
changedec, changedecfix : *cond* $\nrightarrow$ *InputPairs*

---
dom *changedec* = dom *changedecfix* = *argdec*
ran *changedec* $\cup$ ran *changedecfix* $\subseteq$ *decinput* $\leftrightarrow$ *decinput*
$\forall c : cond \mid c \in argdec \bullet$
  changedec $c = \{i_0, i_1 : decinput \mid value\ i_0 \neq value\ i_1 \wedge c\ i_0 \neq c\ i_1\} \wedge$
  changedecfix $c = \{i_0, i_1 : decinput \mid (i_0, i_1) \in changedec\ c \wedge$
    $(\forall othercond : argdec \mid othercond \neq c \bullet othercond\ i_0 = othercond\ i_1)\}$

The set *changedec* c is a set of pairs of data which simultaneously varying the decision and condition c, i.e., condition c equals 0 for one element of the pair and equals 1 for another element. The *changedecfix* c set contains pairs of data which varying the decision and given and only given condition c, i.e., for all other conditions from the decision, the condition value for the first element of the pair coincides with the condition value for the second element. Obviously, *changedecfix* $c \subseteq$ *changedec* c.

For definition of MC/DC (and, later, RC/DC), the *choice* function is used.

---
choice : *InputPairs* $\times$ *InputPairs* $\rightarrow$ *InputPairs*

$\forall a, b : InputPairs \bullet$
$(a \neq \varnothing \Rightarrow choice(a, b) = a) \wedge (a = \varnothing \Rightarrow choice(a, b) = b)$

The arguments are two sets. If the first one is not empty the function just returns it; otherwise, the second set is returned.

Now we can create a formal definition of MC/DC. For each condition in each decision, the aim of this criterion is to have, as a part of the testing data, pairs of input data that vary this condition simultaneously with the decision while, if it is possible, fixing all other conditions. The following Z schema captures MC/DC:

MC_DC

StatementCoverage

$\forall DecModified;\ c : cond \mid c \in argdec \bullet$
  $(testset \times testset) \cap choice(changedecfix\ c, changedec\ c) \neq \varnothing$

Let us prove that it is always possible to choose the testing data, which satisfy MC/DC, i.e., that $choice(changedecfix\ c, changedec\ c) \neq \varnothing$, using the method of the proof by contradiction.

Lemma 1

$MC\_DC;\ c : cond \vdash choice(changedecfix\ c, changedec\ c) \neq \varnothing$

Proof

$choice(changedecfix\ c, changedec\ c) = \varnothing$	[assumption]
$\Leftrightarrow changedec\ c = \varnothing$	[definition of $choice$]
$\Leftrightarrow \neg\ (\exists\ i_0, i_1 : decinput \mid c\ i_0 \neq c\ i_1 \bullet$	[definition of $changedec$]
$\quad value\ i_0 \neq value\ i_1)$	
$\Leftrightarrow \forall\ i_0, i_1 : decinput \mid c\ i_0 = 0 \land c\ i_1 = 1 \bullet$	[logic]
$\quad value\ i_0 = value\ i_1$	
$\Rightarrow \forall\ i_0, i_1 : decinput \mid$	$[decinput0, decinput1 : \mathbb{P}_1\ INPUT]$
$\quad c\ i_0 = 0 \land c\ i_1 = 1 \bullet \exists\ i_2 : decinput \bullet$	
$\quad value\ i_2 \neq value\ i_0 \land value\ i_2 \neq value\ i_1$	
$\Rightarrow \forall\ i_0, i_1 : decinput \mid c\ i_0 = 0 \land c\ i_1 = 1 \bullet$	$[c\ i_2 = 0 \lor c\ i_2 = 1]$
$\quad \exists\ i_2 : decinput \bullet (c\ i_2 = 0 \land value\ i_2 \neq value\ i_1) \lor$	
$\quad (c\ i_2 = 1 \land value\ i_2 \neq value\ i_0)$	
$\Rightarrow \exists\ i_0, i_1, i_2 : decinput \bullet$	$[c\ i_1 = 1 \land c\ i_0 = 0]$
$\quad (c\ i_2 \neq c\ i_1 \land value\ i_2 \neq value\ i_1) \lor$	
$\quad (c\ i_2 \neq c\ i_0 \land value\ i_2 \neq value\ i_0)$	
$\Rightarrow \exists\ n, m : decinput \bullet$	$[n = i_2 \land (m = i_0 \lor m = i_1)]$
$\quad c\ n \neq c\ m \land value\ n \neq value\ m$	
$\Leftrightarrow changedec\ c \neq \varnothing$	[definition of $changedec$]
$\Leftrightarrow choice(changedecfix\ c, changedec\ c) \neq \varnothing$	[definition of $choice$]
$\Rightarrow false$	[contradiction]

Let us consider how the proposed formal definition of MC/DC answers the questions formulated in Section 2.1:

- *How to handle the situation when it is impossible to vary a condition and a decision while holding fixed all other conditions.* According the formal definition of MC/DC, if it is impossible to find such testing data (i.e., *changedecfix c = $\varnothing$*) we can vary the condition and the decision without fixing other conditions (i.e., take testing data from *changedec c*).
- *How to understand multiple occurrences of a condition in a decision.* According to the definition of a decision (in the *Dec* schema), we consider a set (*argdec*) of conditions that make a decision. This means that each condition is considered only once. This approach is more mathematically valid and corresponds with understanding a decision as a function of conditions.

- *How to treat degenerate conditions and decisions.* According to the definition of a decision (again in the *Dec* schema), both of the sets *decinput0* and *decinput1* are non-empty. This means that every decision should take the value of both 0 and 1 and the degenerate decisions are excluded from consideration. The *Dec* schema also ensures that the range of every condition is equal to *Bool*; i.e., every condition should take both 0 and 1 values and thus degenerate conditions are excluded from consideration. The reason for this approach is that degenerate conditions and decisions are always covered by any testing data. So, we consider them as satisfying MC/DC because it does not make demands on such decisions and conditions.
- *How to consider the coupled conditions.* The coupled conditions [4] make problems for selecting the testing data satisfying the MC/DC criterion. However, these problems exist only for weakly coupled conditions. As we show below (see Lemma 2), if one condition A always varies the other condition B then $A = B \lor A = \neg B$, where $\neg$ is formally defined as follows:

$$\neg : cond \rightarrowtail cond$$
$$\forall c : cond \bullet \neg c = c \mathbin{\substack{\circ \\ 9}} \{0 \mapsto 1, 1 \mapsto 0\}$$

So we can consider A and B as entering the same condition into a decision. In other words, strongly coupled conditions as mentioned in [4] do not exist.

Lemma 2

$MC\_DC;\ A, B : cond \vdash$

$(\forall i_0, i_1 : decinput \bullet A\, i_0 \neq A\, i_1 \Rightarrow B\, i_0 \neq B\, i_1) \Rightarrow (A = B \lor A = \neg B)$

Proof

$\neg ((\forall i_0, i_1 : decinput \bullet$	[assumption]
$\quad A\, i_0 \neq A\, i_1 \Rightarrow B\, i_0 \neq B\, i_1) \Rightarrow (A = B \lor A = \neg B))$	
$\Leftrightarrow (\forall i_0, i_1 : decinput \bullet$	[logic]
$\quad A\, i_0 \neq A\, i_1 \Rightarrow B\, i_0 \neq B\, i_1) \land (A \neq B \land A \neq \neg B)$	
$\Rightarrow (\exists i_0, i_1 : decinput \bullet$	
$\quad A\, i_0 = 0 \land A\, i_1 = 1 \land B\, i_0 \neq B\, i_1) \land$	[ran $A = Bool$]
$\quad (A \neq B \land A \neq \neg B)$	

$[\textbf{CASE 1} :\ B\, i_0 = 1,\ B\, i_1 = 0]$

$\Rightarrow A \neq \neg B$	[logic]
$\Leftrightarrow \exists i_2 : decinput \bullet A\, i_2 = B\, i_2$	[logic]

$[\textbf{CASE 1.1} :\ A\, i_2 = 0,\ B\, i_2 = 0]$

$\Rightarrow A\, i_2 \neq A\, i_1$	$[A\, i_1 = 1]$
$\Rightarrow B\, i_2 \neq B\, i_1$	$[A\, i_2 \neq A\, i_1 \Rightarrow B\, i_2 \neq B\, i_1]$
$\Rightarrow B\, i_2 = 1$	$[B\, i_1 = 0]$
$\Rightarrow$ false	[contradiction with **CASE 1.1**]

[**CASE 1.2** : $A\ i_2 = 1,\ B\ i_2 = 1$]

$\Rightarrow A\ i_2 \neq A\ i_0$ $\hfill [A\ i_0 = 0]$

$\Rightarrow B\ i_2 \neq B\ i_0$ $\hfill [A\ i_2 \neq A\ i_0 \Rightarrow B\ i_2 \neq B\ i_0]$

$\Rightarrow B\ i_2 = 0$ $\hfill [B\ i_0 = 1]$

$\Rightarrow$ false $\hfill$ [contradiction with **CASE 1.2**]

[**CASE 2** : $B\ i_0 = 0,\ B\ i_1 = 1$]

$\Rightarrow A \neq B$ $\hfill$ [logic]

$\Leftrightarrow \exists\ i_2 : decinput \bullet A\ i_2 \neq B\ i_2$ $\hfill$ [logic]

[**CASE 2.1** : $A\ i_2 = 0,\ B\ i_2 = 1$]

$\Rightarrow A\ i_2 \neq A\ i_1$ $\hfill [A\ i_1 = 1]$

$\Rightarrow B\ i_2 \neq B\ i_1$ $\hfill [A\ i_2 \neq A\ i_1 \Rightarrow B\ i_2 \neq B\ i_1]$

$\Rightarrow B\ i_2 = 0$ $\hfill [B\ i_1 = 1]$

$\Rightarrow$ false $\hfill$ [contradiction with **CASE 2.1**]

[**CASE 2.2** : $A\ i_2 = 1,\ B\ i_2 = 0$]

$\Rightarrow A\ i_2 \neq A\ i_0$ $\hfill [A\ i_0 = 0]$

$\Rightarrow B\ i_2 \neq B\ i_0$ $\hfill [A\ i_2 \neq A\ i_0 \Rightarrow B\ i_2 \neq B\ i_0]$

$\Rightarrow B\ i_2 = 1$ $\hfill [B\ i_0 = 0]$

$\Rightarrow$ false $\hfill$ [contradiction with **CASE 2.2**]

2.3   A Case Study

The contents of the proposed formal definitions are considered below. Different examples of MC/DC use that have been presented previously (for example, see [4]) have often considered only simple decisions containing two or three conditions. Using MC/DC for such decisions has no great practical use because full searching of all test cases is easy achieved. Furthermore, such examples do not reflect complicated situations, which are typical in realistic practical examples of use of this criterion. We consider a more complex example (but one that is still far from a real practical problem because of space considerations), which takes into account the following factors:

– dependence of the values of the conditions and decisions on input data;
– dependence of the specific decision on its place in the computer program, i.e., on the values of other decisions in the program;
– dependence of the conditions in the specific decision on each other, i.e., the possibility that one condition takes a value depending on the value of other conditions in this decision.

This example uses a computer program fragment, whose graph is given in Figure 2.

In this fragment, the input data x and y are read; let the value of both x and y be between 0 and 100. For this simple example, we could consider *INPUT* as just a pair of values:

$$INPUT == (0\,..\,100) \times (0\,..\,100)$$

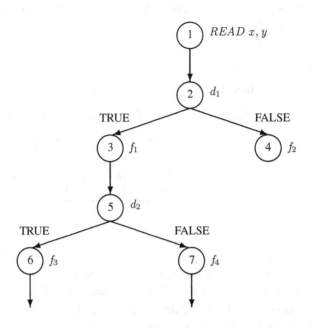

Fig. 2. Flow graph of a program fragment.

Then, depending on the values of x and y, the computation by one from four formulae $f_1 - f_4$ is implemented.

The control flow of this program is determined by two decisions: d_1 and d_2. Let d_1 depends on conditions A_1 and B_1 and d_2 depends on conditions A, B, C, and D, as it is shown below:

$d_1, d_2 : DecModified$
$A, B, C, D, A_1, B_1 : cond$
$x, y : 0 .. 100$

$A(x, y) = 1 \Leftrightarrow x > 20$
$B(x, y) = 1 \Leftrightarrow y < 60$
$C(x, y) = 1 \Leftrightarrow x > 40$
$D(x, y) = 1 \Leftrightarrow y < 80$
$A_1(x, y) = 1 \Leftrightarrow x > 20$
$B_1(x, y) = 1 \Leftrightarrow y > 60$
$d_1.decinput = INPUT$
$d_2.decinput = \{x, y : 0 .. 100 \mid x > 20 \lor y > 60\}$
$d_1.argdec = \{A_1, B_1\}$
$d_2.argdec = \{A, B, C, D\}$
$d_1.value(x, y) = 1 \Leftrightarrow A_1(x, y) = 1 \lor B_1(x, y) = 1$
$d_2.value(x, y) = 1 \Leftrightarrow$
$\quad ((A(x, y) = 1 \land B(x, y) = 1) \lor (C(x, y) = 1 \land D(x, y) = 1))$

For the d_1 decision, $d_1.decinput$ is all possible $INPUT$s and both conditions A_1 and B_1 are independent. The examples of the testing data satisfy the MC/DC criterion for d_1 are given in Table 1.

Table 1. Testing data satisfied the MC/DC criterion for d_1.

num	values			testing data	variations		MC/DC
	A_1	B_1	$d_1.value$	(x, y)	A_1	B_1	
1	1	1	1	$(50, 70)$			
2	1	0	1	$(50, 50)$	*		+
3	0	1	1	$(10, 70)$		*	+
4	0	0	0	$(10, 50)$	*	*	+

For the d_2 decision, the set $d_2.decinput$ is more restricted than the full set of possible $INPUT$s because of the interdependency of $d1$ and $d2$. The members of $d_2.decinput$ are only the input data for which d_1 equals 1, i.e., is TRUE.

The conditions in d_2 are interdependent in pairs. For conditions A and C the situation $(A = 0 \wedge C = 1)$ is impossible because $(C = 1) \Rightarrow (A = 1)$. For conditions B and D the situation $(B = 1 \wedge D = 0)$ is impossible because $(B = 1) \Rightarrow (D = 1)$.

As it is shown in Table 2, only 8 of the 16 combinations of condition values are possible.

Table 2. Testing data satisfied the MC/DC criterion for d_2.

num	values					testing data	variations				MC/DC
	A	B	C	D	$d_2.value$	(x, y)	A	B	C	D	
1	1	1	1	1	1	$(50, 50)$					
2	1	1	1	0	-	impossible					
3	1	1	0	1	1	$(30, 50)$	●	*			+
4	1	0	1	1	1	$(50, 70)$			*	*	+
5	0	1	1	1	-	impossible					
6	1	1	0	0	-	impossible					
7	1	0	1	0	0	$(50, 90)$				*	+
8	1	0	0	1	0	$(30, 70)$		*	*		+
9	0	1	1	0	-	impossible					
10	0	1	0	1	-	impossible					
11	0	0	1	1	-	impossible					
12	1	0	0	0	0	$(30, 90)$					
13	0	1	0	0	-	impossible					
14	0	0	1	0	-	impossible					
15	0	0	0	1	0	$(10, 70)$	●				+
16	0	0	0	0	0	$(10, 90)$					

The combinations $(0, 1, 1, 1)$, $(0, 1, 1, 0)$, $(0, 0, 1, 1)$, and $(0, 0, 1, 0)$ are impossible because of the value of the condition A. Combinations $(1, 1, 1, 0)$, $(1, 1, 0, 0)$, and $(0, 1, 0, 0)$ are impossible because of the value of the condition D. The combination $(0, 1, 0, 1)$ is impossible because of the value of the decision d_1.

Testing data satisfying the MC/DC criterion for the conditions B, C, and D are shown in Table 2 marked as '$*$'.

For the condition A it is impossible to choose similar combinations, i.e., combinations for which the values of A and d_2 are changed and the values of B, C, and D are fixed. So, following the formal definition of MC/DC, in this case for the condition A it is sufficient to take any combinations which vary A and d_2 without fixing other conditions. For example, it is possible to take combinations $(1, 1, 0, 1)$ and $(0, 0, 0, 1)$, for which the values of A and d_2 vary simultaneously (marked '$\bullet$' in Table 2). The testing set satisfying the MC/DC criterion for the decision d_2 consists of five pairs of the input data (marked '$+$' in Table 2).

2.4   The Main Shortcoming of MC/DC

As already mentioned in this paper, the MC/DC criterion is used mainly for testing of safety-critical avionics software [16]. The main aim of MC/DC is testing situations when changing a condition implies a change in a decision. Often a decision can be associated with some safety-critical operation of a system. In such cases, MC/DC requires the testing of situations when changing one condition has some consequence on the operation of the system. A software error in such situations could involve "*non-operation*" (inability to operate on demand) type of failures. Such situations are extremely important and the MC/DC requirements are entirely reasonable.

But, as we show below, this criterion has one substantial shortcoming, not previously mentioned in the literature, namely deficiency of requirements for testing of the "*false actuation*" (operation without demand) type of failures. This could make this criterion insufficient for many safety-critical applications.

The false actuation of a system could be invoked by a software error in situations when changing a condition should not imply changing a decision. We now consider several examples.

Railway points: Consider a railway computer control system and a decision that is responsible for switching over the points by which trains can be routed in one direction to another.

Let there be two tracks (main and reserved); the condition determines track states (which may be either occupied or clear) and the decision determines changing the route from the main track to the reserved track and vice versa. Consider two situations for the non-operation and false actuation types of failures.

The first situation is when the main track becomes occupied (varying the condition) and, therefore, it is necessary to switch over the points to the reserve track (varying the decision). The failure in this situation involves keeping the value of the decision instead of varying it; this means non-operation of the system and could result in a possible crash.

The second situation is when the reserved track becomes occupied (varying the condition) and, therefore, it is necessary to keep the main track as a route (keeping the decision). The failure in this situation involves varying the value of the decision instead of keeping it fixed that means false operation of the system and a possible crash.

Thus, from the safety point of view, these situations are symmetrical and can lead to a crash. Therefore, both types of failures should be considered and both situations should be tested with the same accuracy.

Nuclear reactor protection system: Consider a decision that is responsible for actuating a reactor protection system at a nuclear power plant (i.e., the reactor shutdown) and a condition that describes some criterion for the actuation (e.g., excessive pressure over some specified level). Varying this decision because of variation of the condition should be tested since failure in this situation means the non-operation of the system in case of emergency conditions and can lead to the nuclear accident.

Nevertheless, keeping the value of the decision is also important. The failure in this situation means the false actuation of the system during normal operating and can lead to non-forced reactor shutdown, the deterioration of the physical equipment, and the underproduction of electricity.

The typical architecture of nuclear reactor protection systems (three channels with "2 from 3" logical voting) takes into account this particular problem. The use of three identical channels decreases the probability of the system not operating correctly. However, if it is only required to consider this factor, the "1 from 3" logic is more reliable. The aim of using "2 from 3" voting is to provide protection against false actuation of a system as in this case the false signal from one channel does not lead to system actuation.

Thus, during the reactor protection system software testing, it is necessary to include test cases both for varying and keeping a decision's outcomes.

Planned halt of a computer control system: Sometimes specific situations are possible, when keeping a decision is much more important for safety than varying a decision. Consider a decision that is responsible for a planned (non-emergency) halt of a continuous process control system, e.g., for planned maintenance. Conditions describe when this process is in a safe state allowing switching off of the control system. Again consider two situations.

The first situation is when the state of the process becomes safe (varying the condition) and it is possible to switch system off (varying the decision). The failure in this situation does not have grave consequences and means only a delay of the system halt.

The second situation is when the state of the process becomes unsafe (varying the condition) and the control system should continue in operation (keeping the decision). The failure in this situation means that the system is erroneously switched off. Such a fault leads to loss of control and this is important with respect to safety. So it is important to test the keeping of the value of this decision.

The examples considered above demonstrate that for many cases testing only varying a decision when varying a condition (i.e., using MC/DC) is insufficient from the safety point of view. To eliminate this shortcoming, we propose the use of a new RC/DC criterion in critical applications.

3   RC/DC

3.1   General Definition

The main idea of RC/DC is for future development of MC/DC with the purpose of making it even stronger. In that way, all requirements of MC/DC are valid and a new requirement for keeping the value of a decision when varying a condition is added to the testing regime.

With the objective of ensuring compatibility and continuity with the MC/DC definition, we define RC/DC as follows:

> Every point of entry and exit in the program has been invoked at least once, every condition in a decision in the program has taken on all possible outcomes at least once, every decision in the program has taken all possible outcomes at least once, each condition in a decision has been shown to independently affect the decision's outcome, *and each condition in a decision has been shown to independently keep the decision's outcome.* A condition is shown to independently affect *and keep* a decision's outcome by varying just that condition while holding fixed *(if it is possible)* all other conditions.

The reservation "if it is possible" is used because it is far from always being possible to independently affect or keep the value of a decision. For more accurate consideration and analysis of all possible situations we propose the formal definition of RC/DC using the Z notation in the next section.

3.2   Formal Definition in the Z Notation

For elaboration of the formal definition of RC/DC, we carry out further development of the Z schema, describing the notion of the decision. The *DecReinforced* schema below differs from the *DecModified* schema by adding four new functions: $keep0$, $keep1$, $keep0fix$ and $keep1fix$. Each of these functions connects conditions with pairs of input data.

$DecReinforced$
$DecModified$
$keep0, keep1, keep0fix, keep1fix : cond \nrightarrow InputPairs$

$\mathrm{dom}\ keep0 = \mathrm{dom}\ keep1 = \mathrm{dom}\ keep0fix = \mathrm{dom}\ keep1fix = argdec$
$\mathrm{ran}\ keep0 \cup \mathrm{ran}\ keep1 \cup \mathrm{ran}\ keep0fix \cup \mathrm{ran}\ keep1fix \subseteq decinput \leftrightarrow decinput$
$\forall c : cond \mid c \in argdec \bullet$
$\quad keep0\ c = \{i_0, i_1 : decinput \mid value\ i_0 = value\ i_1 = 0 \wedge c\ i_0 \neq c\ i_1\} \wedge$
$\quad keep1\ c = \{i_0, i_1 : decinput \mid value\ i_0 = value\ i_1 = 1 \wedge c\ i_0 \neq c\ i_1\} \wedge$
$\quad keep0fix\ c = \{i_0, i_1 : decinput \mid (i_0, i_1) \in keep0\ c\ \wedge$
$\qquad (\forall othercond : argdec \mid othercond \neq c \bullet othercond\ i_0 = othercond\ i_1)\} \wedge$
$\quad keep1fix\ c = \{i_0, i_1 : decinput \mid (i_0, i_1) \in keep1\ c\ \wedge$
$\qquad (\forall othercond : argdec \mid othercond \neq c \bullet othercond\ i_0 = othercond\ i_1)\}$

The functions $keep0$ and $keep1$ assign for each condition the subset of pairs of input data that vary the condition but keep the value of the decision equal to 0 (for $keep0$) or 1 (for $keep1$). The difference between the $keep0fix/keep1fix$ functions and the $keep0/keep1$ functions is that, for the $keep0fix/keep1fix$ functions, all the other conditions are kept fixed. This is similar to the difference between $changedecfix$ and $changedec$ in the $DecModified$ schema.

The introduced functions allow formulation of the formal definition of the RC/DC criterion:

RC_DC
MC_DC

$\forall\, DecReinforced;\ c : cond \mid c \in argdec\ \bullet$
(**let** $target0 == choice(keep0fix\ c, keep0\ c);$
$target1 == choice(keep1fix\ c, keep1\ c)\ \bullet$
$(target0 \neq \varnothing \Rightarrow (testset \times testset) \cap target0 \neq \varnothing)\ \wedge$
$(target1 \neq \varnothing \Rightarrow (testset \times testset) \cap target1 \neq \varnothing))$

This criterion contains MC/DC as a component and also includes the requirements for testing of invariability of the decision when a condition varies. In this way, the set of test cases should contain pairs of input data from two sets $target0$ and $target1$, which keep (if it is possible) the value of the decision and fix (if it is possible) all other conditions. If holding other conditions is not possible, the test cases that keep the value of the decision without fixing other conditions should be used.

Let us prove that if the decision does not coincide with the condition or the condition's negation, it is always possible to choose testing data that satisfies RC/DC; i.e., ($target0 \neq \varnothing$) $\vee$ ($target1 \neq \varnothing$).

Lemma 3

$RC\_DC;\ c, value : cond \vdash$

$(value \neq c \wedge value \neq \neg\, c) \Rightarrow (target0 \neq \varnothing \vee target1 \neq \varnothing)$

Proof

$\neg\,((value \neq c \wedge value \neq \neg\, c) \Rightarrow$ [assumption]
$\quad(target0 \neq \varnothing \vee target1 \neq \varnothing))$

$\Leftrightarrow value \neq c \wedge value \neq \neg\, c \wedge target0 = \varnothing \wedge target1 = \varnothing$ [logic]

$\Rightarrow choice(keep0fix\ c, keep0\ c) = \varnothing\ \wedge$ [definition of $target0$ and $target1$]
$\quad choice(keep1fix\ c, keep1\ c) = \varnothing$

$\Leftrightarrow keep0\ c = \varnothing \wedge keep1\ c = \varnothing$ [definition of $choice$]

$\Leftrightarrow \neg\,(\exists\, i_0, i_1 : decinput\ \bullet$ [definition of $keep0$ and $keep1$]
$\quad c\ i_0 \neq c\ i_0 \wedge value\ i_0 = value\ i_1)$

$\Leftrightarrow \forall\, i_0, i_1 : decinput\ \bullet\ c\ i_0 \neq c\ i_1 \Rightarrow value\ i_0 \neq value\ i_1$ [logic]

$\Rightarrow c = value \vee c = \neg\, value$ **[Lemma 2]**

$\Rightarrow false$ [contradiction]

It should be noted that there are decisions that do not keep one of the values (0 or 1) when varying a condition. For example, $A \lor B$ does not keep 0 and $A \land B$ does not keep 1.

3.3 A Case Study

Consider the following criterion of the protection system actuation for the VVER-1000 type [21] nuclear reactor: the system should shut down a reactor in the case of decrease of the pressure in the first circuit to less than $150 \; kg/cm^2$ under the coolant temperature of more than $260°$ centigrade and reactor capacity equal or more than 75% of the rated capacity or decrease of the pressure in the first circuit to less than $140 \; kg/cm^2$ under the coolant temperature more than $280°$ centigrade.

For measurement of each input parameter (pressure p and temperature t), three sensors are used (inputs p_1, p_2, p_3 and t_1, t_2, t_3) with majority voting of inputs.

Hence, 13 conditions are used for determination of necessity of the system actuation:

$$
\begin{array}{lll}
P_{11} = p_1 < 150 & P_{21} = p_2 < 150 & P_{31} = p_3 < 150 \\
P_{12} = p_1 < 140 & P_{22} = p_2 < 140 & P_{32} = p_3 < 140 \\
T_{11} = t_1 > 260 & T_{21} = t_2 > 260 & T_{31} = t_3 > 260 \\
T_{12} = t_1 > 280 & T_{22} = t_2 > 280 & T_{32} = t_3 > 280 \\
NR = N \geq 0.75 N_r
\end{array}
$$

The decision that is responsible for this actuation criterion is:

$$(((P_{11} \land P_{21}) \lor (P_{11} \land P_{31}) \lor (P_{21} \land P_{31})) \land ((T_{11} \land T_{21}) \lor (T_{11} \land T_{31}) \lor (T_{21} \land T_{31})) \land NR) \lor (((P_{12} \land P_{22}) \lor (P_{12} \land P_{32}) \lor (P_{22} \land P_{32})) \land ((T_{12} \land T_{22}) \lor (T_{12} \land T_{32}) \lor (T_{22} \land T_{32})))$$

A general number of all possible combinations of values of the conditions equals $2^{13} = 8192$. Not all combinations are possible because of coupled conditions. Thus, it is impossible to have $P_{i1} = 0 \land P_{i2} = 1$ and also $T_{i1} = 1 \land T_{i2} = 0$. However, the number of possible combinations is still too large to be completely checked during practical testing.

The RC/DC criterion requires a maximum of 6 test cases for each condition (two for varying the decision, two for keeping it 0, and two for keeping it 1). So, not more than 78 combinations are required for this decision. However this number is overestimated since the same combinations can be used for testing different conditions. The minimization of the number of test cases for RC/DC could demand special analysis and be a hard task. Costs of this analysis could exceed the obtained benefit from the minimization. But if further reduction of test cases is not very important, the selection of test cases presents no difficulty.

For testing maintaining the value 0 for the decision during variation of the condition P_{11}, it is sufficient to select combinations of input data, for which $T_{11} = 0$ and $T_{21} = 0$. It ensures that the decision equals 0; therefore any possible values of the other conditions could be fixed. For testing maintaining the value 1 for the decision during variation of the condition P_{11}, it is sufficient to fix, for example, $P_{12} = P_{22} = T_{12} = T_{22} = 1$ and any allowed values of the other conditions.

4   Conclusion and Future Work

In this paper we have proposed and formalized a new Reinforced Condition/Decision Coverage (RC/DC) criterion for software testing, which strengthens the requirements of the well-known Modified Condition/Decision Coverage (MC/DC) criterion [16].

Z schemas have been formulated to provide the formal definition of MC/DC (see also [19]) in the Z notation [1], which accurately capture the particular features of this criterion. However, the MC/DC criterion does not include requirements for testing of "false operation" type failures. Such failures, as we have shown in several examples, can be highly important for safety-critical computer systems.

The proposed RC/DC criterion aims to eliminate this shortcoming and requires the consideration of situations when varying a condition keeps the value of a decision constant. Though the number of required test cases rises, this growth remains linear compared to the number of conditions in a decision, making the approach practicable. We have illustrated application of the RC/DC criterion in the testing of nuclear reactor protection system software. An important area of application of the RC/DC criterion could be using it as a regulatory requirement in standards [18,20].

One direction for future work could be using RC/DC not only for software testing but also for integration testing of a whole computer system, assuming the system specification as a basis. Another important aim is automated generation of test inputs in line with the RC/DC criterion.

References

1. Bowen, J. P. Z: A formal specification notation. In M. Frappier and H. Habrias (eds.), *Software Specification Methods: An Overview Using a Case Study*, Chapter 1. Springer-Verlag, FACIT series, 2001, pp. 3–19.
2. Burton, S. *Towards Automated Unit Testing of Statechart Implementations*. Technical Report YCS319, Department of Computer Science, University of York, UK. September 1999.
3. Chapman, R. Industrial Experience with SPARK. *Proceedings of ACM SIGAda Annual International Conference (SIGAda 2000)*, November 12–16, 2000, Johns Hopkins University/Applied Physics Laboratory, Laurel, MD, USA.
4. Chilenski, J. and Miller, S. Applicability of modified condition/decision coverage to software testing. *Software Engineering Journal*, September 1994, pp. 193–200.
5. Chilenski, J. and Newcomb, P. H. Formal specification tool for test coverage analysis. *Proceedings of the Ninth Knowledge-Based Software Engineering Conference*, September 20–23, 1994, pp. 59–68.
6. DeWalt, M. MCDC. A blistering love/hate relationship. *FAA National Software Conference*, Long Beach, CA, USA, April 6–9, 1999.
7. Dolman, B. *Definition of Statement Coverage, Decision Coverage and Modified Condition Decision Coverage*. WG-52/SC-190 Discussion paper. Paper reference: D004, revision 1. Draft, September 25, 2000.
8. Dupuy, A. and Leveson, N. An empirical evaluation of the MC/DC coverage criterion on the HETE-2 satellite software. *Proceedings of the Digital Aviation Systems Conference (DASC)*, Philadelphia, USA, October 2000.
9. Hayhurst, K. J., Veerhusen, D. S., Chilenski, J. J., and Rierson, L. K. *A Practical Tutorial on Modified Condition/Decision Coverage*, Report NASA/TM-2001-210876, NASA, USA, May 2001.

10. Jasper, R., Brennan, M., Williamson, K., Currier, B., and Zimmerman, D. Test data generation and feasible path analysis. *Proceedings of the 1994 International Symposium on Software Testing and Analysis*, Seattle, WA, USA, August 17–19, 1994, pp. 95–107.
11. Jia, X. *ZTC: A Type Checker for Z Notation. User's Guide. Version 2.03, August 1998.* Division of Software Engineering, School of Computer Science, Telecommunication, and Information Systems, DePaul University, USA, 1998.
12. Li, Y. Y. Structural test cases analysis and implementation. *42nd Midwest Symposium on Circuits and Systems*, 8–11 August, 1999, Volume 2, pp. 882–885.
13. Myers, G. *The Art of Software Testing.* Wiley-Interscience, 1979.
14. Offutt, A. J., Xiong, Y., and Liu, S. Criteria for generating specification-based tests. *Proceedings of the Fifth IEEE International Conference on Engineering of Complex Computer Systems (ICECCS'99)*, Las Vegas, Nevada, USA, October 18–21, 1999, pp. 119–129.
15. Roper, M. *Software Testing.* McGraw-Hill, 1994.
16. RTCA/DO-178B. *Software Considerations in Airborne Systems and Equipment Certification*, RTCA, Washington DC, USA, 1992.
17. Vilkomir, S. A. and Bowen, J. P. *Formalization of Control-flow Criteria of Software Testing.* Technical Report SBU-CISM-01-01, SCISM, South Bank University, London, UK, January 2001.
18. Vilkomir, S. A. and Bowen, J. P. *Application of Formal Methods for Establishing Regulatory Requirements for Safety-Critical Software of Real-Time Control Systems.* Technical Report SBU-CISM-01-03, SCISM, South Bank University, London, UK, 2001.
19. Vilkomir, S. A. and Bowen, J. P. Formalization of software testing criteria using the Z notation, *Proceedings of COMPSAC 2001: 25th IEEE Annual International Computer Software and Applications Conference*, Chicago, Illinois, USA, 8–12 October 2001. IEEE Computer Society Press, 2001, pp. 351–356.
20. Vilkomir, S. A. and Kharchenko, V. S. Methodology of the Review of Software for Safety Important Systems. *Safety and Reliability. Proceedings of ESREL'99 – The Tenth European Conference on Safety and Reliability*, Munich-Garching, Germany, 13–17 September 1999, Vol. 1, pp. 593–596.
21. Voznessensky, V. and Berkovich, V. VVER 440 and VVER-1000. Design Features in Comparison with Western PWRS. *International Conference on Design and Safety of Advanced Nuclear Power Plants*, Tokyo, October 1992, Vol. 4.
22. Zhu, H., Hall P. A., and May, H. R. Software unit test coverage and adequacy. *ACM Computing Surveys*, Vol. 29, No. 4, December 1997, pp. 336–427.

A Comparison of the BTT and TTF Test-Generation Methods

Bruno Legeard[1], Fabien Peureux[1], and Mark Utting[2]

[1] Laboratoire d'Informatique de Franche-Comté, France
{legeard,peureux}@lifc.univ-fcomte.fr,
http://lifc.univ-fcomte.fr/~bztt
[2] Visiting from: The University of Waikato, Hamilton, New Zealand
marku@cs.waikato.ac.nz

Abstract. This paper compares two methods of generating tests from formal specifications. The *Test Template Framework* (TTF) method is a framework and set of heuristics for manually generating test sets from a Z specification. The *B Testing Tools* (BTT) method uses constraint logic programming techniques to generate test sequences from a B specification. We give a concise description of each method, then compare them on an industrial case study, which is a subset of the GSM 11.11 smart card specification.

1 Introduction

Traditionally, formal methods and testing were viewed as opposites, with formal methods concentrating on the analysis of abstract specifications and on ways of proving programs to be completely correct, while testing concentrated on finding some of the errors in informally-developed programs. Testing is almost universally used in industry, while formal methods have relatively little industrial impact outside of a few critical sectors.

The design and use of good test suites is a time-consuming manual craft that typically consumes one quarter or more of the total development effort. There are two main philosophies of test design: in *structural* (or *clear box/white box*) testing tests are derived from the structure of the implementation, while in *functional* (or *black box*) testing tests are derived from the specification of the system, which is typically expressed in natural language.

Recently there has been an increasing interest in combining formal methods and testing, particularly by using formal specifications as the basis for developing more systematic sets of functional (black box) tests [1,2,3,4,5,6,7,8,9]. This field of research, called *specification-based testing*, is seeing an increasing number of publications, and workshops, such as the *Formal Approaches to Testing of Software (FATES'01)* workshop, at the CONCUR'01 conference in Aalborg, Denmark. The attractions of combining formal methods and testing include:

- a formal specification is a precise and consistent description of the functionality of a system and this is an ideal starting point for functional testing;

D. Bert et al. (Eds.): ZB 2002, LNCS 2272, pp. 309–329, 2002.

- the formality of the specification can allow some aspects of test generation to be automated, which may reduce the cost of testing;
- industry interest in testing is an opportunity to introduce formal specifications into industry, which is an initial step to introducing other formal methods.
- generating tests from a formal specification helps to validate the specification, because the tests are concrete instantiations of the specification and because the tests expose differences between the specification and its implementation, which can necessitate changes to either of them.

This paper describes and compares two specification-based testing methods: the TTF (Test Template Framework) method [10,11,12] for Z specifications, and the BTT (B Testing Tools) method [13,14] for B specifications. To make the comparison more concrete, both methods have been applied to a subset of a smart card GSM 11.11 case study and the resulting tests are compared with a set of hand-generated tests from an industrial partner. We chose the TTF method because it is the most widely known of the test-generation methods for Z, and the BTT method because it is more amenable to automation and has been developed in close collaboration with, and partly funded by, the French software industry.

The TTF and BTT methods are quite different, as will be seen. The longer-term goal behind our comparison study is to explore ways of combining the best aspects of the two methods and to develop supporting tools. This comparison is part of the *BZ Testing Tools* project, which is a two-year, eight-person project that is developing specification-based testing tools for both B and Z. The main contributions of this paper are:

- a concise description of the key aspects of each method, using similar vocabulary, and with a minimum of technical detail.
- a common case study using both methods and a comparison between them.
- an analysis of the strengths and weaknesses of each method, particularly with regard to automation.
- identification of several ways in which TTF ideas could be incorporated into the BTT method.

2   The TTF Method for Z

The *Test-Template Framework* (TTF) was invented by Stocks and Carrington [10,11] at the University of Queensland, Australia.

The best paper for an introduction to TTF is the 1994 *Two Paradigms* paper [12]. This identifies the following key issues in testing. In this paper we use the first three of these issues as a framework for comparing the two methods.

1. Selecting test data (which input and state values to test);
2. Providing test oracles (checking the outputs of the implementation);
3. Sequencing of tests (to get the implementation into the correct state to execute a particular test);

4. Defining tests (management of the whole process);
5. Evaluating tests (introspection and meta-results about the tests).

For **selecting test data**, the key idea of the TTF is an elegantly simple Z framework for manually defining a tree[1] of possible tests for each operation of a system. The root of the tree is the *valid input space* of the operation, which is the set of all inputs that satisfy the precondition. Each tree node specifies a set of input values, is called a *test template*, and is written as a Z schema. The children of that node specify various interesting subsets of that set of inputs. The children of a node are generated by applying a test *strategy*, such as the widely used *partition analysis* strategy [2], which puts a predicate into DNF (*disjunctive normal form*), then tests each disjunct. Leaves of the tree specify the actual tests that should be performed on the operation. Normally, leaves are fully instantiated so that they specify a unique test value, but it is possible to stop extending the tree at any time, so a leaf can also specify a set of values to indicate that any one of those values is an acceptable test input.

TTF provides an open-ended set of strategies, mostly adapted from previous researchers. However, it includes two novel strategies which were found to be particularly useful for high-level Z specifications which have a 'flat' structure (such as a simple conjunction of predicates), but use powerful Z toolkit operators like $\lhd, \oplus, \cap$ or *squash*.

The first strategy, *domain propagation*, uses a library of *input domains* for each operator. For example, for binary set operations like $A \cup B$ or $A \cap B$, the suggested input domains are (note that these partition the inputs):

$$A = \{\} \wedge B = \{\}$$
$$A = \{\} \wedge B \neq \{\}$$
$$A \neq \{\} \wedge B = \{\}$$
$$A \cap B = \{\} \wedge A \neq \{\} \wedge B \neq \{\}$$
$$A \neq \{\} \wedge A \subset B$$
$$B \neq \{\} \wedge B \subset A$$
$$A \neq \{\} \wedge A = B$$
$$A \cap B \neq \{\} \wedge A \nsubseteq B \wedge B \nsubseteq A$$

To apply domain propagation to an operator, the disjunction of its input domains is conjoined to the predicate containing the operator. Then the whole schema is transformed to DNF, and false branches are removed. The effect is to propagate the interesting sub-domains of an operator up to the top level of the operation.

The second novel strategy is *specification mutation*. This is inspired by mutation testing of programs, but has different aims. The specification is mutated, and for each mutation a test is derived which will distinguish the behaviours of the mutated and original specifications. The effect is to ensure that the implementation does not implement any of the incorrect specifications.

Oracles are provided very easily in the TTF and can be derived directly from the tree of test templates. Each leaf is simply conjoined with the original

---

[1] In fact, the TTF allows directed acyclic graphs of tests, but trees are most common.

operation schema, then projected onto the output variables, to find the acceptable outputs. Note that when the leaf specifies unique input values, this oracle specifies precisely the allowable outputs of the implementation. However, when the leaf specifies multiple possible inputs, this oracle does not check exact input-output correspondance, but simply checks that the output is correct for that set of inputs.

Sequencing of tests was not originally addressed by the TTF. No algorithms or heuristics for this were defined, except for an example in Stocks' thesis where the sequencing was done manually. More recent work [15] has adapted the Dick-Faivre *'convert into a finite state machine'* approach [2] to the TTF. The states of the finite state machine are quite fine-grained, because they are obtained by projecting every TTF leaf and oracle of every operation onto the system state variables, then normalizing that set of states to produce a set of states S that partition the system state. Then a state machine is built by considering all possible transitions and including those that are feasible (this requires $O(\#S^2 \times \#Ops)$ schemas to be simplified!). Then any unreachable states must be removed from the state machine (the paper defines reachability, but calculates it manually, with no discussion of possible algorithms). In the case study in that paper, test sequences were then generated automatically from the state machine, using an existing tool called ClassBench.

2.1   Current Status

The TTF method was first published internationally in 1993 [16], and it has become quite widely known and referenced since that time. More recent publications from UQ have extended the TTF to support inheritance [17] and interactive systems [18], and added some heuristics that promote greater reuse by distinguishing the state invariant from operation schema. A support tool called Tinman has been developed [19], but it is aimed primarily at providing organizational support and has little support for manipulating predicates, such as simplification or theorem proving. This is better supported by standard Z tools, like Z/EVES, but there is currently no integrated solution that supports both aspects.

Meudec [20] has adapted the TTF domain propagation strategy to work with VDM-SL, and extended it to work better with quantifiers. His thesis discusses possible automation, and concludes that the complex simplification and VDM automatic proving support that is needed may be beyond the state of the art, but that a constraint logic programming implementation looks promising.

3   The BTT Test Generation Method for B

The *B Testing Tools* (BTT) test generation method is a mostly-automated method that uses a customized constraint logic programming solver called CLPS-B [21]. This is a finite domain CLP (Constraint Logic Programming) system with special support for B set-theory operations.

Background: CLPS-B provides good support for reasoning about variables that range over small enumerated sets, like $x \in \{a, b, c\}$. For such variables, it supports about 80% of the set and relation operators of B and performs deep semantic reasoning, similar to a highly automated theorem prover. For variables that range over infinite sets (like $\mathbb{N}$) or over given sets of unknown cardinality, CLPS-B provides less support. For example, it cannot compute the intersection of two such sets, but can reason about membership and cardinality. For test generation purposes, this emphasis on finiteness is often acceptable, since it is common to replace an unbounded set by a small enumerated set of possible test values before testing commences.

Internally, it uses only the operators $\in$, $=$ and $\neq$, and maintains, in disjunctive normal form, a *constraint store* involving these operators. Each disjunct in the constraint store is actually a separate Prolog choicepoint that returns a conjunction of constraints. When new predicates are added to the constraint store, constraint propagation rules deduce all the interesting consequences and simplifications that are possible.

All other B operators are translated into the above three operators. This translation is done on the fly, as each predicate is needed, so that as much as possible is known about each variable, and the possible members of each set, and the translation can be very specific. For example, $A \subseteq B$ is translated into $\forall x : A \bullet x \in B$, so if the predicate $A \subseteq B$ is encountered in an operation precondition when the current CLPS-B state is $A = \{V1, V2\}$ and $B = \{2, 5\}$, then $A \subseteq B$ will be translated into the constraints $V1 \in \{2, 5\}$, $V2 \in \{2, 5\}$.

CLPS-B also uses an ingenious extension of CLP which allows domains to contain variables rather than just constants. For example, it can directly represent states like $x \in \{y, z, 3\} \wedge z \neq 3$ where we do not know whether or not $y = z$ or $y = 3$ etc. This extension dramatically reduces the number of constraint states (disjuncts) that are needed to represent a predicate and makes it easy to preserve the non-determinism of the specification. For example, if an operation has an input parameter X, with the precondition $X \in S$, it is not necessary to know the precise value of the input. Instead we just add the constraint $X \in S$ and execute the operation with X remaining symbolic.

The BTT method starts from a B specification of a single abstract machine. The test designer partitions the operations into *observation* operations (which do not change the state) and the *update* operations. It then goes through two main stages:

1. identifies interesting *boundary values* of each specification variable of finite type.
2. finds sequences of operations that allow those boundary values to be tested.

Generation of Tests: The first stage is done by using the CLPS-B solver to analyze each atomic predicate P in the specification. After this predicate has been put into CLPS-B, the constraint solver can calculate a set of *boundary values* $Lim(P, V)$ for each variable V. For the $A \subseteq B$ example above, the bounds for A would be just $\{2, 5\}$, the bounds for $V1$ would be 2 and 5, while for $x : 0 .. 10$ the bounds for x would be 1 and 10. So, for each state variable

V in the B specification, we obtain a set of all its boundary values, $Lim(V) = \bigcup_P Lim(P, V)$. For example, if the specification contains a state variable k with invariant $k \in \mathbb{N} \wedge k \leq 9$, and an operation contains the statement *if* $k >$ 2 *then* $k := 1$ *else* ..., then the boundary values for k will include at least $\{0, 3, 9\}$.

Sequencing of Tests: The second stage attempts to generate sequences of operations that will take the implementation under test (IUT) from the initial state to a boundary state. That is, for each state variable V, and each of its boundary values $L \in Lim(V)$, the B specification is executed symbolically, using the CLPS-B solver, to find a sequence of operations that reaches a state where $V = L$ is true. The search tries all operations, using best-first search, plus metric functions that measure the distance between the current and desired states. This process does not always terminate, because some boundary states may not be reachable, or the search space may be too big (even though the powerful constraint representation does reduce the number of states, and makes some exponential search spaces become polynomial [22]).

Once a path P to a boundary state has been found, it is used to control the IUT, so that all update operations can be tested at that boundary state. For each update operation Op, the IUT is made to execute the sequence $Init \ ⨾ \ P \ ⨾ \ Op \ ⨾ \ Observ$, where $Observ$ is a sequence of all the observation operations whose preconditions are true.

If Op has input parameters, various possibilities are tested, as follows. While searching for P, the CLPS-B solver is simulating the actual values of the state variables, which are accessible only by the IUT. Note that the IUT may resolve specification non-determinism by choosing a particular state, but the CLPS-B simulation records the set of *all* possible states at each step along P. When the CLPS-B simulation reaches Op, its constraint store is a disjunct of constrained solutions, so one instance of each of those disjuncts is chosen. This is similar to the partition-analysis strategy of TTF.

Generation of Oracles: The expected output of each Op test is obtained (as a value, or as some constraints that describes a set of values) from the CLPS-B solver during the above simulation of $Init \ ⨾ \ P \ ⨾ \ Op \dots$. This is very similar to the oracles of TTF, provided that a separate TTF oracle is defined for each leaf.

3.1   Current State

The BTT test-generation method has been under development at the *Laboratoire d'Informatique de l'Université de Franch-Comté* (LIFC) for several years, and was first used in a large industrial case study in 1999-2000. The techniques used in the test-generation are only now being published, though the underlying CLPS-B technology has been published earlier [22] and used for animation and model-checking of B specifications. The project has recently received significant funding to further develop the method and implement test-generation tools for both B and Z using these techniques.

4   B and Z Specifications of GSM 11.11

GSM is a worldwide standard for digital mobile phone systems, which is widely used (109 countries), particularly in Europe. Its specification is managed by the European Telecommunication Standard Institute (ETSI). In a joint project with the SCHLUMBERGER company, we specified a subset of the GSM specification in B, generated tests from that specification, and compared them with the manually-generated tests from SCHLUMBERGER. The subset specified was the protocol that allows the mobile equipment to communicate with the *Subscriber Identification Module* (SIM). The SIM maintains a fixed hierarchy of directories and files, and allows the mobile equipment to query and update these files, subject to various security checks. The B specification of this subset contained around 30 constants and variables and 12 operations. The results of this project are reported more fully elsewhere [23,24].

For this paper, we use a further simplified version of this GSM subset, with a reduced state and just five operations. The B specification is shown in Appendix A. The key data structure to note is *FILES_CHILDREN*, which defines the fixed file hierarchy that the SIM manages (see Fig. 1).

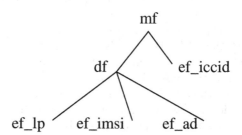

Fig. 1. Tree Structure of considered SIM files

It would probably be possible to adapt the TTF method to work directly on the B specification, but that is beyond the scope of this paper, so we instead translated the B specification into Z for use with the TTF method. The remainder of this section discusses the interesting aspects of this translation. The most interesting difference is that the B machine is written in an *event style* with *guards* to determine when each operation is enabled. This is not directly translatable to Z, but for the purposes of this case study, it made no difference whether an event style or standard state machine style was used, so we translated the B guards into Z preconditions.

The basic sets and constants of the B specification were easily translated into Z. Enumerated sets became Z free types, while *DATA* (which has arbitrary elements of unknown value) became a Z given set (note that we renamed the members of *VALUE* to *yes* and *no*, because *true* and *false* are predicates in Z).

For example:

$$
\begin{aligned}
FILES &::= mf \mid df\_gsm \mid ef\_iccid \mid ef\_lp \mid ef\_imsi \mid ef\_ad \\
MF &== \{mf\} \\
VALUE &::= yes \mid no \\
[DATA] &
\end{aligned}
$$
...

As is common Z style, we built up the state schema in several steps, grouping the variables of the B machine into logical groups. This is useful for framing purposes, when only a subsection of the state changes.

PState

$counter\_chv : COUNTER\_CHV$
$counter\_unblock\_chv : COUNTER\_UNBLOCK\_CHV$
$permission\_session : PERMISSION \to VALUE$
$blocked\_chv\_status : BLOCKED\_STATUS$
$blocked\_status : BLOCKED\_STATUS$
$code\_chv : CODE$

$(counter\_chv = 0 \Leftrightarrow blocked\_chv\_status = blocked)$
$(counter\_unblock\_chv = 0 \Leftrightarrow blocked\_status = blocked)$

State

$PState$
$current\_file : \mathbb{P}\, EF$
$current\_directory : DF \cup MF$
$data : EF \to DATA$

$\#current\_file \leq 1$
$(current\_file = \emptyset$
$\quad \vee \operatorname{dom}(FILES\_CHILDREN \rhd current\_file) = \{current\_directory\})$

The initialization follows directly from the B specification, so is not shown in Z. Note that all state variables are initialized to unique values, except for *data* which can take any arbitrary value.

The *SELECT_FILE* operation was written using a nested if-then-else in B, but we decided to split it into several schemas and follow a more Z-like style that separates the error cases from normal operation. (A disadvantage of this separation is that it required some duplication in the preconditions, but advantages are that the preconditions are expressed in positive form, rather than in negated form like in the B specification, and that we had the opportunity to prove that the disjunction of the separate preconditions gives a total operation.) The constants like *h90FF* in the specification represent hexadecimal response codes.

___ *SELECT_FILE1* _____
$\Delta State$
$ff? : FILES$
$sw! : Status$

___

$(ff? = mf$
$\lor ff? = current\_directory$
$\lor ff?\ \underline{FILES\_CHILDREN}\ current\_directory$
$\lor current\_directory\ \underline{FILES\_CHILDREN}\ ff? \land ff? \in DF$
$\lor (\exists\, dp : FILES \bullet dp\ \underline{FILES\_CHILDREN}\ current\_directory$
$\qquad\qquad \land dp\ \underline{FILES\_CHILDREN}\ ff? \land ff? \in DF))$
$current\_directory' = ff?$
$current\_file' = \emptyset$
$data' = data$
$\Xi PState$
$sw! = h90FF$

___ *SELECT_FILE2* _____
$\Delta State$
$ff? : FILES$
$sw! : Status$

___

$(current\_directory\ \underline{FILES\_CHILDREN}\ ff?$
$\lor (\exists\, dp : FILES \bullet dp\ \underline{FILES\_CHILDREN}\ current\_directory$
$\qquad\qquad \land dp\ \underline{FILES\_CHILDREN}\ ff?))$
$ff? \notin DF$
$current\_file' = \{ff?\}$
$current\_directory' = current\_directory$
$data' = data$
$\Xi PState$
$sw! = h90FF$

___ *SELECT_FILE3* _____
$\Xi State$
$ff? : FILES$
$sw! : Status$

___

$\neg\ \text{pre}\ SELECT\_FILE1$
$\neg\ \text{pre}\ SELECT\_FILE2$
$sw! = h9404$

$$SELECT\_FILE \mathrel{\widehat{=}} SELECT\_FILE1 \lor SELECT\_FILE2 \lor SELECT\_FILE3$$

For the *VERIFY_CHV* operation, just to be different, we used an if-then-else style like the B specification. The frame problem (saying what variables have not changed) is more difficult to solve in Z than B, but we ameliorated it by defining a schema for the unchanged state subset:

$$Others \;\widehat{=}\; State \setminus (counter\_chv, blocked\_chv\_status, permission\_session)$$

BLOCKED
$\Xi State$
$sw! : Status$

$blocked\_chv\_status = blocked$
$sw! = h9840$

VERIFY_CHV_OK
$\Delta State$
$code? : CODE$
$sw! : Status$

$\neg$ pre $BLOCKED$
(**if** $code\_chv = code?$
then
    $counter\_chv' = MAX\_CHV \;\wedge$
    $blocked\_chv\_status' = blocked\_chv\_status \;\wedge$
    $permission\_session' = permission\_session \oplus \{(chv, yes)\} \;\wedge$
    $sw! = h9000 \;\wedge$
    $\theta\, Others = \theta\, Others'$
else
    (**if** $counter\_chv = 1$
    then
        $counter\_chv' = 0 \;\wedge$
        $blocked\_chv\_status' = blocked \;\wedge$
        $permission\_session' = permission\_session \oplus \{(chv, no)\} \;\wedge$
        $sw! = h9840 \;\wedge$
        $\theta\, Others = \theta\, Others'$
    else
        $counter\_chv' = counter\_chv - 1 \;\wedge$
        $blocked\_chv\_status' = blocked\_chv\_status \;\wedge$
        $permission\_session' = permission\_session \;\wedge$
        $sw! = h9804 \;\wedge$
        $\theta\, Others = \theta\, Others'))$

$$VERIFY\_CHV \;\widehat{=}\; VERIFY\_CHV\_OK \lor BLOCKED$$

These two examples illustrate how the B operations was translated into Z quite easily, either by following exactly the same style, or by translating it into a more typical Z style which uses the schema calculus to separate error schemas from the normal operation schemas. The latter style is slightly more convenient when using the TTF, because the individual schemas often appear as branches of the test tree. However, the differences are a matter of style, rather than deep, although the frame problem is often a little more verbose to solve in Z than in B. Note that Abrial [25] gives a translation from any generalized substitution into a pre/postcondition pair, and this is essentially what we used.

The original B specification was type-checked with Atelier B, and all proof obligations were proved (most automatically, one interactively). The translated Z specification was type-checked with Z/EVES and two domain checks were generated.[2] One was proved automatically, the other one involved μ and was too difficult for the automatic tactics of Z/EVES.

5   Test Generation

First we use the TTF method to generate tests from the Z specification. In fact, this was done after the BTT generation, but to eliminate any bias towards generating similar (or different) tests, the TTF test designer (Utting) did not study the BTT-generated tests until after TTF generation was complete.

5.1   Generation Using TTF

In this paper, to save space, we show the structure of the TTF trees informally, rather than defining them in Z as is normal in TTF. [3]

We start with the *SELECT_FILE* operation. The syntactic structure of the *SELECT_FILE* schema is a disjunct of three schemas with distinct effects, so the *cause-effect* heuristic suggests partitioning the valid input space according to these three effects. So our initial TTF tree is:

```
          SF_VIS
         /   |   \
       SF1  SF2  SF3
```

where

$$SF\_VIS \cong \text{pre } SELECT\_FILE$$
$$SF1 \cong \text{pre } SELECT\_FILE1$$
$$SF2 \cong \text{pre } SELECT\_FILE2$$
$$SF3 \cong \text{pre } SELECT\_FILE3$$

---

[2] The Z/EVES domain checks ensure that all functions are called within their domains. Note that there are fewer opportunities for well-formedness proofs in Z, because the state invariant is automatically included in every precondition and postcondition, so all operations automatically preserve the invariant.

[3] The formal definitions of the TTF trees are mostly useful for proving results about the test tree, such as completeness or partitioning properties.

The precondition of *SELECT_FILE1* contains 5 disjuncts, so we use the *partition-analysis* heuristic, which splits a disjunct $A \lor B$ into three mutually disjoint cases: $A \land B$; $A \land \neg B$; $\neg A \land B$). For five disjuncts, this potentially results in $2^5 - 1$ cases, but most of these are contradictory given the definition of *FILES_CHILDREN*, so we get the following non-false cases (several more test cases would be obtained if the directory structure were more complex, with more subdirectories than just *df_gsm*):

$$SF1a \cong [SF1 \mid ff? = current\_directory \land current\_directory \neq mf]$$
$$SF1b \cong [SF1 \mid ff? = current\_directory \land current\_directory = mf]$$
$$SF1c \cong [SF1 \mid ff? = mf \land current\_directory \neq mf]$$

Similarly, using the *partition-analysis* heuristic on *SF2*, we get:

$$SF2a \cong [SF2 \mid current\_directory \ \underline{FILES\_CHILDREN} \ ff? \land ff? \notin DF]$$
$$SF2b \cong [SF2 \mid (\exists p : FILES \bullet p \ \underline{FILES\_CHILDREN} \ current\_directory \land$$
$$p \ \underline{FILES\_CHILDREN} \ ff? \land ff? \notin DF)]$$

The *SF3* schema, when expanded, contains lots of conjuncts (some of which contain nested disjuncts and $\exists$), but we decide that one test instance will suffice for this schema, so we do not partition it further. This gives the following final TTF tree for *SELECT_FILE*:

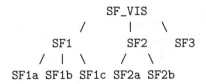

The second step in the TTF test-generation process is to find a *ground instance* of each of these six leaves. This gives us the actual input and initial state values that we want to exercise. For example, for the above leaves, we could choose the following values (variables not mentioned can take any value that satisfies the state invariant):

$$iSF1a \cong [SF1a \mid current\_directory = df\_gsm \land ff? = current\_directory]$$
$$iSF1b \cong [SF1b \mid current\_directory = mf \land ff? = current\_directory]$$
$$iSF1c \cong [SF1c \mid current\_directory = df\_gsm \land ff? = mf]$$
$$iSF2a \cong [SF2a \mid current\_directory = mf \land ff? = ef\_iccid]$$
$$iSF2b \cong [SF2b \mid current\_directory = df\_gsm \land ff? = ef\_iccid]$$
$$iSF3 \cong [SF3 \mid current\_directory = mf \land ff? = ef\_lp]$$

The generation of tests for the other operations is similar. For the *READ_BINARY* operation, it is sufficient to treat each of the three subschemas as TTF leaves. The $i \ldots$ schemas are instantiated versions of these.

$RB1 \mathrel{\widehat{=}} \text{pre } READ\_BINARY1$
$RB2 \mathrel{\widehat{=}} \text{pre } NO\_CURR\_FILE$
$RB3 \mathrel{\widehat{=}} \text{pre } READ\_BINARY3$
$iRB1 \mathrel{\widehat{=}} [RB1 \mid current\_file = \{ef\_iccid\}$
$\qquad \wedge \, permission\_session = (\lambda\, p : PERMISSION \bullet yes)]$
$iRB2 \mathrel{\widehat{=}} [RB2 \mid current\_file = \{\}]$
$iRB3 \mathrel{\widehat{=}} [RB3 \mid current\_file = \{ef\_iccid\}$
$\qquad \wedge \, permission\_session = (\lambda\, p : PERMISSION \bullet no)]$

The $RB1$ and $RB3$ cases are interesting because they contain a dependency between *current_file* and *permission_session*, with many values of each variable that satisfy the dependency. The values shown above were chosen by setting *permission_session* to a simple value (a constant function), but in fact this is not a good heuristic, because the operations of this machine do not provide any way of changing *permission_session*, so these state values are not reachable from the initial state. A better heuristic for choosing values that satisfy these predicates would be to choose values that are as much like the initial state as possible. This would result in using the initial state value for *permission_session* and setting *current_file* to *ef_lp* for $RB1$ and to any other member of EF for $RB3$. This illustrates the danger in instantiating TTF leaves – it is easy to over-restrict the state components so that the state becomes unreachable. This is not a problem when the testing process has complete freedom to set the state variables, or when there are no state variables (only inputs). But in the common case where we are testing a component with private state, it is better to leave the TTF leaves as general as possible.

For the $VERIFY\_CHV$ operation, the *cause-effect* heuristic suggests that we split into the two cases $VERIFY\_CHV\_OK$ and $BLOCKED$. The if-then-else structure of the $VERIFY\_CHV\_OK$ branch suggests the *disjunct* heuristic, which gives three branches, which are specific enough that they need no further analysis. So we get four leaves in total: $VC1$ and $VC2a/b/c$.

$VC \mathrel{\widehat{=}} \text{pre } VERIFY\_CHV$
$VC1 \mathrel{\widehat{=}} [VC \mid blocked\_chv\_status = blocked]$
$VC2 \mathrel{\widehat{=}} [VC \mid blocked\_chv\_status \neq blocked]$
$VC2a \mathrel{\widehat{=}} [VC2 \mid code\_chv = code?]$
$VC2b \mathrel{\widehat{=}} [VC2 \mid code\_chv \neq code? \wedge counter\_chv = 1]$
$VC2c \mathrel{\widehat{=}} [VC2 \mid code\_chv \neq code? \wedge counter\_chv \neq 1]$

Finally, the $STATUS$ operation is so simple that one test will suffice. In practice, it is typically called to check the results of other tests.

$ST\_VIS \mathrel{\widehat{=}} \text{pre } STATUS$

The third stage in the TTF test-generation process is to write a test driver that makes the implementation-under-test reach each of these states, then uses the schemas of the TTF tree (or the original schema) as an oracle to check the correctness of the output from the implementation. Due to limited space, we shall not show this step. One example of such a conversion is given in [15].

Note that TTF generation is a manual process, requiring extensive expertise at manipulating and simplifying schemas. In principle, a prover like Z/EVES could be used to simplify each node of the TTF tree before continuing, but in this case study the predicates were too difficult for the automatic simplification commands of Z/EVES. Typically, we find that considerable expertise with Z/EVES interactive proof is required to get useful simplifications of most nontrivial schemas, and this is too time-consuming to be worthwhile during TTF generation.

5.2   Generation Using BTT

Next we use the BTT method to generate tests from the B specification. After automatic translation of each predicate into constraints, CLPS analyzes the boundary values of each state variable, and returns the boundary values shown in Fig. 2 ($\_$ means 'dont care').

$$counter\_chv = 0$$
$$counter\_chv = 1$$
$$counter\_chv = 3$$
$$counter\_unblock\_chv = 0$$
$$counter\_unblock\_chv = 10$$
$$permission\_session = \{(adm,\_),(always,\_),(chv,no),(never,\_)\}$$
$$permission\_session = \{(adm,\_),(always,\_),(chv,yes),(never,\_)\}$$
$$blocked\_chv\_status = blocked$$
$$blocked\_chv\_status = unblocked$$
$$blocked\_status = blocked$$
$$blocked\_status = unblocked$$
$$code\_chv = a1$$
$$code\_chv = a2$$
$$code\_chv = a3$$
$$code\_chv = a4$$
$$current\_file = \{\}$$
$$current\_file = \{ef\_ad\}$$
$$current\_file = \{ef\_iccid\}$$
$$current\_file = \{ef\_imsi\}$$
$$current\_file = \{ef\_lp\}$$
$$current\_directory = mf$$
$$current\_directory = df\_gsm$$

Fig. 2. BTT Boundary Values for GSM 11.11

No boundary values are obtained for *data*, because its type is not enumerated.
   We classify the *STATUS* and *READ_BINARY* operations as observation operations. With the BTT method, each of the remaining three update operations will be tested at each of these 22 boundary values. In fact, slightly more

than 66 test sequences are generated, because for some operations the CLPS-B solver tries more than one input. For example, in $VERIFY\_CHV$, it tries one value of $code$ that is equal to $code\_chv$ and one that is not equal. An example of one test sequence for the boundary state $current\_directory = df\_gsm$ is:

$$90FF \qquad \leftarrow SELECT\_FILE(df\_gsm); \quad \text{Reach the boundary state}$$
$$90FF \qquad \leftarrow SELECT\_FILE(mf); \qquad \text{The operation being tested.}$$
$$mf, \{\}, 3, 10 \leftarrow STATUS \qquad\qquad\qquad \text{Observation operations}$$

5.3   Comparison of the Tests

It is interesting to compare the BTT boundary states with the 14 TTF leaves: $SF1a$, $SF1b$, $SF1c$, $SF2a$, $SF2b$, $SF3$, $RB1$, $RB2$, $RB3$, $VC1$, $VC2a$, $VC2b$, $VC2c$, ST_VIS. Precise comparison is difficult, because the TTF leaves focus mostly on input values, whereas the BTT boundary values are purely based on the system state. However, we can note the following interesting points:

– TTF has $counter\_chv = 1$ and $\neq 1$, while BTT splits the latter case into $= 0$ and $= 3$.
– BTT has boundary values on the $counter\_unblock\_chv$ and $blocked\_status$ variables while TTF has none, because those variables did not appear within any operations (except for framing equalities). If we had studied all 12 operations, those state components would have been updated or tested, then TTF tests would have been warranted.
– For the $permission\_session$ and $blocked\_chv\_status$ cases, there is a close match between the BTT and TTF tests.
– For the $code\_chv$ and $current\_file$ (non-empty) cases, BTT simply tries all possible values because the types of variables are enumerated sets, whereas TTF has analyzed the way they are used by the specification and clustered them accordingly (in the $current\_file$ case, the clustering is based on the directory tree structure).
– Finally, TTF has *more* cases for $current\_directory$, because it interacts in subtle ways with the $ff?$ input of $SELECT\_FILE$. These interactions are lost in BTT, because input variables are not treated until later in the process, so TTF tests this operation more thoroughly.

BTT generates around 80 tests, which is significantly more than the 14 generated via TTF. This is because the BTT strategy is to test *all* update operations at each boundary state.[4]

On the full subset of the GSM 11.11 (larger state space and 12 operations), 48 boundary values were found, which resulted in 1104 test sequences being generated (this took around 1.5 hours of CPU-time). The resulting tests were compared in detail with the 71 manually-generated tests from SCHLUMBERGER.

---

[4] If desired, it is possible to use heuristics to reduce some of this repetition. For example, update operations that do not refer to a given variable need not be tested on the boundary values of that variable.

The automatically generated tests covered 85% of the manual tests, which is excellent coverage, and 50% of the generated tests were additional to the manually-generated tests.

The GSM 11.11 specification is deterministic and uses quite simple data structures, so the only TTF heuristics we used were the *cause-effect* and *partition-analysis* heuristics. In a second BTT/TTF comparison study using a process scheduler, the state invariant included $active \cap (ready \cup waiting) = \{\}$, where *ready*, *active* and *waiting* were sets of processes. Applying the TTF domain propagation heuristic to both the $\cap$ and $\cup$ operators gave a very elegant and concise partitioning of the scheduler state space, more concise than the boundary values from BTT. From that case study, we conjecture that it might be useful to adapt the domain propagation strategy to work with BTT, to assist in the generation of boundary states.

6 Conclusions

The 'boundary values' of BTT are global across the whole specification, whereas the TTF leaves, which specify similar boundary states, are specific to each operation. This typically results in a lot more tests being generated by BTT than TTF. Some of the repetitive nature of the BTT tests could be reduced via heuristics, as discussed earlier, but generally BTT still produces more tests than TTF.

On the other hand, for a system with a complex state space, there are often several operations that need to be tested on the partitions of that state space. This can result in duplication within the TTF trees for those operations, because the same state-partitioning subtree must be derived for each operation. A variant of the TTF has been proposed [26] which factors this state-related partitioning out into a separate subtree, and this would make the two methods more similar.

The BTT method maintains a clearer distinction between the state variables and the input parameters of each operation than the TTF. This is important because, during testing, the state variables are controlled by the IUT, while the input parameters can be chosen by the test generation method.

TTF builds the entire state machine before commencing testing, and does not give an algorithm for removing unreachable states. In contrast, BTT does not build an explicit state machine, instead it explores the reachable paths one at a time. However, it still has to address the reachability issue, to avoid searching infinitely for a path to an unreachable boundary state. This is an area of ongoing research.

BTT generates more accurate oracles than TTF. In BTT, the oracle is the complete specification of the operation, so it relates output values to the inputs. But in TTF, the oracle is purely a set of output values. If every TTF leaf is fully instantiated, and then has its own specific oracle calculated for that input value, then this is equivalent to the BTT oracles. But if a TTF leaf represents a set of inputs, then the corresponding oracle may be less accurate. For example, in an experimental evaluation of TTF on topological sorting programs, the proposed top-level oracle for non-circular input graphs allows the implementation to always output an empty sequence! [27].[5]

In TTF, full instantiation of leaves can be dangerous. It seems best to leave the TTF leaves as general as possible (rather than instantiating them to specify unique inputs), because this leaves maximum flexibility when converting the test trees into a state machine, and thus into test sequences. Our example illustrated how instantiating too far can easily result in specifying an unreachable test state.

TTF seems to give a more detailed analysis of the interesting partitions of complex operations or predicates. Strategies like domain propagation are good for analyzing predicates involving several variables and complex operators. In comparison, BTT is relatively weak at producing boundary values involving multiple variables, mostly relying on a cross-product style of generation. So it would seem useful to use TTF techniques, such as domain propagation, as an alternative way of generating the BTT boundary values that contain multiple variables. This could be done by applying TTF strategies to the system state, including the invariant, to partition it into semantically interesting subsets. A possible difficulty here is that TTF generates predicates rather than simple boundary values, but it may be possible to adapt the BTT sequence-generation method to search for a path to a state described by a predicate (the issue is finding a good measure function).

Another interesting use of TTF techniques within the BTT method would be to use the TTF leaves as a measure of *coverage* of the BTT-generated tests. That is, what percentage of the TTF leaves are covered by (or have some non-empty intersection with) the states explored by BTT.

Our final conclusion is that BTT is better designed for automation than TTF, but a consequence of this is that it is more limited to specifications that have smaller state spaces, and that are written in a subset of B, whereas TTF handles any state space and all of Z (because it leaves all reasoning to the human!). It is not clear how much automated support for TTF would be possible or desirable. Tinman automates many simple clerical tasks of TTF, but a higher level of automation of the reasoning support would be useful. However, this would certainly require some restrictions on the specification notation, just like the BTT system. It is interesting, and perhaps promising, that after considering the automation of the TTF domain propagation strategy in VDM, Meudec [20] concludes that a CLP approach seems the best way to provide the necessary reasoning support. Our future work will certainly be exploring this.

One of the findings of applying the BTT method to the full GSM subset was that some human interaction is desirable, to customize the choice of boundary values etc. Even if were possible, full automation may not be desirable because it would lead to much larger test sets. But TTF is currently done largely by hand, and practitioners are looking for a little more automation. So it seems worthwhile to combine some of the techniques, to reach a middle ground where automation is available, but can be controlled and customized.

---

[5] They comment later in the paper that they did in fact recalculate a specific oracle for each TTF leaf in the non-circular case.

References

1. John D. Gannon, Paul R. McMullin, and Richard G. Hamlet. Data-abstraction implementation, specification, and testing. *ACM TOPLAS*, 3(3):211–223, 1981.
2. Jeremy Dick and Alain Faivre. Automating the generation and sequencing of test cases from model-based specifications. In J. C. P. Woodcock and P. G. Larsen, editors, *FME '93: Industrial-Strength Formal Methods*, number 670 in LNCS, pages 268–284. Springer-Verlag, April 1993.
3. Michael R. Donat. Automating formal specification-based testing. In Michel Bidoit and Max Dauchet, editors, *TAPSOFT '97: Theory and Practice of Software Development*, volume 1214 of *LNCS*, pages 833–847. Springer, 1997.
4. Susan Stepney. Testing as abstraction. In J. P. Bowen and M. G. Hinchey, editors, *ZUM'95: 9th International Conference of Z Users, Limerick, 1995*, volume 967 of *Lecture Notes in Computer Science*, pages 137–151. Springer-Verlag, 1995.
5. Hans-Martin Hoercher and Jan Peleska. Using formal specifications to support software testing. *Software Quality Journal*, 4(4):309–327, 1995.
6. C. Péraire, S. Barbey, and D. Buchs. Test selection for object-oriented software based on formal specifications. In David Gries and Willem-Paul de Roever, editors, *Programming Concepts and Methods: PROCOMET '98, IFIP TC2/WG2.2, 2.3 International Conference, June 1998, Shelter Island, New York*, pages 385–403. Chapman and Hall, 1998.
7. R. M. Hierons. Testing from a z specification. *Journal of Software Testing, Verification and Reliability*, 7:19–33, 1997.
8. Igor Burdonov, Alexander Kossatchev, Alexander Petrenko, and Dmitri Galter. KVEST: Automated generation of test suites from formal specifications. In Wing et al. [28], pages 605–621. LNCS 1708 (Volume 1).
9. S. Behnia and H. Waeselynck. Test criteria definition for B models. In Wing et al. [28], pages 509–529. LNCS 1708 (Volume 1).
10. Philip Stocks. *Applying formal methods to software testing*. PhD thesis, The Department of Computer Science, The University of Queensland, 1993. Available from `http://athos.rutgers.edu/~pstocks/pub.html`.
11. P. A. Stocks and D. A. Carrington. A framework for specification-based testing. *IEEE Transactions in Software Engineering*, 22(11):777–793, November 1996.
12. D. Carrington and P. Stocks. A tale of two paradigms: Formal methods and software testing. In *Proceedings of the 8th Z User Meeting*, pages 51–68. Springer-Verlag, 1994.
13. L. Py, B. Legeard, and B. Tatibouet. Évaluation de spécifications formelles en programmation logique avec contraintes ensemblistes – application à l'animation de spécification formelles B. In *AFADL'2000, Grenoble, 26-28 Jan 2000*, pages 21–35, 2000.
14. Bruno Legeard and Fabien Peureux. Generation of functional test sequences from B formal specification – presentation and industrial case-study. Submitted to Automated Software Engineering 2001, 2001.
15. L. Murray, D. Carrington, I. MacColl, J. McDonald, and P. Strooper. Formal derivation of finite state machines for class testing. In J. P. Bowen, A. Fett, and M. G. Hinchey, editors, *ZUM'98: The Z Formal Specification Notation*, volume 1493 of *LNCS*, pages 42–49. Springer-Verlag, 1998.
16. P. Stocks and D. Carrington. Test templates: A specification-based testing framework. In *Proceedings of the 15th International Conference on Software Engineering*, pages 405–414. IEEE Computer Society Press, 1993.

17. L. Murray, D. Carrington, I. MacColl, and P. Strooper. Extending test templates with inheritance. In *Proceedings of 1997 Australian Software Engineering Conference (ASWEC'97)*, pages 80–87. IEEE Computer Society, 1997. Also SVRC Technical Report 97–18.

18. I. MacColl and D. Carrington. Extending the TTF for specification-based testing of interactive systems. In *Australasian Computer Science Conference (ACSC99)*, pages 372–381. Springer-Verlag, 1999.

19. L. Murray, D. Carrington, I. MacColl, and P. Strooper. Tinman – a test derivation and management tool for specification-based class testing. In *Technology of Object-Oriented Languages and Systems (TOOLS 32)*, pages 222–233. IEEE Computer Society, 1999. Available as SVRC Technical Report 99-07.

20. Christophe Meudec. *Automatic Generation of Software Test Cases from Formal Specifications*. PhD thesis, Faculty of Science, Queen's University of Belfast, 1997.

21. F. Bouquet, B. Legeard, and F. Peureux. Constraint logic programming with sets for animation of B formal specifications. In *Proceedings of 1st International Conference on Computational Logic (CL'2000). Workshop on Constraint Logic Programming and Software Engineering, LPSE'2000, London, July 2000*, pages 62–81, 2000.

22. Fabrice Bouquet, Bruno Legeard, Fabien Peureux, and Paurent Py. Un système de résolution de contraintes ensemblistes pour l'évaluation de spécifications B. In *Programmation en logique avec contraintes. JFPLC'00. Marseilles, June 2000*, pages 125–144, 2000.

23. B. Legeard and F. Peureux. Generation of functional test sequences from B formal specifications - presentation and industrial case-study. In 16th *IEEE International conference on Automated Software Engineering (ASE'2001)*, 2001.

24. B. Legeard, F. Peureux, and J. Vincent. Automatic generation of functional of test patterns from a formalized smart card model - application to the GSM 11-11 specification. Rapport de fin de contrat (Confidentielle) Tome 1 : 79 pages, Tome 2 : 418 pages, Convention de recherche Schlumberger R&D Smart Card/LIFC, TFC01-01, Juillet 2000.

25. J.-R. Abrial. *The B-Book: Assigning Programs to Meanings*. Cambridge University Press, 1996.

26. I. MacColl and D. Carrington. Extending the test template framework. In *Proceedings of the Third Northern Formal Methods Workshop, Ilkley, UK, Sept. 1998*, 1998.

27. I. MacColl, D. Carrington, and P. Stocks. An experiment in specification-based testing. In *Proceedings of the 19th Australian Computer Science Conference (ACSC'96)*, pages 159–168, 1996. Also SVRC Technical Report 96–05.

28. Jeannette M. Wing, Jim Woodcock, and Jim Davies, editors. *FM'99 – Formal Methods*. Springer-Verlag, 1999. LNCS 1708 (Volume 1).

B Specification of a Fragment of the GSM 11-11 Standard

MACHINE GSM

SETS

$FILES$
$\quad = \{mf, df\_gsm, ef\_iccid,$
$\qquad ef\_lp, ef\_imsi, ef\_ad\};$
$PERMISSION$
$\quad = \{always, chv, never, adm\};$
$VALUE = \{true, false\};$
$BLOCKED\_STATUS$
$\quad = \{blocked, unblocked\};$
$CODE = \{a_1, a_2, a_3, a_4\};$
$DATA$

CONSTANTS

$FILES\_CHILDREN,$
$PERMISSION\_READ,$
$MAX\_CHV,$
$MAX\_UNBLOCK,$
$CODE\_UNBLOCK\_CHV$

DEFINITIONS

$MF == \{mf\};$
$DF == \{df\_gsm\};$
$EF == \{ef\_iccid, ef\_lp, ef\_imsi, ef\_ad\};$
$COUNTER\_CHV == 0..MAX\_CHV;$
$COUNTER\_UNBLOCK\_CHV$
$\quad == 0..MAX\_UNBLOCK;$

PROPERTIES

$FILES\_CHILDREN \in FILES \leftrightarrow FILES \wedge$
$FILES\_CHILDREN = \{(mf, df\_gsm),$
$\quad (mf, ef\_iccid), (df\_gsm, ef\_lp),$
$\quad (df\_gsm, ef\_imsi), (df\_gsm, ef\_ad)\} \wedge$
$PERMISSION\_READ$
$\quad \in EF \longrightarrow PERMISSION \wedge$
$PERMISSION\_READ$
$\quad = \{(ef\_iccid, never), (ef\_lp, always),$
$\qquad (ef\_imsi, chv), (ef\_ad, adm)\}$
$MAX\_CHV = 3 \wedge$
$MAX\_UNBLOCK = 10 \wedge$
$CODE\_UNBLOCK\_CHV \in CODE \wedge$
$CODE\_UNBLOCK\_CHV = a_3$

VARIABLES

$current\_file,$
$current\_directory,$
$counter\_chv,$
$counter\_unblock\_chv,$
$blocked\_chv\_status,$
$blocked\_status,$
$permission\_session,$
$code\_chv,$
$data$

INVARIANT

$current\_file \subseteq EF \wedge$
$card(current\_file) \leq 1 \wedge$
$current\_directory \in DF \cup MF \wedge$
$counter\_chv \in COUNTER\_CHV \wedge$
$counter\_unblock\_chv$
$\quad \in COUNTER\_UNBLOCK\_CHV \wedge$
$code\_chv \in CODE \wedge$
$permission\_session$
$\quad \in PERMISSION \longrightarrow VALUE \wedge$
$blocked\_chv\_status \in BLOCKED\_STATUS \wedge$
$blocked\_status \in BLOCKED\_STATUS \wedge$
$data \in EF \longrightarrow DATA \wedge$
$(counter\_chv = 0)$
$\quad \Leftrightarrow (blocked\_chv\_status = blocked) \wedge$
$(counter\_unblock\_chv = 0)$
$\quad \Leftrightarrow (blocked\_status = blocked) \wedge$
$(current\_file = \emptyset$
$\quad \vee (dom(FILES\_CHILDREN$
$\qquad \rhd current\_file) = \{current\_directory\}))$

INITIALIZATION

$current\_file := \emptyset \;\|$
$current\_directory := mf \;\|$
$counter\_chv := MAX\_CHV \;\|$
$counter\_unblock\_chv := MAX\_UNBLOCK \;\|$
$blocked\_chv\_status := unblocked \;\|$
$blocked\_status := unblocked \;\|$
$permission\_session$
$\quad := \{(always, true), (chv, false),$
$\qquad (adm, false), (never, false)\} \;\|$
$code\_chv := a_1 \;\|$
$data :\in EF \longrightarrow DATA$

OPERATIONS

$sw \longleftarrow$ **SELECT_FILE**(ff) $\hat{=}$
PRE
  $ff \in FILES$
THEN
  IF ((($current\_directory, ff$)
    $\notin FILES\_CHILDREN$)
  $\wedge (\exists(dp) \cdot (dp \in FILES$
  $\wedge ((dp, current\_directory)$
    $\notin FILES\_CHILDREN$
    $\vee (dp, ff) \notin FILES\_CHILDREN)))$
  $\wedge ((ff, current\_directory)$
    $\notin FILES\_CHILDREN)$
  $\wedge (ff \neq mf)$
  $\wedge (ff \neq current\_directory))$
  THEN
    $sw := 9404$
  ELSE
    $sw := 90FF$ $\|$
    IF ($ff \in DF \cup MF$)
    THEN
      $current\_directory := ff$ $\|$
      $current\_file := \emptyset$
    ELSE
      $current\_file := \{ff\}$
    END
  END
END;

$sw, dd \longleftarrow$ **READ_BINARY** $\hat{=}$
PRE
  $current\_file \subseteq EF$
THEN
  IF ($current\_file = \emptyset$)
  THEN
    $sw := 9400$
  ELSE
    IF ($permission\_session[$
      $PERMISSION\_READ[$
        $current\_file]] = \{true\}$)
    THEN
      $sw := 9000$ $\|$
      ANY ff **WHERE** $ff \in current\_file$
      THEN
        $dd := data(ff)$
      END
    ELSE
      $sw := 9804$
    END
  END
END;

$sw \longleftarrow$ **VERIFY_CHV**($code$) $\hat{=}$
PRE
  $code \in CODE$
THEN
  IF ($blocked\_chv\_status = blocked$)
  THEN
    $sw := 9840$
  ELSE
    IF ($code\_chv = code$)
    THEN
      $counter\_chv := MAX\_CHV$ $\|$
      $permission\_session(chv) := true$ $\|$
      $sw := 9000$
    ELSE
      IF ($counter\_chv = 1$)
      THEN
        $counter\_chv := 0$ $\|$
        $blocked\_chv\_status := blocked$ $\|$
        $permission\_session(chv) := false$ $\|$
        $sw := 9840$
      ELSE
        $counter\_chv := counter\_chv - 1$ $\|$
        $sw := 9804$
      END
    END
  END
END;

$cd, cf, cc, cuc \longleftarrow$ **STATUS** $\hat{=}$
BEGIN
  $cd := current\_directory$ $\|$
  $cf := current\_file$ $\|$
  $cc := counter\_chv$ $\|$
  $cuc := counter\_unblock\_chv$
END;

RESET $\hat{=}$
BEGIN
  $current\_file := \emptyset$ $\|$
  $current\_directory := mf$ $\|$
  $permission\_session$
    $:= \{(always, true), (chv, false),$
      $(adm, false), (never, false)\}$
END;
  $\ldots$
END

A Formal Analysis of the CORBA Security Service[*]

David Basin, Frank Rittinger, and Luca Viganò

Institut für Informatik
Albert-Ludwigs-Universität Freiburg
Georges-Köhler-Allee 52
D-79110 Freiburg, Germany
Tel: +49 (0)761 203-8240, Fax: +49 (0)761 203-8242
{basin,rittinge,luca}@informatik.uni-freiburg.de
http://www.informatik.uni-freiburg.de/~{basin,rittinge,luca}

Abstract. We give a formal specification and analysis of the security
service of CORBA, the Common Object Request Broker Architecture
specified by the Object Management Group, OMG. In doing so, we tackle
the problem of how one can apply lightweight formal methods to improve
the precision and aid the analysis of a substantial, committee-designed,
informal specification. Our approach is scenario-driven: we use representative scenarios to determine which parts of the informal specification
should be formalized and verify the resulting formal specification against
these scenarios. For the formalization, we have specified a significant part
of the security service's data-model using the formal language Z. Through
this process, we have been able to sharpen the OMG-specification, uncovering a number of errors and omissions.

1 Introduction

In large, heterogeneous, distributed information systems, *middleware* is often
used to provide a communication medium between different system components,
which may run on different platforms. By acting as a mediator between the different components, and as a uniform interface for the developer, middleware provides access and location transparency and thereby creates the illusion of a single
composite system. Various middleware systems exist, such as OMG's Common
Object Request Broker Architecture (CORBA) [19], Microsoft's COM+ [23],
and Sun's Enterprise Java Beans [28]. In this paper, we focus on CORBA and
use the formal language Z to give a formal analysis of CORBA's security service.

The CORBA security service specification of the OMG [20] consists of more
than 350 pages, detailing interfaces and using natural language and ad hoc diagrams to describe the semantics of the interfaces. The overall specification is
quite complex and this complexity stems from several factors. First, the problem itself is inherently complex. Any specification-conform implementation must

---

[*] This work was partially supported by the Deutsche Forschungsgemeinschaft under
grant BA 1740/5-1 "Formale Sicherheitsarchitekturen".

D. Bert et al. (Eds.): ZB 2002, LNCS 2272, pp. 330–349, 2002.

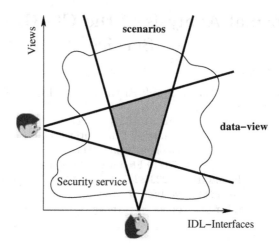

Fig. 1. Reduction of complexity along two dimensions.

achieve multilateral security guarantees for distributed object systems where the objects execute on different platforms in different environments. Second, the solution must be *loose* (or *open*). The OMG, a consortium whose members come from over 800 companies and organizations [18], does not define new technologies, but rather develops standards that encompass the technologies of its members. Hence, any specification of a standard by the OMG must be loose in the sense that it admits as models all implementations given by OMG members. This looseness is achieved, for example, by simply giving signatures for interfaces, where the use of general types does not constrain the values of possible implementations. Finally, the specification has the complexity inherent in any document designed by a political process.

Our research is motivated by the question of whether, despite this complexity and the informal nature of the specification of the CORBA security service, it is possible to reason about the security of systems built using this service. Moreover, whether one can do this in a *lightweight* way [10] without plunging into the black hole of formalizing both CORBA and its security service in their entireties.

What we have accomplished is to formalize part of the OMG-specification for this service, guided by particular security requirements. More specifically, to carry out a formal analysis in a pragmatic way, we have reduced the complexity of the problem along two dimensions, as illustrated in Fig. 1. First, we took a data-oriented view of the system and formalized part of the service's data-model using the formal language Z. Second, we took a scenario-driven approach: we used representative scenarios to determine which parts of the OMG-specification should be formalized, and then verified our formal specification against the requirements of these scenarios. These reductions allowed us to give a partial, but meaningful, formal analysis, without formalizing both CORBA and its security service in their entireties, or managing the additional complexity of employing multi-language formalizations. Through this process, we have been able to sharpen the OMG-specification and uncover a number of errors and omissions. Moreover, as Z supports high-level, readable specifications, we believe that our formal specification can also be of use to different vendors who wish to implement the CORBA security service or compare (or even formally verify) their implementation against the specification.

We have used tool support to improve the precision of our specification and the accuracy of our analysis. We wrote and type-checked our Z-specification using the ZETA-tool [34], which provided substantial assistance not only in finding mistakes in preliminary versions of our Z-specification but also in uncovering a number of errors and omissions in the original OMG-specification. We initially carried out the analysis of the security properties by hand, and afterwards machine-checked some of our proofs, including the ones given here, using HOL-Z, which is an embedding of Z in higher-order logic within the generic theorem prover Isabelle [9,14,22]. In doing so, we have been able to show that some of the security requirements are satisfied, while for others a refinement of our (and thus of the OMG's) specification is needed.

More generally, our work contributes to the understanding of how formal languages like Z can be used to model and analyze complex, informal standards. A common misconception is that formal methods are too heavy in such cases, as they require a complete formalization of the standard, which is of overwhelming complexity due to its size and the need to integrate different views of the system, e.g. static data-models with dynamic event or process-oriented models. We show here that it is possible to proceed with a pragmatic "light touch" in the formalization: One can proceed in a scenario-driven way, and even restrict attention to a particular view of the system (namely the data-view), and still arrive at general, useful results, with limited resources.

The rest of this paper is organized as follows. In § 2, we provide background on CORBA and its security service. In § 3, we discuss, and illustrate by means of examples, our Z-specification of this service. In particular, we consider parts relevant for authentication and access control. In § 4, we provide highlights from the formal analysis of our Z-specification. We conclude in § 5, where we summarize the lessons learned from our research and discuss related and future work.

2   Background on CORBA

2.1   OMG's CORBA

CORBA systems are based on the *Object Management Architecture* (OMA) [21, 24], which consists of an object model that describes the properties of all objects, and a reference model that describes how these objects interact. Fig. 2 shows the four different types of components distinguished by the OMA: the *business applications* (*objects*), the *services*, the *facilities*, and the *Object Request Broker* (ORB). The ORB acts as an object bus (analogous to a hardware bus) that

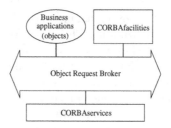

Fig. 2. Object Management Architecture (OMA).

connects the different system components and manages communication between them, thereby ensuring portability, reusability and interoperability. The services provide general interfaces such as transaction management, object location and security, while the facilities provide domain specific interfaces (e.g. for health or financial services). This enables a kind of plug-and-play mechanism. Business applications only implement the business logic and are associated with services or facilities to gain additional functionality.

To achieve platform independence, all components in a CORBA system are known only through their interfaces, which are defined in a programming language independent way using the CORBA-specific *Interface Definition Language* (IDL). The OMG-specification, however, does not always have a clear semantics of what functionality implementations of the interfaces must provide; what semantics there is, is given only by informal prose or ad hoc diagrams. As a result, the OMG-specification is ambiguous in many places, and this problem is complicated by the desire to have a specification open to all member technologies.

We focus here on version 1.2 of the CORBA security service [20]; for more information in general on CORBA consult [19,21,24], and [2,5] for security in particular.

2.2   The CORBA Security Service

There are many different goals that must be achieved to obtain a secure system, and in this paper we focus on *confidentiality* and *integrity*. Together these two goals express that only authorized users can access and modify confidential data.

The protection mechanisms that are employed by the CORBA security service to achieve these goals are:

Identification and authentication: A user's identity is validated, e.g. by name and password, by biometry, or by similar mechanisms.

Authorization and access control: An authenticated user is assigned rights and is only allowed to perform actions for which he has rights.

It is mandatory for security that these protection mechanisms are always used and cannot be bypassed.

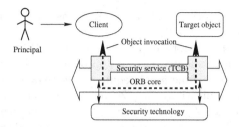

Fig. 3. Integration of the security service into the ORB.

The CORBA security service specification defines interfaces between standard security technologies (such as Kerberos or SSL) and the ORB or the business applications. The architecture of the security service is based on a *Trusted Computing Base* (TCB), which encapsulates all security-related functionality in a small trusted kernel that cannot be bypassed. Given this encapsulation, the business components can then be implemented independently of concrete security mechanisms. This has the advantage that the code implementing the security functionality is minimized, security policies are easier to keep consistent, and the trust between different components is limited to the trust between the TCBs of different domains. Fig. 3 illustrates how the security service is integrated into the ORB.

The actors in a CORBA environment, which are normally human users but can also be software or hardware components, are called *principals*. Principals are known to the system through *security attributes*, which are divided into *identity attributes* and *privilege attributes*. These attributes are assigned to a principal on registration, during which the principal is authenticated and is given its *credentials* (which include the principal's attributes).

Since CORBA is an object system, all actions in the system are carried out through object invocations, whereby the access control mechanisms must check the identity and rights of the initiating principals. The CORBA security model does not commit to one access model but allows designers to choose between many different models, such as role-based access control, access control lists, label-based policies or capabilities-based ones. Some of these models have been formally specified in [3,13]; as a comparison, note that these works specify individual access control models for CORBA while the OMG specification of the CORBA security service, and thus our specification in Z, defines how access control is used within an overall security architecture.

3   Z-Specification of the CORBA Security Service

3.1   The Formal Language Z

Z [11,30,33] is a formal language based on typed set theory and first-order logic with equality. It has been successfully applied, in its original form or using extensions such as Object-Z [29,31], in many case studies in formal modeling and development, e.g. [6,7,15,25].

We have chosen Z for our work as the CORBA security service specification is heavily data-oriented in that approximately half of it is focused on data, interfaces, and their functional requirements (while the remainder focuses on inter-ORB communication). The Z language is well-suited for such data modeling. In particular, Z provides constructs for structuring and compositionally building data-oriented specifications: schemas are used to model the states of the system (*state schemas*, e.g. the schema *Env* in §3.4) and operations on the state (*operation schemas*, possibly with input/output parameters, as in the schema *authenticateSuc* in §3.5), and a *schema calculus* is provided to compose these subspecifications.

We have also chosen Z for its simplicity: Z-specifications are fairly easy to write and understand. This is important for us as we want the results of our formal analysis to be meaningful to those working with middleware who are not trained in formal methods. Another reason for choosing Z is its rapidly growing tool support. We have written and type-checked the Z-specification we describe here using the ZETA-tool [34] and we have machine-checked some of our proofs using the HOL-Z embedding in the generic theorem prover Isabelle [9,14] (we return to this in §4 below).

In what follows, we will assume that the reader is familiar with the syntax and features of Z. We will only explain a few details, as need arises, in the examples; a more complete picture of Z can be found in [11,30].

3.2   Our Specification Approach

Given the size and complexity of the informal CORBA security service specification, we sought an alternative to analyzing the complete service. We approached the formalization problem by selecting scenarios for two of the most important security requirements: principal authentication and secure object invocation (including establishing security associations and access control).

As illustrated in Fig. 4, from these scenarios we extract two kinds of specifications. First, as discussed in this section, we identify what parts of the OMG-specification must be formalized to describe those parts of the security service relevant for each scenario. Second, as discussed in §4, we determine the scenarios' security requirements, which constitute a (partial) specification of properties of the security service. Both of these specifications are then formalized in Z and provide a basis for formally establishing (as we do in §4) that the security service specification has the desired properties.

Due to space restrictions, we cannot give here in full our Z-specification of the CORBA security service, which encompasses more than 50 pages and can be found in [26]. Instead, we provide only a few highlights of our specification and illustrate the methodological aspects of our approach. Note also that most of the Z-definitions (for types, schemas, and requirements) in this paper have been significantly simplified to include only those parts of the specification we actually focus on here.

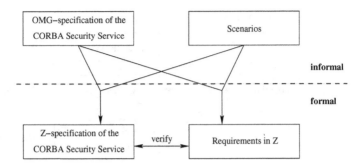

Fig. 4. From informal to formal specifications (and analysis).

Our translation of the informal OMG-specification of the CORBA security service into a Z-specification proceeded in three steps, where we actually translated into Z only those parts of the informal specification that were needed for the chosen scenarios.

In the first step, we defined a mapping from IDL-constructs to Z-constructs.[1] Although it is possible to map each IDL-construct following a fixed scheme, it leads to a clearer and more concise formalization to translate IDL-constructs differently, according to their purposes. For instance, the canonical way to translate operations is to model them as operation schemas, but we can obtain a more concise specification if we model the operations that do not modify the state (i.e. lookup operations) as functions in Z. Similarly, we did not explicitly define a state schema for each of the interfaces, but instead distinguished two kinds of interfaces: (i) interfaces without state information, which serve only to structure the name-space (e.g. PrincipalAuthenticator); (ii) interfaces having state information, which are implemented by objects that populate the object system. For an interface of kind (i), no state schema is created but only operation schemas that (possibly) manipulate the global state of the system. For an interface of kind (ii), we create a state schema that models the datatype of the interface, a state schema that defines functions to manage such objects, and operation schemas that manipulate the state.[2] Fig. 5 illustrates this transformation scheme in the case of the Credentials interface.

In the second step, we transformed all IDL-code into Z, and in the third step we augmented our Z-definitions with the informal descriptions of the interfaces given by the OMG. These two steps leverage off the fact that a Z-schema con-

---

[1] Our mapping, which is defined for the IDL-constructs occurring in the CORBA security service specification of the OMG, has been influenced by the similar mapping given in [16], and could be automated (at least partially) using a compiler similar to the one given there.

[2] Alternatively, we could have used promotion (as described in [33]) to model the operations as a combination of local operations and a promotion schema that connects the local and global state.

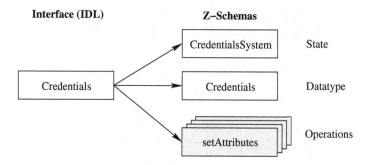

Fig. 5. Translation from IDL to Z-schemas.

sists of two parts: a *declaration* part, where typed variables are declared, and a *predicate* part, where the constraints on the variables are specified. The second step of our translation amounts to mapping IDL-code into the declaration parts of schemas. The predicate parts, defined in step three, allow us to rigorously specify the intended semantics, which is only informally described by the OMG.

In the rest of this section, we motivate and explain the scenarios we chose, provide some of the basic definitions that are needed to understand our specification, and define a formal model of the scenarios.

3.3   Scenarios

As examples, we focus here on two requirements that, as postulated in the OMG-specification, we can impose on the specification, namely (i) that upon authentication, no principal can gain more attributes (i.e. more credentials) than the system allows, and (ii) that a valid security association must be established prior to a secure object invocation. To analyze these two requirements we consider the following two scenarios.

Scenario 1: Principal authentication. Whenever a principal enters a CORBA system for the first time, the principal must authenticate itself to the system. Afterwards, provided the authentication has succeeded, the system assigns identity and privilege attributes to the principal.

Scenario 2: Security association. Before a client can invoke any operation on a target object, there must exist a security association between them. Prior to object invocation, the CORBA environment establishes this association, which is used to establish mutual trust between the client and the target and to negotiate the details of the communication. The process of establishing this association (also known as 'binding') is handled internally and is transparent to the client.

3.4   Basic Definitions

Before we can formalize the interfaces associated with our scenarios, we must define the system environment. This is accomplished by the schema *Env* below, whose declaration part defines: the system time *systemTime* (*UtcT* is a time format defined by the OMG in the CORBA time service and is used in all CORBA services); a set *supportedAuthMethods* of supported authentication methods (the authentication method chosen by the client must be supported by the actual security service implementation); a set *moreSteps* that consists of all authentication methods that require more than one step for authentication; a set *principals* of all registered users of the system; a function *attributes* that maps users to their security attributes; and a function *authenticates* that maps registered principals to their authentication data (*Opaque* is an unstructured sequence of bytes that is used throughout the OMG-specification of the security service when no assumptions about the data are made).

The predicate part of the schema *Env* specifies that the default attribute for each unregistered principal ($s \notin principal$) is equal to the constant *Public*, thereby allowing public access to non-confidential data (if desired).

$\_\_$ *Env* $\_\_$
$systemTime : UtcT$
$supportedAuthMethods : \mathbb{P}\, AuthenticationMethod$
$moreSteps : \mathbb{P}\, AuthenticationMethod$
$principals : \mathbb{P}\, SecurityName$
$attributes : SecurityName \rightarrow \mathbb{P}\, SecAttribute$
$authenticates : SecurityName \rightarrow Opaque$
***
$\mathbf{dom}\ authenticates = principals$
$\forall\, s : SecurityName \mid s \notin principals\ \bullet$
$\qquad \exists\, a : SecAttribute \bullet a.attribute\_type.attribute\_type = Public$
$\qquad\qquad \wedge\ attributes(s) = \{a\}$

As we mentioned above, most of the Z-definitions we give here are incomplete as we have elided details for brevity; the complete definitions can be found in [26]. Note also that the types in the OMG's, and our, specification can be divided into two kinds: (i) basic types that, according to the OMG, are given meaning only by their names, e.g. *SecurityName*; and (ii) complex types such as enumeration types (e.g. *AuthenticationMethod*) or structured types (e.g. *SecAttribute*, which consists of an attribute type, its defining authority, and the attribute's value). Whenever not relevant for our examples, we do not specify the details of these complex types. Moreover note that some of the identifiers may look needlessly complex, e.g. *a.attribute_type.attribute_type*, but they reflect the way the OMG structures its specification and the names chosen there.

In addition to the environment, we also require definitions for some of the basic entities occurring in the specification. As mentioned in § 2.2, all principals are identified by their attributes, which are held in their credentials. These credentials are described by a schema *Credentials*, which defines components for the principal's security attributes (*attributes*). The security attributes are split into privilege attributes (*privileges*), which can be used to decide what the principal can access, and identity attributes, which are assigned by the system after authentication.

```
┌─ Credentials ──────────────────────────────────────────────
│ privileges, attributes : ℙ SecAttribute
├────────────────────────────────────────────────────────────
│ privileges ⊂ attributes
└────────────────────────────────────────────────────────────
```

The following state schema *CredentialsSystem* describes an object that manages all credentials. It includes[3] the schema *Env* to access the system time, it defines a function *credentialsId* that maps credentials identifiers (of basic type *CREDS*) to credentials, and it defines a function *is_valid* that maps each valid credentials identifier to the time t when it expires, whereby t must be greater than the current system time. Note, however, that invalid credentials can be refreshed, which implies that the domain of *is_valid* is only a subset of the domain of *credentialsId*.

```
┌─ CredentialsSystem ────────────────────────────────────────
│ Env
│ credentialsId : CREDS ⇸ Credentials
│ is_valid : CREDS ⇸ UtcT
├────────────────────────────────────────────────────────────
│ dom is_valid ⊆ dom credentialsId
│ ∀ c : CREDS; t : UtcT | is_valid(c) = t • t > systemTime
└────────────────────────────────────────────────────────────
```

The state schema *Current* describes context and technology-specific state information for the current principal, of which only the credentials that will be used for the next invocation are displayed.

```
┌─ Current ──────────────────────────────────────────────────
│ own_credentials : seq CREDS
└────────────────────────────────────────────────────────────
```

3.5   Scenario 1: Principal Authentication

The CORBA security service specification defines the authentication of principals in the interface *PrincipalAuthenticator*. This interface defines two operations: *authenticate* to authenticate a principal in one step, and *continue_authentication*, which specifies the behavior of the system when additional authentication steps are required. (When such steps are necessary, and how they should be

---

[3] Inclusion of a schema S_1 in a schema S_2 means that the declaration part of S_1 is incorporated into the declaration part of S_2 and the constraints on these declarations imposed in the predicate part of S_1 are propagated into the predicate part of S_2. Other forms of schema inclusion are used and explained below.

processed, is technology-dependent and cannot be analyzed without knowledge of the underlying technology; in particular, it is impossible to define a schema describing the operation *continue_authentication* without this knowledge.)

We focus on the case where only one step is needed for the authentication and use the schema *authenticateContinue* to represent the further steps for the authentication and the conditions under which these steps occur. Indeed, we exploit the schema calculus of Z to define the operation *authenticate* as a disjunction of several schemas capturing successful, partial or (in the last three disjuncts) erroneous authentication:

$$authenticate == authenticateSuc \lor authenticateContinue$$
$$\lor\ methodsNotSupported \lor wrongAuthData \lor tooManyAttributes$$

As an example, the operation schema *authenticateSuc*, for successful authentication, is as follows:

```
┌─ authenticateSuc ─────────────────────────────────────────
│ Ξ Env
│ Δ Current
│ Δ CredentialsSystem
│ method? : AuthenticationMethod
│ security_name? : SecurityName
│ auth_data? : Opaque
│ privileges? : seq SecAttribute
│ creds! : Credentials
│ return! : AuthenticationStatus
├──────────────────────────────────────────────────────────
│ method? ∈ supportedAuthMethods
│ method? ∉ moreSteps
│ auth_data? = authenticates(security_name?)
│ ran privileges? ⊆ attributes(security_name?)
│ ∃ c : Credentials; cid : CREDS; t : UtcT •
│     cid ∉ dom credentialsId ∧ c ∉ ran credentialsId
│     ∧ credentialsId' = credentialsId ∪ {cid ↦ c}
│     ∧ is_valid' = is_valid ∪ {cid ↦ t}
│     ∧ creds! = c ∧ c.privileges = ran privileges?
│     ∧ own_credentials' = ⟨cid⟩
│ return! = SecAuthSuccess
└──────────────────────────────────────────────────────────
```

This schema describes the possible transitions of the part of the system state that is represented by the schemas *Current*, *CredentialsSystem* and *Env*. To this end, *Current* (and similarly for *CredentialsSystem*) is included prefixed by a Δ, meaning that two copies, primed and unprimed, of all variables of *Current* are included in *authenticateSuc*; unprimed variables represent the state before the transition, while primed variables define the state after it. Inclusion of a state schema prefixed by Ξ, like for *Env* in this case, works similarly but with the restriction that the values of the variables remain unchanged during the transition, i.e. the values of primed and unprimed variables must be equal. Overall, the predicate part of *authenticateSuc* describes the conditions for successful authentication, and thereby formalizes the semantics of this operation.

A successful authentication requires that the chosen authentication method *method?* is supported by the environment and that, for a one-step authentication, *method?* is not an element of the set *moreSteps*. Furthermore, the correct authentication data for the given security name *security_name?* is checked against the provided data *auth_data?*, and the privileges requested by the client (*privileges?*) must be contained in the attributes that the client is entitled to. The appropriate credentials (including the requested privileges) are then created, added to the set of valid credentials, and set in the *own_credentials* component of the client's *Current*.

To capture the scenario when the authentication is refused by the system because the provided authentication data does not match the required data, we define the schema *wrongAuthData*. Its precondition states that the authentication data is incorrect and therefore indicates failure of the authentication (*return!* = *SecAuthFailure*) in contrast to *authenticateSuc*. No state changes are described by this schema.

$$
\begin{array}{|l}
\hline
\;\text{---}\; wrongAuthData \;\text{---} \\
\hline
\Xi Env \\
security\_name? : SecurityName \\
auth\_data? : Opaque \\
return! : AuthenticationStatus \\
\hline
auth\_data? \neq authenticates(security\_name?) \\
return! = SecAuthFailure \\
\hline
\end{array}
$$

The other schemas that indicate failures, *methodsNotSupported* and *tooManyAttributes*, are defined analogously.

3.6   Scenario 2: Security Association

A security association between communicating parties is employed to establish trust between them and to negotiate the details of the communication, such as the type and strength of the protection used for the messages they exchange. The state/context information of this association is represented on both sides by a *SecurityContext* interface, through which the target object has access to the credentials of the initiator of the invocation.

The Z-formalization of this interface is analogous to that of the *Credentials* interface. To represent the state information of the context, we define a schema *SecurityContext*, which defines components for the credentials, for the target object, and for additional context-specific information (which we do not display in the schema). The state schema *ObjectSystem* is included since it defines components that represent the objects that are currently associated with the system;

in particular, it defines a function *objectId* that maps object identifiers to the actual objects.

```
┌─ SecurityContext ─────────────────────────────────────────────
│ ObjectSystem
│ credentials : ℙ CREDS
│ target : dom objectId
└───────────────────────────────────────────────────────────────
```

A security context can only be created through the operations *init_security_context* and *accept_security_context*, which are defined in the interface *Vault*. But since we are only concerned here with the existence of such contexts, in our examples we elide the parts of the specification that handle their establishment.

The schema *SecurityContextSystem* describes an object that manages all *SecurityContext* objects (i.e. it describes the state of the system of all client and target contexts). It defines a function *securityContextId* that maps identifiers of the basic type $SEC\_CONTEXT$ to security context objects, and a function *is_valid* that maps the identifiers of all valid objects to their expiration date. In addition, it is necessary to define two sets of contexts, one for the client-side contexts and one for the target-side contexts. In the predicate part, these two sets are constrained to be of the same size and to contain all valid contexts in the system.

```
┌─ SecurityContextSystem ───────────────────────────────────────
│ Env
│ securityContextId : SEC_CONTEXT ⇸ SecurityContext
│ is_valid : SEC_CONTEXT ⇸ UtcT
│ clientContexts, targetContexts : ℙ SEC_CONTEXT
├───────────────────────────────────────────────────────────────
│ clientContexts ∪ targetContexts = dom securityContextId
│ #clientContexts = #targetContexts
│ dom is_valid ⊆ dom securityContextId
│ ∀ s : SEC_CONTEXT; t : UtcT | is_valid(s) = t • t > systemTime
└───────────────────────────────────────────────────────────────
```

The next section will show that, although abstracting away details about how the security association is established, these two schemas are sufficient to describe the part of the security service against which we will check the requirements of this scenario, i.e. the existence of two corresponding *SecurityContext* elements.

4 Formal Analysis

Given the two kinds of formal specifications discussed above, we have carried out a formal analysis by proving that the Z-specification of the security service satisfies the Z-specification of the requirements associated with each scenario. Alternatively, when a proof is not possible, we have used the failed attempt to uncover inadequacies in the OMG-specification.

Since we are only using Z for data-modeling, some requirements fall outside of our formal specification and analysis. In general, such an analysis yields three different kinds of results (assuming, as we do here, that the requirements are consistent in the first place):

(i)   the specification adheres to the requirement, which follows by proving a formula that represents the requirement;

(ii) the specification does not adhere to the requirement, but may do so after the specification is refined; and

(iii) the requirement cannot be expressed using our formalization in the first place.

We present two examples that illustrate these three situations; further examples are given in [26].

A Formal Analysis of Scenario 1

The first example of our formal analysis, which illustrates results of kinds (i) and (iii), is to investigate whether principal attributes are assigned correctly upon authentication. Namely, we require from our Z-specification that principals are always authenticated and are assigned correct credentials.[4]

We can split this requirement into two subrequirements, the first of which can be formulated as: The operation *authenticate* must be called for the initiating principal prior to any object invocation on behalf of that principal. This subrequirement cannot be directly expressed in our Z-specification by formalizing properties about schemas that specify state or operations that transform state. Capturing such temporal dependencies is best accomplished by combining Z with another specification language, like a temporal logic, or some visual, temporally-oriented specification formalism like UML's sequence diagrams, where one can express constraints on the *dynamics* of the system, e.g. which sequences of operations are allowed. The importance of such combinations is discussed in more detail in § 5.

Many of the security requirements can, however, be expressed as conditions on the data-model. Consider, for example, the second subrequirement, which states that the credentials assigned to a principal during authentication are correct, in the sense that no privilege attributes are assigned to a particular principal other than those that the system allows for him. We can formalize this in Z as follows:

$$\forall \, Env; \; Current; \; CredentialsSystem \bullet authenticate \Rightarrow$$
$$creds!.privileges \subseteq attributes(security\_name?) \quad (1)$$

---

[4] It is important to note that in this example we are only concerned with the initial assignment of credentials, and therefore privileges, to principals. Since the security service provides operations for a client/principal to copy and modify its credentials, it is another proof obligation (not discussed in this paper) to show that these operations cannot be used to illegally modify a principal's privileges.

Here, quantification over the schemas *Env*, *Current*, and *CredentialsSystem* quantifies over the free variables in the declaration parts of these schemas, which binds the variables in the body of (1). The schemas *Env*, *Current*, and *CredentialsSystem* denote the system state for our example and the quantification formalizes that property (1) must hold for all possible states.

We are now in a position to prove that the specification adheres to this subrequirement, i.e. that (1) holds. To do so, we must show that, for an arbitrary system state (represented by *Env*, *Current*, and *CredentialsSystem*), the operation *authenticate* only returns credentials that are in accordance with the system's policy for the initiating principal. Note that it is only necessary to show this for a successful authentication, as formalized by *authenticateSuc*, since otherwise no credentials are created and set in the principal's *Current*. The essence of the proof is then as follows: one of the conditions in the predicate part of *authenticateSuc* states that

$$\text{ran } privileges? \subseteq attributes(security\_name?)$$

and *c.privileges* = ran *privileges?* (also in the predicate part of *authenticateSuc*) implies that

$$c.privileges \subseteq attributes(security\_name?),$$

so that *creds!* = *c* yields the requirement

$$creds!.privileges \subseteq attributes(security\_name?).$$

As the instances of *Env*, *Current*, *CredentialsSystem* and *authenticate* were arbitrary, the formula (1) follows.

To study the feasibility of the mechanization and automation of our approach, we have carried out this proof and others using HOL-Z [9,14], an embedding of Z in higher-order logic for the Isabelle system [22]. HOL-Z provides a conservative extension of HOL with Z, which we further extended with our formalized theory of the CORBA security service. We then formalized and proved security requirements within this extended theory, i.e. we formalized the lower part of Fig. 4 as

$$\vdash_{\text{HOL + Z + CORBAsec}} \text{Z-Requirements}$$

where CORBAsec stands for our formalization of the security service. Using support for automated reasoning provided by the Isabelle system, proofs like the above can be carried out at a high level and, in some cases, even completely automated. Although hand proofs like that of (1) are straightforward, in general verification can be substantially more complex, and theorem proving plays an important role in aiding and improving its precision.

A Formal Analysis of Scenario 2

The second requirement that we consider concerns the security association for a secure object invocation, and is an example of a result of kind (ii) illustrating how formal reasoning can lead to an improvement of the specification. In order to formulate the requirement, we briefly explain the access control mechanisms in the CORBA security service.

Since CORBA is an object system, to control access to confidential data one must control the invocation of operations on objects. Access control is realized by an operation *access_allowed* that checks the rights of a principal (which is represented through its credentials) to invoke an operation on an interface. The operation *access_allowed* is specified by a relation that contains all tuples of credentials, targets, operation names and interface names for which access is allowed, i.e. if $(c, t, op, i) \in access\_allowed$ then the principal with credentials c is allowed to invoke the operation op on the target object t implementing the interface i. As far as the security service is concerned, prior to an access decision for an operation invocation of a client object on a target object, a security context must have been established between the client and the target.

We formulate the second security requirement as: A client can only invoke an operation on a target if a security association between them has been previously established. More technically, the following formula states that a valid client context must be present in the *clientContexts* component of the state schema *SecurityContextSystem* for each tuple in *access_allowed* before an invocation can occur. Since the actual object invocation is not part of the security service, we can assume that the last action of the security service prior to the invocation is a call of *access_allowed* for an access decision, and that the ORB handles this correctly. Therefore *access_allowed* will be used as a synonym for the actual invocation and, since *access_allowed* is modeled as a relation, the test for membership corresponds to the operation call. Moreover, we require that all security contexts and credentials are valid with respect to the current system time.

$$\exists\, scs : SecurityContextSystem;\ cs : CredentialsSystem;$$
$$sc : scs.clientContexts;\ cl : \text{seq } CREDS;\ t : UtcT;$$
$$i, op : Identifier \bullet$$
$$(\text{ran } cl \subseteq (scs.securityContextId\ sc).credentials$$
$$\wedge\ scs.is\_valid(sc) = t$$
$$\wedge\ (\forall\, c : cl \bullet \exists\, tt : UtcT \bullet cs.is\_valid(c) = tt))$$
$$\Leftrightarrow (cl, (scs.securityContextId\ sc).target, op, i) \in access\_allowed$$

Additionally to the formula above, we must also require that an element exists in *targetContexts* and that it is connected to the one in *clientContexts*, i.e. that they were created by consecutive calls to *init_security_context* and *accept_security_context*.

Our use of formal methods helped us uncover a problem here. Namely, we discovered that we could not formulate this additional requirement since the connection between the elements in *targetContexts* and *clientContexts* is not present in the specification of the OMG and, a fortiori, in ours. Although the OMG-specification in its current form is inadequate, it is possible to extend it, and, correspondingly, refine our Z-specification, so that we can then state and prove the requirement. This is achieved by augmenting the schema *SecurityContextSystem* with a relation *associates* that relates a client-side security context object with the corresponding target-side object:

| *associates* : $SEC\_CONTEXT \leftrightarrow SEC\_CONTEXT$

Having repaired this omission, we can now check that a pair of valid contexts is contained in this relation when access decisions are made. That is, we can now establish that when a security service meets the specification given by the *SecurityContext* schema and the extended *SecurityContextSystem* schema then it satisfies the requirements of our second scenario.

5   Conclusions

We now summarize the lessons learned and discuss related and future work.

Lightweight formal methods. As explained in the introduction, the informal specification given by the OMG is large and complex. A number of groups have carried out partial, informal analyses of the CORBA security service [1,17], but the task of a full formal analysis is rather daunting. Our work indicates that there are reasonable points in-between when formal methods are used in a pragmatic, lightweight fashion.

To carry out a formal analysis, we have reduced the complexity of the problem along two dimensions. First, we took a data-oriented view of the system and formalized the service's data-model in the Z language, and second we took a scenario-driven approach to determine which parts of the OMG-specification should be formalized, and then verified our formal specification against requirements of these scenarios. This reduction allowed us to give a partial, but meaningful, formal analysis, without formalizing both CORBA and its security service in their entireties, or managing the additional complexity of employing multi-language formalizations.

Our formal specification. OMG-specifications are informal and open by design. This is politically attractive (since it subsumes the reference implementations of OMG members) but problematic. Although loose specifications (allowing model/implementation classes) are successfully used, e.g. in algebraic specification, when carried out using natural language, by a political process, too much "looseness" creeps in. This allows future developers and implementors of the security service to interpret the specification in unintended ways that may lead to security-holes in applications built using the service. A formalization of the specification (especially in a language that is relatively easy for non-specialists to understand) helps solve this problem, as it provides a rigorous semantics that can serve as a basis for communication between different developers, implementors and users of the system. Even a partial formalization is of assistance here, e.g. clarifying the data-model, while leaving control semi-formal. Moreover, even with a partial formalization, it is still possible, and becomes interesting, to formally establish security properties and thus demonstrate that the specification is "tight" enough.

In the formal specification process (and even more so in the formal analysis process), our experience indicates that appropriate tool support is a necessity. Tool support improves the understanding of, and increases the confidence in, the specification and analysis. Indeed, it is difficult to write a Z-specification of more than a handful of schemas without any typographical errors or inconsistencies. The use of the type-checker ZETA [34] enabled us not only to correct many such mistakes in preliminary versions of our Z-specification but also to discover a number of errors and omissions in the original OMG-specification.

Our formal analysis. A formal specification like ours provides the basis for a formal verification of many of the multilateral security properties of the security service and of applications built upon it. Rather than completely analyzing the entire service, a Herculean task due to its complexity, we adopted a scenario-driven analysis and, as examples, focused in this paper on two security scenarios: principal authentication and secure object invocation. (Other scenarios and requirements are analyzed in [26].) Our examples illustrate how our analysis yields three different kinds of results: (i) the Z-specification adheres to the requirement; (ii) the Z-specification does not adhere to the requirement; and (iii) the requirement cannot be expressed using our formalization in the first place. In other words, we have been able to show that some of the security requirements are satisfied, while for others a refinement of our (and thus of the OMG's) specification is needed. Hence our formal analysis suggests how the OMG-specification can be improved both by removing subtle errors (results of the second kind) and omissions (results of the third kind).

Our proofs of results of the first two kinds were carried out by hand and also machine-checked using the HOL-Z embedding in the generic theorem prover Isabelle [9,14]. Results of kind (iii), on the other hand, identify a shortcoming of our approach.

Z and other formal methods. Our examples show how the use of Z can improve the precision and analysis of an informally specified standard like the CORBA security service. As mentioned above, Z is well-suited for data-oriented modeling, e.g. for modeling the states of the system and the operations on them, and hence well-suited for our approach where we define such a model. The decision to focus on the data-model bounds the overall complexity, but has, as its main drawback, the limitation that there are security scenarios and properties that fall outside of this model as they require formalizing other views, for which other formal methods are better suited than Z. Examples include event-oriented or process-oriented views like those formalizable using process algebras like CSP [27] or temporal logics, as was done in [4,12], where the model-checker SPIN is used to validate parts of the CORBA specification. These methods allow one to express explicit temporal constraints and analyze sequences of actions for safety and liveness properties, which is important for security, since it may happen that different components or operations taken by themselves are well-defined but their correct use requires their proper sequencing with events from the environment. As future work, we plan not only to analyze further scenarios within the data-oriented view, but also to analyze other views by enhancing and combining our approach with event-oriented or process-oriented formalisms, and investigating to what extent one can still do this in a lightweight way. Existing embeddings of such formalisms, such as the HOL-CSP embedding in Isabelle [8,32], will provide a good starting point, with HOL serving as a common semantic basis for combining our Z-specification with a process-oriented CSP-specification.

Important related work is also that involving the use of Object-Z [29], which combines object oriented and temporal extensions to Z. We strongly considered the use of this language for our work, especially since it has been applied in case studies of specifications of CORBA services [15,25]. In the end, we chose Z for two reasons. First, Z is an ISO standard, which carries weight with standardization bodies like the OMG. Second, and more importantly, Z is supported by many tools for modeling and reasoning, such as ZETA and HOL-Z, which both provided a basis for our work. We are currently working on linking ZETA and HOL-Z and providing further support for automating reasoning like ours.

References

1. J. Arndt and T. Österdahl. Network security in distributed systems using CORBA. Technical report, Ericsson Hewlett-Packard Telecommunications AB, 1998.
2. B. Blakley. *CORBA Security*. Addison-Wesley-Longman, 1999.
3. G. Brose. A Typed Access Control Model for CORBA. In F. Cuppens, Y. Deswarte, D. Gollmann, and M. Waidner, editors, *Proceedings of ESORICS 2000*, LNCS 1895, pages 88–105. Springer-Verlag, 2000.
4. G. Duval. Specification and verification of an object request broker. In *Proceedings of the 20th International Conference on Software Engineering (ICSE'98)*, pages 43–52. IEEE Computer Society Press, 1998.
5. D. Gollmann. *Computer security*. John Wiley & Sons, 1999.
6. J. Hall. Using Z as a specification calculus for object oriented systems. In D. Bjorner, C. A. R. Hoare, and H. Langmaack, editors, *Proceedings of VDM'90*, LNCS 428, pages 290–318. Springer-Verlag, 1990.
7. I. Hayes. *Specification Case Studies*. Prentice-Hall International, 1993.
8. HOL-CSP.
   http://www.informatik.uni-freiburg.de/~wolff/doc/isa_doc/CSP/.
9. HOL-Z. http://www.informatik.uni-freiburg.de/~wolff/doc/isa_doc/HOL-Z/home.html.
10. D. Jackson and J. Wing. Lightweight formal methods. *IEEE Computer*, 1996.
11. J. Jacky. *The way of Z*. Cambridge University Press, 1997.
12. M. Kamel and S. Leue. Formalization and Validation of the General Inter-ORB Protocol (GIOP) using Promela and Spin. *Software Tools for Technology Transfer*, 2:394–409, 2000.
13. G. Karjoth. Authorization in CORBA Security. *Journal of Computer Security*, 8(2–3):89–108, 2000.

14. Kolyang, T. Santen, and B. Wolff. A structure preserving encoding of Z in Isabelle/HOL. In J. von Wright, J. Grundy, and J. Harrison, editors, *Proceedings of TPHOLs'96*, LNCS 1125, pages 283–298. Springer-Verlag, 1996.
15. D. Kreuz. Formal specification of CORBA services using Object-Z. In *Proceedings of 2nd IEEE International Conference on Formal Engineering Methods (ICFEM'98)*. IEEE Computer Society Press, 1998.
16. D. Kreuz. *Formale Semantik von Konnektoren*. PhD thesis, Technische Universität Hamburg-Harburg, Germany, 1999. In German.
17. U. Lang. CORBA Security. Master's thesis, Royal Holloway University of London, 1997.
18. OMG. http://www.omg.org.
19. OMG. *CORBA/IIOP 2.2*. OMG, 1998.
20. OMG Security Working Group. CORBAservices Security Service Specification v1.2. Technical report, Document: formal/98-12-17, OMG, 1998.
21. R. Orfali, D. Harkey, and J. Edwards. *Instant CORBA*. Addison-Wesley-Longman, 1998.
22. L. C. Paulson. *Isabelle: a generic theorem prover*. LNCS 828. Springer-Verlag, 1994.
23. D. Platt. *Understanding COM+*. Microsoft Press, 1999.
24. A. Pope. *The CORBA Reference Guide*. Addison-Wesley-Longman, 1998.
25. J.-C. Real. Object-Z specification of the CORBA repository service. Technical Report 351, Université Libre de Bruxelles, 1997.
26. F. Rittinger. Formale Analyse des CORBA Sicherheitsdienstes. Diplomarbeit, Institut für Informatik, Albert-Ludwigs-Universität Freiburg, Germany, 2000. In German.
27. A. Roscoe. *The Theory and Practice of Concurrency*. Prentice Hall International, 1997.
28. G. Seshadri. *Enterprise Java Computing*. Cambridge University Press, 1999.
29. G. Smith. *The Object-Z Specification Language*. Kluwer Academic Publishers, 1999.
30. J. Spivey. *The Z Notation: A Reference Manual*. Prentice Hall International, 2nd edition, 1992.
31. S. Stepney. *Object Orientation in Z*. Springer-Verlag, 1990.
32. H. Tej and B. Wolff. A corrected failure-divergence model for CSP in Isabelle/HOL. In J. Fitzgerald, C. Jones, and P. Lucas, editors, *Proceedings of FME'97*, LNCS 1313, pages 318–337. Springer-Verlag, 1997.
33. J. Woodcock and J. Davis. *Using Z*. Prentice Hall International, 1995.
34. ZETA. http://uebb.cs.tu-berlin.de/zeta/.

Type Synthesis in B and the Translation of B to PVS

Jean-Paul Bodeveix and Mamoun Filali

IRIT - Université Paul Sabatier
118, route de Narbonne, F-31062 Toulouse
{bodeveix,filali}@irit.fr
tel: 05 61 55 69 26 fax: 05 61 68 47

Abstract. In this paper, we study the design of a typed functional semantics for B. Our aim is to reuse the well known logical frameworks based on higher order logic, e.g., Isabelle, Coq and PVS as proving environments for B. We consider type synthesis for B and study a semantics and some of its composition mechanisms by translation to PVS.

Keywords: B, semantics, logical frameworks, type theory, PVS.

1   Introduction

In this paper, we study the design of a typed functional semantics for B [Abr96]. Our aim is to reuse the well known logical frameworks based on higher order logic, e.g., Isabelle [Pau92], Coq [BBC+97] and PVS [COR+95] as proving environments for B. Usually, the proposed semantics for B-like formal methods (Z and VDM) are set-based and even if they are encoded in higher order logic, they ignore the functional aspects: functions are encoded as specific relations. In fact, they build the semantics from a basic type and a membership relation [Gor94, Pau93]. Our approach consists in statically recognizing functional objects so that they can be translated to functions of the target language and their properties easily recognized for proof purposes.

The paper is organized as follows: Section 2, we present the basics of higher order logic and set theory. Section 3 reviews some embeddings of formal methods. Section 4 presents type synthesis for B. Section 5 describes a semantics for B by translation to PVS. Section 6 considers some B composition mechanisms and their translation.

2   Higher Order Logic and Set Theory

2.1   Higher Order Logic

Higher order logic is a specialization of typed lambda-calculus with some distinguished symbols to encode at least the boolean type and logical constants,

D. Bert et al. (Eds.): ZB 2002, LNCS 2272, pp. 350–369, 2002.
© Springer-Verlag Berlin Heidelberg 2002

connectives, and quantifiers. Models are restricted so that these symbols have their intended meaning. As a consequence of this definition many variants have been proposed, like HOL, Isabelle/HOL [Pau92] and PVS. So the typed lambda-calculus is the core of all these languages. It is defined by a type language $\tau ::= \tau \rightarrow \tau \mid \tau \times \tau \mid \mathcal{B}$ where $\mathcal{B}$ is a set of basic types, a term language over a set x of variables and a set c of typed constants $t ::= x \mid c^\tau \mid \lambda x^\tau t \mid (t\, t)$, together with typing rules defining the well-formedness conditions of the terms. One rule is associated to each term constructor:

$$\frac{}{\Gamma, x : \tau \vdash x : \tau} \text{ (variable)} \qquad \frac{}{\Gamma \vdash c^\tau : \tau} \text{ (constant)}$$

$$\frac{\Gamma, x : \tau_1 \vdash t : \tau_2}{\Gamma \vdash \lambda x^{\tau_1}.t : \tau_1 \rightarrow \tau_2} \text{ (abstraction)} \qquad \frac{\Gamma \vdash t_1 : \tau \rightarrow \tau' \quad \Gamma \vdash t_2 : \tau}{\Gamma \vdash (t_1\, t_2) : \tau'} \text{ (application)}$$

2.2   Set Theory

Set theory allows the use of higher order objects as sets and relations with no more than first order logic. Furthermore, operators such as set inclusion, generalized union or intersection avoid the systematic use of quantifiers. However, the unrestricted use of set constructors leads to paradox, such as Russel's set expression $\{x \mid x \notin x\}$ which cannot specify a set. Set construction must be controlled. This is the objective of Zermelo-Fraenkel axioms.

Set membership is considered as a primitive relation. Then the theory is defined by the following five axioms schemata expressed in first order logic over the atoms $x \in y$ and $x = y$:

$$A \subseteq B \equiv \forall x \in A.\; x \in B$$
$$A = B \Leftrightarrow A \subseteq B \wedge B \subseteq A \quad \text{(extensionality)}$$
$$A \in \bigcup(C) \Leftrightarrow \exists B \in C.\; A \in B \quad \text{(generalized union)}$$
$$A \in \wp(B) \Leftrightarrow A \subseteq B \quad \text{(powerset)}$$
$$A = \emptyset \vee \exists B \in A.\; \forall x \in B.\; x \notin A \quad \text{(foundation)}$$
$$b \in \{y \mid x \in A \wedge \Phi(x,y)\} \Leftrightarrow \exists x \in A.\; \Phi(x,b) \wedge \exists! z.\; \Phi(x,z) \quad \text{(replacement)}$$

Set theory is more difficult to mechanize than type theory: it is of lower level and does not incorporate information hiding or representation hiding mechanisms. For example, the couple $\langle a, b \rangle$ is defined by $\{\{a\}, \{a, b\}\}$ and $3 = \{0, 1, 2\}$.

2.3   Set Theory in B

A simplified set of axioms forms the basis of the B method. The main difference between Zermelo-Fraenkel axioms and those on which B relies on is that the foundation axiom, allowing to prove the non-existence of some sets, has been replaced by typing conditions. In fact, in B, the collection $\mathcal{S}$ of valid sets is inductively defined from a given collection $\mathcal{B}$ of basic sets and a collection $\mathcal{L}$ of labels used to build records [Abr96] :

$$\mathcal{S} ::= \mathcal{B} \mid \mathcal{S} \times \mathcal{S} \mid \mathcal{P}(\mathcal{S}) \mid \{l_1 : \mathcal{S}, \ldots, l_n : \mathcal{S}\}_{l_i \in \mathcal{L}} \mid \{\mathcal{V} \mid \mathbb{P}\}$$

where $\mathbb{P}$ denotes the collection of first order predicates on variables of $\mathcal{V}$, atomic formulas being restricted to equality and set membership. Then, the choice axiom is translated to a function `choice` which, applied to a non-empty set returns an unspecified element of the set.

A predicate used in the definition of a set must be well typed. The type language is a restriction of the set language which excludes sets defined by predicates. It is inductively defined by:

$$\mathcal{T} ::= \mathcal{T} \times \mathcal{T} \mid \{l_1 : \mathcal{T}, \ldots, l_n : \mathcal{T}\}_{l_i \in \mathcal{L}} \mid \mathcal{P}(\mathcal{T}) \mid \mathcal{B}$$

Note that contrarily to higher order logic, functional types are not introduced: a function is a particular relation, i.e. a subset of the Cartesian product of its domain and range satisfying some properties.

3   Embeddings of Formal Methods

In this section, we review some existing works on formal methods embeddings within logical frameworks based on higher order logic. It should be mentioned that we have not considered dedicated tools and environments [DT00a,DT00b].

3.1   Partiality in HOL

In most proof assistants, functions are total while in set theory, functions are particular relations which are first specialized in partial functions and then in total functions. Thus, in most systems, partial functions must be encoded or eliminated. Several solutions have been proposed [MS97] :

- functions returning an *arbitrary* value defined with a choice operator over the whole type (HOL [GM94]),
- under-specified functions with non exhaustive pattern matching (HOL),
- non-terminating functions in systems based on domain theory (HOLCF [Reg95]),
- pre-conditions associated with type correctness conditions generated to ensure the effective arguments belong to the domain (PVS [COR+95,MB97]),
- extra-arguments containing the proof of domain membership (Coq [DDG98]).

The closest solution to that of B is the one proposed by PVS: the domain of a function is characterized by a pre-condition predicate. In B, the call to an operation leads to the generation of a proof obligation expressing that the operation precondition is satisfied by the caller. However, no proof obligation is generated for partial functions. In fact, the correctness of the code is ensured by the existence of an implementation. In PVS, there is no distinction between operations and functions. A type correctness condition is systematically generated to ensure that the effective arguments of the caller function belong to the domain of the function. It follows that in PVS some potential errors can be detected sooner but the correctness of the design can be proved twice.

3.2   Z Embedding in Higher Order Logic

The Z / HOL approach [BG94] is qualified as a simple "shallow" embedding of the Z notation into the HOL logic. A fragment of the Z notation is supported through a HOL theory containing the definition of the set theoretic operators of Z and theorems about them. For instance, the well known Z notations SCHEMA, ::, P, ><, ↔ are introduced with the following generic definitions:

```
⊢ SCHEMA decs body = CONJL decs ∧ CONJL body
⊢ x :: s = x IN s
⊢ x ↦ y = x,y
⊢ dom R = {x | ∃y. (x,y) IN R}
⊢ P X = {Y | Y SUBSET X}
⊢ X >< Y = {(x,y) | x IN X ∧ y IN Y}
⊢ X ↔ Y = P (X >< Y)
⊢ X ⇸ Y = {f | f IN (X ↔ Y) ∧
      (∀x y1 y2. (x ↦ y1) IN f ∧ (x ↦ y2) IN f ⇒ (y1 = y2))}
⊢ f ^^ x = @y. (x,y) IN f
```

Moreover, a number of theorems about the Z operators are provided in order to simplify assertions resulting from Z schemas. For instance, we have:

```
⊢ dom (X UNION Y) = (dom X) UNION (dom Y)
⊢ dom {x ↦ y} = {x}
⊢ {x ↦ y} ^^ x = y
```

Given the declaration in HOL of all base sets used in a Z specification, the translation is straightforward. The verification of Z declarations relies on HOL type checking. Hence, functions are not recognized during the translation and are translated to HOL relations, as defined above. Along these lines, [KB95] have developed Z-in-Isabelle: a deep embedding of Z in the generic theorem proving tool Isabelle [Pau92]. They have not only given a HOL semantics to the Z notation but also formalized the deductive system of Z. Such a work is important since the user interacts with the proof tool through Z dedicated rules. He does not need to know the basic inference rules of the underlying meta-logic. Starting from the core inference rules of Z, they have derived a large set of sound (thanks to the underlying framework) new inference rules. [KSW96] have also implemented a deep embedding of Z on top of the logical framework Isabelle. Unlike the previous works, they provide support to use Z schemas as first class objects. Meta rules are provided for reasoning over schemas. Indeed, they provide a schema calculus. A salient feature of their approach is to avoid the systematic expansion of schemas. Thus, their proof terms remain small.

However, it should be said that, in general, the resulting expression cannot be decided automatically and as noted by the authors the user must have some knowledge of the underlying proof tool in order to give appropriate hints for simplifications or for proof completion. From our point of view, this is not the main drawback of such an approach. We rather believe that in general when translating to low level definitions and then making simplifications we obtain a formula that is hardly recognized by its author. It follows that it is hard to interact with the underlying proof tool.

In our proposal, type checking is performed at translation time and annotates functional expressions so that they can be translated to functions.

3.3   B Embedding in Higher Order Logic

The formalisation of B in Isabelle/HOL proposed in [Cha98] describes a semantic embedding of substitutions and machines. His goal is to produce a formally checked proof obligation generator. The proof obligations of some composing mechanisms like INCLUDES have been formalized and validated. However, he does neither consider B typing aspects nor B expression language.

The PBS approach [BFM99] is an attempt to mix the B method and the PVS language. In fact, it consists in taking from B the substitution language, the methodological aspects for software specification and development, i.e., invariants and refinements, and from PVS the term language, the type language, the modularity and proving aspects. The goal was to formalize the B substitution and refinement theory and to validate its associated proof rules.

4   Type Synthesis in a B Machine

4.1   The Extended Type Language

Besides ensuring the wellformedness of set expressions occurring in a B machine, type checking has here a second objective: it provides type annotations for variables. This knowledge is required for typed systems, such as PVS, that do not support type synthesis for lambda-variables. However, the type language proposed by Abrial [Abr96] is not sufficiently accurate to allow a good translation to a λ-calculus based formalism: it does not incorporate the function space operator $\rightarrow$. Consequently, we propose an extended type language that preserves information about partial and total functions, together with a translation to Abrial's type language for type checking purpose. It is performed by the function $t\uparrow$ which computes the base type of an extended type t. Let $\mathcal{ET}$ be the extended type language. It is inductively defined as follows:

$$\mathcal{ET} ::= \mathcal{B} \mid a..b \mid \{l_i : \mathcal{ET}\} \mid \mathcal{ET} \rightarrow \mathcal{ET} \mid \mathcal{ET} \nrightarrow \mathcal{ET} \mid \mathcal{ET} \times \mathcal{ET} \mid \mathcal{P}(\mathcal{ET}) \mid \text{seq}(\mathcal{ET})$$

Note that all B set expressions could be declared as part of the extended type language. However, we have only kept set expressions of which translation to the type language of higher order logic is efficiently supported.

4.2 Supertype Computation

The type language introduced in paragraph 4.1 is too precise for type checking the application of set operators which do not differentiate functions from sets. Thus, we define the function $t{\uparrow}$ which associates the maximal set containing the set associated to the type t. The resulting set is expressed in the B type language defined in section 2.3.

$id{\uparrow}= id$	$\mathbb{N}{\uparrow}= \mathbb{Z}$	$i_1..i_2{\uparrow}= \mathbb{Z}$
$E_1 \rightarrow E_2{\uparrow}= \mathcal{P}(E_1{\uparrow} \times E_2{\uparrow})$	$E_1 \nrightarrow E_2{\uparrow}= \mathcal{P}(E_1{\uparrow} \times E_2{\uparrow})$	$seq(t){\uparrow}= \mathcal{P}(\mathbb{Z} \times t{\uparrow})$
$\{l_i : t_i\}{\uparrow}= \{l_i : t_i{\uparrow}\}$	$E_1 \times E_2{\uparrow}= E_1{\uparrow} \times E_2{\uparrow}$	$\mathcal{P}(t){\uparrow}= \mathcal{P}(t{\uparrow})$

4.3 Synthesis of the Type of Variables

In the B book, type checking is specified by deduction rules over type equality constraints. Constraint solving is performed via the B prover and is not guaranteed to be fully automatic. Furthermore, the type synthesized for variables belongs to the minimal type language of section 2.3. The proposed type inference rules target the extended language and recognize partial or total functions.

In the following, we assume that the expression has been normalized: $\Rightarrow$ and $\forall$ are eliminated and negation is over atomic formulae. Then, within an existential formula $\exists x \,.(p_1 \wedge \ldots \wedge p_n)$, the type of the variable x is defined by the atomic predicate p_i containing its first occurrence, if it has one the following shapes:

- $x = y$: the type of x is the type of y,
- $x \subseteq e$: the type of x is the type of e,
- $x \in e$: the type of x is a set of $\mathcal{ET}$ containing e.

For example, the formula $\exists x. (y > 0 \wedge x \subseteq \mathbb{N})$ is transformed into $\exists x : \mathcal{P}(\mathbb{N}). (y > 0 \wedge x \subseteq \mathbb{N})$ so that all variables become typed at their introduction point. Note that the three identified typing patterns can be extended to synthesize the type of variables inside tuples, as it is allowed by B.

4.4 Type Synthesis Rules

Type checking and synthesis is performed through the backward application of a set of rules defining the transition relation $\triangleright$ over judgements $\Gamma \vdash \omega$ where:

- Γ is an environment containing variable declarations $x : t$. A variable may be either typed or untyped if $t =?$.
- ω is a wellformedness assertion.
  - If ω is a sole formula P, it asserts it is a predicate.
  - If ω is $e :?$, it asserts e is an expression of unknown type.
  - If ω is $e : t$, it asserts e is an expression of type t.

The transition relation partially defined in figure 1 specifies the transformation of the environment and of the expression during the type-checking process. Applied backward, rules can be read as follows:

- The existential quantifier rule adds the untyped variable x to the environment, checks the body P. As a result, P is transformed into P' and x becomes typed in the environment. Then the existential formula is build again, but with type annotations.
- Logical connector rules give a non commutative semantics to conjunction: the first occurrence of an untyped variable must type it. So, the first conjunct transforms P into P' and produces an intermediate environment Γ'' used to type check Q. However, disjunction and negation cannot type variables of Γ. For this purpose, the typed part of Γ, noted $\overline{\Gamma}$, is passed to sub-predicates.
- Variable annotation rules recognize the three mentioned shapes and update the environment. Note that the type checking of sub-expressions is performed within the context $\overline{\Gamma}$: the free variables of sub-expressions must be typed.
- The lambda-calculus rules section is an adaptation of simply typed lambda calculus rules to the context of B, taking into account partiality. These rules can be optimized to improve type information. First, within $\lambda x : t.\ (p\ |\ e)$, the predicate p can be implied by the type annotation t so that the function becomes total. Second, the type of f in the application $f(x)$ may be declared as functional, so that considering the super type becomes useless. The synthesized type for the result is then more precise.
- Even if all set operators were declared within the environment, specific rules would be needed to manage overloaded operators (for example, $\times$ for both set and integer products). Furthermore, the writing of specific rules avoids the introduction of generic types and the need for type unification. However, type comparison must take into account anonymous type variables introduced to type empty sets and sequences.

It is important to note that as our type language is more precise than the B type language, type comparisons use the base type $t\uparrow$ of a type annotation t. Thus, set operators can be applied to functions as well.

4.5   The Typing of B Expressions

The set of rules presented in the previous section must be extended to take into account B operators. For each binary operator $\circ$, we have the following rule:

$$\frac{\Gamma \vdash u_1 :? \rhd \Gamma \vdash u_1 : t_1 \quad \Gamma \vdash u_2 :? \rhd \Gamma \vdash u_2 : t_2 \quad \text{condition}}{\Gamma \vdash u_1 \circ u_2 :? \rhd \Gamma \vdash u_1 \circ u_2 : t_1\ \overline{\circ}\ t_2}$$

where $\overline{\circ}$ is the abstraction over types of $\circ$ and `condition` checks that $\overline{\circ}$ is defined over the types t_1 and t_2. When read backward, it says that typing the application of the operator o to u_1 and u_2 requires to compute the type t_i of each u_i. Then, if o applies to arguments of type t_1 and t_2 (which is controlled by the condition), the result is typed $t_1\overline{\circ}t_2$. For instance, we have

$$\frac{\Gamma \vdash u_1 :? \rhd \Gamma \vdash u_1 : t_1 \quad \Gamma \vdash u_2 :? \rhd \Gamma \vdash u_2 : t_2 \quad t_1\uparrow = \mathbb{Z} \wedge t_2\uparrow = \mathbb{Z}}{\Gamma \vdash u_1 + u_2 :? \rhd \Gamma \vdash u_1 + u_2 : \mathbb{Z}}$$

Existential quantifier

$$\frac{\Gamma, x :? \vdash P \rhd \Gamma', x : t \vdash P'}{\Gamma \vdash \exists x.\ P \rhd \Gamma' \vdash \exists x : t.\ P'}$$

Logical connectors

$$\frac{\Gamma \vdash P \rhd \Gamma'' \vdash P'\quad \Gamma'' \vdash Q \rhd \Gamma' \vdash Q'}{\Gamma \vdash P \wedge Q \rhd \Gamma' \vdash P' \wedge Q'} \qquad \frac{\overline{\Gamma} \vdash P \rhd \overline{\Gamma} \vdash P'\quad \overline{\Gamma} \vdash Q \rhd \overline{\Gamma} \vdash Q'}{\Gamma \vdash P \vee Q \rhd \Gamma \vdash P' \vee Q'}$$

$$\frac{\overline{\Gamma} \vdash P \rhd \overline{\Gamma} \vdash P'}{\Gamma \vdash \neg P \rhd \Gamma \vdash \neg P'}$$

Annotation of variables

$$\frac{\overline{\Gamma} \vdash e :? \rhd \overline{\Gamma} \vdash e' : u\quad \vdash u \subseteq \mathcal{P}(t)}{\Gamma, x :? \vdash x \in e \rhd \Gamma, x : t \vdash x \in e} \qquad \frac{\overline{\Gamma} \vdash e :? \rhd \overline{\Gamma} \vdash e' : t}{\Gamma, x :? \vdash x = e \rhd \Gamma, x : t \vdash x = e'}$$

$$\frac{\overline{\Gamma} \vdash e :? \rhd \overline{\Gamma} \vdash e' : u\quad \vdash u \subseteq \mathcal{P}(t)}{\Gamma, x :? \vdash x \subseteq e \rhd \Gamma, x : t \vdash x \subseteq e'}$$

Lambda calculus rules

$$\frac{}{\Gamma, x : t \vdash x :? \rhd \Gamma \vdash x : t}\ \text{(variable)}$$

$$\frac{\overline{\Gamma}, x :? \vdash p \rhd \overline{\Gamma}, x : t_1 \vdash p'\quad \overline{\Gamma}, x : t_1 \vdash e :? \rhd \overline{\Gamma}, x : t_1 \vdash e' : t_2}{\Gamma \vdash \lambda x.\ (p \mid e) :? \rhd \Gamma \vdash \lambda x : t_1.(p' \mid e') : t_1 \nrightarrow t_2}\ \text{(abstraction)}$$

$$\frac{\overline{\Gamma} \vdash f :? \rhd \overline{\Gamma} \vdash f : t\quad \overline{\Gamma} \vdash x :? \rhd \overline{\Gamma} \vdash x : u\quad \vdash t\!\uparrow = \mathcal{P}(t_1 \times t_2)\quad \vdash u\!\uparrow = t_1}{\Gamma \vdash f(x) :? \rhd \Gamma \vdash f(x) : t_2}\ \text{(application)}$$

Some set operators

$$\frac{\overline{\Gamma} \vdash x :? \rhd \overline{\Gamma} \vdash x : t_1\quad \overline{\Gamma} \vdash e :? \rhd \overline{\Gamma} \vdash e : \mathcal{P}(t_2)\quad \vdash t_1\!\uparrow = t_2\!\uparrow}{\Gamma \vdash x \in e \rhd \Gamma \vdash x' \in e'}$$

$$\frac{\overline{\Gamma} \vdash e_1 :? \rhd \overline{\Gamma} \vdash e_1' : t_1\quad \overline{\Gamma} \vdash e_2 :? \rhd \overline{\Gamma} \vdash e_2' : t_2\quad \vdash t_1\!\uparrow = t_2\!\uparrow}{\Gamma \vdash e_1 \subseteq e_2 \rhd \Gamma \vdash e_1' \subseteq e_2'}$$

$$\frac{}{\Gamma \vdash \emptyset :? \rhd \Gamma \vdash \emptyset : \mathcal{P}(\_)}$$

Fig. 1. Type synthesis rules

Such rules can be strengthened to give a better upper approximation of the resulting type. With respect to the preceding rule, we can state the strengthenings illustrated by the following abstract interpretation [CC77] of $+$:

$\mp$	$\mathbb{Z}$	$\mathbb{N}$	$\mathbb{N}_1$
$\mathbb{Z}$	$\mathbb{Z}$	$\mathbb{Z}$	$\mathbb{Z}$
$\mathbb{N}$	$\mathbb{Z}$	$\mathbb{N}$	$\mathbb{N}_1$
$\mathbb{N}_1$	$\mathbb{Z}$	$\mathbb{N}_1$	$\mathbb{N}_1$

Of course, we can make better approximations of $+$ by considering a lattice of intervals [Cou00]. These techniques apply to the other domains used by B.

5   Semantics of a B Machine

In this section, we propose a semantics of B machines defined by a translation to the PVS language. The latter has been chosen mainly for two reasons:

- Its interaction with the user is close to that of B where typing and proving are mixed. They both generate proof obligations that must be discharged.
- The units of modularity of B and PVS are close. It is easy to translate a B machine to a PVS theory.

We define the semantics of a B machine through the following semantic functions translating B constructs to PVS:

- $[\![\_]\!]_t$ translates a B set expression to a PVS type.
- $[\![\_]\!]_e$ translates a B expression to a PVS expression.
- $[\![\_]\!]_p$ translates a B predicate to a PVS expression.
- $[\![\_]\!]_s$ translates a B substitution to a PVS expression.
- $[\![\_]\!]_o$ translates a B operation to a PVS function.

Note that this paper does not address the comparison between the proposed semantics and the one defined in the B book. Such a comparison should relate the checks as defined in the B book and those performed by PVS on the translated machine. It relies on the precise semantics of the PVS type system and its composition mechanisms. It follows that such a comparison deserves by itself a separate study and is beyond the scope of this paper.

5.1   The Target Language

First, we present two aspects of the PVS language: the subset of the type language used to translate B expressions types, and the basic constructs of the expression language.

Mapping of the extended type language. The translation of the B extended type language (4.1) to PVS is introduced by the function $[\![\_]\!]_t$ which is inductively defined as follows:

$$[\![A]\!]_t = A_{\text{pvs}}$$
$$[\![a..b]\!]_t = \{i : int \mid [\![a]\!]_e \leq i \ \& \ i \leq [\![b]\!]_e\}$$
$$[\![\{l_i : T_i\}]\!]_t = [\#l_i : [\![T_i]\!]_t \#]$$
$$[\![T_1 \rightarrow T_2]\!]_t = [[\![T_1]\!]_t \rightarrow [\![T_2]\!]_t]$$
$$[\![T_1 \nrightarrow T_2]\!]_t = [\#\text{dom} : set[[\![T_1]\!]_t], \text{fun} : [(\text{dom}) \rightarrow [\![T_2]\!]_t]\#]$$
$$[\![T_1 \times T_2]\!]_t = [[\![T_1]\!]_t, [\![T_2]\!]_t]$$
$$[\![\mathcal{P}(T)]\!]_t = set[[\![T]\!]_t]$$
$$[\![seq(T)]\!]_t = \text{list}[[\![T]\!]_t]$$

The PVS expression language. The core of the PVS expression language is described by the following grammar. We have mainly the usual expressions of lambda calculus: abstraction, application and usual abstract data types constructors as tuples. An originality of PVS is the declaration of functions over subtypes: $\lambda(x : T \mid E) : E$. Such a function can only be applied to arguments of type T and satisfying the predicate E. This condition being undecidable in general, a proof obligation is generated by PVS.

$$E::=\lambda(x : T \mid E) : E \mid E(E) \mid (E, \ldots , E) \mid \ldots$$

5.2   B Machine Translation Schemata

The semantics of a B machine is defined by its translation to a PVS theory according to figure 2. This translation is performed by a tool which synthesizes all the type information required by the target language. Moreover, we suppose that the source B machine has been type-checked. In this paragraph, an isolated machine is considered and machine composition operators are ignored:

- A parameterized machine is translated to a parameterized theory where constraints become assumptions over the theory parameters. As in B, they must be verified when instantiating the theory.
- An abstract set declaration is translated to a finite set of integers. An enumerated set declaration is translated to a PVS enumeration and a set containing all the elements of the type. Note that type declarations have been added in order to declare the PVS sets representing B sets. These types will be used to declare objects of these sets.
- Constants are represented by PVS constants. Their type is synthesized from the PROPERTIES section of B.
- Properties are translated to a PVS axiom. It should be noted that the type declaration of constants allows to simplify the axiom formula.
- Variables are grouped into a record which defines the state type. As for constants, the type of each field is synthesized from the INVARIANT clause.
- Assertions being consequences of the invariant, they are translated to a PVS theorem stating that each state satisfying the invariant also satisfies the assertion.
- The initialization is translated to a function returning a subset of the state space.
- Each operation is translated to a PVS function. Its parameters are the state space and the parameters of the B operation. It returns a set of tuples containing the state space and the results of the B operation. The precondition of the B operation is translated to a subtyping constraint over the parameters of the PVS function. The invariant preservation is translated to a subtyping judgement. It generates a proof obligation corresponding to the one required by B for invariant preservation.

Remark: The VALUES clause of implementation machines allows the instantiation of deferred abstract sets. Such a mechanism is actually not supported in PVS. We have solved this problem by representing all abstract sets as finite sets of integers. Consequently, PVS type-checking will be less accurate than B with respect to the use of abstract sets. Furthermore, the VALUES clause is translated to a PVS axiom.

MACHINE $M(P_i)$	$M[P_i: \ldots]$: THEORY
	BEGIN
CONSTRAINTS C_i	ASSUMING C_i: ASSUMPTION ENDASSUMING
SETS S_i	S_i: finite_set[int]
	S_iType: TYPE = (S_i)
CONSTANTS K_i	K_i: ... % type synthesis
PROPERTIES A_i	properties: AXIOM A_i
VARIABLES V_i	state: TYPE = [# V_i: ... #] % type synthesis
INVARIANT I	invariant(st: state): bool = I
ASSERTION T	assertion: THEOREM $\forall$st:(invariant) T
INITIALISATION S	initialisation: subset[invariant] = S
OPERATIONS	
$r_i \leftarrow op_i(a_i) =$	op_i(st:(invariant), a_i:... | P_i)
PRE $P_i(a_i, V_i)$	:set[state, ...] =
THEN O_i	O_i
END	JUDGEMENT op_i HAS_TYPE ...
...	...
END	END

Fig. 2. Translation of a B machine

5.3   B Machines and PVS Theories

In this section, we condider the translation of B composition clauses to PVS theory importations. In the following, we suppose that all the B visibility checks have been performed before translation. We rely on the PVS system only for verifying the constraints of the component machines. Let $I_k(\ldots), U_k(p_k^U)$ and $S_k(p_k^S)$ be parameterized B machines and M the following machine:

```
MACHINE M(p)
INCLUDES I_k(p_k^I), i_k.I_k(q_k^I)
USES U_k, u_k.U_k
SEES S_k, s_k.S_k
...
END
```

The following rules are applied for the translation to PVS:

– Formal parameters of used machines become formal parameters of the using machine and actual parameters of included machines must be passed to the machines that use them. Thus, the machine M is translated to the following PVS theory, parameterized by those of M and of the used and seen machines:

```
M[p:..., p_k^U :..., u_k-q_k^U]: THEORY
BEGIN
IMPORTING ...
...
END
```

This theory imports the theories associated to the component machines.
– In B, objects of unprefixed component machines can be referenced implicitly. The behaviour is the same in PVS after importing the theory associated to the component. Potential naming conflicts are the same in B and PVS. So, if no conflict occurs in B, no conflict will occur in PVS.

```
IMPORTING   I_k[p_k^I], U_k[p_k^U], S_k[p_k^S]
```

– B allows renaming of component machines that can be inserted several times, thus avoiding naming conflicts. Then, reference to the component objects must be prefixed by the renaming identifier. The same behaviour is allowed in PVS if the identifier is that of a theory.

```
i_k: THEORY = I_k[q_k^I]    u_k: THEORY = U_k[q_k^U]    s_k: THEORY = S_k[q_k^S]
```

5.4   Semantics of B Substitutions

The B method gives two isomorphic semantics of a substitution:

– a conjunctive predicate transformer mapping a predicate over the *after* state space to a predicate over the *before* state space: the image predicate identifies states from where an execution will terminate in a state satisfying the after predicate.
– a couple (precondition p, before-after relation r) which satisfies the constraint $\forall x: \neg p(x) \Rightarrow \forall y: r(x, y)$.

The type $(t_1 \times t_2) \to$ bool of the before-after relation r is isomorphic to the type of functions from t_1 to sets of elements of t_2. This reading is closest to a functional view of substitutions and makes easier the writing of compositions. Furthermore, we encode the precondition p as a constraint over the argument of the function. As in B, the caller must ensure that the precondition is satisfied. So, PVS, as well as B, will produce a proof obligation. The following equation defines our translation of B substitutions expressed as a couple or as a predicate transformer to a PVS partial function:

$$[\![(p, r)]\!]_s = \lambda(x : t_1 \mid p) : \{y : t_2 \mid r(x, y)\}$$
$$[\![S]\!]_s = \lambda(x : t_1 \mid \text{pre}(S)) : \{y \mid \text{rel}(S)(x, y)\}$$

The translation of a B substitution is specified by the function $[\![\_]\!]_s^T$ of type $T \to \text{set}[T]$ where T is the type of the state space. Given the translation $[\![\_]\!]_p$ of B predicates and the translation $[\![\_]\!]_e$ of expressions, it is inductively defined as follows[1]:

$$[\![P \mid S]\!]_s^T = \lambda(x : T \mid [\![P]\!]_p^T) : \; [\![S]\!]_s^T(x)$$

$$[\![P \Rightarrow S]\!]_s^T = \lambda(x : T) : \; \textbf{if } [\![P]\!]_p^T(x) \textbf{ then } [\![S]\!]_s^T(x) \textbf{ else } \emptyset$$

$$[\![@y^U.S]\!]_s^T = \lambda(x : T) : \bigcup_{y \in U} [\![S]\!]_s^{T \times U}(x, y)$$

$$[\![S_1 [] S_2]\!]_s^T = \lambda(x : T) : \; [\![S_1]\!]_s^T(x) \cup [\![S_2]\!]_s^T(x)$$

$$[\![S_1 \; ; \; S_2]\!]_s^T = \lambda(x : T) : \bigcup_{y \in [\![S_1]\!]_s^T(x)} [\![S_2]\!]_s^T(y)$$

$$[\![S_1 \parallel S_2]\!]_x^T = \lambda(x : T) : \; \{ \, y : T \mid \quad \exists(y_1, y_2 : T) : y_1 \in [\![S_1]\!]_s^T(x)$$
$$\wedge \; y_2 \in [\![S_2]\!]_s^T(x)$$
$$\wedge \; y = \text{merge}(y_1, \text{sup}(S_1), y_2, \text{sup}(S_2), x) \, \}$$

$$[\![x_1, \ldots, x_n := e_1, \ldots, e_n]\!]_s^T = \lambda(x : T) : \{x \textbf{ with } [\ldots, x_i := [\![e_i]\!]_e, \ldots]\}$$

where:

the function sup calculates the support of a substitution S, i.e., the variables that are syntactically modified by S,

and the function $\text{merge}(y_1, \text{sup}(S_1), y_2, \text{sup}(S_2), x)$ returns the state x where the fields in S_1 (resp. S_2) have been updated by those of y_1 (resp. y_2).

5.5   Semantics of Operations

We associate to the following operation declaration:

$$v_1, \ldots, v_n \longleftarrow op(p_1, \ldots, p_k) \triangleq P \mid S$$

the PVS function $[\![op]\!]_o$ defined by:

$$[\![op]\!]_o(st : (\text{invariant}), p_i : T_i \ldots p_k : T_k \mid [\![P]\!]_p) : \text{set}[\text{machine\_state}, R_1, \ldots, R_n]$$
$$= [\![S]\!]_s$$

where the types T_i of parameters are synthesized from the precondition P and the types R_j of results are synthesized from the substitution S.

The preservation of the invariant by an operation is specified by a PVS subtyping judgement.

5.6   Semantics of B Expressions

The semantics of B expressions in higher order logic is defined by the function $[\![\_]\!]_e$, mapping B expressions to PVS expressions. The definition of this function

---

[1] We use the PVS notation r **with** [(f) := e] which denotes the record r where the field f has been updated by the expression e.

raises two problems connected to the fact that set theory is untyped. First, the proposed type language allows the identification of functional relations and thus a specialized translation of such relations. Then, relational operators must be either evaluated at translation time or overloaded in an appropriate PVS theory. Type conversions must be provided to reuse general operators when no specialization exists. For example, composition can be defined efficiently at a functional level while it needs a quantifier at the relational level. However, the reverse of a function can reuse the definition for relations. Second, the term language of B defines operators that are usually constructors of the type language: Cartesian product and function space operators must also be defined in PVS as set operators.

Functional overloading of set operators. The static analysis performed by type synthesis identifies some partial or total functions. These functions are appropriately encoded in PVS, which implies the overloading of set operators on functions. These functions are unfold at translation time when the resulting expression is more compact and readable.

Partial functions are encoded by a dependent record containing a set (the function domain) and a total function from that set. Given a coercion to functions, partial functions may be applied to arguments using the same syntax as for usual (total) functions.

The coercion to binary relations will be implicitly applied if a relational function is called with a partial function as argument.

Total functions are encoded by usual PVS functions. Given the total functions $f : A \to B$, $g : B \to C$ and $h : D \to E$, some set operators can be specialized or statically evaluated:

$$\mathrm{dom}(f) = A$$
$$\mathrm{ran}(f) = B$$
$$f \; ; \; g = g \circ f$$
$$f \lhdplus \{x_i \mapsto t_i\} = f \text{ with } [x_i := t_i]$$
$$f \lhdplus h = \lambda(x : A \cup D) : \textbf{if } x \in D \textbf{ then } h(x) \textbf{ else } f(x)$$

Total functions are special cases of partial functions. Thus, a conversion is defined to coerce a total function to a partial one and inherit some functions defined in the **partial** theory. Other functions may be specialized. For example, overloading a total function by a (partial or total) function over the same domain results in a total function.

Synthesis of functional types. Along with these overloading theories, type checking integrates rules dedicated to the synthesis of functional types. For instance, we have the following rule:

$$\frac{\overline{\Gamma} \vdash A :? \rhd \overline{\Gamma} \vdash A' : t_A \quad t_A\!\uparrow = \mathcal{P}(t_1) \quad \overline{\Gamma} \vdash B :? \rhd \overline{\Gamma} \vdash B' : t_B \quad t_B\!\uparrow = \mathcal{P}(t_2)}{\Gamma \vdash A \to B :? \rhd \Gamma \vdash A' \to B' : \mathcal{P}(t_1 \nrightarrow t_2)}$$

Remark:

- If t_A is syntactically equal to some $\mathcal{P}(t_1)$, it is useless to compute its base type and thus we get a more precise type: t_1 is taken as the type of elements of A.
- If $A = t_1$ the functional type becomes total.

Definition of set operators. Cartesian products and function space operators which are usually defined by the underlying type language must also be defined in the term language to support the translation of B expressions, e.g., $f \in A \times B$. As an example, the following theory introduces the PVS definitions for Cartesian product, partial and total functions.

```
setops[T1,T2:TYPE]:THEORY
BEGIN
  *(S1:set[T1],S2:set[T2]): set[[T1,T2]] =
  { x1x2: [T1,T2] | S1(x1x2'1) & S2(x1x2'2)}

  partial_functions(S1:set[T1],S2:set[T2]): set[set[[T1,T2]]] =
  {R : set[[T1,T2]] | FORALL (x:T1,y1,y2:T2):
     R(x,y1) & R(x,y2) => y1 = y2}
  ...
END setops
```

These definitions are used to translate B expressions such as $f \in A \to B$. However, the typing information attached to f can be used to simplify this expression. It reduces to **true** when no loss of information occurs between the set language and the type language for the representation of the domain. For instance, the B expression $\forall f.(f \in \mathbb{N} \to \mathbb{N} \Rightarrow ...)$ is translated to:

```
FORALL (f:[nat -> nat]): ...
```

where the membership predicate is fully taken into account by the type information and becomes useless.

6   Composition Mechanisms

In this section, we consider the definition of machine states and the semantics of operation calls. We will be mainly concerned by the sharing aspects of the B composition clauses: USES and SEES. It should be noted that we take as axioms the so called "architectural conditions" stated in the B book and completed by [PR98,Rou99,DBMM00]. Actually, such axioms concerning proof modularity assert that proof obligations can be done once for all. Consequently, when developing a refinement of an existing machine, we do not reconsider the proof of machines relying on the specification of the refined machine.

6.1   Building the State Space

As in [Age96], we represent the state of a machine by a record. This record contains fields for the machine variables and for the component machines. However, in our case, such a record is constrained by the machine invariant. Thus, it exactly encodes the state space of a B machine.

For building the state space, we first build an unconstrained record as follows:

- machine variables are represented by the fields l_i of the record.
- if the component includes, sees, or imports other machines, we have a field f_{m_i} for each such machine. They are typed by the invariant associated to the component machine.
- if the component is a refinement or an implementation, we add a field s for the refined component. It is used to express the refinement relation within the invariant. Then, invariant preservation expresses operation refinement.

The unconstrained record associated to a machine has the following shape:

$$\text{machine\_state} : \textbf{TYPE} \ = [\# \ f_{m_0} : ([\![m_0.\text{invariant}]\!]_e), \ldots, f_{m_m} : ([\![m_m.\text{invariant}]\!]_e),$$
$$l_0 : T_0, \ldots, l_l : T_l, \%\text{local variables}$$
$$s : I_s \ \%\text{refined machine state}$$
$$\#]$$

In a second step, we translate the machine invariant I to a predicate over the unconstrained record. This predicate is used to define the machine state as a subtype of the record.

$$\text{invariant:pred[machine\_state]} = [\![I]\!]_p$$
$$\text{state} : \textbf{TYPE} = (\text{invariant})$$

Remark: In B the assertion clause allows us to state a predicate A that is implied by the invariant. Such a property is proved once and can be reused for proving invariants or preconditions. In the PVS framework, it is easy to state such a clause first by a theorem asserting the implication and by a tighter subtyping of the state space as follows:

$$\text{ASSERTION\_CLAUSE} : \textbf{THEOREM}$$
$$\text{FORALL} \ (st : (\text{invariant})) : [\![\text{assertion}]\!]_p(st)$$
$$\text{typed\_machine\_state} : \textbf{TYPE} = ([\![I \wedge A]\!]_p)$$

6.2   Semantics of Operation Calls

In this section, we present the translation of an operation call. We suppose that the called operation belongs to some component machine m. Such an operation is called with the projection of the state over the component m. The result is used to update the state of the composed machine.

Calling without sharing. The semantics of a call to the operation op of machine m is the following:

$$[\![v_1, \ldots, v_n \longleftarrow m.op(p_1, \ldots, p_k)]\!] =$$
$$\lambda(st) : \{st' \mid \exists m', r_1, \ldots, r_n : (m', r_1, \ldots, r_n) \in [\![m.op]\!](m(st), p_1, \ldots, p_k)$$
$$\wedge \; st' = st \text{ with } [\; f_m := m', v_1 := r_1, \ldots, v_n := r_n]\}$$

Specification stage sharing. As illustrated by figure 3, there exists a compound machine MO which includes all the machines M1, ... Mi using the shared machine Ms. Moreover, Ms must be directly included by MO also.

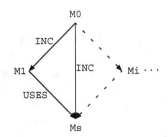

Fig. 3. Specification stage sharing

With respect to the state representation in PVS, the state of the shared machine Ms is in fact duplicated in all the included machines M1, ... Mn and thus MO contains $n + 1$ replica of Ms. However, coherence is ensured by working only on the version directly included in MO. More precisely, when an operation of some Mi is called, the state space provided for the call is the state of Mi updated by the state of Ms. The semantics of a call to the operation op of machine m is the following:

$$[\![v_1, \ldots, v_n \longleftarrow m_i.op(p_1, \ldots, p_k)]\!]_o =$$
$$\lambda(st) : \{st' \mid \exists m', r_1, \ldots, r_n :$$
$$(m', r_1, \ldots, r_n) \in [\![m.op]\!](m(st \text{ with } [\; f_{m_s} := m_s(st)]), p_1, \ldots, p_k)$$
$$\wedge \; st' = st \text{ with } [\; f_{m_i} := m', v_1 := r_1, \ldots, v_n := r_n]\}$$

Implementation stage sharing. Unlike with the USE clause where the top machine $M0$ was the unique writer of the shared machine Ms, with the SEES clause, we only know that a seen machine should be imported only once somewhere in the development. Moreover, when the top machine $M0$ imports a machine Mi that sees $M2$, $M0$ does not necessarily see $M2$.

We implement this kind of sharing at the level of the top machine $M0$. To each shared component, i.e., seen component by one of $M0$'s components, we dedicate a field in the record associated to $M0$. When calling an operation on one of M's components that sees or imports a shared component, as in the previous section, we provide to the operation a record updated by the record associated to the shared component. However, since $M0$'s components do not access Ms necessarily in a read mode, after the call, we update the records associated to the shared components. To such a call, we give the following semantics:

$$
\begin{aligned}
&[\![v_1, \ldots, v_n \longleftarrow m_i.op(p_1, \ldots, p_k)]\!]_o = \\
&\lambda(st) : \{ st' \mid \exists m', r_1, \ldots, r_n : \\
&\qquad (m', r_1, \ldots, r_n) \in [\![m.op]\!](m(st \textbf{ with } [\ f_{m_s} := m_s(st)]), p_1, \ldots, p_k) \\
&\qquad \wedge st' = st \textbf{ with } [\ f_{m_i} := m', v_1 := r_1, \ldots, v_n := r_n, \\
&\qquad\qquad f_{m_s} := m_s(m')]\}
\end{aligned}
$$

7   Conclusion

In this paper, we have given a functional semantics to B machines. Our main goal was to map directly B functional constructs to the corresponding ones of the underlying logic, namely the PVS one. We have used the subtyping feature of PVS to translate preconditions. With respect to operations (or Z schemas), most of the earlier works have adopted an expansion approach. By giving a functional semantics to operations and operation calls, we hope to avoid that. We believe that such an approach is important first to keep proof terms small and close to their original expression and second in synthesizing(guessing) the needed operation properties. As in B, we have proof obligations to ensure that the operations are called correctly. We hope that through this translation the interaction with the underlying proof tool will be natural and that it will be easy to reuse the PVS proof environment, e.g., cooperation of powerful decision procedures, proof subsumption[2]. Lastly, we have sketched a functional semantics for the sharing clauses of B.

We are currently implementing the B to PVS translator. We have already implemented the typing step. The latter has been built on top of the B platform developed by [Mar01].

This work can be extended with respect to several aspects:

– This study has highlighted a target functional kernel close to PVS. We have shown how PVS could be used as a proof engine for B. In order to validate

---

[2] PVS recognizes that some proof obligations are the same and does not ask to prove them twice.

the proposed translations, such a kernel must be defined with a well defined semantics.
- As suggested in 4.5, type synthesis can be improved by using abstract interpretation techniques. The increased typing accuracy can help in simplifying proof obligations.
- dedicated proof tactics should be developed in order to support specific B concepts or operators and their properties.

References

[Abr96]   J.R. Abrial. *The B-Book Assigning programs to meanings*. Cambridge University Press, 1996.

[Age96]   Agerholm, S. Translating specifications in VDM-SL to PVS. In J. Grundy J. Von Wright and J. Harisson, editors, *Proceeding of the 9th International Conference On Theorem Proving in Higher Order Logics*, volume 1125 of *Lecture Notes in Computer Science*, pages 1–16, Turku, Finland, 1996. Springer-Verlag.

[BBC⁺97]   B. Barras, S. Boutin, C. Cornes, J. Courant, J.C. Filliatre, E. Giménez, H. Herbelin, G. Huet, C. Muñoz, C. Murthy, C. Parent, C. Paulin, A. Saïbi, and B. Werner. The Coq Proof Assistant Reference Manual – Version V6.1. Technical Report 0203, INRIA, August 1997. http://coq.inria.fr.

[BFM99]   J.-P. Bodeveix, M. Filali, and C. Munoz. A formalization of the B method in Coq and PVS. In *FM'99 – B Users Group Meeting – Applying B in an industrial context : Tools, Lessons and Techniques*, pages 32–48, 1999.

[BG94]   J. Bowen and M. Gordon. Z and HOL. In *8th Z User Meeting (ZUM'94)*. BCS FACS, June 1994.

[CC77]   P. Cousot and R. Cousot. Abstract interpretation: a unified lattice model for static analysis of programs by construction or approximation of fixpoints. In *Conference Record of the Fourth Annual ACM SIGPLAN-SIGACT Symposium on Principles of Programming Languages*, pages 238–252, Los Angeles, California, 1977. ACM Press, New York, NY.

[Cha98]   P. Chartier. Formalisation of B in Isabelle/HOL. In *Proc. Second B International Conference, Montpellier, France*, 1998.

[COR⁺95]   S. Crow, S. Owre, J. Rushby, N. Shankar, and S. Mandayam. A Tutorial Introduction to PVS. In *Workshop on Industrial-Strength Formal Specification Techniques, Boca Raton*, http://www.csl.sri.com/pvs, April 1995.

[Cou00]   P. Cousot. Abstract interpretation: Achievements and perspectives. In *Proceedings of the SSGRR 2000 Computer & eBusiness International Conference*, Compact disk paper 224 and electronic proceedings http://www.ssgrr.it/en/ssgrr2000/proceedings.htm, L'Aquila, Italy, July 31 – August 6 2000. Scuola Superiore G. Reiss Romoli.

[DBMM00]   T. Dimitrakos, J. Bicarregui, B. Matthews, and Maibaum. Compositional structuring in the B-method: A logical viewpoint of the static context. In J.-P. Bowen, S. Dunne, A. Galloway, and S. King, editors, *ZB'2000: Formal Specification and Development in Z and B*, volume 1878 of *Lecture Notes in Computer Science*. Springer-Verlag, September 2000.

[DDG98]   C. Dubois and V. Donzeau-Gouge. A step towards the mechanization of partial functions: domains as inductive predicates. In *In Workshop on mechanization of partial functions, CADE 15*, July 1998.

[DT00a]   D. Duffy and I. Toyn. Reasoning inductively about Z specifications via unification. In *Proceedings International Conference of Z and B Users, ZB2000*, volume 1878 of *Lecture Notes in Computer Science*. Springer-Verlag, May 2000.

[DT00b]   D. Duffy and I. Toyn. Typechecking Z. In *Proceedings International Conference of Z and B Users, ZB2000*, volume 1878 of *Lecture Notes in Computer Science*. Springer-Verlag, May 2000.

[GM94]   M.J.C. Gordon and T.F. Melham. *Introduction to HOL*. http://www.cl.cam.ac.uk/Research/HVG/HOL. Cambridge University Press, 1994.

[Gor94]   M.J.C. Gordon. Merging hol with set theory: preliminary experiments. Technical Report 353, University of Cambridge Computer Laboratory, 1994.

[KB95]   I. Kraan and P. Baumann. Implementing Z in Isabelle. In Bowen and Hinchey, editors, *ZUM'95: The Z formal specification notation*, number 967 in Lecture Notes in Computer Science, pages 355–373. Springer-Verlag, 1995.

[KSW96]   Kolyang, T. Santen, and B. Wolff. A structure preserving encoding of Z in Isabelle/HOL. In J. von Wright, J. Grundy, and J. Harrison, editors, *Theorem Proving in Higher Order Logics — 9th International Conference*, volume 1125 of *Lecture Notes in Computer Science*, pages 283–298. Springer Verlag, 1996.

[Mar01]   G Mariano. The Bcaml project. Technical report, INRETS, http://www3.inrets.fr/Public/ESTAS/Mariano.Georges, 2001.

[MB97]   S. Maharaj and J. Bicarregui. On verification of VDM specification and refinement with PVS. In *proceedings of the 12th IEEE International Conference in Automated Software Engineering*, pages 280–289, 1997.

[MS97]   O. Müller and K. Slind. Treating partiality in a logic of total functions. *The Computer Journal*, 40(10):1–12, 1997.

[Pau92]   L.C. Paulson. Introduction to Isabelle. Technical report, Computer laboratory, university of Cambrige, 1992.

[Pau93]   Lawrence C. Paulson. Set theory for verification: I. From foundations to functions. *Journal of Automated Reasoning*, 11(3):353–389, 1993.

[PR98]   M.-L. Potet and Y. Rouzaud. Composition and refinement in the B-method. volume 1393 of *Lecture Notes in Computer Science*, pages 46–65. Springer-Verlag, 1998.

[Reg95]   F. Regensburger. HOLCF: Higher Order Logic of Computable Functions. In E. T. Schubert, P. J. Windley, and J. Alves-Foss, editors, *Proceedings of the 8th International Workshop on Higher Order Logic Theorem Proving and Its Applications*, number 971 in Lecture Notes in Computer Science, Aspen Grove, Utah, 1995. Springer-Verlag.

[Rou99]   Y. Rouzaud. Interpreting the B-method in the refinement calculus. In J. Wing, J. Woodcock, and J. Davies, editors, *FM'99*, volume 1708 of *Lecture Notes in Computer Science*, pages 411–430. Springer-Verlag, Sep 1999.

"Higher-Order" Mathematics in B

Jean-Raymond Abrial[1], Dominique Cansell[2], and Guy Laffitte[3]

[1] Consultant, Marseille France jr@abrial.org
[2] LORIA, Metz France Dominique.Cansell@loria.fr
[3] INSEE, Nantes France Guy.Laffitte@insee.fr

Abstract. In this paper, we investigate the possibility to mechanize the proof of some real *complex* mathematical theorems in B [1]. For this, we propose a little *structure language* which allows one to encode mathematical structures and their accompanying theorems. A little tool is also proposed, which translates this language into B, so that Atelier B, the tool associated with B, can be used to prove the theorems. As an illustrative example, we eventually (mechanically) prove the Theorem of Zermelo [6] stating that any set can be well-ordered. The present study constitutes a complete reshaping of an earlier (1993) unpublished work (referenced in [4]) done by two of the authors, where the classical theorems of Haussdorf and Zorn were also proved.

1 Introduction

Is it interesting at all of proving the theorem of Zermelo? In fact, one might indeed wonder why, as designers of algorithms and system engineers, we are interested in such a fancy theorem, and also why we are interested in proving it mechanically with Atelier B. Is this theorem of any interest in the formal development of, say, an electronic circuit, or a transmission protocol, or a real-time monitor ? Certainly not. The answer is rather that we are not so much interested in the theorem itself than in the *approach of its proof*. Mathematics is unique: our experience has led us to believe that a complex proof in "higher-order" mathematics is very much of the same nature as one done in some more elementary domain. We think that there is something to be learned from exercising ourselves in a field that is completely different from the one we are used to in our every day work.

There is, however, another important reason for engaging in this effort. The small "structure language" proposed in this paper constitutes in fact a framework, within which a collection of generic theories (and theorems) can be developed, accumulated and eventually re-used in some real system engineering projects. This will considerably enhance the possibilities already offered today by the B Language. Before starting a project, the idea might then be to use that structure language to develop (or use off the shelf) some *static* theories (and theorems) that are relevant to the problem at hand. Due to an obvious lack of space, we shall not really develop these ideas in this paper, only slightly touching on them towards the end of this presentation.

From what we have just said, it seems that our approach and goal are very close to those aimed at by such formalisms and tools as PVS [3] or Isabelle [5]. This is certainly the case. There is, however, an important distinction to be

D. Bert et al. (Eds.): ZB 2002, LNCS 2272, pp. 370–393, 2002.

made. The mentioned systems (and others) make heavy use of the, so-called, "higher-order" logic. We do not. Rather, we construct the proposed formalism as a *super-structure* based on first order predicate logic *extended with set theory*. In doing so, we follow the classical tradition of mathematicians. One might think, however, that this approach is less "powerful" than others. This is not our opinion (hence the choice of a non-trivial example entirely conducted within the proposed formalism). In fact, in our proposal, the higher-order aspect is achieved by the use of set theory, which offers the possibility to *quantify over all the subsets of a set* (the proposed example will illustrate this on a number of occasions). Such quantifications give indeed the possibility to climb up to "higher-orders" in a way that is always framed.

But to begin with: what do we mean exactly by a *complex* proof? Is it a difficult proof? a long proof? a tricky proof? A little of those certainly but, above all, this is a proof that needs to be *structured* in order to be intelligently communicated to others (a proof, being a rational argument, is, we must remember, an act of communication). By "structured", we do not only mean the process by which a proof is decomposed into small pieces (lemmas) that are all put together in a final simple argument. Structuring also means for us that the proof in question relies on some results that pertain to *other* theories situated outside the very field where the theorem to be proved is stated.

For instance, in the case of the Theorem of Zermelo, we shall rely, not only on a theory of well-ordering and of a related morphism (able to transport a well-ordering into another one), but also on a theory of transfinite induction, itself based on a theory of fixpoint, etc. The proof of the Theorem of Zermelo is *complex* for that very reason, in other words because you have to be familiar with various *mathematical structures* in order to discover and understand it. This contrasts with the kind of problems that you may find in Sunday newspapers, where the difficulty rather lies in the logical intricacy of the very statement of the problem: such puzzles are *not* the sort of things we are interested in, they are not complex, they are just tricky.

2   Mathematical Structure

A mathematical structure (as presented in Bourbaki [2]) corresponds essentially to the definition of a certain *generic universe of discourse*, within which some mathematical theorems can be stated and proved. A structure comprises a number of parts, which we shall present in what follows on a small example: partial-ordering.

A partial-order structure is first introduced by a certain set s: the, so-called, *carrier set* of the structure. Then you have a certain relation r: the, so-called, *component* of the structure. This relation is a binary relation built on s, which is reflexive, transitive, and antisymmetric: this constitutes the, so-called, *axioms* of the structure. Finally, you may *prove* some other results depending on these axioms: these are the *theorems* of the structure. The statements of the axioms and theorems are expressed by means of a certain formal language, which, in the most general case, is that of *set theory*. Notice that some structures may contain more components (sometimes also more carriers): for instance, the algebraic structure of group has three components, namely the binary operation, the inverse, and the neutral element. Practically, a structure can thus be given the following concrete shape with a number of optional clauses. Next to the general form, we show how the small partial-order structure we have mentioned above can be "encoded" (without any theorems).

```
STRUCTURE
   < name >
SETS
   < list of set names >
COMPONENTS
   < list of component names >
AXIOMS
   < list of predicates >
THEOREMS
   < list of predicates >
END
```

```
STRUCTURE
   Partial_order
SETS
   s
COMPONENTS
   r
AXIOMS
   r ∈ s ↔ s ;
   id (s) ⊆ r ;        /*  injectivity   */
   r ∩ r⁻¹ ⊆ id (s) ;  /*  antisymetry   */
   r ∘ r ⊆ r           /*  transitivity  */
END
```

$$r \in s \leftrightarrow s \,;$$
$$\text{id}\,(s) \subseteq r \,; \quad /* \; injectivity \; */$$
$$r \cap r^{-1} \subseteq \text{id}\,(s)\,; \quad /* \; antisymetry \; */$$
$$r \circ r \subseteq r \quad /* \; transitivity \; */$$

A structure is said to be *generic* in that the carrier sets are completely abstract and independent of each other. A structure is thus a template. The primary interest of a mathematical structure resides in the encapsulation of a well defined body of mathematical knowledge. But this is not enough. In fact, there exists another complementary and fundamental interest, which is the following: a structure *can be re-used* in other structures either as such, or, far more interestingly, by means of the mechanism of *instantiation*.

3   Proof Obligations Associated with a Structure

Before presenting the notion of instantiation, we have to make precise what is to be proved with regards to the predicates that one may find in the THEOREMS clause of a structure. Suppose we have a simple structure of the shape shown on the right (for simplification, the lists contained in some clauses have been reduced to a single element), then we simply have to prove that the theorem $T(s, c)$ holds under the axioms $A(s, c)$, formally: $A(s, c) \Rightarrow T(s, c)$. As can be seen, the identifier s is free in this statement. It can be considered denoting a *set meta-variable*. The presence of such a meta-variable makes this statement "generic".

```
STRUCTURE
   S
SETS
   s
COMPONENTS
   c
AXIOMS
   A(s, c)
THEOREMS
   T(s, c)
END
```

In other words, it stands for a *family of statements*. The identifier c is also free in this statement: it is a term meta-variable. Such meta-variables could thus be replaced by any set and any term. We shall see in the next section how such replacements are performed.

4   Instantiating a Mathematical Structure

Re-using a mathematical structure $\mathcal{S}$ in another one, $\mathcal{T}$, allows one to assume in the latter the theorems of the former without re-proving them. This is done however after possibly "specializing" the structure $\mathcal{S}$ and, in some cases, after proving the resulting axioms. Roughly speaking, we can think of our structure $\mathcal{S}$ as being *parameterized* by its carrier sets s and components c.

Suppose that, besides the previous structure $\mathcal{S}$, we are given an additional structure $\mathcal{T}$ with carrier set t and component d. The process of instantiating the structure $\mathcal{S}$ in $\mathcal{T}$ consists of exhibiting a set term $\sigma(t,d)$, defined in terms of the set and component of $\mathcal{T}$, and also a similar component term $\gamma(t,d)$ able to instantiate the components c of S. The instantiation consists thus of "repainting" $\mathcal{S}$ with $\sigma(t,d)$ and $\gamma(t,d)$ and to "invoke" it as $\mathcal{S}(\sigma(t,d),\gamma(t,d))$[1]. Such an invocation *stands for a certain predicate*, namely the conjunction of the instantiated axioms of $\mathcal{S}$. When situated in the AXIOMS clause of the structure $\mathcal{T}$, this invocation says that $\mathcal{T}$ contains, among its own axioms, those, properly instantiated, of $\mathcal{S}$. And when situated in the THEOREMS clause, it asserts that $\mathcal{T}$ is "of the structure" $\mathcal{S}$ (a fact that has to be ensured by proving in $\mathcal{T}$ that the instantiated axioms of $\mathcal{S}$ do hold). Here is an example of the first case, where it is shown that well-ordering is a simple extension of partial-ordering:

```
STRUCTURE
    Well_order
SETS
    s
COMPONENTS
    r
AXIOMS
    Partial_order(s, r) ;
    ∀t · ( t ∈ ℙ₁(s)
          ⇒
              ∃x · ( x ∈ t ∧ t ⊆ r[{x}] ) )
END
```

In other words, a well-ordering built on s, is a partial-ordering together with the extra property stating that every non-empty subset t of s possesses a least element x. Notice that the potential theorems proved in the structure *Partial_order* can be expanded and assumed in the structure *Well_order* without being locally re-proved: we have thus factorized the proof effort that has already been done elsewhere.

Here is another example showing an extension of well-ordering. We have two generic sets, s and t, and, besides the relation q, which is supposed to well-order t (this is an axiom), we have an extra component f, which is supposed to be a *total and injective* function from s to t. We now have another "structure call", this time in the THEOREMS clause, which asserts that the relation $f^{-1} \circ q \circ f$ well-orders s. What has to be proved thus is that the relation $f^{-1} \circ q \circ f$ satisfies the instantiated axioms of the structure

```
STRUCTURE
    Well_order_transportation
SETS
    s, t
COMPONENTS
    q, f
AXIOMS
    Well_order(t, q) ;
    f ∈ s ↣ t
THEOREMS
    Well_order(s, f⁻¹ ∘ q ∘ f)   √
END
```

Well_order (this is easy, hence the "tick"). Once this is done, we can assume locally the theorems, if any, of the structure $Well\_order(s, r)$ with r replaced by $f^{-1} \circ q \circ f$.

---

[1] Notice that the meta-variables t and d should not be captured in these replacements.

5   Mathematical Construct

In the previous sections we have explained how to build structures and how to call them in various contexts. The structure is certainly an interesting device, which is able to help the formalist in organizing his mathematical discourse, but clearly this is not enough. A mathematical text (as found in a textbook) also contains some *definitions*, which allows the mathematician to extend at will his way of expressing formal concepts. In this section, we shall present a device that helps writing such definitions.

In the previous example, a structure, called $Well\_order\_transportation$, has shown how a well-ordering relation q, built on t, can be transported into another one, this time built on s, by means of a total and injective function f. The problem with this $Well\_order\_transportation$ structure is that we have to look *inside* it to know the exact form of the new well-ordering relation on s. This might be necessary, should we need to use it in further mathematical developments. But, most of the time, we are not interested at all in the precise definition of this well-ordering relation on s, we are rather interested *in its very existence and unicity only*.

More precisely, it would be amply sufficient for us to have a name for it and just to know that it corresponds to a well-ordering on s. For this, we propose a second formal device, the, so-called, *mathematical construct*. This is a device that looks very much like the *mathematical structure* presented above: it has a number of generic carrier sets, components, axioms and theorems of exactly the same form as a structure. But, besides this, it also *returns* one (or several) constants, together with a number of useful theorems concerned with the relationship between them and the components of the construct. On the right is an example, where we have transformed the previous structure into a construct called $Transport\_w\_o$.

```
CONSTRUCT
    Transport_w_o
SETS
    t, s
COMPONENTS
    q, f
RETURNS
    r
AXIOMS
    Well_order(t, q) ;
    f ∈ s ↣ t ;
DEFINITION
    r ∈ s ↔ s ;
    r = f⁻¹ ∘ q ∘ f
THEOREMS
    Well_order(s, r)   √
END
```

As can be seen, a new clause, the RETURNS clause, contains the list of *returned constants* (in the present case, the single constant r). A DEFINITION clause has also been added, which contains, as its name indicates, the precise definition of the returned constant. It is made of two parts: (1) the typing of the constant, which takes the form of a set membership, and (2) one or more predicates supposed to define the constant non-ambiguously. We shall come back on this very point in section 6, where a number of corresponding proof obligations will be stated.

Note that, besides the returned constants, some *hidden constants* can be introduced as well (in a construct but also in a structure). They are local, and their rôle is to help in defining other (possibly returned) constants and express more easily some of the axioms or lemmas (defined in the last paragraph of this section). Such constants are introduced in a CONSTANTS clause. Each local constant is defined (like the returned constants) in a DEFINITION clause.

As in a structure, we can also call a construct in another one. More precisely, a "construct call" stands for the constant (or list of constants) returned by the construct. Such construct calls are used in DEFINITION clauses only to give the definitions of some constants declared in the calling construct. This is done by means of an equality whose left member is a (list of) local constant name(s) and whose right member is a construct call (examples of this will be given in section 7). Once the instantiated axioms of the called construct are proved, then its instantiated theorems can, as usual, be assumed locally.

It might be convenient to also have some local lemmas (like in a book of mathematics) that are not externally visible as are the theorems. The lemmas might be useful to help proving other lemmas or theorems. Lemmas are introduced in a LEMMAS clause. Of course, lemmas can use the local constants.

6   Summary of Proof Obligations

In this section we summarize the various proof obligations associated with structures and constructs. Each proof obligation corresponds to a formal statement that has to be proved under a number of hypotheses, namely:

- the axioms of the structure or construct,
- the previous definitions,
- the previous theorems or lemmas,
- the instantiated axioms and theorems of the structures or constructs previously called.

Here are these proof obligations:

(1) THEOREMS or LEMMAS **clauses:** Predicates situated in such clauses generate one proof obligation each. Only genuine predicates are involved here. Structure calls, found in such clauses, are taken care of in (4) below.

(2) DEFINITION **clauses** : Each such clause is made of the typing of a certain (possibly multiple) constant followed by one or several predicates aimed at completely and unambiguously defining it. The corresponding constant should then be proved *to exist and be unique*. Only genuine definitional predicates are involved here, construct calls are not concerned since the corresponding existence and unicity has already been proved locally within the called construct itself. Note, however, that other obligations are required for such calls. These are studied below.

(3) **Structure calls in** AXIOMS **clauses:** Such calls do not generate any proof obligations. The corresponding instantiated axioms and theorems are simply generated and can be used as hypotheses in further proof obligations.

(4) **Structure calls in** THEOREMS **or** LEMMAS **clauses:** The various instantiated axioms of the called structure generate one proof obligation each. The corresponding instantiated theorems are then generated and can be used as hypotheses in further proof obligations.

(5) **Construct calls in** DEFINITION **clauses:** The various instantiated axioms of the called constructs generate one proof obligation each. The corresponding instantiated theorems are then generated and can be used as hypotheses in further proof obligations.

(6) **Non-circularity:** Finally, a global test should be performed, making sure that no circularity occurs in the various structure and construct calls.

7 An Example

We now have enough material to describe our formal strategy with the aim of eventually proving the Theorem of Zermelo. With this example, we would like to illustrate our proposal on a non-trivial case. Our intention is also to show how the proposed framework allows the construction of a complex proof by a sort of "backward reasoning" starting from the goal and progressively elaborating the various intermediary results that are needed until one reaches some completely obvious statements. In that respect, the development of this example shows a certain *mathematical methodology* at work, as advocated in [7].

The reader will probably notice that there exists certain similarities between this approach, used here to design *complex proofs*, and that used in the design of *complex systems*. This only means that complexity, wherever it lies, seems to demand the same sort of ingredients. Here is the well known statement of the Theorem of Zermelo, which we propose to prove in what follows:

$$\boxed{\textit{Every set can be well-ordered}}$$

7.1 Formalizing the Problem in a Construct

The possibility expressed in this statement leads us to define a construct associated with a generic set s, and returning a relation r, which well-orders s. The purpose of the proof is precisely to "construct" that relation r.

7.2 Necessity of Structuring the Set s

We have absolutely no idea how such a well-ordering r on s could be constructed. This is because the set s has no pre-defined "structure". We even suspect that without any particular property of s, the problem is unsolvable. Suppose, conversely, that the problem is solved.

In this case, every non-empty subset of s has, by definition, a least element (since r well-orders s), an element that can thus be *well identified* (even if the subset in question is infinite). This is the very property we are going to start from: given a non-empty subset of s, we shall assume that there exists a *well identified* element of it. This can be formalized by means of a certain component, c, in our construct, component which is a total function (the, so-called, "choice function") mapping any non-empty subset a of s to an element $c(a)$ of a: this axiom corresponds to one form of the, so-called, "axiom of choice".

```
CONSTRUCT
    Zermelo
SETS
    s
COMPONENTS
    c
RETURNS
    r
AXIOMS
    c ∈ ℙ₁(s) → s ;
    ∀a · ( a ∈ ℙ₁(s)  ⇒  c(a) ∈ a )
...
DEFINITION
    r ∈ s ↔ s ;
    r = ...
THEOREMS
    Well_order(s, r)
END
```

We can thus now rephrase our problem as follows (which seems more reasonable than the previous statement):

> *Every set, equipped with a "choice function", can be well-ordered*

7.3   Moving the Problem

Clearly, this new statement of the problem is richer than the previous one, but we still have difficulties to construct the well-ordering on s. We could certainly think of linking each element of a non-empty subset a of s with the element $c(a)$, but we would not obtain a partial-order in general (it is easy to construct a counter-example with a set s made of three elements and a particular choice function c on it).

Since the choice function c deals with subsets of s, the idea is then to transfer the research of a well-ordering on s to that of a well-ordering on a certain set t (to be constructed) of subsets of s, and, quite naturally, to propose the inclusion relation q on t as our potential well-ordering on t (there is a clear advantage in doing so, due to the fact that the relation q is already a partial order).

The next step is then to use the mentioned construct $Transport\_w\_o$, which will be able to transport our well-ordering q into the well-ordering r we are looking for. But to do this, we first need a total injection f from s to t. All this can be formalized as shown on the left. As can be seen, we have introduced three local constants t, q, and f. The set t is just typed for the moment, we do not know yet how to construct it more precisely. The relation q is completely determined: it is the inclusion relation on members of t (these are sets). Finally the constant f is declared as a total function, but, as for t, not yet constructed. Then we state in a LEMMAS clause the two lemmas to be proved, concerning the well-ordering of q on t and the total injectivity of f. Provided these are proved (which is far from being the case yet), then the instantiated axioms of the construct call $r = Transport\_w\_o(t, s, q, f)$ can be proved (hence the $\sqrt{}$). Consequently, we can expand the theorem of this construct yielding our main theorem (thus also ticked).

```
CONSTRUCT
  Zermelo
. . .
CONSTANTS
  t, q, f
DEFINITION
  t ∈ ℙ(ℙ(s)) ;
  t = . . .
DEFINITION
  q ∈ t ↔ t ;
  ∀ (a, b) · ( a, b ∈ t × t
             ⇒
             (a, b ∈ q  ⇔  a ⊆ b))
DEFINITION
  f ∈ s → t ;
  f = . . .
LEMMAS
  Well_order(t, q) ;
  f ∈ s ↣ t
DEFINITION
  r ∈ s ↔ s ;
  r = Transport_w_o(t, s, q, f)   √
THEOREMS
  Well_order(s, r)   √
END
```

7.4   Proving the Well-Ordering on t

In order to prove that q well-orders t (this is our first lemma), we have to prove that every non-empty subset a of t has a least element with respect to set inclusion (we remind the reader that t is a set of subsets of s). The least element of a set of subsets, if it exists, is simply its generalized intersection. We therefore have to prove, as an extra lemma, that the generalized intersection in question is a member of a. This has the consequence that the well-ordering of t is now proved (hence the new tick) since q, as already observed, is a partial order.

CONSTRUCT
  Zermelo
· · ·
LEMMAS
  $\forall a \cdot (a \in \mathbb{P}_1(t) \;\Rightarrow\; \mathsf{inter}\,(a) \in a)$;
  $Well\_order(t, q)$; $\checkmark$
  $f \in s \rightarrowtail t$
DEFINITION
  $r \in s \leftrightarrow s$;
  $r = Transport\_w\_o(t, s, q, f)$ $\checkmark$
THEOREMS
  $Well\_order(s, r)$ $\checkmark$
END

CONSTRUCT
  Zermelo
· · ·
CONSTANTS
  n, t, q, f
DEFINITION
  $n \in \mathbb{P}(s) \rightarrow \mathbb{P}(s)$;
  $n(s) = s$;
  $\forall a \cdot (a \in \mathbb{P}(s) \wedge a \neq s$
        $\Rightarrow$
        $n(a) = a \cup \{c(s - a)\})$
· · ·
END

A consequence of this new lemma (still to be proved) is that t is made of a series of embedded subsets of s (since, given two members of t, one is necessarily included in the other). We wonder then how we could characterize two "successive" members of t. This is where the choice function c may enter into the scene. For this, we introduce a (local) constant n (for next) which maps any proper subset of s into another one, which is just enlarging the former with the "chosen" element of its complement (and in the case of s, n maps it to itself).

7.5   Giving more Shape to t

The embedding of the members of t leads us to consider that the image ot t under n is included in t. If we suppose that t contains the empty set, then it appears as something like: $\emptyset, n(\emptyset), n(n(\emptyset)), n(n(n(\emptyset))), \ldots$

The set t seems to be a construction like that of the natural numbers, where each member of t "represents" one of these numbers. But, clearly, if we limit ourselves to such a construction, we certainly have no chance of finding a *total injection* f from s to t, as, in general, t will not be big enough (think of s as being $\mathbb{P}(\mathbb{N})$). In order to enlarge t a "little" more, the revolutionary idea (of Cantor) is to make the generalized union of any subset of t also a member of t (notice that this has the consequence of making the empty set a member of t).

CONSTRUCT
  Zermelo
· · ·
LEMMAS
  $\forall x \cdot (x \in t \;\Rightarrow\; n(x) \in t)$;
  $\forall a \cdot (a \in \mathbb{P}(t) \;\Rightarrow\; \mathsf{union}(a) \in t)$;
  $\forall a \cdot (a \in \mathbb{P}_1(t) \;\Rightarrow\; \mathsf{inter}\,(a) \in a)$;
  $Well\_order(t, q)$; $\checkmark$
  $f \in s \rightarrowtail t$
DEFINITION
  $r \in s \leftrightarrow s$;
  $r = Transport\_w\_o(t, s, q, f)$ $\checkmark$
THEOREMS
  $Well\_order(s, r)$ $\checkmark$
END

For the moment, we do not know whether such a set t does exist, we have just exhibited a number of wishful properties of it. As can be seen, we have added two extra lemmas. The first one claims that t is closed under the "next" function n, and the second one claims that t is closed under generalized union.

7.6 Defining the Injection between s and t

Clearly, the set t takes more and more shape, but we have left the injective function f rather unexplored for the moment. It is time to propose some concrete realization for it. To achieve the total injectivity of f, we have to map each member z of s to a member $f(z)$ of t in such a way that a *distinct* member z' of s is mapped to a member $f(z')$ of t that is distinct from $f(z)$.

The idea of Zermelo is extremely "simple": he maps each member z of s to the *largest* member, $f(z)$, of t which *does not* contain z. The set $f(z)$ is thus defined to be the union of such members of t, which thus belongs to t according to the second closure property of t. As a consequence, $n(f(z))$ will then necessarily contain z, for otherwise $f(z)$ would *not* be the largest member of t not containing z. This means, according to the definition of n, that $c(s - f(z))$ is exactly z. Thus, for z' distinct from z, we have $c(s - f(z)) \neq c(s - f(z'))$, which clearly implies $f(z) \neq f(z')$ and thus proves the injectivity of f.

The previous informal proof, as for all the proofs of this paper, has been mechanically conducted with Atelier B. We can thus now "tick" the lemma concerning the injectivity of f.

7.7 Finalizing the Construct *Zermelo*

Now there remains the finalization of the construction of t. This will be done in a separate construct called *Transfinite*. The idea is to define t as the *smallest* set of subsets of s that is closed under n and under generalized union. This will give us automatically our two first lemmas. What we only then require from this construct is a theorem corresponding to our third lemma. Here is thus the final text of the construct *Zermelo*:

```
CONSTRUCT
    Zermelo
SETS
    s
COMPONENTS
    c
RETURNS
    r
AXIOMS
    c ∈ ℙ₁(s) → s ;
    ∀a · ( a ∈ ℙ₁(s)  ⇒  c(a) ∈ a )
CONSTANTS
    n, t, q, f
DEFINITION
    n ∈ ℙ(s) → ℙ(s) ;
    n(s) = s ;
    ∀a · ( a ∈ ℙ(s) ∧ a ≠ s  ⇒  n(a) = a ∪ {c(s − a)} )
```

DEFINITION
$\quad t \in \mathbb{P}(\mathbb{P}(s))$;
$\quad t = Transfinite(s, n)$
DEFINITION
$\quad q \in t \leftrightarrow t$;
$\quad \forall (a, b) \cdot (a, b \in t \times t \ \Rightarrow \ (a, b \in q \ \Leftrightarrow \ a \subseteq b))$
DEFINITION
$\quad f \in s \rightarrow t$;
$\quad \forall z \cdot (z \in s \ \Rightarrow \ f(z) = \mathsf{union}(\{x \mid x \in t \ \wedge \ z \notin x\}))$
LEMMAS
$\quad \forall x \cdot (x \in t \ \Rightarrow \ n(x) \in t)$; $\checkmark$
$\quad \forall a \cdot (a \in \mathbb{P}(t) \ \Rightarrow \ \mathsf{union}(a) \in t)$; $\checkmark$
$\quad \forall a \cdot (a \in \mathbb{P}_1(t) \ \Rightarrow \ \mathsf{inter}(a) \in a)$; $\checkmark$
$\quad Well\_order(t, q)$; $\checkmark$
$\quad f \in s \rightarrowtail t$ $\checkmark$
DEFINITION
$\quad r \in s \leftrightarrow s$;
$\quad r = Transport\_w\_o(t, s, q, f)$ $\checkmark$
THEOREMS
$\quad Well\_order(s, r)$ $\checkmark$
END

7.8   Constructing the Transfinite Set t

The $Transfinite$ construct should give us a set t with theorems corresponding to the three first lemmas of the construct $Zermelo$. Clearly this construct is parameterized by a set s and a function such as n. Notice that we don't know yet which axioms are required for n, we know only that the definition of n in the construct $Zermelo$ should imply the axioms in question.

CONSTRUCT
$\quad Transfinite$
SETS
$\quad s$
COMPONENTS
$\quad n$
RETURNS
$\quad t$
AXIOMS
$\quad n \in \mathbb{P}(s) \rightarrow \mathbb{P}(s)$;
$\quad \ldots$
DEFINITION
$\quad t \in \mathbb{P}(\mathbb{P}(s))$;
$\quad t = \ldots$
THEOREMS
$\quad \forall x \cdot (x \in t \ \Rightarrow \ n(x) \in t)$;
$\quad \forall a \cdot (a \in \mathbb{P}(t) \ \Rightarrow \ \mathsf{union}(a) \in t)$;
$\quad \forall a \cdot (a \in \mathbb{P}_1(t) \ \Rightarrow \ \mathsf{inter}(a) \in a)$
END

CONSTRUCT
$\quad Transfinite$
$\quad \ldots$
CONSTANTS
$\quad g$
DEFINITION
$\quad g \in \mathbb{P}(\mathbb{P}(s)) \rightarrow \mathbb{P}(\mathbb{P}(s))$;
$\quad \forall a \cdot (\ a \in \mathbb{P}(\mathbb{P}(s))$
$\qquad \Rightarrow$
$\qquad g(a) = n[a] \ \cup \ \mathsf{union}[\mathbb{P}(a)])$
$\quad \ldots$
LEMMAS
$\quad g(t) \subseteq t$
THEOREMS
$\quad \forall x \cdot (x \in t \ \Rightarrow \ n(x) \in t)$; $\checkmark$
$\quad \forall a \cdot (a \in \mathbb{P}(t) \ \Rightarrow \ \mathsf{union}(a) \in t)$; $\checkmark$
$\quad \forall a \cdot (a \in \mathbb{P}_1(t) \ \Rightarrow \ \mathsf{inter}(a) \in a)$
END

The first two theorems of our construct state that t is closed under n and under generalized union. To prove this, we define a local function g mapping a set typed as t (that is $\mathbb{P}(\mathbb{P}(s))$) into a set typed in the same way. More precisely, this function g maps a set a (of subsets of s) into its own image under n, unioned with the image of $\mathbb{P}(a)$ under generalized union. Now, provided we locally prove that $g(t)$ is included into t (this is thus a local lemma to be proved), then we have the theorems in question. This yields the following more precise text for the construct $Transfinite$, where the two theorems are now ticked.

Proving Well-Ordering

We certainly have not enough material to locally prove our last theorem stating that any non-empty subset of t has a least element, but we could try to do so, in order, precisely, to discover what are the missing assumptions. Let a be a non-empty subset of t. We have to prove that the generalized intersection, inter (a), of a belongs to a. The idea is to form the set b of members x of t that are all subsets of inter (a)

$$b = \{\, x \mid x \in t \,\wedge\, x \subseteq \mathsf{inter}\,(a) \,\}$$

Clearly b is a subset of t. Consequently, its generalized union, union (b), is also a member of t (closure theorem). As a consequence, and according to a basic property of generalized union, union (b) belongs to b, and we have:

$$\mathsf{union}\,(b) \subseteq \mathsf{inter}\,(a)$$

If union (b) is equal to s then we have succeeded because inter (a) is then equal to s (since inter $(a) \subseteq s$) and thus a is equal to $\{s\}$ and then inter $(a) \in a$. We can thus suppose that union (b) is a proper subset of s. Remembering what the concrete value of the function n was in the construct $Zermelo$, we are tempted to form $n(\mathsf{union}\,(b))$. But we lack any property for n locally at the moment. For this, we now axiomatize the function n as follows (notice that we have avoided using the choice function c of the construct $Zermelo$ since that function is not known to the construct $Transfinite$, hence the rôle of the existential quantification):

AXIOMS
  $n \in \mathbb{P}(s) \rightarrow \mathbb{P}(s)$;
  $n(s) = s$;
  $\forall a \cdot (\, a \in \mathbb{P}(s) \,\wedge\, a \neq s \;\Rightarrow\; \exists x \cdot (\, x \in s - a \,\wedge\, n(a) = a \cup \{x\}\,))$
  $\ldots$

We note in passing that it will not be difficult to prove this axiom from within the construct $Zermelo$ since the term $c(s - a)$ will provide us with a witness x needed for proving the existential quantification. Now, clearly, $n(\mathsf{union}\,(b))$ does not belong to b since union (b) is *strictly* included in $n(\mathsf{union}\,(b))$, and is also, by definition, the largest member of b. We have thus:

$$\neg\; n(\mathsf{union}\,(b)) \subseteq \mathsf{inter}\,(a)$$

As a consequence, and according to a classical property of generalized intersection, we have

$$\neg\; \forall y \cdot (\, y \in a \;\Rightarrow\; n(\mathsf{union}\,(b)) \subseteq y \,)$$

or equivalently

$$\exists y \cdot (y \in a \ \wedge \ \neg \, n(\text{union}\,(b)) \subseteq y)$$

Since the existence of such a y is asserted, we can assume that we have it in the rest of the proof. We notice that both $n(\text{union}\,(b))$ and y belong to t. If we now suppose, as an extra property of t, that it is *totally ordered* by inclusion, we thus have the following, since the other inclusion has just been eliminated (notice, the *strict* inclusion we use):

$$y \subset n(\text{union}\,(b))$$

But we also have $\text{inter}\,(a) \subseteq y$ since y belongs to a. And we remind the reader that we also have $\text{union}\,(b) \subseteq \text{inter}\,(a)$. Finally, we also have $n(\text{union}\,(b)) = \text{union}\,(b) \cup \{x\}$ for some x not belonging to $\text{union}\,(b)$. To summarize, we have thus the following embedding:

$$\text{union}\,(b) \ \subseteq \ \text{inter}\,(a) \ \subseteq \ y \ \subset \ \text{union}\,(b) \cup \{x\}$$

Clearly, there is not very much "room" between $\text{union}\,(b)$ and y. We have:

$$\text{union}\,(b) \ = \ \text{inter}\,(a) \ = \ y \ \subset \ \text{union}\,(b) \cup \{x\}$$

Thus $\text{inter}\,(a)$, being equal to y, indeed belongs to a as y does. This completes the informal proof. The construct $Transfinite$ is thus now as follows:

CONSTRUCT
  $Transfinite$
SETS
  s
COMPONENTS
  n
RETURNS
  t
AXIOMS
  $n \in \mathbb{P}(s) \to \mathbb{P}(s)$;
  $n(s) = s$;
  $\forall a \cdot (a \in \mathbb{P}(s) \ \wedge \ a \neq s \ \Rightarrow \ \exists x \cdot (x \in s - a \ \wedge \ n(a) = a \cup \{x\}))$
CONSTANTS
  g
DEFINITION
  $g \in \mathbb{P}(\mathbb{P}(s)) \to \mathbb{P}(\mathbb{P}(s))$;
  $\forall a \cdot (a \in \mathbb{P}(\mathbb{P}(s)) \ \Rightarrow \ g(a) = n[a] \ \cup \ \text{union}\,[\mathbb{P}(a)])$
DEFINITION
  $t \in \mathbb{P}(\mathbb{P}(s))$;
  $t = Fixpoint(\mathbb{P}(s), g)$
LEMMAS
  $g(t) \subseteq t$;
  $\forall (x, y) \cdot (x \in t \ \wedge \ y \in t \ \Rightarrow \ x \subseteq y \ \vee \ y \subseteq x)$
THEOREMS
  $\forall x \cdot (x \in t \ \Rightarrow \ n(x) \in t)$; $\checkmark$
  $\forall a \cdot (a \in \mathbb{P}(t) \ \Rightarrow \ \text{union}(a) \in t)$; $\checkmark$
  $\forall a \cdot (a \in \mathbb{P}_1(t) \ \Rightarrow \ \text{inter}\,(a) \in a)$ $\checkmark$
END

As can be seen, we have added what we have learned from the previous informal proof, namely the axiomatization of n (in the AXIOMS clause) and the total ordering of t (in the LEMMAS clause). This was the (not too heavy) price to be paid in order to prove the well-ordering of t, which is now ticked.

What remains for us to do is the proof of the two lemmas and, of course, to give the precise definition of t. For this, we shall use another more general construct, called $Fixpoint$, providing us with what we need. This construct will directly give us the first lemma, and also an extra result that will help us to prove locally the second one. So our strategy is now to leave the construct $Transfinite$ as it is for the moment. We shall then define the construct $Fixpoint$ in the next section, and then come back in section 7.10 to finalize the construct $Transfinite$.

7.9   Constructing *Fixpoint*

The skeleton for the $Fixpoint$ construct is shown below.

```
CONSTRUCT
    Fixpoint
SETS
    s
COMPONENTS
    h
RETURNS
    t
AXIOMS
    h ∈ ℙ(s) → ℙ(s) ;
    ...
CONSTANTS
    a
DEFINITION
    a ∈ ℙ(ℙ(s)) ;
    a = { x | x ∈ ℙ(s) ∧ h(x) ⊆ x }
DEFINITION
    t ∈ ℙ(s) ;
    t = inter (a)
THEOREMS
    h(t) ⊆ t
END
```

We have just mentioned the theorem we need in the construct $Transfinite$, namely the closure property of the function h (called g in the construct $Transfinite$). As already explained, the idea is to take for t the smallest set closed under h. For this, we have defined a local constant a denoting the set of subsets x of s making $h(x)$ included in x, and we have defined t as the generalized intersection of a. This is not sufficient yet to prove that t is closed under h, since nothing indeed guarantees that the set t belongs to a. The theorem of Tarski (used by mathematicians long before it was named as such) establishes our result provided the function h is monotonic. In fact, this theorem says more, namely that t is the *least fixpoint* of h (but this reciprocal property, easily proved from the direct part, is not needed here). Formally:

```
CONSTRUCT
    Fixpoint
...
AXIOMS
    h ∈ ℙ(s) → ℙ(s) ;
    ∀ (a,b) · ( a ∈ ℙ(s) ∧ b ∈ ℙ(s) ∧ a ⊆ b ⇒ h(a) ⊆ h(b) )
...
END
```

In order to prove (mechanically) the theorem of Tarski, we need two classical properties of generalized intersection, properties that we can state as theorems in an ad-hoc structure called *Inter*. The properties in question establish that inter(a) is the greatest lower bound of a. They are easy to establish (mechanically). Formally:

```
STRUCTURE
  Inter
SETS
  s
COMPONENTS
  a
AXIOMS
  a ∈ ℙ₁(ℙ(s))
THEOREMS
  ∀n · ( n ∈ a  ⇒  inter(a) ⊆ n ) ;   √
  ∀m · ( m ⊆ s  ∧  ∀n · ( n ∈ a  ⇒  m ⊆ n )  ⇒  m ⊆ inter(a) ) )   √
END
```

Equipped with these theorems, the theorem of Tarski is easily (mechanically) established. Here is thus the final shape of construct *Fixpoint*:

```
CONSTRUCT
  Fixpoint
SETS
  s
COMPONENTS
  h
RETURNS
  t
AXIOMS
  h ∈ ℙ(s) → ℙ(s) ;
  ∀ (a,b) · ( a ∈ ℙ(s)  ∧  b ∈ ℙ(s)  ∧  a ⊆ b  ⇒  h(a) ⊆ h(b) )
CONSTANTS
  a
DEFINITION
  a ∈ ℙ(ℙ(s)) ;
  a = {x | x ∈ ℙ(s)  ∧  h(x) ⊆ x }
DEFINITION
  t ∈ ℙ(s) ;
  t = inter (a)
LEMMAS
  Inter(s, a)   √
THEOREMS
  h(t) ⊆ t ;   √
  ∀p · ( p ∈ ℙ(s)  ∧  h(p) ⊆ p  ⇒  t ⊆ p )   √
END
```

Note that the call of the structure *Inter* in the LEMMAS clause requires proving that a is not empty (which is easy as it contains s). As can be seen, we have added a theorem in the construct *Fixpoint*. It will allow us to prove properties of t *by induction*. It is very easily (mechanically) proved by using the theorems of the structure *Inter*. It can be read as follows: provided you can prove that a certain subset p of s is closed under h, then the set t is included in p.

7.10   Revisiting the Construct *Transfinite*

We are now ready to finalize our development by proving the remaining unproved lemmas of construct *Transfinite*.

Previous State

Here is the situation we left in section 7.8 concerning the construct *Transfinite*. We can now add some more ticks since the instantiation of the axiom of construct *Fixpoint* is easily provable (the function g is clearly monotone), and the first lemma of *Transfinite* is now easily proved by instantiating the first theorem of *Fixpoint*.

CONSTRUCT
   Transfinite
$\ldots$
DEFINITION
   $t \in \mathbb{P}(\mathbb{P}(s))$;
   $t = Fixpoint(\mathbb{P}(s), g)$ $\checkmark$
LEMMAS
   $g(t) \subseteq t$; $\checkmark$
   $\forall (x, y) \cdot (x \in t \wedge y \in t \Rightarrow x \subseteq y \vee y \subseteq x)$
$\ldots$
END

Proving the Total Ordering of t

The *last* unproved lemma of *Transfinite* (and, in fact, of our overall proof effort) concerns the total ordering of the set t. For this we shall use the instantiation of the induction theorem of the construct *Fixpoint*. From this instantiation, we can easily prove the following more convenient lemma called the *transfinite induction lemma*:

CONSTRUCT
   Transfinite
$\ldots$
LEMMAS

$$\forall p \cdot \left(\begin{array}{l} p \in \mathbb{P}(\mathbb{P}(s)) \ \wedge \\ \forall a \cdot (a \in p \Rightarrow n(a) \in p) \ \wedge \\ \forall b \cdot (b \in \mathbb{P}(p) \Rightarrow \text{union}(b) \in p) \\ \Rightarrow \\ t \subseteq p \end{array} \right) ; \quad \checkmark$$

$\ldots$
END

Proving the total ordering of t, that is

$$\forall (x, y) \cdot (x \in t \wedge y \in t \Rightarrow x \subseteq y \vee y \subseteq x)$$

is equivalent to proving:

$$\forall x \cdot (x \in t \Rightarrow \forall y \cdot (y \in t \Rightarrow x \subseteq y \vee y \subseteq x))$$

that is:

$$t \subseteq \{x \mid x \in t \ \wedge \ \forall y \cdot (y \in t \ \Rightarrow \ x \subseteq y \ \vee \ y \subseteq x)\}$$

For proving this, we instantiate p in the transfinite induction lemma with the set $\{x \mid x \in t \ \wedge \ \forall y \cdot (y \in t \ \Rightarrow \ x \subseteq y \ \vee \ y \subseteq x)\}$. Proving the first and last antecedents is easy. For proving the second antecedent, we assume:

$$x \in t \ \wedge \ x \neq s \ \wedge \ \forall y \cdot (y \in t \ \Rightarrow \ x \subseteq y \ \vee \ y \subseteq x) \qquad \textbf{Hyp1}$$

Notice that we have added the hypothesis $x \neq s$ since in the case that $x = s$ holds we have succeeded (the next statement holding trivially). We have to prove

$$\forall y \cdot (y \in t \ \Rightarrow \ n(x) \subseteq y \ \vee \ y \subseteq n(x))$$

that is, equivalently

$$t \subseteq \{y \mid y \in t \ \wedge \ n(x) \subseteq y \ \vee \ y \subseteq n(x)\}$$

For proving this, we now instantiate p in the transfinite induction lemma with the set $\{y \mid y \in t \ \wedge \ (n(x) \subseteq y \ \vee \ y \subseteq n(x))\}$. Again, proving the first and third antecedents is easy. For proving the second antecedent, we assume:

$$y \in t \ \wedge \ y \neq s \ \wedge \ (n(x) \subseteq y \ \vee \ y \subseteq n(x)) \qquad \textbf{Hyp2}$$

Notice again that we have added the hypothesis $y \neq s$ since in case $y = s$ holds we have succeeded (the next statement holding trivially). We have to prove:

$$n(x) \subseteq n(y) \ \vee \ n(y) \subseteq n(x)$$

The proof is by cases, as suggested in **Hyp2**. If $n(x) \subseteq y$ holds then we have succeeded (since $y \subseteq n(y)$). We assume then:

$$y \subseteq n(x) \qquad \textbf{Hyp3}$$

Instantiating y with $n(y)$ in **Hyp1** yields:

$$x \subseteq n(y) \ \vee \ n(y) \subseteq x \qquad \textbf{Hyp4}$$

Again, the proof is by cases. If $n(y) \subseteq x$ holds then we have succeeded (since $x \subseteq n(x)$). We assume then:

$$x \subseteq n(y) \qquad \textbf{Hyp5}$$

Instantiating y with y in **Hyp1** yields:

$$x \subseteq y \ \vee \ y \subseteq x \qquad \textbf{Hyp6}$$

In case $x \subseteq y$ holds, **Hyp3** yields $x \subseteq y \subseteq n(x)$, hence $x = y \ \vee \ n(x) = y$ (remember the definition of $n(x)$ when x is not equal to s (as stipulated in **Hyp1**): $n(x) = x \cup \{z\}$ for some z not in x). In both cases, we have $n(x) \subseteq n(y)$. In case $y \subseteq x$ holds, **Hyp5** yields $y \subseteq x \subseteq n(y)$, hence $y = x \ \vee \ n(y) = x$ (again remember the definition of $n(y)$ when y is not equal to s (as stipulated in **Hyp2**)). In both cases, we have $n(y) \subseteq n(x)$.

This completes the proof of the total ordering of t, *and thus the proof of the Theorem of Zermelo*: all lemmas and theorems have indeed been (informally) ticked and also (mechanically) proved with Atelier B. Next we present the final version of the construct *Transfinite*.

CONSTRUCT
  Transfinite
SETS
  s
COMPONENTS
  n
RETURNS
  t
AXIOMS
  $n \in \mathbb{P}(s) \to \mathbb{P}(s)$;
  $n(s) = s$;
  $\forall a \cdot (a \in \mathbb{P}(s) \wedge a \neq s \Rightarrow \exists x \cdot (x \in s - a \wedge n(a) = a \cup \{x\}))$
CONSTANTS
  g
DEFINITION
  $g \in \mathbb{P}(\mathbb{P}(s)) \to \mathbb{P}(\mathbb{P}(s))$;
  $\forall a \cdot (a \in \mathbb{P}(\mathbb{P}(s)) \Rightarrow g(a) = n[a] \cup \text{union}\,[\mathbb{P}(a)])$
DEFINITION
  $t \in \mathbb{P}(\mathbb{P}(s))$;
  $t = Fixpoint(\mathbb{P}(s), g)$ $\checkmark$
LEMMAS
  $g(t) \subseteq t$; $\checkmark$
  $\forall p \cdot \begin{pmatrix} p \in \mathbb{P}(\mathbb{P}(s)) \wedge \\ \forall a \cdot (a \in p \Rightarrow n(a) \in p) \wedge \\ \forall b \cdot (b \in \mathbb{P}(p) \Rightarrow \text{union}\,(b) \in p) \\ \Rightarrow \\ t \subseteq p \end{pmatrix}$; $\checkmark$
  $\forall (x, y) \cdot (x \in t \wedge y \in t \Rightarrow x \subseteq y \vee y \subseteq x)$ $\checkmark$
THEOREMS
  $\forall x \cdot (x \in t \Rightarrow n(x) \in t)$; $\checkmark$
  $\forall a \cdot (a \in \mathbb{P}(t) \Rightarrow \text{union}(a) \in t)$; $\checkmark$
  $\forall a \cdot (a \in \mathbb{P}_1(t) \Rightarrow \text{inter}\,(a) \in a)$ $\checkmark$
END

7.11   The Overall Structure of the Proof

Clearly, this proof is *complex*, although no individual lemma and theorem in it are particularly difficult. This is because it appeals to various relatively independent mathematical theories as shown in the following diagram where the complete *structure* of the proof is presented:

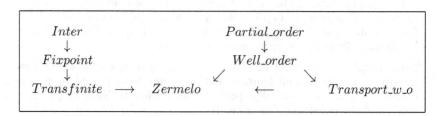

8   Other Approaches

In this section, we shall compare our approach with that of PVS and Isabelle. This will be done by means of examples, which we have extracted from the corresponding documentation.

8.1   PVS

Next we present a sample of a PVS text showing a theory where partial and total orderings are defined.

```
orderings[t: TYPE]: THEORY
BEGIN
 x, y, z: VAR t
 pp, qq: VAR PRED[t]
 <=: VAR PRED[[t,t]]
 reflexive?(<=): bool = FORALL x: x<=x)
 antisymmetric?(<=): bool = (FORALL x, y: x <= y AND y <= x IMPLIES x = y)
 transitive?(<=): bool = (FORALL x, y, z: x <= y AND y <= z IMPLIES y <= x)
 partial_order?(<=): bool = reflexive?(<=) AND
                        antisymmetric?(<=) AND transitive?(<=)
 linear?(<=): bool = (FORALL x, y: x <= y OR y <= x)
 total_order?(<=): bool = partial_order?(<=) AND linear?(<=)
END orderings
```

In our proposed formalism, this would correspond to the following structures:

STRUCTURE
$\qquad Partial\_order$
SETS
$\qquad s$
COMPONENTS
$\qquad r$
AXIOMS
$\qquad r \in s \leftrightarrow s$;
$\qquad \mathrm{id}\,(s) \subseteq r$;
$\qquad r \circ r \subseteq r$;
$\qquad r \cap r^{-1} \subseteq \mathrm{id}\,(s)$
END

STRUCTURE
$\qquad Total\_order$
SETS
$\qquad s$
COMPONENTS
$\qquad r$
AXIOMS
$\qquad Partial\_order(s,r)$;
$\qquad s \times s \subseteq r \cup r^{-1}$
END

We now present a PVS text importing the previous one and defining a sorted array:

```
sorto [domain, range: type,
       (IMPORTING orderings[t})
       d_order: total_order?[domain]),
     r_order: (partial_order?[range])]: THEORY
  BEGIN
   Array_type: TYPE = ARRAY[domain->range]
   A, B, C: VAR Array_type
   sorted?(A): bool =
     (FORALL (x, y: domain): (d_order(x, y) AND x/=y)
            IMPLIES
              NOT r_order(A(y), A(x)))
  END sorto
```

Next is how this would be encoded in our formalism. We also present another structure showing how one can write (and easily prove) that the composition of two sorted functions is also a sorted function.

STRUCTURE
  Sorting
SETS
  s, t
COMPONENTS
  f, p, q
AXIOMS
  $f \in s \to t$;
  $Total\_order(s, p)$;
  $Partial\_order(t, q)$;
  $f \circ p \subseteq q \circ f$
END

STRUCTURE
  Sorting_compose
SETS
  s, t, u
COMPONENTS
  f, g, p, q, r
AXIOMS
  $Sorting(s, t, f, p, q)$;
  $Sorting(t, u, g, q, r)$
THEOREMS
  $Sorting(s, u, (g \circ f), p, r)$
END

Isabelle

Next is an Isabelle encoding of various well known algebraic structures, namely monoids, semi-groups, groups, and abelian groups. This is followed by a number of instantiations: a boolean structure and the product of two groups.

```
theory Group = Main:
subsection {* Monoids and Groups *}
consts
  times:: "'a => 'a => 'a"      (infixl "[*]" 70)
  invers:: "'a => 'a"
  one:: 'a
axclass monoid < "term"
  assoc:      "(x [*] y) [*] z = x [*] (y [*] z)"
  left_unit:  "one [*] x = x"
  right_unit: "x [*] one = x"
axclass semigroup < "term"
  assoc: "(x [*] y) [*] z = x [*] (y [*] z)"
axclass group < semigroup
  left_unit:    "one [*] x = x"
  left_inverse: "invers x [*] x = one"
axclass agroup < group
  commute: "x [*] y = y [*] x"
subsection {* Abstract reasoning *}
  theorem group_right_inverse: "x [*] invers x = (one::'a::group)"
  theorem group_right_unit: "x [*] one = (x::'a::group)"
subsection {* Abstract instantiation *}
  instance monoid < semigroup
  instance group < monoid
subsection {* Concrete instantiation *}
  defs (overloaded)
    times_bool_def:   "x [*] y == x ~= (y::bool)"
    inverse_bool_def: "invers x == x::bool"
    unit_bool_def:    "one == False"
  instance bool:: agroup
subsection {* Lifting and Functors *}
  defs (overloaded)
```

```
    times_prod_def: "p [*] q == (fst p [*] fst q, snd p [*] snd q)"
instance *:: (semigroup, semigroup) semigroup
```

Here is the equivalent encoding in our formalism:

STRUCTURE
  Monoid
SETS
  s
COMPONENTS
  t, i, o
AXIOMS
  $Algebra(s, t, i, o)$;
  $\forall(x, y, z) \cdot \left(\begin{array}{c} x, y, z \in s \times s \times s \\ \Rightarrow \\ t(t(x, y), z) = t(x, t(y, z)) \end{array} \right)$;
  $\forall x \cdot (x \in s \Rightarrow t(o, x) = x)$;
  $\forall x \cdot (x \in s \Rightarrow t(x, o) = x)$
THEOREMS
  $Semigroup(s, t, i, o)$
END

STRUCTURE
  Algebra
SETS
  s
COMPONENTS
  t, i, o
AXIOMS
  $t \in s \times s \rightarrow s$;
  $i \in s \rightarrow s$;
  $o \in s$
END

STRUCTURE
  Semigroup
SETS
  s
COMPONENTS
  t, i, o
AXIOMS
  $Algebra(s, t, i, o)$;
  $\forall(x, y, z) \cdot \left(\begin{array}{c} x, y, z \in s \times s \times s \\ \Rightarrow \\ t(t(x, y), z) = t(x, t(y, z)) \end{array} \right)$
END

STRUCTURE
  Group
SETS
  s
COMPONENTS
  t, i, o
AXIOMS
  $Semigroup(s, t, i, o)$;
  $\forall x \cdot (x \in s \Rightarrow t(o, x) = x)$;
  $\forall x \cdot (x \in s \Rightarrow t(i(x), x) = o)$
THEOREMS
  $\forall x \cdot (x \in s \Rightarrow t(x, i(x)) = o)$;
  $\forall x \cdot (x \in s \Rightarrow t(x, o) = x)$;
  $Monoid(s, t, i, o)$
END

STRUCTURE
  Agroup
SETS
  s
COMPONENTS
  t, i, o
AXIOMS
  $Group(s, t, i, o)$;
  $\forall (x, y) \cdot \left(\begin{array}{c} x, y \in s \times s \\ \Rightarrow \\ t(x, y) = t(y, x) \end{array} \right)$
END

STRUCTURE
  Bool
COMPONENTS
  t
AXIOMS
  $t \in \{0, 1\} \times \{0, 1\} \rightarrow \{0, 1\}$;
  $\forall (x, y) \cdot \left(\begin{array}{c} x, y \in \{0, 1\} \times \{0, 1\} \\ \Rightarrow \\ (x \neq y) \Leftrightarrow t(x, y) = 1 \end{array} \right)$
THEOREMS
  $Agroup(\{0, 1\}, t, \mathrm{id}(\{0, 1\}), 0)$
END

CONSTRUCT
  Group_product
SETS
  s_1, s_2
COMPONENTS
  $t_1, i_1, o_1, t_2, i_2, o_2$
RETURNS
  t, i, o
AXIOMS
  $Group(s_1, t_1, i_1, o_1)$;
  $Group(s_2, t_2, i_2, o_2)$
DEFINITION
  $t \in (s_1 \times s_2) \times (s_1 \times s_2) \to s_1 \times s_2$;
  $\forall (x_1, y_1, x_2, y_2) \cdot \left(\begin{array}{c} x_1, y_1, x_2, y_2 \in s_1 \times s_1 \times s_2 \times s_2 \\ \Rightarrow \\ t((x_1, x_2), (y_1, y_2)) = t_1(x_1, y_1), t_2(x_2, y_2) \end{array} \right)$
DEFINITION
  $i \in s_1 \times s_2 \to s_1 \times s_2$;
  $\forall (x_1, x_2) \cdot \left(\begin{array}{c} x_1, x_2 \in s_1 \times s_2 \\ \Rightarrow \\ i(x_1, x_2) = i_1(x_1), i_2(x_2) \end{array} \right)$
DEFINITION
  $o \in s_1 \times s_2$;
  $o = o_1, o_2$
THEOREMS
  $Group(s_1 \times s_2, t, i, o)$
END

9   Converting into B (for B Users)

The translation of our structure language into B is not very difficult although sometimes delicate. Each structure or construct is translated into a B "machine". The set clause is readily translated into a B set clause. The component and constant clauses are put together and translated into a single B constant clause. The axioms clauses are put together and translated into a single B properties clause. Finally, the lemmas and theorems clauses are all put together and translated into a single B assertions clause.

Structure and construct *calls* are all instantiated within the tool. Calls situated in axioms clauses have their instantiated axioms and theorems put in the B properties clause. When these calls are situated in theorems or lemmas clauses their instantiated axioms and theorems are put in the B assertions clauses, but the fact that the instantiated theorems are implied by the instantiated axioms is put in the B properties clause (which makes the theorems trivially proved as soon as the axioms are).

To summarize, a translation results in a number of B machines dealing with constants only. During the translation process, a few technicalities are encountered, which we shall not cover here due to the lack of space.

Once the B translation is performed, the various tools of Atelier B can come into play: lexical analysis, parsing, type-checking, proof obligation generation, automatic and interactive proofs.

10   Towards a Practical Library of Structures and Constructs

The previous examples are probably not characteristic of what we need in our practical formal model constructions. In what follows, we formalize and give theorems concerning a useful theory, namely that of the reflexive and transitive closure of a graph. It is then used in the definition of the tree structure built on a set s. It is simply defined by means of the "top", t, of the tree and of its "father" function f. The fact that the tree f does not contain one or more separate circuits is formalized by saying that the entire set s belongs to the image of the singleton $\{t\}$ under the transitive closure of the converse of f.

CONSTRUCT
Closure
SETS
s
COMPONENTS
r
RETURNS
c
AXIOMS
$r \in s \leftrightarrow s$
CONSTANTS
g
DEFINITION
$g \in \mathbb{P}(s \times s) \to \mathbb{P}(s \times s)$;
$\forall h \cdot (\ h \in \mathbb{P}(s \times s)$
$\Rightarrow$
$g(h) = \mathsf{id}\,(s) \cup (h \circ r)\,)$
DEFINITION
$c \in s \leftrightarrow s$;
$c = Fixpoint(s \times s, g)$
THEOREMS
$\mathsf{id}\,(s) \subseteq c$;
$c \circ r \subseteq c$;
$c \circ c \subseteq c$;
$$\forall p \cdot \begin{pmatrix} p \in s \leftrightarrow s \ \wedge \\ \mathsf{id}\,(s) \subseteq p \ \wedge \\ p \circ r \subseteq p \\ \Rightarrow \\ c \subseteq p \end{pmatrix}$$
END

STRUCTURE
Tree
SETS
s
COMPONENTS
t, f
AXIOMS
$t \in s$;
$f \in s - \{t\} \longrightarrow s$
CONSTANTS
c
DEFINITION
$c \in s \leftrightarrow s$;
$c = Closure(s, f^{-1})$
AXIOMS
$\{t\} \times s \subseteq c$
LEMMAS
Partial_order(s, c)
THEOREMS
$$\forall u \cdot \begin{pmatrix} u \in \mathbb{P}(s) \ \wedge \\ t \in u \ \wedge \\ f^{-1}[u] \subseteq u \\ \Rightarrow \\ s \subseteq u \end{pmatrix}$$
END

11   Conclusion

In this paper, we have presented an important extension of the B language. It allows one to express higher-order formalizations and proofs. It has been illustrated on a non-trivial example. The proposed language is certainly not yet completely stable. Moreover, it has to be experimented further in some practical developments. We are also thinking about a number of generalizations consisting of using structure and construct calls in wider contexts (i.e. not only at the outermost level of clauses).

References

1. J.R. Abrial. *The B-Book:Assigning Programs to Meanings*. Cambridge University Press (1996).
2. N. Bourbaki. *Théorie des ensembles*. Hermann, Paris, 1970.
3. S. Owre, N. Shankar, J. M. Rushby, and D. W. J. Stringer-Calvert *PVS language reference version 2.3*. Technical report, SRI International, September 1999.
4. L. C. Paulson and K. Grabczewski. *Mechanizing set theory: Cardinal arithmetic and the axiom of choice*. Journal of Automated Reasoning, 17(3):291–323, December 1996.
5. Markus Wenzel. *Using axiomatic type classes in Isabelle. part of the Isabelle distribution*. Technical report, TU München, February 2001.
6. E. Zermelo. *Neuer Beweis für die Möglichkeit einer Wohlordnung*. Mathematische Annalen, 65:107–128, 1908.
7. A. J. M. van Gasteren. *On the Shape of Mathematical Arguments*, volume 445 of *Lecture Notes in Computer Science*. Springer-Verlag, 1990.
8. ClearSy. *Atelier B (version 3.6)*. 2001.

ABS Project: Merging the Best Practices in Software Design from Railway and Aircraft Industries

Pierre Chartier

RATP
7, square Félix Nadar
94684 Vincennes Cedex
tel: 0149578981 — fax: 0149578723
Pierre.Chartier@ratp.fr

Abstract. The design of safety critical systems requires specific methods and tools to reach the safety level required. In the Railway Industry the B Method supported by the Atelier B have been used with success for several years now as shown by the emblematic METEOR metro system. In the Aircraft Industry the use of synchronous declarative languages like Lustre supported by the SCADE tool improves the quality of softwares and saves costs.

The aim of the ABS (Atelier B-SCADE) project is to design a software engineering tool by merging the best practices from both domains:

- On one hand, the B language allows the specification of high-level properties, such as functional and safety ones, through invariant clauses. The B method guarantees the conformity of the code to the requirements by the refinement proof obligations.
- On the other hand, SCADE is very close to the equations used to define critical real time systems. Its user-friendly graphical interface allows non software engineering experts to design such a system. SCADE integrates a simulator and a code generator certified compliant with regards to DO178B standard.

The ABS tool is composed of a generic part and a system specific part. The generic part is independent of the system to design and is based on three items:

- the automatic translation from Lustre to B (the result is called B1) based on rules which are essentially syntactic and reduce drastically the risk of translation errors,
- the proof rules and tactics written by the B expert in order to get an automatic proof between the high-level properties (item 2 below) and the B1 level (equivalent to the Lustre/SCADE counterpart, item 1 below),
- the implementation and proof rules written by the B expert in order to get an automatic B0 implementation from the B1 level.

The system specific part is based on two items:

1. the SCADE system description written by the system expert, actually based on a component library dedicated to this kind of systems,

D. Bert et al. (Eds.): ZB 2002, LNCS 2272, pp. 394–395, 2002.

2. the B abstract specification written jointly by the B expert and the system expert describing the high-level properties of this kind of systems. This specification is written only once for each family of systems (for example in the railway domain: interlocking systems, automatic train protection systems, etc.).

The system specific part of ABS will be tested on the design of an interlocking system.

At the end, the ABS tool will provide a double chain of code generation: a certified chain with SCADE and a proved chain with the Atelier B. Besides the high quality of each chain, the double chain offers other interesting benefits: for example the use of a 2 out-of 2 architecture with a different code on each processor, the validation of each code by the other, etc. The main point is that once the B expert has done his job, the ABS tool will allow non software engineering experts to design high quality software systems.

Generalised Substitution Language and Differentials

James Blow and Andy Galloway

High Integrity Systems Engineering,
Department of Computer Science,
University of York,
UK
{jrb,andyg}@cs.york.ac.uk

Abstract. Embedded continuous control systems can be thought of as implementing complex (piecewise and pipelined) differential functions. Each 'piece' of the function may be preconditioned with a 'domain of applicability', which prescribes the circumstances the piece was designed to handle. The preconditions often involve rate of change, i.e. differentials, as well as range constraints. In this paper we present an adaptation of the substitution calculus which can be used to reason about such systems. Our approach is based on generalising the traditional view that a component is a fragment of a sequential programme. We consider a component to be an autonomous transformation which is 'clocked' to perform its computation at regular intervals, over and over again. In the case of such a component we can generalise the notion of weakest precondition to traces (sequences of values) of inputs and outputs. In our approach we characterise such traces by 'step' predicates over adjacent elements in the trace. We also generalise our calculus to cover n^{th} order differentials. Since analysis can be performed at a comparable complexity to regular wp, our techniques are a powerful tool in the validation of continuous control systems.

1 Introduction

The 'Practical Formal Specification' (PFS) project has been running for approximately five years with the aim of applying formal methods to the problem of specifying and implementing embedded control software. It is funded jointly by the UK Ministry of Defence, Rolls-Royce plc and BAE SYSTEMS.

The focus of the project is to provide, for industries producing control software, the enabling technology to meet Defence Standard 00-55 [15] – which mandates the use of formal methods for safety critical software. Examples of the kind of product in the target domain are full authority engine controllers and flight control systems for aircraft.

The classes of system under consideration are unlike the staple of examples in the formal methods literature. The core software requirements hail from continuous mathematical analysis of the embedding system. The solutions to the

D. Bert et al. (Eds.): ZB 2002, LNCS 2272, pp. 396–415, 2002.

differential equations, which model the physics of the real world, give rise to PID (proportional, integral, differential) calculations which monitor the state of the environment and, in real time, adjust the controlled variables in the environment to achieve a desired effect.

Different control models (for different situations) are then combined with mode selection logic. Fault monitoring and failure management requirements are then added to the model. The result is a set of requirements that:

- has little richness in data (most data are boolean or fixed-point numbers)
- is deterministic (only one outcome is prescribed for each situation)
- is highly explicit (describes the 'how' rather than the 'what')

As such we have found the usual levers for abstraction, in formalisms that lead to executable code, difficult to apply to the domain[1] [14,12,13].

In addition, control system requirements are complex. They involve transformation of input values and state into output values and new state. They involve timing properties – the computations must be performed at a stable frequency in order not to invalidate the continuous modelling. Increasingly, the requirements also involve distribution, communication and asynchrony. In [10,11] we presented the use of Generalised Substitution Language within the context of a process algebra, which we are now using to validate complex requirements. However, the work presented in this paper concentrates on an orthogonal problem.

Of the conventional abstraction mechanisms available in model-based formalisms (such as B, Z) the easiest to apply to the domain is that of specifying sub-components within assumptions (preconditions) and then weakening the precondition through design. The 'firewalling' of components with assumptions is also one of the most valuable aspects when it comes to validation. By demonstrating that components are only ever supplied parameters which conform to their preconditions (as we effectively do in the B method), the engineers confidence in correctness of the requirements is significantly increased.

However, since the component requirements hail from differential models of the environment, the 'handling assumptions' which precondition the components' intended use often involve constraints on rate of change of a variable as well as the range a variable can take. Clearly the 'flat' style preconditions of Generalised Substitution Language (GSL)[2] – a predicate over a single instance of the input variables and before state – are insufficient to represent such characteristics. Moreover, in order to prove that components' preconditions are preserved by their context or usage, we would like a substitution calculus which also embraces differential information.

In what follows we present a first step towards such a calculus. An adaptation of the Generalised Substitution theory is outlined which allows to characterise

---

[1] Note that abstract requirements *are* embraced by the continuous analysis. Requirements such as performance, physical limits (such as the maximum working temperatures of the materials) and physical optimals (desired control parameters to minimise maintenance etc.) are all taken into account when engineering the control requirements.

and manipulate first-order (subsequently generalised to n^{th}-order) pre and post conditions. These are specified (for assumed to be regularly clocked operations), as 'step' (or sequences of step) predicates involving several instances of the relevant variables. A 'macro' Generalised Substitution Language is then overviewed which allows the specifier to construct substitutions preconditioned with first-order (or n^{th}-order) constraints and to compose them together in sequence. The combined theory allows us to derive preconditions for systems constituted by synchronously composed components. The theories are illustrated with examples including deriving preconditions for proportional, integral and differential components, as well as for a complete PID specification.

Our hope is that whilst these are early results, they will open the debate on this important aspect of formalism for control software.

The paper is structured as follows: sections 2 and 3 discuss the generalisation of weakest precondition calculus to synchronously composed operations; section 4 introduces our characterisation of a first-order condition; sections 5 and 6 explain how regular GSL can be used to derive preconditions from first-order postconditions; section 7 includes proportional, integral and differential examples; section 8 generalises the theory to n^{th}-order; section 9 introduces the macro substitution language and provides the PID example; sections 10 and 11 discuss related work and record our conclusions respectively.

2   Generalised Substitution Language, wp and Synchronously Composed Operations

We can think of a generalised substitution (operation) as a component (program fragment, procedure or autonomous process) which has inputs (i) and outputs (o) and some special 'outputs' (x) which are also 'inputs' (state). See figure 1.

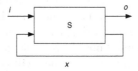

Fig. 1. Example Component with State x and Operation S

The substitution has a termination condition trm(S) and feasibility condition fis(S) which can both be expressed as predicates over the inputs. In addition, the substitution can be said to have a behaviour prd(S) relating the outputs to the inputs.

Traditionally, we view the behaviour of the component as one which given some inputs i in state x, updates x and assigns o according to its trm, fis and prd. Indeed, weakest precondition calculi (wp or wlp depending upon ones insistence on termination) will calculate the largest set (weakest characterising predicate) of input values such that if we know we have inputs and state in this set, then we are guaranteed to arrive in some designated set of outputs and state values (as characterised by some predicate). B's substitution calculus is a theory for deriving weakest preconditions on operations specified using the Generalised Substitution Language.

The traditional view is inspired by the notion that a component is a fragment of a sequential programme. However, a potentially powerful generalisation of this view comes from considering the component to be an autonomous transformation (say a function on a chip, a parallel process, or cyclically scheduled procedure) which is 'clocked' to perform its computation at regular intervals, over and over again. Our view of a 'programme' under this interpretation is a synchronous composition of operations, invoked repeatedly, in which every operation output is uniquely named (and therefore set once only during each repetition).

In the case of the 'clocked' component we can generalise the notion of weakest precondition to 'traces'[2] of inputs and outputs. Thus, given a set of output traces (characterised symbolically in some way), we want to be able to derive the largest set (again characterised symbolically) of input traces such that if we present one of the traces in the input set we are guaranteed, given the behaviour of the component, to produce a trace in the designated output set. This prompts the question: which classes of traces are we interested in, and how do we represent and manipulate them symbolically?

3   Standard Generalised Substitution Language and Traces

For standard GSL, in fact, we can already be said to be working on traces – albeit on a restricted class of traces. By calculating $[S]Q$, we are using Q as an acceptance criterion for each element in the output trace. Thus Q characterises the set of traces in which every element in the trace satisfies Q. The weakest precondition $[S]Q$ for a standard generalised substitution S yields another acceptance criterion for inputs. Thus if every element in the input sequence satisfies $[S]Q$ then every output in the sequence will correspondingly satisfy Q and any trace characterised by the $[S]Q$ acceptance criterion is guaranteed to produce a trace characterised by the Q acceptance criterion.

If there is state, then there is the additional condition that the after state from each input presentation becomes the before state for each immediately following input presentation. We need the notion of a state invariant. If we have a predicate on the state variables, Inv, and condition on the inputs P, such that $P \wedge Inv \Rightarrow [S](Q \wedge Inv)$ and we know that the initial state satisfies Inv, then we also have a precondition (though not necessarily the weakest) on traces. To guarantee output traces whose elements' acceptance criteria is Q, we must guarantee input traces whose elements' acceptance criteria is P.

---

[2] By 'traces' we mean sequences of associations of variables to values.

4   First Order Differentials and Traces

The above deals with a restricted class of traces which can be characterised by an acceptance predicate over each of the trace's elements (in effect, zeroth order differentials). However, there are more interesting classes of traces to be considered. For instance, traces that also take into account first-order differentials. In other words, traces characterised by acceptance predicates which not only describe the set of values each individual element make take, but also how values are permitted to change from trace element to element.

For example, if we know that a component is clocked at a particular frequency, we might want to describe a desired output trace with particular range and rate of change constraints: 'all the outputs are greater then 10 and they must not increase or decrease by more than 5 each invocation of the component's operation'.

Moreover, we are interested in discovering the acceptance predicate that describes the widest set of input traces, for the component, that guarantees one of the desirable output traces.

The simplest, least redundant, acceptance predicate is a 'step' predicate[3]. By 'step' predicate we mean one which constrains two arbitrary adjacent elements in a trace. To be able to reason about steps we need variables with different decorations.

We adopt a prefix-prime notation[4] to denote values in the first element of a 'step'[5] (for example, $'o$) and an unprimed variable to denote values in the second element of the 'step'[6] (for example, o).

We can therefore encode our example constraint introduced above as:

$$o > 10 \land o - 'o < 5 \land o - 'o > -5$$

We will also need a way to adapt substitutions on unprimed variables to operate on primed variables. We adopt the notation $'[S]_y Q$ to mean the substitution S with all unprimed free variables (including preconditions, guards, assignment variables and assigning expressions), except state variables y, replaced with their primed equivalent, applied to the postcondition Q. $'[S]Q$ therefore denotes the special case where there are no exceptions. For example, $'[x > 5 \mid x := x + 2]Q$ is equivalent to $['x > 5 \mid 'x := 'x + 2]Q$.

Finally, we adopt the notation $'P$ to mean the predicate P with all unprimed free variables replaced with their primed equivalent.

---

[3] Other characterisations could have been chosen, such as sequences. However, the sequence representation is less natural to the engineers and is a less efficient encoding since it will encode any trace, not just the class we are interested in, and in a more verbose way.

[4] We adopt the prefix use of a prime so as not to overload the postfix use of primes, which has well defined meanings in both control theory and discrete formal methods.

[5] We can also read this as the 'previous' value of the variable.

[6] We can also read this as the 'current' value of the variable.

5 Deriving Preconditions from First Order Postconditions

Recall that in terms of traces characterised by zeroth order constraints, by calculating $[S]Q$, we are using Q as an acceptance criterion for each output and deriving $[S]Q$, the acceptance criteria for each input.

If we have state, we also need a predicate on the state variables, Inv. For a given predicate P, if we prove $P \wedge Inv \Rightarrow [S](Q \wedge Inv)$ for traces with output acceptance criteria Q, then we know input acceptance criteria P is sufficient to guarantee traces characterised by Q.

We present the following argument to extend the theory to traces characterised by *first order* constraints (and later in Section 8, n^{th} order constraints).

Consider a single component as shown in Figure 1. Assume we have a first order postcondition given by Q, a predicate over o and $'o$. Dealing first with the second element of the step, i.e. the 'current' invocation of the operation giving the 'current' output: $[S](Inv \wedge Q)$ yields a predicate on x, i and $'o$ (i.e. the unprimed state variables and input variables, and the primed output variables). This effectively gives us the weakest precondition, on the before state and input of the 'current' operation invocation and the output from the 'previous' operation invocation, to guarantee the postcondition Q (and maintain the invariant). However, the operation substitution is defined on unprimed variables and we need it to operate on primed variables to manipulate the output from the 'previous' operation invocation.

To link the before state of the 'current' invocation of the operation to the after state of the 'previous' invocation, we must now substitute the component state variables by their primed equivalent:

$$[x := 'x][S](Inv \wedge Q)$$

This is now the postcondition which the 'previous' invocation of the operation (adapted to primed variables) must establish. However we must also maintain the invariant on the 'previous' operation invocation[7], giving:

$$'[S]('Inv \wedge [x := 'x][S](Inv \wedge Q))$$

To give the final precondition, in terms of the input variables, we universally quantify over the primed state variables of the operation:

$$\forall \, 'x \bullet ('Inv \Rightarrow '[S]('Inv \wedge [x := 'x][S](Inv \wedge Q)))$$

---

[7] For an example of the need to preserve the invariant on *both* invocations of the operation, consider an operation with no inputs, a single output o and state variable x, where o and x are natural. The operation's specification is $o := x \, || \, x = 1 - x$. Assuming the post condition $o - 'o > 0$, and choosing the invariant $x = 0$ and initialisation $x := 0$, we can derive an erroneous result by not insisting the invariant is maintained by the first invocation of the operation. The initialisation establishes the invariant. Every pair of invocations reestablishes the invariant. The precondition is therefore *true*. However, the output is clearly not guaranteed to increase.

Alternatively, if we have a candidate precondition P defined over the input variables i, and $'i$ we obtain the following proof obligation:

$$\vdash P \wedge {}'Inv \Rightarrow {}'[S]({}'Inv \wedge [x := {}'x][S](Inv \wedge Q))$$

In the above we have used invariant preservation over consecutive operation calls. However, if we are able to preserve the invariant over a single operation call, then two consecutive calls will also preserve it. Consequentially, we can separate our proof obligation above into the following two proofs[8]:

$$\vdash P \wedge Inv \Rightarrow [S]Inv$$

$$\vdash P \wedge Inv \Rightarrow {}'[S]_x[S]Q$$

We therefore have a preondition given by:

$$\forall x \bullet Inv \Rightarrow ([S]Inv \wedge {}'[S]_x[S]Q) \tag{1}$$

6   Proof Obligations for First-Order Theory

We are now in a position to state proof obligations for the first-order theory[9].

Proof Obligation 1: Initialisation. We need to prove that the invariant is established when the system is initialised. Given an initialisation substitution *Init*:

$$\vdash [Init]Inv \tag{2}$$

Proof Obligation 2: State. If we have a candidate precondition P on free variables i and $'i$, a postcondition Q on free variables o and $'o$, and an invariant *Inv* constraining state variable(s) x, we must prove:

$$\vdash P \wedge Inv \Rightarrow ([S]Inv \wedge {}'[S]_x[S]Q) \tag{3}$$

Proof Obligation 3: No State. If we have no state variable, there is no need to involve the invariant and rename the state variables. Proof Obligation 3 simplifies to:

$$\vdash P \Rightarrow {}'[S][S]Q \tag{4}$$

---

[8] The invariant is still needed as an antecedent for the step case since in general $'[S]_x[S]Q$ will contain free occurances of the state variables.

[9] In what follows we ignore contextual information e.g. those given by constraints and properties clauses in B method.

7   Examples

The components of a control system are invoked at a particular frequency, mapping their inputs (and their integrals/differentials) to outputs (and their integrals/differentials) and updating their internal state (belief of the state of the embedding system) in a time insignificant to the invocation frequency (the synchrony hypothesis). The component requirements often involve pipelines of proportional (scaling function on inputs - aka gain functions), integrator and differentiator sub-components. In the examples below we focus on these three types of subcomponent, as shown in figure 2, and highlight the application of Proof Obligations 3 and 4 (we do not consider initialisation issues within these examples).

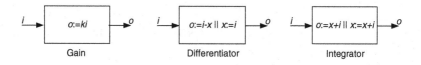

Fig. 2. Main Components Types

Proportional. Consider first a Gain component, as in Figure 2. It is a function which maps outputs directly in some relationship to inputs. It does not have state and in general has the form $o = f(i)$, where o is an output variable of the component, i an input variable, and f the functionality of the component. In our example we choose f to be $\times k$ and therefore have a component specification given by the substitution $o := i * k$. We assume a step postcondition given by $o - 'o < k$, i.e. o must not increase by more than k.

Since we have no state, we can use Proof Obligation 4 to prove that a candidate precondition P is sufficient to guarantee the postcondition:

$$\vdash P \Rightarrow {}'[S][S]Q$$

We are able to derive the weakest precondition as follows:

$$[{}'o := k * {}'i][o := k * i](o - {}'o < k)$$
$$\equiv [{}'o := k * {}'i](k * i - {}'o < k)$$
$$\equiv k * i - k * {}'i < k$$

Which simplifies to $i - {}'i < 1$ (i.e. i must not increase).

Note that since the component has no state and assigns the outputs as a constant function of the inputs, we are able to derive a precondition which has the same order of differential as the postcondition. For traces with output acceptance criteria $o - {}'o < k$, we need to prove any candidate input acceptance criteria P is stronger than $i - {}'i < 1$:

$$\vdash P \Rightarrow i - {}'i < 1$$

Differential. We now consider a differential component as shown in Figure 2. In such a component, state, denoted by x, is used to record the previous input and subtract the current value from the previous value. In general, we have $o = i - x$ and $x = i$, where i and o once again represent the component's input and output. Our specification is therefore given by the substitution: $o := i - x \parallel x := i$.

In our example, we assume that we have a flat postcondition $o < k$. Since we now have state, we need to consider the invariant. We choose the invariant (Inv) to be that x is an integer. We now derive the precondition, using (1) in Section 5:

$$\forall x \bullet (x \in \mathbb{Z} \Rightarrow ([o := i - x \parallel x := i] x \in \mathbb{Z} \wedge$$
$$[{}'o := {}'i - x \parallel x := {}'i][o := i - x \parallel x := i] o < k))$$

$$\equiv$$

$$\forall x \bullet (x \in \mathbb{Z} \Rightarrow (i \in \mathbb{Z} \wedge$$
$$[{}'o := {}'i - x \parallel x := {}'i](i - x) < k))$$

$$\equiv$$

$$\forall x \bullet (x \in \mathbb{Z} \Rightarrow (i \in \mathbb{Z} \wedge (i - {}'i) < k))$$

Note that we have derived a step (first order) precondition from a flat (zeroth order) postcondition. Our proof obligation, based on Proof Obligation 2 and assuming a candidate precondition P, is therefore given by:

$$\vdash P \wedge x \in \mathbb{Z} \Rightarrow (i \in \mathbb{Z} \wedge i - {}'i < k)$$

Furthermore, if P implies $i \in \mathbb{Z}$, then our proof obligation simplifies to P is stronger than $i - {}'i < k$:

$$\vdash P \Rightarrow i - {}'i < k$$

Integral. The final example considers an integrator component. An integrator uses state to accumulate input values. More precisely, the output and state of the component are the sum of the current input and the current value of the state. An integrator (a linear integrator) has the specification $o := x + i \parallel x := x + i$. We take the post-condition to be $o - {}'o < k$ and again choose the invariant (Inv) as $x \in \mathbb{Z}$:

$$\forall x \bullet (x \in \mathbb{Z} \Rightarrow([o := i + x \| x := i + x]x \in \mathbb{Z} \wedge$$
$$['o := 'i + x \| x := 'i + x][o := i + x \| x := i + x](o - 'o < k)))$$

$$\equiv$$

$$\forall x \bullet (x \in \mathbb{Z} \Rightarrow(i + x \in \mathbb{Z} \wedge ['o := 'i + x \| x := 'i + x](i + x - 'o < k)))$$

$$\equiv$$

$$\forall x \bullet (x \in \mathbb{Z} \Rightarrow(i + x \in \mathbb{Z} \wedge (i + 'i + x - 'i - x) < k))$$

The precondition is therefore given by:

$$\forall x \bullet (x \in \mathbb{Z} \Rightarrow (i + x \in \mathbb{Z} \wedge i < k))$$

Observe how the desired output differential has been (mostly) 'absorbed' by the wp effect of the component, yielding an input trace characterisation without a differential constraint (other than that both instances of the input variables are of the right type ($\mathbb{Z}$)).

If we have a candidate precondition P,then we have the following proof obligation:

$$\vdash P \wedge x \in \mathbb{Z} \Rightarrow x + i \in \mathbb{Z} \wedge i < k$$

Moreover, if P implies $i \in \mathbb{Z}$ then the proof obligation simplifies to P is stronger than $i < k$:

$$\vdash P \Rightarrow i < k$$

Summary of Results. The results of the examples are summarised in Table 1, where $\Sigma(x_1, \ldots, x_n)$ denotes a predicate Σ over free variables $x_1, \ldots, x_n$. We can see how the form of the postcondition is preserved by the wp effect on the gain component, how a differential postcondition is 'absorbed' and so on. Importantly, the table also introduces two results which we have not yet discussed. These are the effect of wp on an integral component with a flat postcondition and the wp effect on a differential component with a step postcondition.

For integral components, we are able to maintain a first-order postcondition by controlling 'flat' (zeroth order) inputs. However, there is generally no way to limit inputs to ensure a flat postcondition. This can be achieved, however, if the postcondition states that the output is multiple of some k, in which case the precondition is that the input is a multiple of the same k. The simplest case is $k = 1$, so the postcondition is $o \in \mathbb{Z}$ and the precondition is $i \in \mathbb{Z}$. Therefore, the wp effect on an integral with a flat postcondition is *false* unless the precondition and the postcondition can be formulated in terms of multiples of some k.

We have seen how a flat (zeroth order) postcondition and a differential component give rise to a step (first order) precondition. In other words, the wp effect on such a component results in a precondition which is of one order higher than the postcondition. Similarly, a second order precondition should be derived from a first order postcondition. However, the theory which we have presented so

Table 1. Results for First Order Differentials on Selected Component Types

Precondition	Component	Postcondition
$P(i, {}'i)$	Gain	$Q(o, {}'o)$
$P(i)$	Gain	$Q(o)$
$P(i, {}'i, {}''i)$	Diff	$Q(o, {}'o)$
$P(i, {}'i)$	Diff	$Q(o)$
$P(i)$	Int	$Q(o, {}'o)$
Multiple_of_k(i)/$false$	Int	$Q(o)$

far has only been able to deal with zeroth order and first-order constraints. In Section 8, we address this limitation by extending the theory to be able to characterise and manipulate constraints involving arbitrary orders of differentials.

8 n^{th} Order Differential Constraints

So far we have used step predicates to characterise traces involving first order differentials. To represent acceptance criteria for traces involving n^{th} order differentials we need constraints which range over $n + 1$ elements. For example, a constraint which states that the 'acceleration' of the variable cannot be more than 6, would need to range over three adjacent elements in a trace.

n^{th} Order Variable Decorations. Given that we now want to reason about n^{th} order pre and post conditions, ranging over several elements of a trace we need to build on the earlier prefix-prime notation. We will still have an unprimed variable, such as x referring to the most recent element in the step sequence and a primed variable, such as $'x$, denoting the second most recent element in the step sequence. However, we now have $''x$ denoting the third most recent element in the step sequence, and so on, until the $(n + 1)^{th}$ most recent element in the step sequence, given by $\overbrace{'\cdots'}^{n}x$. By step sequence we mean the sequence of elements needed to characterise the relevant n^{th} order constraint.

We are now able to represent the restriction that an output can not 'accelerate' by more than 6 units:

$$(o - {}'o) - ({}'o - {}''o) < 6$$

In other words, we are now in a position to be able to reason about traces involving n^{th} order differentials by predicates involving the free variables:

$$x, {}'x, {}''x, \ldots \overbrace{'\cdots'}^{n-1}x, \overbrace{'\cdots'}^{n}x$$

Again, we need to adapt substitutions and predicates to their primed equivalents. We interpret $\overbrace{'\cdots'}^{k}[S]_y\ Q$ as S with all free variables $\overbrace{'\cdots'}^{l}x$ replaced by $\overbrace{'\cdots'}^{l+k}x$, except state variables y, applied to Q. We assume $\overbrace{'\cdots'}^{k}P$, for predicate P, to be defined similarly.

For example, $''[x < 5 \Longrightarrow x := x + 2]\ Q$ is equivalent to $['' x < 5 \Longrightarrow '' x := '' x + 2]\ Q$.

n^{th} Order Precondition Derivation and Proof Obligations. Given the extra notation it is now possible to extend the first-order theory. Firstly, we will consider the second-order case, and then generalise to n^{th} order.

Recall the weakest precondition formulae for zeroth-order and first-order constraints:

$$\forall x \bullet (Inv \Rightarrow ([S]Inv \wedge [S]Q))$$
$$\forall x \bullet (Inv \Rightarrow ([S]Inv \wedge '[S]_x[S]Q))$$

The first order version was derived by applying the substitution twice (one for unprimed variables and once for primed variables), maintaining the invariant after each application. For the second order formulae, we need to apply the substitution three times maintaining the invariant on each application:

$$Inv \Rightarrow ([S]Inv \wedge ''[S]_x\,'[S]_x[S]Q) \tag{5}$$

Since we want the precondition in terms of i, $'i$ and $''i$, we must also universally quantify over the state variable $''x$ to give the *second order* precondition:

$$\forall x \bullet (Inv \Rightarrow ([S]Inv \wedge ''[S]_x\,'[S]_x[S]Q)) \tag{6}$$

The second order version of the proof obligation, assuming a candidate P over i, $'i$, and $''i$, is therefore:

$$\vdash P \wedge Inv \Rightarrow ([S]Inv \wedge ''[S]_x\,'[S]_x[S]Q)$$

Repeating the above process n times, we obtain the following general forms for the precondition formula and proof obligation for n^{th} *order* predicates:

$$\forall x \bullet (Inv \Rightarrow([S]Inv \wedge \overbrace{'\cdots'}^{n}[S]_x\overbrace{'\cdots'}^{n-1}[S]_x \ldots '[S]_x[S]Q)) \tag{7}$$

$$\vdash P \wedge Inv \Rightarrow ([S]Inv \wedge \overbrace{'\cdots'}^{n}[S]_x\overbrace{'\cdots'}^{n-1}[S]_x \ldots '[S]_x[S]Q) \tag{8}$$

The nth order precondition formula and proof obligation shown above are verbose. For concise presentation we adopt the following recursive representation for the precondition formula, where k denotes the order under consideration.

$$\{\!|S|\!\}^k_x\ Q \equiv \overbrace{'\ldots'}^{k}[S]_x\{\!|S|\!\}^{k-1}_x\ Q) \tag{9}$$

where the base case is given by:

$$\{\!|S|\!\}^0_x\ Q \equiv [S]Q \tag{10}$$

Finally, quantifying over the state variables and adding the initial invariant we have:

$$\langle\!|S|\!\rangle^k_{Inv_x}\ Q \equiv \forall x \bullet (Inv \Rightarrow ([S]Inv \wedge \{\!|S|\!\}^k_x\ Q))) \tag{11}$$

where $\langle\!|S|\!\rangle^k_{Inv_x}$ denotes the kth-order precondition formula for the substitution, S (with invariant Inv on state variables x) acting on Q.

Special Case. If we have no state then 9 above simplifies to:

$$\{\!|S|\!\}^k\ Q \equiv \overbrace{'\ldots'}^{k}[S]\{\!|S|\!\}^{k-1}\ Q \tag{12}$$

where the base case is given by:

$$\{\!|S|\!\}^0\ Q \equiv [S]Q \tag{13}$$

and the kth-order precondition is given by:

$$\langle\!|S|\!\rangle^k\ Q \equiv \{\!|S|\!\}^k\ Q \tag{14}$$

8.1   Differential Example

To illustrate the higher order theory we present an example which incorporates higher order differentials. Given the differential component again as shown in figure 2, the aim is to derive a second-order precondition ranging over free variables i, $'i$, $''i$.

Recall the Differential example in Section 4. From the flat postcondition $o < k$ we were able to derive the step precondition $i - 'i < k$ (for simplicity ignore the type information). In this example we shall use this precondition as our postcondition (by relabelling i and $'i$ with o and $'o$, respectively).

Since (up to) a second-order precondition is required, we use the second-order theory. Setting $k = 2$ in (11), we obtain:

$$\langle\!\langle S \rangle\!\rangle^2_{Inv_x} = \forall x \bullet (Inv \Rightarrow ([S]Inv \wedge ''[S]_x\,'[S]_x[S]Q))$$

If *Diff* is the specification of the differentiator (as employed earlier) we derive the precondition as follows:

$$\forall x \bullet (x \in \mathbb{Z} \Rightarrow ([o := i - x \parallel x := i]x \in \mathbb{Z} \wedge$$
$$''[\mathit{Diff}]_x\,'[\mathit{Diff}]_x[o := i - x \parallel x := i](o - 'o < k)))$$

$$\equiv$$

$$\forall x \bullet (x \in \mathbb{Z} \Rightarrow (i \in \mathbb{Z} \wedge$$
$$''[\mathit{Diff}]_x['o := 'i - x \parallel x := 'i](i - x - 'o < k)))$$

$$\equiv$$

$$\forall x \bullet (x \in \mathbb{Z} \Rightarrow (i \in \mathbb{Z} \wedge$$
$$[''o := ''i - x \parallel x := ''i](i - 'i - 'i + x < k)))$$

$$\equiv$$

$$\forall x \bullet (x \in \mathbb{Z} \Rightarrow (i \in \mathbb{Z} \wedge$$
$$(i - 'i - 'i + ''i < k)))$$

We therefore have the following proof obligation for a candidate precondition P:

$$\vdash P \wedge x \in \mathbb{Z} \Rightarrow i \in \mathbb{Z} \wedge i - 2 * 'i + ''i < k$$

If P implies $i \in \mathbb{Z}$ and then the proof obligation simplifies to:

$$\vdash P \Rightarrow i - 2 * 'i + ''i < k$$
$$\equiv \quad \vdash P \Rightarrow (i - 'i) - ('i - ''i) < k$$

This means that rate of change of rate of the input must be less than k, i.e. i must not 'accelerate' at a rate greater than k.

9   Extended Substitution Theory

So far we have generalised the notion of weakest precondition to traces, where each trace is characterised by n^{th} order differential constraints. However, the development of the theory has been restricted to single components. E.g. the integrator and differentiator components in Section 4.

In practise, however, components such as these are not used in isolation. They are the building blocks for large, complex subsystems, such as those found in aeroengine and flight control systems. Consequentially, we need a theory which not only permits component composition, but also the propagation of desirable traces from outputs, through the intermediate components, into inputs. The theory is presented in terms of the earlier n^{th} order definitions. Therefore all formulae can ultimately be rewritten in terms of standard generalised substitutions.

9.1   The 'Macro' Substitution Language

Real systems are composed from more than one subcomponent. In addition, we wish to precondition systems, sub-systems and their components with 'handling conditions', which specify the context in which they were designed to be used. These handling conditions need to be n^{th}-order constraints. Therefore, rather than ensuring the condition over every invocation of an operation[10] – as for regular preconditions – we want to ensure them over several invocations. Moreover, for a composed system and postcondition, we wish to derive a system precondition which guarantees the postcondition as well as all the embedded handling assumptions. To achieve this we employ a 'macro' substitution language, which permits n^{th}-order preconditioning and composition. The 'macro' language has four rules corresponding to its four syntactic forms. We use S_m and T_m to range over macrosubstitutions and S over regular substitutions.

Standard Substitution Rule. For a component specified by a standard substitution S, with state x, invariant Inv and a postcondition Q, for a kth-order precondition:

$$[\![S]\!]^k_{Inv_x} Q \equiv \langle\!\langle S \rangle\!\rangle^k_{Inv_x} Q \tag{15}$$

Precondition Rule. If P ranges over $i, \ldots, \overbrace{\text{'}\cdots\text{'}}^{n} i$ and is an assumption which provides the 'handling context' of component S_m, then we want to 'separate' the precondition from the macrosubstitution and conjoin it to the component's derived precondition. We use a $\upharpoonright$ to denote this form of preconditioning.

$$[\![P \upharpoonright S_m]\!]^k_{Inv_x} Q \equiv P \wedge [\![S_m]\!]^k_{Inv_x} Q \tag{16}$$

Invariant Decoration Rule. To delineate the components of the system with their own state, we allow 'macro' substitutions to carry invariant and state annotations[11]:

$$[\![(S_m)_{Inv_x}]\!]^k Q \equiv \wedge [\![S_m]\!]^k_{Inv_x} Q \tag{17}$$

---

[10] It would be impossible to ensure an nth order constraint ($n > 0$) over a single operation invocation.

[11] For this presentation we assume the static semantic constraint that invariant decorations may not be nested i.e. there is no need for a rule for the form $[\![(S_m)_{Inv1_x}]\!]^k_{Inv2_y} Q$. This will be the subject of future publications.

Sequential Composition Rule. Sequential composition of 'macro' substitutions is defined as the application of the first 'macro' substitution to the result of applying the second 'macro' substitution to the postcondition:

$$[\![S_m \, \text{\fontsize{5}{6}\selectfont ;} \, T_m]\!]^k Q \equiv [\![S_m]\!]^k [\![T_m]\!]^k Q \qquad (18)$$

9.2   Example – A PID Control Loop

The goal with this example is to show the propagation of step predicates through a pipeline of components.

A PID Loop is a standard structure within control theory. It is used predominantly in embedded systems for maintaining precise control in a continually changing environment. We regard the component in Figure 3 as the system which we are analysing. However, in reality the component would be a very small section of a much larger system.

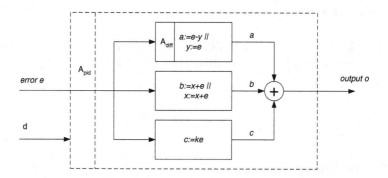

Fig. 3. PID Loop

The component shown in Figure 3 can be regarded as having four subcomponents – a gain (proportional), an integrator, a differentiator and a sum. The first three act independently, but in parallel, to produce outputs a, b and c. These are then summed by the sum subcomponent to produce the PID component's output, o.

The component has two inputs - an 'error' input, denoted by e, and a 'logical' input, denoted by d. The 'error' input represents the difference between a measured value of a variable and some desirable value and is used in the calculation of the component's output. The logical input is *not* used in any calculations, but is needed to 'contextualise' the requirements.

The differentiator is assumed to be an 'off-the-shelf' component which should only be used under certain circumstances. Consequentially, it has a 'handling condition' (precondition) scoping its correct usage. The component is specified for a certain range of input values, and should not be used for rapidly fluctuating input values. The assumption is given by $e > min \wedge e < max \wedge e - {}'e < \delta$, and is denoted by A_{diff} in figures 3 and 4. In other words, the error input must be within some maximum and minimum values, and that the rate of change of the error signal must be less than the constant δ.

We also assume that the PID has been developed using a particular mathematical model of the environment. Such a model might make assumptions on the 'cleanliness' of the signals e and d (such as rate of change constraints), as well as catering for a particular operational scenario (e.g. high pressure turbine speed above idle), say, by the range d can take. We refer to these handling conditions as A_{pid}, and use an example definition of A_{pid}:

$$A_{pid} \equiv e - {}'e < \sigma \wedge {}'e - e < \sigma \wedge d \geq \gamma$$

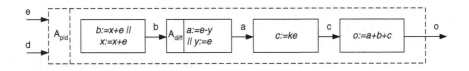

Fig. 4. Sequential Representation of a PID Loop

Considering the dependencies of the specified calculation, it is possible to represent the system sequentially. The calculation of a, b and c are independent (i.e. they do not rely on each other). However, o is dependent on a, b and c. Therefore, as long as o is calculated last it does not matter about the order of the other calculations. We can therefore represent figure 3 by figure 4 and specify the system in terms of the macro-substitution language:

$$A_{pid} \upharpoonright ((Int)_{Inv1_x} \, {}^\circ_\circ \, (A_{diff} \upharpoonright Diff)_{Inv2_y} \, {}^\circ_\circ \, Gain \, {}^\circ_\circ \, Sum)$$

where $Inv1$ is the chosen invariant for the integrator acting on its state variable x, and $Inv2$ is the chosen invariant for the differentiator acting on its state variable y. It is now possible to derive a second order precondition[12] for an example postcondition $(o - {}'o <)\alpha$, such that all the embedded handling constraints are ensured. The details of the derivation can be found at [1].

---

[12] A second-order order precondition will be sufficient since the postcondition is first-order and there is a single differentiator in the system which we know will add one differential order to the postcondition.

The derived precondition is:

$$e - {}'e < \sigma \wedge {}'e - e < \sigma \wedge d \geq \gamma \wedge$$
$$\forall x \bullet (x \in \mathbb{Z} \Rightarrow (x + e \in \mathbb{Z} \wedge$$
$$e > min \wedge e < max \wedge e - {}'e < \delta \wedge$$
$$\forall y \bullet (y \in \mathbb{Z} \Rightarrow (e \in \mathbb{Z} \wedge$$
$$(2 + k) * (e - {}'e) + {}''e < \alpha))))$$

Which, given a candidate precondition P which implies $e \in \mathbb{Z}$, requires a proof of:

$$\vdash P \Rightarrow e - {}'e < \sigma \wedge {}'e - e < \sigma \wedge d \geq \gamma \wedge$$
$$e > min \wedge e < max \wedge e - {}'e < \delta \wedge$$
$$(2 + k) * (e - {}'e) + {}''e < \alpha$$

10   Related Work

The lack of examples of work which are aimed at solving the sorts of problems encountered in the development of control systems highlights a particular 'gap' in formal methods research. Some attempts at tackling the issues involved result in solutions which are complex, and arguably impractical for industrial usage. For example, [8].

Our approach has not been to try and devise a new formal method, but to ask how we can extend existing techniques so that we are able to reason *more meaningfully* about the types of systems we are interested in. Building on an existing well established approach also has the advantage of being able to adapt associated tool support.

Our focus has been on GSL[2] and weakest precondition calculus [9]. However trying to model systems in terms of the sequences of input and output values of a component is not novel in itself. In the work of Broy [3,5,6,4,7], a modular method is used for the specification of distributed systems. Communication histories between components are represented by 'streams', where a stream is a finite or infinite sequence of message or action. Components can then be modelled as functions or relations (stream processing functions) on streams and assertions can be made about the communication channels using temporal logic. E.g. a history assertion refers to the stream of values sent via a channel. Within this framework, a system therefore corresponds to a network of stream processing functions.

While such an approach has benefits (e.g. its modularity), it may still be too complex for practical control systems development. Moreover, we are not aware of any work which extends the weakest precondition approach to traces in general, and to traces involving differential constraints in particular.

11 Conclusions

Practical Application of the Techniques. PFS [12,14] advocates an approach in which control system requirements including 'handling constraints' are specified using a combination of diagrams and tables. Specification healthiness conditions then allow the engineer to reason formally about the consistency of the requirements. The techniques have been used within Rolls-Royce, to specify requirements for a real engine design – albeit with limited use of formal proof.

Confidence in system validity is increased by demonstrating that given the environmental constraints understood by the engineers – specified as pre and postconditions on the system – the system precondition is sufficient to guarantee the postcondition and ensure all embedded handling constraints. Until now, the composition semantics was based on 'flat' (zeroth order) pre and postconditions. However, our aim now is to apply the theory presented in this paper as a basis for reasoning about higher order conditions.

PFS has provided some support for modularity. Engineers are encouraged to hide local state within subsystems as well as 'firewall' components with handling constraints which must guarantee the components embedded handling constraints. However, we believe value could be gained from trying to apply the Abstract Machine Notation's structuring mechanisms to the domain, and thus further work is needed.

Another aspect of further work is to employ the higher order theory for analysing distributed control systems. The aim will be to integrate the work presented in this paper with the use of GSL in a process algebraic context [10, 11].

Static Code Analysis. Since the techniques presented in this paper build on weakest precondition calculi, we believe that they can also be used for program verification using static code analysis.

SPARK Ada is a subset of Ada tailored towards high integrity programming. It has an associated tool set, the SPARK Examiner, which can be used to generate verification conditions (VC's) for formal proof of the SPARK code. The VC's are derived from annotations (which are ignored by the compiler) in the SPARK code. Annotations include **pre**, which is used to specify the precondition of a procedure or function, and **post**, which is used to specify the postcondition of a procedure or function. The proof obligations generated are based on standard weakest precondition calculus. To be able to generate proof obligations containing n^{th} order predicates several adaptations to SPARK Ada and the tool set would be necessary. Firstly, it would be necessary to extend the annotations to cover differential information using the prime notation introduced in this paper. In addition, structuring annotations would need to be provided for the macro-substitution style to allow macro-preconditions and delineate between sections of code with their own state variables and invariants. Finally, the macro-substitution theory would need to be implemented by the static analyser.

Acknowledgments. We would like to acknowledge the contribution of an anonymous reviewer who provided a simplification to the theory presented in the submitted version of this paper.

References

1. http://www.cs.york.ac.uk/~jrb/appendix.ps.
2. J-R Abrial. *The B Book – Assigning Programs to Meanings*. Cambridge University Press, 1996.
3. M. Broy. Compositional Refinement of Interactive Systems. *Journal of the ACM*, 44(6):850–891, November 1997.
4. M. Broy. The Specification of System Components by State Transition Diagrams. Technical Report TUM-I9729, Technische Univeritat Munchen, 1997.
5. M. Broy. Compositional Refinement of Interactive Systems Modelled by Relations. In *Compositionality: The Significant Difference*, number 1536 in Lecture Notes in Computer Science, pages 130–149, 1998.
6. M. Broy. A Logical Basis for Component-Based Systems Engineering. International Summer School, Marktoberdorf, July-August 1999.
7. M. Broy. From States to History. In *International Summer School, Marktoberdorf*, 2000.
8. Z. Chaochen, A.P. Ravn, and M.R. Hansen. An Extended Duration Calculus for Hybrid Real Time Systems. In *Hybrid Systems*, Lecture Notes in Computer Science, pages 36–59, 1993.
9. E.W. Dijkstra. Guarded Commands, Nondetermincy and Formal Derivation of Programs. *Communications of the ACM*, 18:453–457, August 1975.
10. A. Galloway. Communicating Generalised Substitution Language. In *Proceedings of the International Conference on Perspectives of System Informatics, PSI'01*, 2001.
11. A. Galloway and J. Blow. Multi Layered Domain Specific Formal Languages. In *Proceedings of the Workshop on Formal Specification of Computer Based Systems, FSCBS'01*, April 2001.
12. A. J. Galloway, T. J. Cockram, and J. A. McDermid. Experiences with the Application of Discrete Formal Methods to the Development of Engine Control Software. In *Proceedings of DCCS (Distributed Computer Control Systems) 98*. IFAC - International Federation of Automatic Control, 1998.
13. J. Blow, A. Galloway, J.A. McDermid, M. Dowding and T. Cockram. The Industrial Use of a Formal Method in a Gas Turbine Engine Electronic Control System. In *Proceedings of the Workshop on Formal Specifications of Computer Based Systems, FSCBS'00*, April 2000.
14. John McDermid and Andy Galloway et al. Towards Industrially Applicable Formal Methods: Three Small Steps, and One Giant Leap. In *The International Conference on Formal Engineering Methods (ICFEM) 1998*. IEEE Press, 1998.
15. UK Ministry of Defence. *Defence Standard 00-55 - The Procurement of Safety Critical Software in Defence Equipment*. 1997.

Communicating B Machines

Steve Schneider and Helen Treharne

Department of Computer Science, Royal Holloway,
University of London, Egham, Surrey, TW20 0EX, UK.

{steve,helen}@cs.rhul.ac.uk

Abstract. This paper describes a way of using the process algebra CSP to enable controlled interaction between B machines. This approach supports compositional verification: each of the controlled machines, and the combination of controller processes, can be analysed and verified separately in such a way as to guarantee correctness of the combined communicating system. Reasoning about controlled machines separately is possible due to the introduction of guards and assertions into description of the controller processes in order to capture assumptions about other controlled machines and provide guarantees to the rest of the system. The verification process can be completely supported by different tools. The use of separate controller processes facilitates the iterative development and analysis of complex control flows within the system. The approach is motivated and illustrated with a non-trivial running example.

Keywords: B-Method, CSP, Combining Formalisms, Concurrency.

1 Introduction

This paper introduces a new approach to combining B machines concurrently in a verifiable way. It provides an alternative approach to using the B-Method at a system level and forces a specifier to make explicit the flow of control (i.e. order of execution) of operations in a B machine. The benefit of our approach is the ability to make use of existing tool support for all aspects of the verification, and ultimately a direct route to code for the B machines using supporting tools.

The paper builds on previous work [13,14,15] using the process algebra CSP [6] to describe controllers for B machines, in order to express and reason about complex flows of control in a natural way. Previous work has been concerned with a single *controller* process P encapsulating a single flow of control for a B *machine* M; M can be a single machine or be comprised of a hierarchy of machines. We also focused on methods for proving that a controller is consistent with its underlying machine. In this paper we consider how a collection of such combinations (of B machines M_i and their controllers P_i) can interact. The controllers play a key role in enabling communication between the machines. Furthermore, we propose an architecture (pictured in Figure 1) which requires all interaction between machines to be through the controllers. The architecture

D. Bert et al. (Eds.): ZB 2002, LNCS 2272, pp. 416–435, 2002.

is therefore appropriate both when controlled machines are distributed across a network and when they are executed on the same processor. This enables deadlock and divergence freedom of a *combined communicating system* to be verified in a compositional way, simply by analysing smaller parts of the system: the individual *controlled machines* $P_i \parallel M_i$, and the parallel combination of only the P_i (without the M_i). Overall correctness follows automatically from Lemma 1 and Theorem 2 presented in Section 4.2. These results form the main technical contributions of the paper.

In practice, we find that the correctness of particular controlled machines within the system rests on the behaviour of other controlled machines, and aspects of the rest of the system's behaviour must be incorporated into the verification. We achieve this by extending the language of controllers to include *guards* (which block unwanted inputs) and *assertions* (which diverge on unexpected communications). These guards and assertions are used in a similar style to reply guarantee predicates [7]. In particular, guards on an input channel of a controller P are used to describe the inputs expected from the rest of the system, and thus capture the assumptions about the process' environment. This enables P to be analysed in the absence of the rest of the system. Assertions on outputs to the rest of the system describe what the process itself should guarantee. Assertions are also used to encapsulate the expectations on inputs to a controller P from its associated machine M. This enables the combination of controllers to be analysed together, independently of the machines they control, and hence entirely at the level of process algebra.

The development of a combined communicating system proposed typically involves the following steps:

1. Define the individual B machines,
2. Give CSP controllers for them that describe the flow of control for their use,
3. Prove consistency between the B machines and their controllers,
4. Prove deadlock freedom of the combination of the controllers,
5. Refine and implement the machines and controllers independently.

Steps 1–4 of the process will be illustrated by a running example in the paper.

This paper is organised as follows: Section 2 introduces the CSP controller language and semantics; Section 3 describes the running example used to motivate and illustrate the approach; Section 4 is concerned with the consistency results which underpin the approach, and the use of assertions and guards; and Section 5 ends with a discussion. The paper assumes familiarity with AMN; further details can be found in [1].

2   CSP Controllers and B Machines

2.1   Notation

Communicating Sequential Processes (CSP) is a language for describing processes of concurrent systems and their patterns of interactions. The unit of interaction is the atomic *event* which processes perform and on which they may

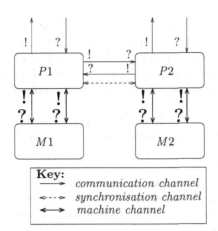

Fig. 1. An architecture for concurrent B machines

synchronise. Events can be unstructured (such as *start*), or they can have some structure, generally of the form of a channel name c and some values v that are passed along a channel. Thus the occurrence of $c.3.5$ may be understood as the passing of the values 3 and 5 along the channel c. The *occurrence* of events is atomic. The set of all events is denoted Σ.

We will use a subset of CSP to describe the controllers for B machines. The language we use is based on the language in [14,15] and is given by the following pseudo-BNF rule;

$$P ::= a \rightarrow P \mid c?x\langle E(x)\rangle \rightarrow P \mid d!v\{E(v)\} \rightarrow P \mid e?v!x\{E(x)\} \rightarrow P \mid$$

$$P_1 \ \Box \ P_2 \mid P_1 \ \sqcap \ P_2 \mid \sqcap_{x|E(x)} P \mid if \ b \ then \ P_1 \ else \ P_2 \ end \mid S(p)$$

where $a \in \Sigma$ and is a *synchronisation event*, c is a *communication channel* accepting inputs, d is a communication channel sending output values, e is a *machine channel*, x represents all data variables on a channel, v represents all data values being passed along a channel, $E(x)$ is a predicate on x (it may be elided, in which case it is considered to be *true*), b is a boolean expression, and p is a process expression.

All the terms in the above grammar have established CSP semantics and can be modelled in a CSP model checker such as FDR [9] without any extensions. We will explain each expression shortly but before doing so we need to clarify our use of channels.

In our language we want to distinguish communication between CSP processes from interaction with B machines, as shown in Figure 1. Note that there is no sharing of state between CSP processes and B machines.

Synchronisation events and communication channels serve as communication mediums between two CSP controllers and they have *no* correspondence with B operations. Conversely, a machine channel corresponds to a B operation, with the same types for inputs and outputs. The machine channel comprising of input and output also admits degenerate cases, in which either or both the $?$ and $!$ may be dropped. These cases are all treated as special instances of the machine channel, and do not require separate treatment. For example, a machine channel with no input and outputs will simply correspond to a B operation with no parameters or return values.

The expression, $a \rightarrow P$, means that a process is prepared to engage in the event a, and subsequently behave as P. It is used to provide a way of synchronising on atomic events.

Another expression which makes use of the prefix operator $(\rightarrow)$ is the expression $c?x\langle E(x)\rangle \rightarrow P$. It denotes a process that is initially prepared to accept any value x on the communication channel c that meets the guard predicate $E(x)$, after which it behaves as the process P (which may depend on the value of x) but it will not accept any x that fails to meet that predicate. Guards can be modelled in FDR using standard CSP syntax.

The expression $d!v\{E(v)\} \rightarrow P$ is initially prepared to perform the event $d.v$ (i.e. pass values v along the communication channel d). If the assertion $E(v)$ is true, then its subsequent behaviour is that of P; if $E(v)$ is false then it *diverges*. Divergent behaviour will be discussed below. Assertions are simply coded in FDR as follows;

$$d!v\{E(v)\} \rightarrow P = d!v \rightarrow (if\ E(v)\ then\ P\ else\ DIV)$$

where DIV represents a divergent process.

The expression $e?v!x\{E(x)\} \rightarrow P$ is initially prepared to allow a process to interact on machine channel e. This channel will be used to communicate with a B machine via its corresponding operation $x \longleftarrow e(v)$, so it provides v as input to the B machine (indicated by the $?$), and accepts x as output (indicated by the $!$). If the value x it receives meets the predicate $E(x)$ then it behaves as the process P which may depend on the value of x, otherwise it diverges. Observe that the CSP semantics of this term will be the same as an event communicating over an output and input channel in standard CSP (for example $e!v?x\{E(x)\}$).

The external choice, $P_1 \square P_2$ is initially prepared to behave either as P_1 or as P_2, with the choice being made on occurrence of the first event. The choice of the first event is made by the environment of the choice. Conversely, the choice $P_1 \sqcap P_2$ chooses internally whether to behave as P_1 or as P_2, and its environment has no control over the way the choice is resolved. Indexed internal choice $(\sqcap)$ chooses a value x such that is meets the predicate $E(x)$ and then behaves as the process P which may depend on the value of x. Another form of choice is controlled by the value of a boolean expression in an *if* expression.

$S(p)$ is a process name where p is an expression. Each process expression contains a recursive call, $S(p)$. For example, a process which manages a set of values can be described by a recursive family indexed by sets:

$$Set(S) = in?x\langle x \notin S\rangle \rightarrow Set(S \cup \{x\})$$
$$\square\ (\textstyle\bigsqcap_{v \in S}\ out!v \rightarrow Set(S - \{v\}))$$

Observe the use of the guard on the input channel *in* to block the input of any value which is already in the set S; and that some arbitrary member of S is selected for output.

Note that the difference with the above language and the one used in our previous work is the inclusion of an output communication channel (!), internal choice ($\sqcap$), and indexed internal choice ($\sqcap$). Furthermore, the use of the input channel ? and the atomic event a have been restricted to describing communication between CSP controllers, and they no longer correspond to underlying B operations. We have also augmented machine and communication channels with assertions and guards. This rationalisation of the notation is necessary because we need to deal with concurrency cleanly. In this paper we have also restricted the language to non-terminating controllers; but the inclusion of terminating loops is discussed in [16].

In addition to the language for controllers, CSP provides a number of other operators, including *parallel composition*: $P_1 \parallel P_2$. There is also an indexed form $\parallel_i P_i$. A parallel combination executes P_1 and P_2 concurrently, requiring that they synchronise on events in both their alphabets, and allowing independent performance of events outside their alphabets. In this paper the alphabet of a process will be all the events that it can perform. This allows messages to pass along channels.

For example, if the processes *Copy* and *Copy2* are defined as follows

$$Copy = in?x \rightarrow out!x \rightarrow Copy$$
$$Copy2 = out?y \rightarrow down!y \rightarrow Copy2$$

then *Copy* $\parallel$ *Copy2* can input values v on *in*, have both components synchronise on *out.v* which passes the value to *Copy2*, and then have *Copy2* independently output v on *down*.

2.2   Semantics

CSP processes are identified with the observations that can be made of them: thus the semantics of a CSP process will be a set of observations. The precise form of the observations will describe the CSP model. The *traces model* uses traces as observations. The *stable failures model* uses traces along with subsequent refusals. The *failures/divergences model* uses traces, divergences, and failures. We briefly describe them here. A fuller explanation can be found in [11].

A *trace tr* of a process P is a finite sequence of events that it may be observed to engage in. The *traces model* identifies a process with its set of traces.

A *divergence* of a process P is a sequence of events tr such that P reaches a divergent state (which may be thought of as entering a non-terminating loop, or in specification terms as a specification which allows any behaviour) during the performance of the sequence of events tr. A process is *divergence-free* if it has no divergences. Divergence denotes undesirable behaviour, and it is generally useful to establish that a process is divergence-free.

A *refusal* of a process P is a set X of events that P might be initially prepared to refuse. A *stable failure* of a process P is a trace/refusal pair (tr, X) such that

P can initially perform the sequence of events tr, and reach a non-divergent state in which every event in X is refused. The stable failures of P is denoted $\mathcal{F}_{SF}\,[\![P]\!]$.

If for some tr $(tr, \Sigma) \in \mathcal{F}_{SF}\,[\![P]\!]$ then P can reach a state in which no event at all is possible, and we say that P has a *deadlock*. If there is no such stable failure, then P is *deadlock-free*.

The CSP model-checker FDR allows checks for deadlock and divergence freedom to be made automatically for processes.

3   Example

We describe and motivate the approach of this paper through an illustrative example. This section illustrates the use of the above control language in specifying CSP controllers to drive underlying B machines. Each individual controller acts as an interface for its underlying machines and provides a possible pattern of execution. The aim is to compose a collection of controlled machines in order to specify a large combined communicating system.

For the purposes of presentation we first describe the machines and then the associated controllers. In practice, we develop both alongside each other, since both impact on each other's specification as we shall see in Section 3.2 when we discuss the particular controllers. The B-Toolkit [10] and FDR source files for the example can be downloaded[1].

3.1   Machines

Consider two B machines, *CustomerData* and *CheckoutData*, which are defined in Figures 2 and 3 respectively. The purpose of these machines is to capture information about customers who are shopping and process them through checkout queues in a supermarket. The separation of customers and checkout processing illustrates our approach of modelling different parts of the system separately.

The customer information is captured in the *CustomerData* machine. It introduces the variable *customer* to track whether an individual customer is either *shopping* or *paying* (i.e. the customers' statuses). The set of all possible customers, *CUSTOMERS* is declared in a separate context machine, *Types*, as shown in Figure 4. Similarly, the set defining the customers' statuses, *STATUSES*, is declared in the same context machine. This allows more than one machine to use static information and is typical of B developments.

Four operations are offered by *CustomerData*: *beginshopping* allows a customer to become a shopper; *whatstate* queries the status of a customer; *proceedtocheckout* updates a customer from being one who is shopping to one who is waiting to pay; and *finished* removes a customer from being considered by the system (once goods have been paid for).

The queue of customers is tracked in the *CheckoutData* machine[2]. This machine has a number of checkout counters, some of which will be open with a

---

[1] http://www.cs.rhul.ac.uk/home/helen/papers/zb2002/sources.tar.gz

[2] originally inspired by an example in [12]

MACHINE *CustomerData*
SEES *Types*
VARIABLES *customer*
INVARIANT *customer* ∈ *CUSTOMERS* ↛ *STATUSES*
INITIALISATION *customer* := ∅

OPERATIONS
  beginshopping (*cust*) ≙
   PRE *cust* ∈ *CUSTOMERS* ∧ *customer* (*cust*) ≠ *paying* **THEN**
    customer (*cust*) := *shopping*
   END ;
  ss ⟵ **whatstate** (*cust*) ≙
   PRE *cust* ∈ *CUSTOMERS* **THEN**
    IF *customer* = ∅ **THEN** *ss* := *idle* **ELSE**
     ss := *customer* (*cust*)
    END
   END ;
  proceedtocheckout (*cust*) ≙
   PRE *cust* ∈ *CUSTOMERS* ∧ *customer* (*cust*) = *shopping* **THEN**
    customer (*cust*) := *paying*
   END ;
  finished (*cust*) ≙
   PRE *cust* ∈ *CUSTOMERS* ∧ *customer* (*cust*) = *paying* **THEN**
    customer := { *cust* } ⊲ *customer*
   END
END

Fig. 2. *CustomerData* Machine

queue of customers. It introduces two variables, *opencounters* and *queues*. The variable *opencounters* keeps track of the counters that are currently open. The set of all possible counters, *COUNTERS*, is again defined in the separate context machine, *Types*. Only counters that are open can have a queue of customers associated with them and this information is captured by the *queues* variable. The invariant of the machine provides constraints on a queue of customers, stating that customers should only ever appear in at most one queue, and in at most one position.

Six operations are offered by the machine. The operation *do_open* opens a new counter provided that not all counters are already open. The operation *do_close* closes a counter and removes any customers who are queueing at that counter. The operation *serve* processes the customer at the head of the queue at a given counter. The operation *join* allows a new customer to join the queue at a particular open counter. The operation *drop_out* allows a customer to leave a queue. The operation *choose* non-deterministically selects an open counter whenever possible.

3.2 Controllers

In order to compose two B machines into a combined communicating system we introduce CSP controllers for each of them which communicate with one another. To be consistent with the B machines the controllers must ensure that operations are always called within their preconditions. In this section we present the con-

MACHINE *CheckoutData*
SEES *Types* , *Bool_TYPE*
VARIABLES *queues* , *opencounters*
INVARIANT
opencounters $\subseteq$ *COUNTERS* $\wedge$
queues $\in$ *opencounters* $\rightarrow$ iseq (*CUSTOMERS*) $\wedge$
$\forall$ (*cc* , *dd*) . (*cc* $\in$ *COUNTERS* $\wedge$
dd $\in$ *COUNTERS* $\wedge$ *cc* $\neq$ *dd* $\Rightarrow$
ran (*queues* (*cc*)) $\cap$ ran (*queues* (*dd*)) = $\varnothing$)
INITIALISATION
queues := $\varnothing$ || *opencounters* := $\varnothing$

OPERATIONS
rep_co $\longleftarrow$ **do_open** $\widehat{=}$
  BEGIN
  IF *opencounters* $\neq$ *COUNTERS* **THEN**
    ANY *co*
    WHERE *co* $\notin$ *opencounters*
      $\wedge$ *co* $\in$ *COUNTERS* **THEN**
      opencounters := *opencounters* $\cup$ { *co* } ||
      queues (*co*) := [ ] || *rep_co* := *co*
    END
  ELSE
    rep_co := *def_counter*
  END
  END ;
rep_set $\longleftarrow$ **do_close** $\widehat{=}$
  BEGIN
  IF *opencounters* $\neq$ $\varnothing$ **THEN**
    ANY *co*
    WHERE *co* $\in$ *opencounters* **THEN**
      rep_set := ran (*queues* (*co*)) ||
      queues := { *co* } $\lhd$ *queues* ||
      opencounters := *opencounters* $-$ { *co* }
    END
  END
  END ;
cu , *rep* $\longleftarrow$ **serve** (*co*) $\widehat{=}$
  PRE *co* $\in$ *COUNTERS* **THEN**
  IF *co* $\in$ dom (*queues*) $\wedge$
    queues (*co*) $\neq$ [ ] **THEN**
    cu := first (*queues* (*co*)) ||
    queues (*co*) := tail (*queues* (*co*)) ||
    rep := *TRUE*
  ELSE
    rep := *FALSE* ||
    cu :$\in$ *CUSTOMERS*
  END
  END ;

join (*co* , *cu*) $\widehat{=}$
  PRE *cu* $\in$ *CUSTOMERS* $\wedge$
    co $\in$ *COUNTERS* $\wedge$
    co $\in$ *opencounters* $\wedge$
    cu $\notin$ $\bigcup$ *cc* .
    (*cc* $\in$ dom (*queues*) |
      ran (*queues* (*cc*)))
  THEN
    queues (*co*) := *queues* (*co*) $\leftarrow$ *cu*
  END ;
drop_out (*cu*) $\widehat{=}$
  PRE *cu* $\in$ *CUSTOMERS* **THEN**
    ANY *co*
    WHERE *co* $\in$ *COUNTERS* $\wedge$
      cu $\in$ ran (*queues* (*co*)) **THEN**
      LET *ss*
      BE *ss* = *queues* (*co*) **IN**
        queues (*co*) :=
        ss $\uparrow$ *ss* $^{-1}$ (*cu*) $-$ *1* $\frown$
        ss $\downarrow$ *ss* $^{-1}$ (*cu*)
      END
    END
  END ;
out_co , *rep* $\longleftarrow$ **choose** $\widehat{=}$
  BEGIN
  IF *opencounters* = $\varnothing$ **THEN**
    rep := *FALSE* ||
    out_co := *def_counter*
  ELSE
    ANY *cc*
    WHERE *cc* $\in$ *opencounters*
    THEN
      rep := *TRUE* ||
      out_co := *cc*
    END
  END
  END
  END

Fig. 3. *Checkout* Machine

trollers which drive the *CustomerData* and *CheckoutData* machines. Specifying
the controllers, *CustomerProc* and *CheckoutProc* below, allows us to build the
system $((CustomerProc \parallel CustomerData) \parallel (CheckoutProc \parallel CheckoutData))$.
Figure 5 illustrates the overall architecture of the whole system and highlights
all the channels involved.

CustomerProc is a controller which deals with customer requests, and is given
in Figure 6. The process *CustomerProc* is defined in terms of a parameterised

MACHINE *Types*
SETS *COUNTERS* ; *CUSTOMERS* ; *STATUSES* = { *shopping* , *paying* , *idle* }
CONSTANTS *def_counter*
PROPERTIES *COUNTERS* = *1 .. 10* ∧ *CUSTOMERS* = *1 .. 100* ∧
def_counter ∈ *COUNTERS*
END

Fig. 4. *Types* Machine

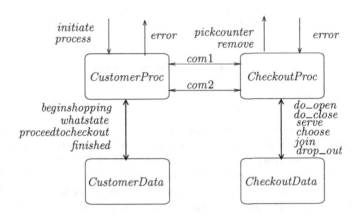

Fig. 5. Example Architecture

recursion. The parameter *custset* holds the set of customers currently paying. Initially, there will be no such customers and the set is empty. Note that the state carried around in this controller is more abstract than the state in the underlying B machine. In practice, our aim is to minimise the amount of state in the CSP controller. However, some state may be necessary in order to determine the flow of control in a process and/or for use in specifying assertions and guards, as is the case with *custset*.

There are three main paths of execution which are controlled by external choice. The first path accepts customers provided they are not already paying. This is ensured by the guard of the communication channel *initiate*, (*cc* ∈ *CUSTOMERS* − *custset*). We proceed after accepting an appropriate customer to observe a communication of that customer value along the *beginshopping* channel which corresponds to the underlying B operation assigning his/her status to be *shopping*. Allowing only certain customers means that the precondition of the *beginshopping* operation will be discharged when invoked. It is true that the above constraint on customers could have been modelled as a *select* statement in B but our philosophy is never to block B operations, so that we can implement the operations in ANSI C within the B-Toolkit. The role of guards and assertions will be examined in more detail in Section 4.3.

The second path processes a customer. If the customer is already shopping then the underlying B state of that customer is updated followed by a message

$CustomerProc = Q1(\varnothing)$

$Q1(custset) \quad = (initiate?cc\langle cc \in CUSTOMERS - custset\rangle \rightarrow beginshopping\,?\,cc \rightarrow$
$\qquad\qquad Q1(custset))$

$\qquad\quad \square\ (process\ ?cc \rightarrow whatstate\,?\,cc\,!\,ss \rightarrow$
$\qquad\qquad\qquad if(ss == shopping)\ then$
$\qquad\qquad\qquad\quad (proceedtocheckout\,?\,cc \rightarrow com1!cc\{cc \notin custset\} \rightarrow$
$\qquad\qquad\qquad\quad Q1(\{cc\} \cup custset))$
$\qquad\qquad\qquad else$
$\qquad\qquad\qquad\quad (error!cc \rightarrow Q1(custset)))$

$\qquad\quad \square\ (com2?cc\langle cc \in custset\rangle \rightarrow finished\,?\,cc \rightarrow Q1(custset - \{cc\}))$

Fig. 6. Customer Process

being passed along *com1* communicating to *CheckoutProc* that the customer can now join a queue. Note that the assertion denotes the assumption that the customer is not already paying. Note also that the parameter in the recursive call updates the *custset* to include the new customer. In the case where the customer is not shopping an error is reported and no underlying B state is affected.

Observe that a request to process a customer could occur even before he/she has begun shopping and thus impacting on the specification of the *whatstate* operation. Initially, we had used a simple assignment and a precondition stating that $cust \in dom(customer)$. The operation in Figure 2 is far more robust and allows the operation to be called with any customer as its input. In order to achieve this particular style of specification we augment the datatype *STATUSES* to include an extra enumeration, *idle*, as shown in Figure 4.

The third path records the fact that a customer has finished shopping. The particular customer, whose status is to be updated, will be communicated along *com2*. The guard on the channel reflects our understanding that the customer will be one who is currently paying. Here again the guard is used to discharge the associated precondition in the *finished* operation. The set substraction in the recursive call keeps the set of paying customers up to date.

CheckoutProc is a controller which controls the opening and closing of supermarket counters, and the processing of customers in queues associated with those counters. The process *CheckoutProc* is defined, in Figure 7, in terms of a parameterised recursion. The parameter *custque* holds the set of customers currently in a queue. Initially, there are no such customers.

There are five main possible execution paths in *CheckoutProc*. The first path effectively invokes the underlying B operation to open a counter if at all possible.

The second path closes a counter. The assertion on the output set *out_que* serves to affirm our belief that the customers in the set are queueing customers.

The third path illustrates how the choice of which counter will be serviced next is resolved by the controller. The information regarding counters is not captured in the controller (at this stage, though it may be included in a refinement at a later stage) and so it could be the case that an empty counter is passed as

$CheckoutProc = Q2(\varnothing)$

$Q2(custque) = (do\_open \, ! \, out\_co \rightarrow Q2(custque))$

$\square \; (do\_close \, ! \, out\_que\{out\_que \subseteq custque\} \rightarrow Q2(custque - out\_que))$

$\square \; (\bigcap_{d \in COUNTERS} pickcounter!d \rightarrow$

$\quad serve \, ? \, d \, ! \, cust \, ! \, rep\{(cust \in custque) \vee (rep == false)\} \rightarrow$

$\quad if \; (rep == true) \; then$

$\quad\quad (com2!cust \rightarrow Q2(custque - \{cust\}))$

$\quad else$

$\quad\quad Q2(custque))$

$\square \; (com1?cust\langle cust \in (CUSTOMERS - custque)\rangle \rightarrow$

$\quad choose \, ! \, out\_co \, ! \, rep \rightarrow$

$\quad if \; (rep == true) \; then$

$\quad\quad join \, ? \, out\_co \, ? \, cust \rightarrow Q2(custque \cup \{cust\})$

$\quad else$

$\quad\quad do\_open \, ! \, out\_co \rightarrow join \, ? \, out\_co \, ? \, cust \rightarrow Q2(custque \cup \{cust\}))$

$\square \; (remove?cc \rightarrow$

$\quad if \; (cc \in custque) \; then$

$\quad\quad drop\_out \, ? \, cc \rightarrow com2!cc \rightarrow Q2(custque - \{cc\})$

$\quad else$

$\quad\quad error!cc \rightarrow Q2(custque))$

Fig. 7. Checkout Process

an input to the B operation *serve*. Therefore, the B operation needs to be robust enough to allow for this possibility. It achieves this by outputting a report value, indicating success or failure of the operation. The report value is used to control the flow of execution so that either *com2* communicates that the customer was served successfully or no such communication occurs. Note that there is only one assertion on the output values of the machine channel *serve*. This is essential since the *cust* value may not always be in the queueing set.

The fourth path allows a new customer to join a queue. A synchronisation along the communication channel *com1* provides the particular customer wishing to join a queue. The guard on the input channel means that we never want to follow this path of execution if the customer is already in a queue. Once a customer is received we then proceed to choose a counter whose queue the customer can join. If there are no open counters then we must open one before the customer is allowed to join its queue. Otherwise he/she is simply appended to the queue of the chosen counter. The operation *join(co, cu)* corresponding to $join \, ? \, out\_co \, ? \, cust$ has a precondition stating that the customer can only join provided he/she is not already present in any of the queues of the open counters. The guard on the communication channel *com1* ensures that the *cust* value meets that precondition.

We originally specified the *do_open* operation with no parameters. During the development of the fourth main execution path we realised that an output parameter was necessary. Consider the following scenario, after performing a *choose* operation the control process reads the *rep* variable in order to decide whether an open counter has been successfully chosen or whether no counters are open and so a failure is reported by the operation. In the latter case we then proceed to non-deterministically choose to open a counter using the *do_open* operation and then allow the customer to join a queue. The problem is that the counter being passed to the *join* operation may not be the same as the one opened by the *do_open* operation. This inconsistency problem is solved by augmenting the *do_open* operation with an output parameter and passing this output value to the *join* operation. The repercussions of this change on process $Q2$ is that the *do_open* operation in the first branch of the choice needs to be consistent in its signature. In the first path the value being passed along the output channel is never used but this is allowable.

The fifth path of process $Q2$ either removes a customer from the queue or reports an error. If the customer is successfully removed from the queue then $Q2$ forces a synchronisation to occur on channel *com2* so that the information about the particular customer is also updated, as we described above.

The above example demonstrates that specifying how the operations are used in controllers forces us to consider their interface and robustness very carefully. If the operations are not robust, that is they contain preconditions, then those preconditions must be discharged by the CSP controller. Furthermore, the controllers themselves enable us to clearly visualise the flow of information between controlled machines in a combined communicating system. In the example all interactions between *CustomerProc* and *CheckoutProc* have been along machine channels *com1* and *com2* passing values. However, in general the language, presented in Section 2.1, also allows pure synchronisations (such as a *start* event).

4   Consistency of Combined Communicating Systems

In this section we discuss how consistency, i.e. divergence and deadlock freedom, of combined communicating systems can be established in a compositional way. There are two separate steps involved in the verification process (steps 3 and 4 in the overall development process outlined in Section 1). First, in Section 4.1, we check that each individual controlled machine is divergence-free. Second, we check deadlock freedom of the combination of the individual controllers *without* their underlying machines. In the case of our example, these steps involve checking that (*CustomerProc* $\|$ *CustomerData*) and (*CheckoutProc* $\|$ *CheckoutData*) are both divergence-free, and then checking that (*CustomerProc* $\|$ *CheckoutProc*) is deadlock-free. These two steps can be supported by the B-Toolkit and FDR respectively. The results, presented in Section 4.2, verify that these two separate steps are enough to ensure deadlock and divergence freedom of a combined communicating system, in our case ((*CustomerProc* $\|$ *CustomerData*) $\|$ (*CheckoutProc* $\|$ *CheckoutData*)) divergence- and deadlock-free.

4.1   Consistency of Single Controlled Machine

When we consider a controller P for a machine M, we have previous results which establish when the CSP controller is appropriate for the B machine: it must not call operations which fail to meet their preconditions. This means that within the failures/divergences semantics for P and M, the parallel combination $(P \parallel M)$ must be divergence-free. The following is the essence of the result from [13,15] giving a sufficient condition to guarantee divergence freedom:

Theorem 1. *If CLI (control loop invariant) is a predicate such that*

$$CLI \wedge I \Rightarrow [\{BBODY_{S(p)}\}] CLI$$

for each $BBODY_{S(p)}$ *in* P *then* $(P \parallel M)$ *is divergence-free.*

The theorem states that the invariant CLI need not necessarily hold after each individual operation but must hold at each recursive call of a controller. As in [13,15], we define a maximum sequence of operations, $\{BBODY_{S(p)}\}$, to be the translation of a process expression associated with $S(p)$ up to and including reaching a recursive call. If such a sequence of operations can establish the CLI then we know that all the operations are called within their preconditions and terminate. Thus if the CLI holds for all such process expressions within P then we have a way of checking whether any machine M acting under its controller P can diverge.

The above result also holds for this paper's CSP controller language, which allows P to have additional channels which are independent of M; and to have assertions and guards on the values passed along channels. In order to ensure this we have to provide extra cases in the translation of process expressions to their corresponding sequences of operations and simplify some existing cases. For example, the translation of communication channels which have no underlying operations from M are outlined as follows;

$$\{c?x\langle E(x)\rangle \to P\} = \textbf{ANY } x \textbf{ WHERE } E(x) \textbf{ THEN } \{P\} \textbf{ END}$$
$$\{d!v\{E(v)\} \to P\} = \textbf{PRE } E(v) \textbf{ THEN } \{P\} \textbf{ END}$$

Indexed internal choice $\bigsqcap_{x|E(x)} P$ is also translated using an **ANY** clause, and is akin to a guard on a communication channel. The fully formal treatment can be found in [16].

Note that the translation rules above reveal why the annotations of CSP channels were termed *assertions* and *guards*. We wanted to model predicates on inputs as guards which when violated would be identified in the CSP analysis in the second step of verification when checking deadlock freedom. On the other hand we wanted to model predicates on outputs as assertions which would be discharged (in step one) when checking the individual controlled machines, $P_i \parallel M_i$, using the above result.

In our example, an appropriate CLI for $(CustomerProc \parallel CustomerData)$ is $custset = customer^{-1}[\{paying\}]$. It states that the set carried in the parameter of the process will match the set of customers who are paying, and this will be true at the start of the loop and hold at each recursive call. A suitable *CLI* for (*CheckoutProc* $\parallel$ *CheckoutData*) is $custque = \bigcup cc.(cc \in dom(queues) \mid ran(queues(cc)))$. It states that the set carried in the parameter will match the set of all customers in all of the queues. Again this will be true at the start of the loop and hold at each recursive call.

4.2   Consistency of Multiple Controlled Machines

Here we consider two B machines M_1 and M_2 under the control of their respective CSP controllers P_1 and P_2, whose overall architecture has been previously depicted in Figure 1. Formally, we have the following constraints on the alphabets of the controllers and machines:

- $\alpha(M_1) \subseteq \alpha(P_1)$
- $\alpha(M_2) \subseteq \alpha(P_2)$
- $\alpha(M_1) \cap \alpha(P_2) = \varnothing$
- $\alpha(M_2) \cap \alpha(P_1) = \varnothing$

Thus each M_i communicates only with its associated P_i, and the various P processes communicate with each other, and with the environment of the combination, on separate channels not involving the M machines.

The results below verify that in order to check deadlock and divergence freedom of the overall combination, only two steps are needed, i.e. it is sufficient to firstly check divergence-freedom of each controlled machine separately, and then to check deadlock freedom only of the combination of the CSP controllers.

Lemma 1. *If $P_1 \parallel M_1$ is divergence-free, and $P_2 \parallel M_2$ is divergence-free, then $(P_1 \parallel M_1) \parallel (P_2 \parallel M_2)$ is divergence-free.*

Proof. This follows immediately from the semantics of parallel composition in the failures/divergences model: parallel composition preserves divergence-freedom.

As noted above, previous work provides techniques for establishing divergence freedom of a $P_i \parallel M_i$.

Theorem 2. *If $(P_1 \parallel M_1)$ is divergence-free, and $(P_2 \parallel M_2)$ is divergence-free, and $(P_1 \parallel P_2)$ is deadlock-free in the stable failures model, then $(P_1 \parallel M_1) \parallel (P_2 \parallel M_2)$ is deadlock-free in the stable failures and failures/divergences model.*

Theorem 2 extends to any number of B machines to be combined in parallel. Thus, Theorem 2 is proved for the general case in the appendix A. The overriding principle is that each B machine should have its own CSP controller, and any interaction it has with its controller is private between them, and no other process participates in it. Communication between controlled machines takes place between the CSP controllers using channels which are independent of the B machines. Using this scheme, all that needs to be checked to establish overall divergence freedom is that each $P_i \parallel M_i$ is divergence-free; and all that then needs to be checked to establish overall deadlock freedom is that the parallel combination $\|_i P_i$ is deadlock-free. The above results also extend to divergence-free controllers which do not have underlying B machines, as we shall see in Section 4.4.

4.3   Role of Assertions and Guards

Assertions and guards allow the overall system to be split into pieces that can each be independently verified. They are used to carry the assumptions and guarantees to decompose the proof obligations on the overall system to proof obligations on small combinations of controlled machines within the system. Moreover, they make explicit the designer's understanding of why the controlled machines should behave correctly when combined.

The assertions on outputs (along machine and communication channels) give conditions that any output value must satisfy. Since failure to meet such an assertion results in a divergence, a successful divergence freedom check of a $P_i \| M_i$ ensures that any output from P_i meets the assertion. Making such assertions too strong however will result in $P_i \| M_i$ failing to be divergence-free. For example, if we had included, in Figure 7, the simpler assertion $cust \in custque$ on the $serve$ communication channel then ($CheckoutProc \| CheckoutData$) would not be divergence-free; since we could not always guarantee that the output provided by the $serve$ operation in $CheckoutData$ met the assertion required by $CheckoutProc$.

Using assertions on machine and communication channels means that values failing such assertions will be ignored when $\|_i P_i$ is analysed for deadlocks in the stable failures model, since a divergence never reaches a stable state in order to contribute to a deadlock.

Guards on inputs on communication channels are also used when checking a particular $P_i \| M_i$ for possible divergences. Without the guards on inputs to the P_is, some inputs which are not possible within the system as a whole might give rise to a divergence within some particular $P_i \| M_i$ when considered separately. Observe that without the guards and assertions in the example, the machines ($CustomerProc \| CustomerData$) and ($CheckoutProc \| CheckoutData$) would not have been divergence-free. For example, if we had omitted the guard on the $cust$ input value along $com1$ in the fourth branch of the choice in $Q2$ (Figure 7) then the precondition of the $join$ operation can be violated on some inputs.

Guards must also be taken into account in the deadlock freedom checks of the combined controllers. Making a guard too strong may result in a deadlock when checking $\|_i P_i$.

If the output assertion on a communication channel between two controllers is the same as (or stronger than) the input guard, then both assertion and guard may be dropped from the controllers once the verification process is complete. In other words, having proven that all messages passed around the system meet the appropriate conditions, they do not require checking at run-time and so they can be dropped from the controller implementations. In the case of $CustomerProc$ this also means that we no longer need to keep track of the set of paying customers in the recursion and as a result we obtain a simpler controller.

$$REG = process?x \rightarrow com1?y \rightarrow REG$$
$$\square\; remove?x \rightarrow com2?y \rightarrow REG$$
$$\square\; pickcounter?x \rightarrow com2?y \rightarrow REG$$

Fig. 8. Regulator Process (REG)

Note that we have not included guards on input values along machine channels in our language. M is not analysed independently of P, so guards capturing properties of inputs from P do not need to be given explicitly.

4.4 Developing Controllers

In Section 3.2 we developed two controllers. In fact an analysis using FDR shows that ($CustomerProc \parallel CheckoutProc$) is not deadlock-free. In retrospect this is not surprising given the complex interactions that occurs between the controllers, and some further development of the controllers was necessary.

This parallel combination of processes can deadlock because $CustomerProc$ can follow a path starting with a *process* event which leads to a point where it must synchronise with $CheckoutProc$ on $com1$ to proceed. However, $CheckoutProc$ can follow a path starting with *remove* or *serve* which lead to a point where it must synchronise with $CustomerProc$ on $com2$ to proceed. If $CustomerProc$ and $CheckoutProc$ both follow these paths at the same time, then they can reach a state in which they cannot synchronise on either $com1$ or $com2$.

Therefore, if we are to guarantee deadlock freedom of the controllers, one solution is to introduce a *regulator* process REG in parallel, which ensures that only one of these paths is entered at any time. Once one process has entered such a critical path, the other process will be blocked from entering a critical path of its own until the relevant com has occurred. An appropriate REG for our example is given in Figure 8 and we can establish that ($CustomerProc \parallel CheckoutProc \parallel REG$) is deadlock-free (now step two of the verification process is finished). The REG process has no assertions, no guards and hence is divergence-free. It also has no underlying B machines, or equivalently (to apply Theorem 2) it can be considered as having a vacuous B machine M_0 with no operations. Thus $REG = REG \parallel M_0$, which is therefore divergence-free. Therefore, all the controlled machines in our example pass step one of the verification process. In turn the overall example is divergence- and deadlock-free.

In general, if the REG process includes assertions and guards, they would be treated in exactly the same way as for controllers with underlying B machines. We would need to prove that REG was divergence-free using our previous result; any assertions and guards would have to hold simply by virtue of the CSP description.

Observe that the regulator process forces a synchronisation on the *pickcounter* communication channel. In the original version of $CheckoutProc$, in Figure 7, we started the branch which serves a customer with a communication along the *serve* machine channel. The introduction of the regulator process meant that we had to force the synchronisation on a communication channel and not on a machine channel, and so we introduced the *pickcounter* channel. This is important since we do not allow two controllers to control the same underlying B operation. Our approach is to couple a single controller and a machine together.

5   Discussion

The B-Method provides supports for the specification and development of software system requirements in terms of a collection of operations which the system must implement. It is ultimately necessary to have some way of describing the flow of control which directs operation calls. We have outlined one way in this paper. Other approaches include the direct introduction of actions (operations with blocking guards) within the B description [2]; or the use of a process algebra to describe a flow of control explicitly as a prelude to translation into B [5]. Both these approaches support the specification of such systems using tools. However, the main difference between their approaches and ours is the lack of a complete development path to executable code using existing tool support (due to *select* not being implementable). Our B machines can use the B-Toolkit to generate executable code. Preliminary investigations also show that CSP controllers can be naturally expressed in Java [16].

In this paper we have shown how the complex interactions between B machines and flow of control of their operations can be naturally expressed using CSP with no sharing of state between the two models. Through the observations made when presenting the paper's example it is clear that our approach to describing combined communicating systems is iterative and compositional. This compositionality is achieved by the addition of assertions and guards to the CSP controller language, and the distinct architecture used to build combined communicating systems. The advantage of compositionality is the ability to maximise the extent to which we can verify controllers and their machines separately.

A further benefit of our approach is the ability to refine each controller and machine separately whilst maintaining divergence and deadlock freedom. For example, we could refine the *CheckoutProc* controller to keep track of the open counters explicitly, and thus refining the internal choice to be a choice over the set of open counters.

In this paper we have focused on verifying deadlock freedom of controllers. Using tool support in the verification process was useful and highlighted the fact that it is not straightforward to specify deadlock-free controllers. Other requirements on the communication patterns of combined communicating system can also be checked using FDR. These are generally specified either in terms of the process algebra itself, or as a predicate on the traces of the system. This is the subject of current research.

Acknowledgements. Thanks to Anna Fukshansky, Martin Green and Craig Saunders for comments on an earlier draft of this paper.

References

1. Abrial J. R.: *The B Book: Assigning Programs to Meaning*, CUP (1996).
2. Abrial J. R.: *Extending B without Changing it (for Developing Distributed Systems)*. In H. Habrias, editor, Proc. of the 1st B Conference, Nantes, France (1996).
3. Butler M. J.: *A CSP Approach to Action Systems*, D.Phil Thesis, Programming Research Group, Oxford University (1992).
4. Butler M. J.: *An Approach to the Design of Distributed Systems with B AMN*. In J. Bowen, M. Hinchey D. Till, editors, ZUM'97, Springer (1998), pp 223-241.
5. Butler M. J.: *csp2B: A Practical Approach to Combining CSP and B*, In J.M.Wing, J. Woodcock, J. Davies, editors, FM'99 World Congress, Springer (1999).
6. Hoare C. A. R.: *Communicating Sequential Processes*, Prentice Hall (1985).
7. Jones C. B.: *Specification and Design of (parallel) Programs*. In R.E.A. Mason, editor, Information Processing '83. IFIP, North Holland (1983).
8. Morgan C. C.: *Of wp and CSP*. In W.H.J. Feijen, A.J.M. van Gasteren, D. Gries and J. Misra, editors, *Beauty is our business: a birthday salute to Edsger W. Dijkstra*. Springer (1990).
9. Formal Systems (Europe) Ltd.: *Failures-Divergences Refinement: FDR2 User Manual* (1997), `http://www.formal.demon.co.uk`
10. Neilson D., Sorensen I. H.: *The B-Technologies: a system for computer aided programming*, B-Core (UK) Limited, Kings Piece, Harwell, Oxon, OX11 0PA (1999), `http://www.b-core.com`
11. Schneider S.: *Concurrent and Real-time Systems: The CSP approach*, Wiley (2000).
12. Schneider S.: *The B-Method: An Introduction*, Palgrave, 2001.
13. Treharne H., Schneider S.: *Using a Process Algebra to control B OPERATIONS*. In K. Araki, A. Galloway and K. Taguchi, editors, IFM'99, York, Springer (1999).
14. Treharne H., Schneider S.: *How to drive a B Machine*. ZB2000, York, LNCS 1878, Springer, September (2000).
15. Treharne H.: *Controlling Software Specifications*. PhD Thesis, Royal Holloway, University of London (2000).
16. Treharne H., Schneider S.: *Communicating B Machines (full version)*. Technical Report, RHUL (2001).

A   Proof of Theorem 2

Theorem 2 (general case). If $(P_i \parallel M_i)$ is divergence-free for each i, and $\parallel_i P_i$ is deadlock-free in the stable failures model, then $\parallel_i (P_i \parallel M_i)$ is deadlock-free in the stable failures and failures/divergences model.

The proof of this theorem requires some preliminary results about the failures of B machines M, and of controllers P.

We first establish that for any operation e of M, given any input v_0, there is some output w_0 that is not refused. This result relies on the fact that operations of M do not block execution.

Lemma 2. *Given an operation $w \longleftarrow e(v)$ of M: if tr is a non-divergent trace of M and (tr, X) is a failure of M, then for any input value v_0 there is some output value w_0 such that $e.v_0.w_0 \notin X$.*

Proof. We use the failures/divergences semantics for action systems (and thus B machines) given in [8], extended to include inputs and outputs in [3,15].

Given (tr, X) as a failure of M, then (by definition) $\neg[tr](\bigvee_{e \in X} g_e)$ (where tr is considered as the sequential composition of the operations listed in it, and $g_e = \neg[e]false$ expresses that e is enabled).

Given $e.v_0.w_0$, the guard $g_{e.v_0.w_0}$ is calculated for the operation e with input v_0 and output coerced to w_0:

$$g_{e.v_0.w_0} = \neg[w \longleftarrow e(v_0)[w = w_0]]false$$
$$= \neg[w \longleftarrow e(v_0)](w \neq w_0)$$

Now assume for a contradiction that $e.v_0.w_0 \in X$ for all possible values of w_0 (of output type T, say), then we have

$$\neg[tr](\bigvee_{e.v_0.w_0 \in X} g_{e.v_0.w_0}) \iff \neg[tr](\neg(\bigwedge_{w_0 \in T} \neg g_{e.v_0.w_0}))$$
$$\iff \neg[tr](\neg(\bigwedge_{w_0 \in T} [w \longleftarrow e(v_0)](w \neq w_0)))$$
$$\iff \neg[tr](\neg[w \longleftarrow e(v_0)](\bigwedge_{w_0 \in T} (w \neq w_0)))$$
$$\iff \neg[tr](\neg[w \longleftarrow e(v_0)] \, false) \qquad (\text{since } w \in T)$$

However, operation e does not block, which means that $\neg[w \longleftarrow e(v_0)]false$ is true, and so we obtain $\neg[tr]true$. On the other hand, however, tr is a non-divergent trace of M, which means that $[tr]true$, yielding a contradiction, and the result follows.

The next lemma states that if a controller process can refuse some output w from a B machine on a machine channel e (with input v), then it can refuse all possible outputs. It uses the notation $\{|\ CV\ |\} = \{c.v.w \mid c.v \in CV \wedge c.v.w \in \Sigma\}$.

Lemma 3. *Given a CSP controller P and one of its failures (tr, X), if CV is a set of machine channels e and input values v such that $\forall e.v \in CV\ .\ \exists w\ .\ e.v.w \in X$, then $(tr, X \cup \{|\ CV\ |\}) \in \mathcal{F}_{SF}\,[\![P]\!]$.*

Proof. This follows by structural induction on the clauses of the controller. The proof requires consideration of the failures semantics of each language construct, and makes use of the fact that none of them enable a controller to partially block output from a B machine (i.e. no guards on such outputs).

Proof of Theorem 2. We prove the contrapositive: that if there is a deadlock (tr, Σ) of the entire system $\|_i (P_i \parallel M_i)$ then it must also be a deadlock of $\|_i P_i$. The entire system is divergence-free by Lemma 1, so there is a deadlock in the failures/divergences semantics of the system if and only if there is a deadlock in its stable failures semantics. We consider the stable failures semantics.

Consider $(tr, \Sigma) \in \mathcal{F}_{SF}\,[\![\|_i (P_i \parallel M_i)]\!]$. Then there must be refusal sets $X_{P_i} \subseteq \alpha(P_i)$, $X_{M_i} \subseteq \alpha(M_i)$, for each i, such that $\bigcup_i (X_{P_i} \cup X_{M_i}) = \Sigma$, and such that for each i

$-\ (tr \restriction \alpha(P_i), X_{P_i}) \in \mathcal{F}_{SF}\, [\![P_i]\!]$
$-\ (tr \restriction \alpha(M_i), X_{M_i}) \in \mathcal{F}_{SF}\, [\![M_i]\!]$

Consider M_i. Define CV to be the set containing all channels $c.v$ made up of an operation name c and an input value v. By Lemma 2, for each $c.v \in CV$ there is some value w such that $c.v.w \notin X_{M_i}$. Thus for each $c.v \in CV$ there is some w such that $c.v.w \in X_{P_i}$, (since $c.v.w$ does not appear in the alphabets of any other machines). By Lemma 3 it follows that $(tr \restriction \alpha(P_i), X_{P_i} \cup \{|\ CV\ |\}) \in \mathcal{F}_{SF}\, [\![P_i]\!]$. But $\alpha(M_i) \subseteq \{|\ CV\ |\}$, and so $X_{M_i} \subseteq \{|\ CV\ |\}$. Thus $(tr \restriction \alpha(P_i), X_{P_i} \cup X_{M_i}) \in \mathcal{F}_{SF}\, [\![P_i]\!]$.

So it follows that

$$(tr, \bigcup_i (X_{P_i} \cup X_{M_i})) \in \mathcal{F}_{SF}\, [\![\, \|_i\, P_i]\!]$$

i.e. $(tr, \Sigma) \in \mathcal{F}_{SF}\, [\![\, \|_i\, P_i]\!]$. This means that $\|_i\, P_i$ has a deadlock in the stable failures model.

Hence, if $\|_i\, P_i$ is deadlock-free in the stable failures model, then $\|_i\, (P_i \parallel M_i)$ is deadlock-free in the stable failures model and the failures/divergences model.

Synchronized Parallel Composition of Event Systems in *B*

Françoise Bellegarde, Jacques Julliand, and Olga Kouchnarenko

Laboratoire d'Informatique de l'Université de Franche-Comté
16, route de Gray, 25030 Besançon Cedex France
Ph:(33) 3 81 66 64 52, Fax:(33) 3 81 66 64 50
{bellegar,julliand,kouchna}@lifc.univ-fcomte.fr,
http://lifc.univ-fcomte.fr

Abstract. A large system typically is or can be decomposed as a composition of components. Usually, these components have to cooperate so, their composition is a synchronized parallel composition. Components are often reactive systems. In the *B* method, each component is an event system. Then, two development paradigms – refinement and component composition – can be used. To provide both paradigms we have a compositionality result of a synchronized parallel composition with respect to refinement. We make use of this result to get an efficient approach to verify the refinement of a synchronized parallel composition between components. Therefore, our proposal allows introducing a second development paradigm in *B*, the component paradigm.

1   Introduction

It is crucial to master the complexity of system specification and development when large systems are considered. Such systems can be specified as a composition of reactive components. So, the complexity of the system is mastered through modularity.

A way to have modules in the *B* method is to decompose the application into separate machines. There are many works on structured development using decomposition into machines and refinement in the framework of the *B* method (see, for example [7,9,14]). Machines are open modules which interact by the authorized operation invocations. However, operation invocations do not provide the parallelism between components. Fortunately, simultaneity can be described as an extension of the simultaneous multiple assignment to generalized substitution in the framework of the *B* event systems.

The *B* event systems can be used to describe the components. However, *B* event systems are closed systems, i.e., they are specified as isolated systems. So, the interactions between components have to be described independently. This is the idea used for example in [8] or in [15] to develop distributed and concurrent systems where the interactions are specified in *CSP*. The tool *csp2B* [8] generates the *B* machine to constrain the execution order between the operation invocations of the modules.

D. Bert et al. (Eds.): ZB 2002, LNCS 2272, pp. 436–457, 2002.

In the paper we propose to specify reactive components by B event systems, and their interactions separately as follows.

- A pair of event names indicates synchronization between two events belonging to two different components.
- Further synchronization between the components is achieved by constraining the activation of events of a component by a predicate over the variables of another component. This way of enhancing the synchronization is also used in [11].

The whole system is then a rearrangement of separate parts, i.e., the components and the interactions. This arrangement is specified by a synchronized parallel composition between components. The key point is to conciliate this specification with refinement.

The approach to verify the refinement is justified by a powerful compositionality theorem. Finite state components can have labeled transition systems as models [6]. This allows us to require no variant in the specification of finite state refined components. For that, we have defined [5] refinement at the model level. The main advantages of such a behavioral semantics are:

- a possibility of an algorithmic verification of the refinement,
- a model based demonstration of the compositionality result,
- a model based verification of dynamic properties of the components and the whole systems.

2   Background

B event systems have been introduced by J.-R. Abrial in [1] for developing distributed systems. Fired events act on the variables while preserving the invariant. An event is a guarded action which is written as a generalized substitution GS. Let p be a predicate, and let x be a state variable. Generalized substitutions can be given by the following grammar:

$$GS ::= x := Exp$$
$$\mid \textbf{SELECT } p \textbf{ THEN } GS \textbf{ END};$$
$$\mid \textbf{ANY } x \textbf{ WHERE } p \textbf{ THEN } GS \textbf{ END};$$
$$\mid GS \| GS$$
$$\mid GS[]GS$$

The notation $[]$ is used for a non-deterministic choice between two actions. An event is enabled only when its guard holds. The guard of an event e, written $grd(e)$, is computed from its generalized substitution GS as $grd(GS)$ in the following way.

$$grd(x := Exp) \qquad\qquad\qquad\qquad = true$$
$$grd(\textbf{SELECT } p \textbf{ THEN } GS \textbf{ END};) \qquad = p \wedge grd(GS)$$
$$grd(\textbf{ANY } x \textbf{ WHERE } p \textbf{ THEN } GS \textbf{ END};) = \exists x.\ p \wedge grd(GS)$$
$$grd(GS_1 \| GS_2) \qquad\qquad\qquad\qquad = grd(GS_1) \wedge grd(GS_2)$$
$$grd(GS_1 [] GS_2) \qquad\qquad\qquad\qquad = grd(GS_1) \vee grd(GS_2)$$

These events provide a means of specifying reactive systems since events are only enabled in certain states. Recall that fired events always terminate.

The paper is organized as follows. Section 3 introduces the B event system components and their communication specifications. We see how to generate their composition. In Section 4 we define the behavioral semantics of components by labeled transition systems. Moreover, we define the compositional hiding and replication operations at the semantic level. In Section 5 we recall the algorithmic definition of B event system refinement already given in our previous works [5,4]. We use this definition in Section 6 to demonstrate compositionality theorem for parallel composition. This theorem allows us to develop a verification of a parallel composition refinement based on the refinement of the components. Section 7 gives some perspectives on property verification, and we conclude in Section 8.

Fig. 1. First scenario

3 Event System Components

In our approach the components are separate event systems M and M', and their parallel composition is defined by a synchronization specification. In this specification two events e of M and e' of M' can be declared synchronized under a feasibility condition. Moreover, any single event e of M or e' of M' can be interleaved in the composition under a feasibility condition. The feasibility conditions are predicates without quantifier on the variable values of other machines.

For example, let us specify a leverage device composed of a pulley and a rail. The pulley lifts the carrier device vertically while the rail pulls it horizontally. We observe only the moves of the pulley and the rail. At the abstract level, we consider the moves according to the two following scenarios:

Fig. 2.   Second scenario

1. The rail moves the device backward. In this back position, the pulley may lift it up. In the up position, the rail may move it forward, so that a carried piece can be put on a deposit belt. At this point, the move can be reversed (see Fig. 1).
2. The rail and the pulley moves can be synchronized, so that the backward and the up moves are done simultaneously. So, in an up and back position, the rail may move the carrier device forward. However, the forward and down moves cannot be done simultaneously because it could hit the feed belt (see Fig. 2).

We specify two event systems, the *Rail* and the *Pulley* as shown in Fig. 3. The synchronization description RP allows specifying the two above scenarios. This means that the F and B moves for the rails are not constrained, and the other moves are constrained. The D and U moves for the pulley are authorized without synchronization in a backward position. However, they are forbidden

```
SYSTEM Rail                      SYSTEM Pulley
VARIABLES HPos                   VARIABLES VPos
INVARIANT                        INVARIANT
HPos ∈ {back, front}             VPos ∈ {down, up}
INITIALIZATION: HPos := back     INITIALIZATION: VPos := down
EVENTS:                          EVENTS:
F ≙SELECT HPos = back            U ≙SELECT VPos = down
    THEN HPos := front END;          THEN VPos := up END;
B ≙SELECT HPos = front           D ≙SELECT VPos = up
    THEN HPos := back END;           THEN VPos := down END;
END                              END


RP = {D when HPos = back, U when HPos = back, F, B, (B, U) when HPos = front}
```

Fig. 3. Pulley and Rail event systems

in a front position, except for (B, U) *when HPos = front*, i.e. an upward move synchronized with the backward move of the rail.

3.1 Component Composition Definition

Let M and M' be two event systems to be composed in parallel under the synchronization specification DS. This specification is a set of event names of either M or M' as well as pairs e, e' where e is an event of M, and e' is an event of M'. The guards of these events or pairs of events can be reinforced by a predicate p over variables of both systems, written e *when* p, e' *when* p', or (e, e') *when* p''. We write this composition $M \|_{DS} M'$.

Let M and M' be as follows.

```
SYSTEM M                         SYSTEM M'
VARIABLES V                      VARIABLES V'
INVARIANT I                      INVARIANT I'
INITIALIZATION: GS_0             INITIALIZATION: GS'_0
EVENTS:                          EVENTS:
e ≙GS;                           e' ≙GS';
...                              ...
END                              END
```

Definition 1. *The event system $M \|_{DS} M'$ is defined as follows:*

SYSTEM $M\|_{DS}M'$
VARIABLES $V \cup V'$
INVARIANT $I \wedge I'$
INITIALIZATION: $GS_0\|GS_0'$
EVENTS:
$e \; \hat{=}$**SELECT** p **THEN** GS **END**;
$e' \; \hat{=}$**SELECT** p' **THEN** GS' **END**;
$(e, e') \; \hat{=}$**SELECT** p'' **THEN** $GS\|GS'$ **END**;
$\ldots$
END

The event system $Rail\|_{RP}Pulley$ is shown in Fig. 4.

SYSTEM $Rail\|_{RP}Pulley$
VARIABLES $HPos, VPos$
INVARIANT
$HPos \in \{back, front\} \wedge VPos \in \{down, up\}$
INITIALIZATION: $HPos, VPos := back, down$
EVENTS:
F $\hat{=}$**SELECT** $HPos = back$
   THEN $HPos := front$ **END**;
B $\hat{=}$**SELECT** $HPos = front$
   THEN $HPos := back$ **END**;
U $\hat{=}$**SELECT** $HPos = back$ **THEN**
   SELECT $VPos = down$ **THEN** $VPos := up$ **END**
   END;
D $\hat{=}$**SELECT** $HPos = back$ **THEN**
   SELECT $VPos = up$ **THEN** $VPos := down$ **END**
   END;
(B,U) $\hat{=}$**SELECT** $HPos = front$ **THEN**
    SELECT $VPos = down$ **THEN** $VPos := up$ **END**
    $\|$ **SELECT** $HPos = front$ **THEN** $HPos := back$ **END**
    END;
END

Fig. 4. Rail and Pulley parallel composition event system

We define also the hiding of an event system M' into M under the synchronization specification DS, written $M/_{DS}M'$.

Definition 2. *The event system $M/_{DS}M'$ is defined as follows:*

SYSTEM $M/_{DS}M'$
VARIABLES $V \cup V'$
INVARIANT $I \wedge I'$
INITIALIZATION: $GS_0 \| GS_0'$
EVENTS:
$e \,\hat{=}$**SELECT** p **THEN** GS **END**;
$(e, e') \,\hat{=}$**SELECT** p'' **THEN** $GS \| GS'$ **END**;
$\ldots$
END

Figure 5 shows the event system $Pulley/_{RP}Rail$, and Figure 6 shows the event system $Rail/_{RP}Pulley$.

SYSTEM $Pulley/_{RP}Rail$
VARIABLES $HPos, VPos$
INVARIANT
$HPos \in \{back, front\} \wedge VPos \in \{down, up\}$
INITIALIZATION: $HPos, VPos := back, down$
EVENTS:
U $\hat{=}$**SELECT** $HPos = back$ **THEN**
   SELECT $VPos = down$ **THEN** $VPos := up$ **END**
   END;
D $\hat{=}$**SELECT** $HPos = back$ **THEN**
   SELECT $VPos = up$ **THEN** $VPos := down$ **END**
   END;
(B,U) $\hat{=}$**SELECT** $HPos = front$ **THEN**
    SELECT $VPos = down$ **THEN** $VPos := up$ **END**
    $\|$ **SELECT** $HPos = front$ **THEN** $HPos := back$ **END**
    END;
END

Fig. 5. Pulley hidden by Rail

Notice that it is not a designer's task to write the event systems resulting from a parallel composition or a hiding under a synchronization description. They could be automatically generated, but it is not needed for the refinement verification (see Section 6.2). The designer only specifies the components, the synchronization description, and the algebraic expression for a parallel composition.

Let $\widehat{DS}$ be a synchronization specification obtained by removing all the feasibility conditions of the elements of DS. The partial hiding and replications of an event system is used in parallel composition by the following equality:

$$M\|_{DS}M' = (M/_{DS}M')\|_{\widehat{DS}}(M'/_{DS}M) \tag{1}$$

SYSTEM $Rail/_{RP}Pulley$
VARIABLES $HPos, VPos$
INVARIANT
$HPos \in \{back, front\} \land VPos \in \{down, up\}$
INITIALIZATION: $HPos, VPos := back, down$
EVENTS:
F $\hat{=}$**SELECT** $HPos = back$
   THEN $HPos := front$ **END**;
B $\hat{=}$**SELECT** $HPos = front$
   THEN $HPos := back$ **END**;
(B,U) $\hat{=}$**SELECT** $HPos = front$ **THEN**
     SELECT $VPos = down$ **THEN** $VPos := up$ **END**
     $\|$ **SELECT** $HPos = front$ **THEN** $HPos := back$ **END**
     END;
END

Fig. 6. Rail hidden by Pulley

The validity of Equality 1 follows from Definitions 1 and 2. Equality 1 allows us to obtain a parallel composition relying simply on event synchronization. Therefore, it simplifies the behavioral semantics of a parallel composition as defined in Section 4.3.

For example, $\widehat{RP} = \{D, U, F, B, (B, U)\}$, and we have $Rail\|_{RP}Pulley = (Rail/_{RP}Pulley)\|_{\widehat{RP}}(Pulley/_{RP}Rail)$.

In the next section we interpret behaviors of components and of their composition as finite labeled transition systems.

4 Event System Component Composition as a Labeled Transition System

4.1 Behavioral Semantics of Components

The behavior of a component is viewed as labeled transition system. A labeled transition system TS is defined by $\langle S, Act, \rightarrow, l \rangle$ where:

- S is a set of states,
- Act is a set of labels,
- $\rightarrow \subseteq S \times Act \times S$ is a transition relation,
- l an interpretation of each state s as a predicate on the variables and their values to evaluate the propositional subformulae of properties.

Let $V \stackrel{\text{def}}{=} \{x_1, \dots, x_n\}$ be a finite set of the *variables* of a component. Let s be a state. The semantics of s over V is a mapping that associates with each variable x_i its value v_i.

Usually, for verification purposes, the set of all the propositions holding in a state s is considered to be a label of s. We consider one of these propositions

$l(s)$ defined by $x_1 = v_1 \wedge \cdots \wedge x_n = v_n$. Then, we know that a predicate p holds on s when $l(s) \Rightarrow p$ holds. So, any *invariant* I, formulating a safety property, holds iff $\forall s \in S, l(s) \Rightarrow I$. The state space of the labeled transition system is the set of states which satisfy the invariant. Let s be a state, and let e be an event enabled in s, i.e. its guard $grd(e)$ is satisfied in s, or $l(s) \Rightarrow grd(e)$ holds.

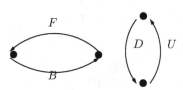

Fig. 7. Rail and Pulley transition systems

Event names are labels of transitions in Act. The events affect state variables. The transition $s \xrightarrow{e} s'$ is the result of the transformation of the state s into the state s' by the event e, i.e. $l(s) \Rightarrow \neg[GS_e]\neg l(s')$ where GS_e is the generalized substitution of e, and $\neg[GS_e]\neg p$ (see [8,6]) represents the weakest precondition under which it is possible for GS_e to establish p.

The labeled transition systems of the examples of the pulley and the rail (see Fig. 3) are given in Fig. 7.

4.2 Behavioral Semantics of a Component Composition

We define the labeled transition system corresponding to the parallel composition of two components. These transition systems are deduced respectively from Definitions 1 and 2.

Let M and M' be two components composed as $M\|_{DS}M'$. Let $TS = \langle S, Act, \rightarrow, l \rangle$ and $TS' = \langle S', Act', \rightarrow', l' \rangle$ be their respective transition systems. We define the transition system corresponding to $M\|_{DS}M'$ as the tuple $\langle S_{||}, Act_{||}, \rightarrow_{||}, l_{||} \rangle$ where

Fig. 8. Transition system of the $Rail\|_{RP}Pulley$

- $S_{||} = S \times S'$. An element of $S_{||}$ is denoted by $s\|s'$.
- $Act_{||} = Act \cup Act' \cup Act \times Act'$,
- $l_{||}(s\|s') = l(s) \wedge l'(s')$,
- $\rightarrow_{||}$ is the set of transitions defined as follows:
  1. $s\|s' \xrightarrow{a}_{||} q\|s'$ iff $a \in Act$, a *when* $p \in DS$, $l(s) \wedge l'(s') \Rightarrow p$, and $(s \xrightarrow{a} q)$.
  2. $s\|s' \xrightarrow{b}_{||} s\|q'$ iff $b \in Act'$, b *when* $p \in DS$, $l(s) \wedge l'(s') \Rightarrow p$, and $(s' \xrightarrow{b}' q')$.
  3. $s\|s' \xrightarrow{(a,b)}_{||} q\|q'$ iff $a \in Act$, $b \in Act'$, (a,b) *when* $p \in DS$, $l(s) \wedge l'(s') \Rightarrow p$, $(s \xrightarrow{a} q)$, and $(s' \xrightarrow{b}' q')$.

The labeled transition system of the rail and pulley parallel composition (see Fig. 4) is given in Fig. 8. Notice how the graph includes the two scenarios of the application (see Fig. 1 and Fig. 2). The nondeterminism allows having both scenarios in one graph.

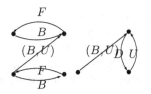

Fig. 9. Transition systems of the $Rail/_{RP}Pulley$ and of the $Pulley/_{RP}Rail$

We deduce the transition system corresponding to $M/_{DS}M'$ as the tuple $\langle S_/, Act_/, \rightarrow_/, l_/ \rangle$ where

- $S_/ = S \times S'$. An element of $S_/$ is denoted by $s\|s'$.
- $Act_/ = Act \cup Act \times Act'$,
- $l_/(s\|s') = l(s) \wedge l'(s')$,
- $\rightarrow_/$ is the set of transitions defined as follows:

  1. $s\|s' \xrightarrow{a}_/ q\|s'$ iff $a \in Act$, a when $p \in DS$, $l(s) \wedge l'(s') \Rightarrow p$, and $(s \xrightarrow{a} q)$.

  2. $s\|s' \xrightarrow{(a,b)}_/ q\|q'$ iff $a \in Act$, $b \in Act'$, (a,b) when $p \in DS$, $l(s) \wedge l'(s') \Rightarrow p$, $(s \xrightarrow{a} q)$, and $(s' \xrightarrow{b} ' q')$.

The mutually hidden transition systems of the event systems $Rail/_{RP}Pulley$ and $Pulley/_{RP}Rail$ are shown in Fig. 9. Notice how the $Rail/_{RP}Pulley$ transition system replicates the rail transitions three times, hiding the free pulley moves. Similarly, the $Pulley/_{RP}Rail$ transition system replicates the pulley transitions twice hiding the free rail moves. Moreover, the rail and pulley parallel composition transition system is the union of these mutually hidden transition systems.

In the next section we define the synchronized parallel composition on transition systems.

4.3   Synchronized Parallel Composition of Transition Systems

In order to synchronize two transition systems TS_1 and TS_2, we have to be able to restrict the interleaving of the moves; some of them being synchronized. For that, we introduce a synchronization set $Synch$ which indicates, among the elements from $Act_1 \cup Act_2 \cup Act_1 \times Act_2$, the authorized moves. For example, given $Act_1 = \{a, b\}$ and $Act_2 = \{c, d, e\}$, a possible synchronization set is $Synch = \{a, d, (a, c), (b, e)\}$.

This synchronization set $Synch$ is a model for a synchronization specification DS. Actually, we syntactically have the set equality $Synch = \widehat{DS}$.

Definition 3. (Parallel composition) *Let $TS_1 = \langle S_1, Act_1, \rightarrow_1, l_1 \rangle$ and $TS_2 = \langle S_2, Act_2, \rightarrow_2, l_2 \rangle$ be two transition systems. The parallel composition of TS_1 and TS_2 is $TS_1\|_{Synch}TS_2 \stackrel{def}{=} \langle S, Synch, \rightarrow, l \rangle$ where S is the set of the $s_1\|s_2$ ($s_1 \in S_1$ and $s_2 \in S_2$), the transition set $Synch \subseteq Act_1 \cup Act_2 \cup Act_1 \times Act_2$ is a synchronization set, and $l(s_1\|s_2) = l_1(s_1) \wedge l_2(s_2)$. The transition relation $\rightarrow$ is defined by combining individual actions in parallel as follows.*

$$[PAR1] \quad \frac{s_1 \xrightarrow{a}_1 s_1'}{s_1\|s_2 \xrightarrow{a} s_1'\|s_2} \; if \; a \in Synch \qquad [PAR2] \quad \frac{s_2 \xrightarrow{a}_2 s_2'}{s_1\|s_2 \xrightarrow{a} s_1\|s_2'} \; if \; a \in Synch$$

$$[PAR3] \quad \frac{s_1 \xrightarrow{a}_1 s_1' \quad s_2 \xrightarrow{b}_2 s_2'}{s_1\|s_2 \xrightarrow{(a,b)} s_1'\|s_2'} \; if \; (a, b) \in Synch$$

This definition means that all moves of parallel composition are moves of either TS_i or of TS_j, or synchronized moves, as indicated in the synchronization set.

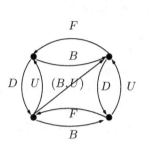

Fig. 10. Transition system of the $Rail\|_{\widehat{RP}}Pulley$

According to the semantics of parallel composition (see Section 4.2), the transition system $TS\|_{\widehat{DS}}TS'$ is a model of the event system $M\|_{DS}M'$ iff $DS = \widehat{DS}$. This is not the case for $Rail\|_{RP}Pulley$ event system since $RP \neq \widehat{RP}$. For that, the reader may compare Fig. 8 and Fig. 10. However, the model of the $Rail\|_{RP}Pulley$ event system is given by the parallel composition of the mutually hidden transition systems under the synchronization set $\widehat{RP}$. Then, by construction, the parallel composition is additive.

In a refinement design we need to examine the parallel and hiding compositionality properties with respect to the refinement. In the next section we recall the algorithmic definition of refinement for B event systems we gave in [5,4].

5   Event System Refinement

In the refinement of abstract event system components, new events can be introduced. These new events must refine the substitution *skip*. A component is refined by applying the standard technique of data refinement to its states. This is specified by the refined event system gluing invariant I_2 which expresses how the variables of the refined system relate to the variables of the abstract system. An event e_2 refines an event e_1 iff the generalized substitution GS_2 of e_2 data-refines[1] the generalized substitution GS_1 of e_1.

Moreover, there must be no new deadlock in the refined system and the new events must not take control forever, thus not preventing the old events to occur. These conditions guarantee behavior similarity between the event systems. Abrial and Mussat [2] give proof obligations for refinement verification. A variant expression, required from the user, gives a termination condition for the cycles of new events. In our verification approach, we propose an algorithmic verification of the refinement. When this verification is effective, no variant is required. This is the case for finite state event systems.

For example, we observe in more detail the pulley and the rail to distinguish the starting points of their diverse movements. So doing, the rail and pulley event systems refine into the event systems as shown in Fig. 11. The new events SU and SD for the pulley, and SF and SB for the rail specify the beginning of refined events U and D for the pulley, and F and B for the rail. The refined events U and D of the pulley, and F and B for the rail have their guards reinforced by refinement. For example, $VPos_2 = to\text{-}up$ implies $VPos = down$ as

---

[1] Let I_1 be the invariant of the abstract event system, GS_2 data-refines GS_1 iff $I_1 \wedge I_2 \Rightarrow [GS_2]\neg[GS_1]\neg I_2$

described by the gluing invariant. Actually, the gluing parts of the invariant are the equivalences.

SYSTEM $Rail_2$	**SYSTEM** $Pulley_2$
REFINES $Rail$	**REFINES** $Pulley$
VARIABLES $HPos_2$	**VARIABLES** $VPos_2$
INVARIANT	**INVARIANT**
$HPos_2 \in \{back, front, to\text{-}back, to\text{-}front\} \wedge$	$VPos_2 \in \{down, up, to\text{-}down, to\text{-}up\} \wedge$
$HPos_2 \in \{back, to\text{-}front\} \Leftrightarrow HPos = back \wedge$	$VPos_2 \in \{down, to\text{-}up\} \Leftrightarrow VPos = down$
$HPos_2 \in \{front, to\text{-}back\} \Leftrightarrow HPos = front$	$\wedge VPos_2 \in \{up, to\text{-}down\} \Leftrightarrow VPos = up$
INITIALIZATION: $HPos_2 := back$	**INITIALIZATION:** $VPos_2 := down$
EVENTS:	**EVENTS:**
F $\hat{=}$**SELECT** $HPos_2 = to\text{-}front$	U $\hat{=}$**SELECT** $VPos_2 = to\text{-}up$
THEN $HPos_2 := front$ **END**;	**THEN** $VPos_2 := up$ **END**;
B $\hat{=}$**SELECT** $HPos_2 = to\text{-}back$	D $\hat{=}$**SELECT** $VPos_2 = to\text{-}down$
THEN $HPos_2 := back$ **END**;	**THEN** $VPos_2 := down$ **END**;
SF $\hat{=}$**SELECT** $HPos_2 = back$	SU $\hat{=}$**SELECT** $VPos_2 = down$
THEN $HPos_2 := to\text{-}front$ **END**;	**THEN** $VPos_2 := to\text{-}up$ **END**;
SB $\hat{=}$**SELECT** $HPos_2 = front$	SD $\hat{=}$**SELECT** $VPos_2 = up$
THEN $HPos_2 := to\text{-}back$ **END**;	**THEN** $VPos_2 := to\text{-}down$ **END**;
END	**END**

Fig. 11. Pulley and Rail refined event systems

In the next section we express the refinement semantics as a relation between transition systems.

5.1   Refinement Semantics

Let $TS_1 = \langle S_1, Act_1, \rightarrow_1, l_1 \rangle$ and $TS_2 = \langle S_2, Act_2, \rightarrow_2, l_2 \rangle$ be two interpreted transition systems giving the behavioral semantics of a system at two levels of refinement. The refinement introduces *new* actions, so $Act_1 \subseteq Act_2$. Our goal is to verify that TS_2 is a *refinement* of TS_1.

First, we define a binary relation $\mu \subseteq S_2 \times S_1$ allowing us to express the relation between states of two consecutive interpreted transition systems.

Definition 4. (Glued states)[2] *Let I_2 be the gluing invariant. The states $s_2 \in S_2$ and $s_1 \in S_1$ are glued, written $s_2 \mu s_1$, if $l_2(s_2) \wedge l_1(s_1) \Rightarrow I_2$.*

Refinement as a simulation. In this section we recall the definition of the refinement relation η as a restriction of μ (see [5,4] for more detail).

---

[2] In previous papers [5,3,4], we defined μ to be a function by requiring the invariant $l_2(s_2) \wedge I_2 \Rightarrow l_1(s_1)$. Then, μ^{-1} gives a partition of the state space of refined systems allowing a modular verification.

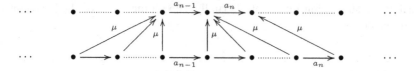

Fig. 12. Path refinement

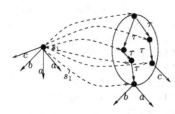

Fig. 13. External non-determinism preservation

The relation η satisfies the following requirements:

1. In order to describe the refinement, we keep the transitions of TS_2 labeled over Act_1 but the *new* ones (from $Act_2 \setminus Act_1$) introduced by the refinement are considered as *non observable* τ moves. We write τ/a to indicate that a τ move covers the action a. Let $Act_{1\tau} \stackrel{\text{def}}{=} Act_1 \cup \{\tau\}$. For $a \in Act_{1\tau}$, and states s, s', we note $s \stackrel{a}{\Longrightarrow} s'$ when there exists $n \geq 0$ such that $s \stackrel{\tau^n a}{\longrightarrow} s'$. Here the occurrence of an action a determines the end of a path of silent moves which all correspond to the activation of new events. This is different from [13] where τ-moves are also allowed after an a. So, the relation η obeys either a strict or a stuttering transition requirement (see Fig. 12).

2. The refinement does not authorize infinite *new* paths composed only with transitions labeled by new actions in TS_2. Since the number of the new actions is finite, traces of infinite τ-paths are ω-regular by the pumping lemma. So, no infinite τ-paths means no infinite τ-iterations (see Fig. 12).

3. The refinement has to maintain the *external non-determinism* though the *internal non-determinism* may be reduced (see Fig. 13).

In [5] we proposed the following definition of a simulation relation between transition systems which expresses the refinement semantics.

Fig. 14. Stuttering

Definition 5. *Let* $TS_1 = \langle S_1, Act_1, \rightarrow_1, l_1 \rangle$ *and* $TS_2 = \langle S_2, Act_{1\tau}, \rightarrow_2, l_2 \rangle$ *be respectively a transition system and its refinement. Let a be in Act_1. The relation $\eta \subseteq S_2 \times S_1$ is defined as the greatest binary relation included into μ and satisfying the following clauses:*

1. (strict transition refinement)
$$(s_2 \; \eta \; s_1 \wedge s_2 \stackrel{a}{\rightarrow}_2 s_2') \Rightarrow (\exists s_1'. \; s_1 \stackrel{a}{\rightarrow}_1 s_1' \wedge s_2' \; \eta \; s_1')$$
2. (stuttering transition refinement) (see Fig. 14)
$$(s_2 \; \eta \; s_1 \wedge s_2 \stackrel{\tau}{\rightarrow}_2 s_2') \Rightarrow (s_2' \; \eta \; s_1)$$

3. **(lack of new deadlock)** $(s_2\ \eta\ s_1 \wedge s_2 \not\rightarrow_2) \Rightarrow (s_1 \not\rightarrow_1)^3$

4. **(non τ-divergence)** $s_2\ \eta\ s_1 \Rightarrow \neg\ (s_2 \overset{\tau}{\rightarrow}_2 s_2' \overset{\tau}{\rightarrow}_2 s_2'' \overset{\tau}{\rightarrow}_2 \cdots \overset{\tau}{\rightarrow}_2 \cdots)$

5. **(non-determinism)** *(see Fig. 15)*
$(s_1 \overset{a}{\rightarrow}_1 s_1' \wedge s_2\ \eta\ s_1) \Rightarrow (\exists\ s_2', s_2'', s_1'.\ s_2'\ \eta\ s_1 \wedge s_2' \overset{a}{\rightarrow}_2 s_2'' \wedge s_1 \overset{a}{\rightarrow}_1 s_1'' \wedge s_2''\ \eta\ s_1'')$

For example, Figure 16 shows the refinement relation between the abstract and refined pulley transition systems, the rail being similar.

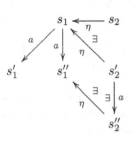

Fig. 15. Non-determinism

We have shown in [5] that η is a τ-simulation. When the relation η holds between states of TS_2 and TS_1, we verify that TS_2 *refines* TS_1, written $TS_2 \sqsubseteq TS_1$. It is well known that a simulation can be computed iteratively when one of the systems is finitely-branching. Since this computation is a depth-first search enumeration of the reachability graphs of the refined system, its order is $\mathbb{O}(|\rightarrow_2| + | S_2 |)$. When we have an effective computation of η, we can verify refinement without proving the decrease of a user provided variant.

6   Refinement and Component Composition

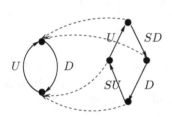

Fig. 16. Refinement relation for the Pulley

In this section we present the main result allowing a design combining a component composition with refinement. This result is an extension of our compositionality theorem in [3] to take into account the synchronization constraints. It is the basis of an efficient component composition refinement verification we present in Section 6.2.

6.1   Parallel Compositionality Refinement Theorem

In the following theorem, η denotes the refinement relation linking transition systems as well as their parallel and hiding compositions. Let TS_2 be a refinement of TS_1, and TS_4 be a refinement of TS_3. Theorem 1 is essential for the component composition refinement verification.

Theorem 1. (Parallel Compositionality) *Let $s_i \in S_i$. Let $s_2\ \eta\ s_1$, and $s_4\ \eta\ s_3$. Let Synch be the synchronization set of $TS_1\|_{Synch}TS_3$, and Synch' be the synchronization set of $TS_2\|_{Synch'}TS_4$. Let*

- *[SH1]: $\forall a \in Act_1 \cup Act_3.\ a \in Synch \Leftrightarrow a \in Synch'$;*
- *[SH2]:*

---

3 We note $s \not\rightarrow$ when $\forall s', s''.(s' \overset{a}{\rightarrow} s'') \in \rightarrow \Rightarrow s \neq s'$.

1. $\forall a.\ a \in Act_1, \forall b.\ b \in Act_3.\ (a,b) \in Synch\ \Leftrightarrow\ (a,b) \in Synch';$
2. $\forall a.\ a \in Act_1, \forall b.\ b \in Act_{3_\tau} \wedge \tau/_b.\ (a,b) \in Synch'\ \Rightarrow\ a \in Synch;$
3. $\forall a.\ a \in Act_{1_\tau} \wedge \tau/_a, \forall b.\ b \in Act_3.\ (a,b) \in Synch'\ \Rightarrow\ b \in Synch;$
- $[SH3]$: $\forall a.\ a \in Act_{1_\tau} \cup Act_{3_\tau}.\ \tau/_a \Rightarrow a \in Synch'$

be Synchronization Hypotheses. If they hold, we have $s_2\|s_4\ \eta\ s_1\|s_3$.

The proof is in Appendix A. We only give here an intuition about Synchronization Hypotheses. First, $[SH1]$ and $[SH2].1$ imply that $Synch \subseteq Synch'$. In other words, every "old" event free in $Synch$ is also free in $Synch'$ and "old" events are synchronized between them in $Synch$ and $Synch'$ in the same way. If an "old" event is synchronized with a "new" event (cf. $[SH3]2$ and $[SH3]3$) then this "old" event appears free in $Synch$. Hypothesis $[SH3]$ forces that every "new" event appears free in $Synch'$.

These Synchronization Hypotheses are a new methodological aspect of a synchronized components refinement design. This compositionality result permits the modular design of the system into separate components.

Consequence 1 Let $s_i \in S_i$. Let $s_2\ \eta\ s_1$. Let $Synch$ be the synchronization set of $TS_1\|_{Synch}TS_3$, and $Synch'$ be the synchronization set of $TS_2\|_{Synch'}TS_3$. We have $s_2\|s_3\ \eta\ s_1\|s_3$ under the following Synchronization Hypotheses:

- $[SH1]$: $\forall a \in Act_1 \cup Act_3.\ a \in Synch\ \Leftrightarrow\ a \in Synch';$
- $[SH2]$:
  1. $\forall a.\ a \in Act_1, \forall b.\ b \in Act_3.\ (a,b) \in Synch\ \Leftrightarrow\ (a,b) \in Synch';$
  2. $\forall a.\ a \in Act_{1_\tau} \wedge \tau/_a, \forall b.\ b \in Act_3.\ (a,b) \in Synch'\ \Rightarrow\ a \in Synch;$
- $[SH3]$: $\forall a.\ a \in Act_{1_\tau}.\ \tau/_a \Rightarrow a \in Synch'$.

Obviously, a similar consequence holds if TS_3 is composed to the right with TS_2 and TS_1. In the next section we use Theorem 1 for an efficient algorithmic verification of the refinement of component parallel composition.

6.2   Component Parallel Composition Refinement Verification

The refinement verification of a composition is based on the separate refinement verification of its components. In this section we present an approach based on the refinement verification of components partially hidden and replicated by a synchronization with another component.

Let M_1 and M_3 be two components to be composed under the synchronization set DS. Let TS_1^h and TS_3^h be the transition systems corresponding respectively to $M_1/_{DS}M_3$ and $M_3/_{DS}M_1$. Let M_2 and M_4 be two components refining respectively M_1 and M_3. The components M_2 and M_4 are composed under the synchronization set DS'. Let TS_2^h and TS_4^h be the transition systems corresponding respectively to $M_2/_{DS'}M_4$ and $M_4/_{DS'}M_2$.

The first verification approach consists in verifying that:

1. $TS_2^h \sqsubseteq TS_1^h$ and that $TS_4^h \sqsubseteq TS_3^h$,
2. *[SH1]*, *[SH2]* and *[SH3]* hold between $\widehat{DS}$ and $\widehat{DS'}$.

This is justified as follows:

- The above items satisfy the assumptions of Theorem 1. Therefore, we have $(TS_2^h \|_{\widehat{DS'}} TS_4^h) \sqsubseteq (TS_1^h \|_{\widehat{DS}} TS_3^h)$.
- The transition system $(TS_2^h \|_{\widehat{DS'}} TS_4^h)$ denotes the semantics of the machine $(M_2 /_{DS'} M_4) \|_{\widehat{DS'}} (M_4 /_{DS'} M_2)$ (see Section 4.3). In the same way, the transition system $(TS_1^h \|_{\widehat{DS}} TS_3^h)$ denotes the semantics of the machine $(M_1 /_{DS} M_3) \|_{\widehat{DS}} (M_3 /_{DS} M_1)$.
- Therefore, we conclude that the component composition $M_1 \|_{DS} M_3$ refines the component composition $M_2 \|_{DS'} M_4$ by Equality 1.

For example, we want to compose the refined systems $Pulley_2$ and $Rail_2$. We consider the synchronization as detailed moves of the two scenarios (see Fig. 1 and 2) taking into account their starting points. The synchronization description RP becomes RP':

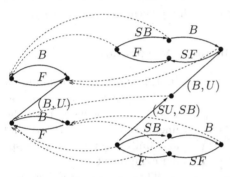

Fig. 17. Refinement relation for the rail hidden by the pulley, $Rail_2 /_{RP'} Pulley_2 \sqsubseteq Rail /_{RP} Pulley$

$\{D$ when $HPos_2 = back,$
SD when $HPos_2 = back,$
U when $HPos_2 = back,$
SU when $HPos_2 \in \{back, to - back\},$
F when $VPos_2 \in \{down, up\},$
SF when $VPos_2 \in \{down, up\},$
B when $VPos_2 \in \{down, up\},$
SB when $VPos_2 \in \{down, up\},$
(B, U) when $HPos_2 = to - back,$
(SU, B) when $HPos_2 = to - back,$
(SU, SB) when $HPos_2 = front\}$

This means that the D, SD and U moves for the pulley are constrained to happen at the rail back position. For the pulley, the beginning move SU is constrained to occur at the *back*, *to − back*, and *to − front* positions. For the rail, the F, SF, B and SB moves are constrained to occur at the *up* and the *down* pulley position.

The moves (B, U), (SU, B) and (SU, SB) are synchronized: SU and SB can occur simultaneously only at the rail front position while the other moves can happen at the rail *to-back* position.

First, the synchronization descriptions $\widehat{RP}$ and $\widehat{RP'}$ satisfy the *[SH1]*, *[SH2]* and *[SH3]* Synchronization Hypotheses. Second, we have verified that the transition system of the $Pulley_2 /_{RP'} Rail_2$ event system refines the transition system of the $Pulley /_{RP} Rail$ event system. Also, the transition system of the $Rail_2 /_{RP'} Pulley_2$ event system refines the transition system of the $Rail /_{RP} Pulley$

event system. Figure 17 and Figure 18 show the refinement relations between the mutually hidden pulley and rail transition systems. Notice that the event system $Rail_2/_{RP'}Pulley_2$ replicates the $Rail_2$ event system 3 times, and that the $Pulley_2/_{RP'}Rail_2$ event system replicates the $Pulley_2$ event system 3 times.

This refinement verification approach can be useful if the sum of the sizes of the state and transition spaces of the systems under hiding is significantly smaller than the sum of the sizes of the state and transition spaces of the whole systems. This can be the case when the refined synchronization specification does not induce a replication.

For our running example this verification is expensive because of the many replications for the pulley and the rail. We have 13 states and 15 transitions to verify that $(Rail_2/_{RP'}Pulley_2) \sqsubseteq (Rail/_{RP}Pulley)$ plus 10 states and 10 transitions to verify that $(Pulley_2/_{RP'}Rail_2) \sqsubseteq (Pulley/_{RP}Rail)$.

This approach can be too restrictive in practice since hiding is likely to introduce deadlocks which disappear in parallel composition. So, we are thinking about a second approach combining algorithmic verification and proof.

6.3 Perspective on Mixed Refinement Verification

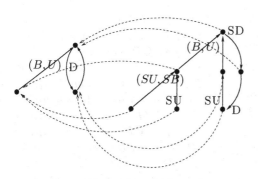

Fig. 18. Refinement relation for the pulley hidden by the rail, $Pulley_2/_{RP'}Rail_2 \sqsubseteq Pulley/_{RP}Rail$

In this section we examine when the refinement verification of the whole components implies the refinement of the partially hidden and finitely replicated corresponding components.

The second verification approach consists in verifying that:

1. $TS_2 \sqsubseteq TS_1$ and that $TS_4 \sqsubseteq TS_3$,
2. for a free or synchronized event e, to each feasibility condition $(e \text{ when } p) \in DS$ corresponds the feasibility condition $(e \text{ when } p') \in DS'$ such that

a) $I \wedge I' \wedge p' \wedge [GS'_e] \Rightarrow p \wedge grd(GS_e)$,
b) $I \wedge I' \wedge p \wedge grd(GS_e) \Rightarrow (p' \wedge grd(GS'_e)) \vee \bigcup_{ne \in NewEvents} (grd(GS'_{ne}) \wedge p'_{ne}))$,
where I and I' are the invariants of the abstract and refined systems;
3. $[SH1]$, $[SH2]$ and $[SH3]$ hold between $\widehat{DS}$ and $\widehat{DS'}$.

We only give here informal arguments for proof. The conjunction of the conditions of the second item with the sufficient conditions of the B refinement for an event e imply the sufficient conditions of the B refinement for e in the hidden event systems. We have shown in [4] that the sufficient conditions of the B event refinement (see [2]) imply that Clauses 1, 2 and 3 of Definition 5 hold. Moreover, the Synchronization Hypothesis $[SH3]$ and the first item imply that

Clause 4 of Definition 5 holds. Clause 5 is obtained from the B event refinement sufficient conditions and the Synchronization Hypotheses. So, the refinement of the parallel composition is verified.

For our example we have 6 states and 6 transitions for each algorithmic refinement verification, and the proofs about the feasibility conditions and the deadlock freeness. The second approach is very interesting for multiple parallel composition of a same machine under different synchronization specifications. It is less restrictive since it checks the lack of new deadlock directly on the whole composed system.

7   Perspectives on Property Verification

On the one hand, refinement as a simulation relation leads to preservation (see, for example, [4]) of $PLTL$ properties. On the other hand, many works exploit parallel composition of components for an efficient model-checking of their composition properties. In particular, we intend to use the verification method of Lind-Nielsen et. al. in [11] for an efficient backward reachability analysis based on a dependency analysis between components. It seems to us that the strong dependency existing between mutually hidden components could be exploited for an efficient reachability analysis of their parallel composition, i.e. a reachability analysis of a parallel composition of the whole systems thanks to Equality 1.

We would like to further exploit this dependency between mutually hidden components. Indeed, we have studied a class of $PLTL$ properties called modular, when holding on each element of a transition system partition, hold on the whole system [10,12]. Notice that these properties include the B modalities in the case of finite systems. These results have been used for a partition of a refined transition system obtained by a particular gluing relation μ (see Definition 4). This allows a separate verification of a modular property on each element of the partition. We call this verification process modular. We would like to also use the separate verification approach for mutually hidden components. We cannot directly use the class of modular $PLTL$ properties since mutually hidden components are not distinct. Nevertheless, we hope to obtain other results because of the additive nature of the parallel composition of the mutually hidden components.

Moreover, we conjecture that the central monotonicity result of [11] applies to a multiple parallel composition of components because of Equality 1. This way we want to exploit the structure of the composition for a reuse purpose using, for example, the assumption-guarantee paradigm as in [16].

8   Conclusion

This paper proposes to introduce components paradigm into B event systems refinement design. It allows mixing between the refinement and components design paradigms. It is very convenient for specifying and verifying large embedded reactive systems such as automatic trains, flight control systems and communication protocols. It also applies to smaller systems such as cell phones, chip cards and car driving assistance systems.

Having both paradigms brings flexibility. For example, adding the carrier device into the $Pulley\|_{RP}Rail$ event system can be obtained either by refinement of the pulley or the rail, or by another component to synchronize in a parallel composition with the $Pulley\|_{RP}Rail$ event system.

We could have chosen to introduce synchronization and communication features using known syntax or calculus such as CSP. However, we remain in the framework of the B event systems to specify the synchronizations for a parallel composition of components. A synchronization is simply a pair of events. Replication and partial hiding of components are generated by feasibility conditions as guards which come to strengthen the proper guards of the events in a parallel composition.

The synchronization can be completed by a description of input and output events. Also, sharing variables between components could be useful for message passing communication. These are important perspectives for future works. Moreover, we are currently limited to a finite parallel composition of components. Parallel composition of an arbitrary number of components is a very difficult but interesting direction.

The renaming operation for a given component – a very useful operation for reuse – is not mentioned in the paper, but it is easy to implement in our framework. For example, it is obvious that the $Pulley$ and the $Rail$ event systems could be viewed as renaming of the same component, a moving device. Going further in the development, the press device which closes to grip a piece on the feed belt and opens to let it go on the deposit belt, is also a renaming of the moving component.

We make use of the compositionality result of a synchronized parallel composition with respect to refinement to get an efficient approach to verify the refinement of a synchronized parallel composition between components. Actually, we verify the refinement of a synchronized parallel composition by verifying separately the refinement of each component. This works if required assumptions between feasibility conditions and Synchronization Hypotheses at the abstract and refined levels hold. Moreover, we can use an algorithmic verification of the refinement for finite state components. This avoids all the variant requirements for design and proof of the finite state components refinement. For infinite state components, the B proof obligations for event system refinement and a variant are required.

The benefit of the refinement approach for verification is that the properties which hold in an abstraction are preserved by refinement. Moreover, we want to make use of the component design for an efficient property verification in the abstract component event systems.

We extend the B system by a synchronized parallel composition operation – an efficient way towards introducing the component paradigm in B.

Acknowledgment. We are grateful to the reviewers for their careful work and their valuable and insightful comments and suggestions.

References

1. J.-R. Abrial. Extending B without changing it (for developing distributed systems). In *1st Conference on the B method*, pages 169–190, Nantes, France, November 1996.
2. J.-R. Abrial and L. Mussat. Introducing dynamic constraints in B. In *Second Conference on the B method, France*, volume 1393 of *LNCS*, pages 83–128. Springer Verlag, April 1998.
3. F. Bellegarde, C. Darlot, J. Julliand, and O. Kouchnarenko. Reformulate dynamic properties during B refinement and forget variants and loop invariants. In *Proc. First Int. Conf. ZB'2000, York, Great Britain*, volume 1878 of *LNCS*, pages 230–249. Springer-Verlag, September 2000.
4. F. Bellegarde, C. Darlot, J. Julliand, and O. Kouchnarenko. Reformulation: a way to combine dynamic properties and B refinement. In *In Proc. Int. Conf. Formal Method Europe'01, Berlin, Germany*, volume 2021 of *LNCS*, pages 2–19. Springer Verlag, March 2001.
5. F. Bellegarde, J. Julliand, and O. Kouchnarenko. Ready-simulation is not ready to express a modular refinement relation. In *Proc. Int. Conf. on Fundamental Aspects of Software Engineering, FASE'2000*, volume 1783 of *LNCS*, pages 266–283. Springer-Verlag, April 2000.
6. D. Bert and F. Cave. Construction of finite labelled transition systems from B abstract systems. In T. Santen W. Grieskamp and B. Stoddart, editors, *2nd international conference on Integrated Formal Methods, IFM 2000*, volume 1945 of *LNCS*, pages 235–255, Germany, November 2000. Springer-Verlag.
7. P. Bontron and M.-L. Potet. Automatic construction of validated B components from structured developments. In *Proc. First Int. Conf. ZB'2000, York, Great Britain*, volume 1878 of *LNCS*, pages 127–147. Springer-Verlag, September 2000.
8. M. J. Butler. csp2B: A practical approach to combining CSP and B. *Formal Aspects of Computing*, 12:182–198, 2000.
9. M. J. Butler and M. Waldén. *Program Development by Refinement (Case Studies Using the B Method)*, chapter Parallel Programming with the B Method. Springer, 1999.
10. J. Julliand, P-A. Masson, and H. Mountassir. Vérification par model-checking modulaire des propriétés dynamiques introduites en B. *Technique et Science Informatiques*, To appear, 2001.
11. J. Lind-Nielsen, H. R. Andersen, H. Hulgaard, G. Behrmann, K. Kristoffersen, and K. G. Larsen. Verification of large state/event systems using compositionality and dependency analysis. *Formal Methods in System Design*, 18(1):5–23, January 2001.
12. P.-A. Masson, H. Mountassir, and J. Julliand. Modular verification for a class of PLTL properties. In T. Santen W. Grieskamp and B. Stoddart, editors, *2nd international conference on Integrated Formal Methods, IFM 2000*, volume 1945 of *LNCS*, pages 398–419. Springer-Verlag, November 2000.
13. R. Milner. *Communication and Concurrency*. Prentice Hall Int., 1989.
14. E. Sekerinski. *Program Development by Refinement (Case Studies Using the B Method)*, chapter Production Cell. Springer, 1999.
15. H. Treharne and S. Schneider. Using a process algebra to control B OPERATIONS. In *IFM'99 1st International Conference on Integrated Formal Methods*, pages 437–457, York, 1999. Springer-Verlag.

16. Y.-K. Tsay. Compositional verification in linear-time temporal logic. In J. Tiuryn, editor, *Proc. 3rd Int. Conf. on Foundations of Software Science and Computation Structures (FOSSACS 2000)*, volume 1784 of *LNCS*, pages 344–358. Springer Verlag, 2000.

A   Proof of Theorem 1

To prove the theorem, we show that $\mathcal{S}$ verifies the clauses of Definition 5, where

$$\mathcal{S} \stackrel{\text{def}}{=} \{(s_1\|s_3, s_2\|s_4) \mid s_2 \ \eta \ s_1 \ \wedge \ s_4 \ \eta \ s_3\}$$

using Lemmas 1, 2, 3, and 4.

From now on, we suppose that $(s_1\|s_3, s_2\|s_4) \in \mathcal{S}$. So, $(s_2\|s_4) \ \eta \ (s_1\|s_3)$, i.e. $s_2 \ \eta \ s_1$ and $s_4 \ \eta \ s_3$.

Lemma 1. *If $(s_2\|s_4 \stackrel{\alpha}{\Rightarrow})$ then $(s_1\|s_3 \stackrel{\alpha}{\Rightarrow})$.*

Proof. There are the following cases.

1. The strict transition refinement case.

    Assuming that $s_2\|s_4 \stackrel{\alpha}{\rightarrow} \tilde{s}'$, we must prove that there exists $\tilde{s}$ such that $s_1\|s_3 \stackrel{\alpha}{\rightarrow} \tilde{s}$ and $\tilde{s}' \ \eta \ \tilde{s}$. There are two cases:

    a) $\alpha = a$ where $a \in Synch'$ may be an action in either TS_2 or TS_4. Consider only the first case ($a \in Act_2$), the second one being similar.

       Then, we have $\tilde{s}' = s_2'\|s_4$ where $s_2 \stackrel{a}{\rightarrow}_2 s_2'$ because of *[PAR1]*. Then, since TS_2 refines TS_1, there exists s_1' such that $s_1 \stackrel{a}{\rightarrow}_1 s_1'$ and $s_1' \ \eta \ s_1$ by the strict transition refinement clause. Therefore, we get the result by taking $\tilde{s} = s_1'\|s_3$ again by *[PAR1]*, because of *[SH1]*.

    b) $\alpha = (a, b)$ where $(a, b) \in Synch'$, a is an action in TS_2, and b is an action in TS_4.

       Then, by *[PAR3]*, we have $\tilde{s}' = s_2'\|s_4'$ where $s_2 \stackrel{a}{\rightarrow}_2 s_2'$ and $s_4 \stackrel{b}{\rightarrow}_4 s_4'$. Then, since TS_2 refines TS_1, there exists s_1' such that $s_1 \stackrel{a}{\rightarrow}_1 s_1'$ and $s_2' \ \eta \ s_1'$ by the strict transition refinement clause. Similarly, since TS_4 refines TS_3, there exists s_3' such that $s_3 \stackrel{b}{\rightarrow}_3 s_3'$ and $s_4' \ \eta \ s_3'$. Therefore, we get the result by taking $\tilde{s} = s_1'\|s_3'$ again by *[PAR3]*, because of *[SH2]*.

    c) $\alpha = (a, \tau)$ where $(a, b) \in Synch'$, a is an action in TS_2, and b is a new action in TS_4 covered by τ.

       Then, by *[PAR3]*, we have $\tilde{s}' = s_2'\|s_4'$ where $s_2 \stackrel{a}{\rightarrow}_2 s_2'$ and $s_4 \stackrel{\tau}{\rightarrow}_4 s_4'$. Then, since TS_2 refines TS_1, there exists s_1' such that $s_1 \stackrel{a}{\rightarrow}_1 s_1'$ and $s_2' \ \eta \ s_1'$ by the strict transition refinement clause. Since TS_4 refines TS_3 we have $s_4' \ \eta \ s_3$ by the stuttering transition refinement clause. Therefore, we get the result by taking $\tilde{s} = s_1'\|s_3$ by *[PAR1]*, because of *[SH2]*.

    d) $\alpha = (\tau, b)$ where $(a, b) \in Synch'$, a is a new action in TS_2 covered by τ, and b is an action in TS_4. The proof is similar to the previous case.

2. The stuttering transition refinement case.

   Assuming that $s_2\|s_4 \xrightarrow{\tau} \tilde{s}'$, we must prove that $\tilde{s}' \; \eta \; s_1\|s_3$.

   There are two cases:

   a) The silent τ move corresponds to an action $a \in Synch'$. It may be a new action in either TS_2 or TS_4. Consider only the first case, the second one being similar.

      Then, we have $\tilde{s}' = s_2'\|s_4$ where $s_2 \xrightarrow{a}_2 s_2'$ because of [PAR1]. Then, since TS_2 refines TS_1, and $\tau \diagup_a$, we have $s_2' \; \eta \; s_1$ by the stuttering transition refinement clause. Therefore, $s_2'\|s_4 \; \eta \; s_1\|s_3$.

   b) The silent τ move corresponds to a move $(a,b) \in Synch'$, a is an action in TS_2, and b is an action in TS_4.

      Then, by [PAR3], we have $\tilde{s}' = s_2'\|s_4'$ where $s_2 \xrightarrow{a}_2 s_2'$ and $s_4 \xrightarrow{b}_4 s_4'$. Then, since TS_2 refines TS_1, and a is covered by τ, we have $s_2' \; \eta \; s_1$ by the stuttering transition refinement clause. Similarly, since TS_4 refines TS_3, and b is covered by τ, we have $s_4' \; \eta \; s_3$. Therefore, $s_2'\|s_4' \; \eta \; s_1\|s_3$.

Lemma 2. If $s_2\|s_4 \not\rightarrow$, then $s_1\|s_3 \not\rightarrow$.

Proof. If $s_2\|s_4 \not\rightarrow$ then no rule of Definition 3 applies, so neither [PAR1], nor [PAR2], nor [PAR3] applies.

1. [PAR1] does not apply.
   - $s_2 \not\rightarrow$, then $s_1 \not\rightarrow$ because of refinement. So, [PAR1] does not apply for $s_1 \|s_3$.
   - $s_2 \xrightarrow{a}_2$ but $a \notin Synch'$. Then, rule [PAR1] does not apply for $s_1\|s_3$ by [SH1].
2. [PAR2] does not apply. The proof is similar.
3. [PAR3] does not apply.
   - either $s_2 \not\rightarrow$, or $s_4 \not\rightarrow$. Then either $s_1 \not\rightarrow$, or $s_3 \not\rightarrow$, because of refinement. So, [PAR3] does not apply for $s_1 \|s_3$.
   - $s_2\|s_4 \xrightarrow{(a,b)}$ where $a \in Act_1$, $b \in Act_2$, but $(a,b) \notin Synch'$. Then, rule [PAR3] does not apply for $s_1\|s_3$ since $(a,b) \notin Synch$ by [SH2].
   - $s_2\|s_4 \xrightarrow{(a,b)}$ but $(a,b) \notin Synch'$ where $a \in Act_1$, $b \in Act_{2_\tau}$ and $\tau \diagup_b$. Then, by [SH3], we have $\tau \in Synch'$. Then, rule [PAR2] can be applied for $s_2\|s_4$. Contradiction.
   - $s_2\|s_4 \xrightarrow{(a,b)}$ but $(a,b) \notin Synch'$ where $a \in Act_{1_\tau}$ and $\tau \diagup_a$, $b \in Act_2$. The proof is similar to the previous case.

Lemma 3. $\neg \, (s_2\|s_3 \xrightarrow{\tau} s_2'\|s_3 \xrightarrow{\tau} s_2''\|s_3\| \xrightarrow{\tau} \cdots \xrightarrow{\tau} \cdots)$

Proof. It is an immediate consequence of the non τ−divergence transition refinement clause in the refined system.

Lemma 4. If $s_1\|s_3 \xrightarrow{\alpha} \tilde{s}_1'$, then $\exists \tilde{s}_2', \tilde{s}_2'', \tilde{s}_1''$ such that $\tilde{s}_2' \; \eta \; s_1\|s_3$, $\tilde{s}_2' \xrightarrow{\alpha}_2 \tilde{s}_2''$, $s_1\|s_3 \xrightarrow{\alpha}_1 \tilde{s}_1''$, and $\tilde{s}_2'' \; \eta \; \tilde{s}_1''$.

Proof. There are three cases derived from $s_1 \| s_3 \xrightarrow{\alpha} \tilde{s}'_1$.

1. $\alpha = a$ where $a \in Synch$ may be an action in either TS_1 or TS_3. Consider only the first case ($a \in Act_1$), the second one being similar.

   By *[PAR1]*, we have $\tilde{s}'_1 = s'_1 \| s_3$ where $s_1 \xrightarrow{a} s'_1$. Then, by the nondeterminism transition refinement clause, $\exists s'_2, s''_2, s''_1$ such that $s'_2 \; \eta \; s_1$, $s'_2 \xrightarrow{a}_2 s''_2$, $s_1 \xrightarrow{a}_1 s''_1$, and $s''_2 \; \eta \; s''_1$. Since $a \in Synch$ we have $a \in Synch'$ by *[SH1]*. Therefore, again by *[PAR1]*, we get the result with $\tilde{s}'_2 = s'_2 \| s_4$, $\tilde{s}''_2 = s''_2 \| s_4$, and $\tilde{s}''_1 = s''_1 \| s_3$.

2. $\alpha = (a, b)$ where $(a, b) \in Synch$, a is an action in TS_1, and b is an action in TS_3.

   By *[PAR3]*, we have $\tilde{s}'1 = s'_1 \| s'_3$ where $s_1 \xrightarrow{a}_1 s'_1$, and $s_3 \xrightarrow{b}_3 s'_3$. Then, by the nondeterminism transition refinement clause, $\exists s'_2, s''_2, s''_1$ such that $s'_2 \; \eta \; s_1$, $s'_2 \xrightarrow{a}_2 s''_2$, $s_1 \xrightarrow{a}_1 s''_1$, and $s''_2 \; \eta \; s''_1$. Again, by the nondeterminism transition refinement clause, $\exists s'_4, s''_4, s''_3$ such that $s'_4 \; \eta \; s_3$, $s'_4 \xrightarrow{b}_4 s''_4$, $s_3 \xrightarrow{b}_3 s''_3$, and $s''_4 \; \eta \; s''_3$.

   By *[SH2]*, $(a, b) \in Synch'$. By rule *[PAR3]*, we can conclude with $\tilde{s}''_1 = s''_1 \| s''_3$, and $\tilde{s}''_2 = s''_2 \| s''_4$.

Global and Communicating State Machine Models in Event Driven B: A Simple Railway Case Study

Antonis Papatsaras and Bill Stoddart

University of Teesside
School of Computing and Mathematics,
Middlesbrough, England.
antonis@papatsaras.com - bill@tees.ac.uk

Abstract. We present a case study of a simple railway system to investigate and compare two ways of modelling a system in "event driven B". We are interested in the specification of a system as a global model as well as the formulation of a distributed state machine model where individual components exchange information by means of shared events. In this paper we investigate the issues of "parameter hiding" and "scaling" as well as the parameterisation of events of the communicating components of such systems. We use two methods for expressing a class of components; we either create indexed B machines that can be instantiated or we represent the state of all components within a given class by means of a function.

1 Introduction

In this paper we are interested in event based system models, known in B as Abstract Systems. We see the development of a system from two different approaches; as a **global** mathematical specification - Abrial's approach - and also in terms of **interacting** components which communicate by means of shared events. Each of these components are specialised to carry out a certain task but they also share information with the rest of the system.

The idea of a global system model has emerged in B with Abrial's paper "Extending B without changing it" [1] and also in [5] and [6]. The use of a model based specification language for developing distributed system models in terms of interacting components has been developed in [11], [9] and [10], where diagrammatic state machine notation is used to express behavioural properties of components and also in [4] where the CSP process algebra style of notation is used. A rather different approach which combines CSP and B is given in [7].

We model a simple railway system as an illustrative case study. The system consists of a track of eight sections, eight sets of signals - one for each section - and two trains. We first define the global model where the whole system is described in a single machine in B [3]. We examine the invariant of such system and investigate its events. Then we specify the same railway system in terms of

D. Bert et al. (Eds.): ZB 2002, LNCS 2272, pp. 458–476, 2002.
© Springer-Verlag Berlin Heidelberg 2002

communicating components. We split it into small parts, the signal machine, the train, and the speed controller machine. We add some stations to examine the behaviour of a train when it is approaching them. Then combine all these small parts into a system machine.

In such a system we realise that some parts may have the same properties. For instance our model consists of eight sections and eight signals. All signals operating in the same way. We should therefore think of a signal as one class of object that we could instantiate eight times. Thus we are facing the question; how could we deal with repetition, otherwise scaling, in such a distributed development in B. Another aspect of scaling for a development technique that relies on diagrams is that these diagrams can become too complex to express on a single page. In such cases it is sometimes possible to break a system down into a number of scenarios. For our railway system we have different scenarios when a train approaches a station and when it is free running. We show how we adapt our diagram notation to take account of this.

Finally we consider how to express our state machine model as a refinement of the global model.

2   Informal Description of the Railway

The model describes a railway system with a single loop of track and no points. Two trains are travelling in one direction only. The track is divided into eight sections two of which are marked as stations. A train will fit completely within a single section. A train traveling at full speed cannot stop within the space of a single section, but can stop within two sections. Two trains cannot exist within the same section.

Each section has an associated signal, which may be red, amber or green, and which is seen by the train when it clears the previous section. Signals affect the speed of trains as follows:

- green: no speed restriction
- amber: slow, the train must adjust its speed so that it can stop in the following section if required to do so.
- red: the train must stop within this section.

A train "clears a section" when its last carriage leaves the section. This is the point at which various signals on the railway system change.

Trains stop at all stations. A trains which is stopped at a station is said to be "docked".

Our study of this railway system requires the ability to be able to see the system from two different points of view; as a global abstract system and as a distributed state machine model consists of communicating components.

3   Global Abstract Model

Here the railway is described as a whole system without taking into account the functionality of each or its components separately. Our formal descriptions of the railway system are written in an extended form of GSL (General Substitution Language).

We define three sets to describe the entities trains, signals and speeds. Then we decide that the stations on our system are located at section 3 and section 7.

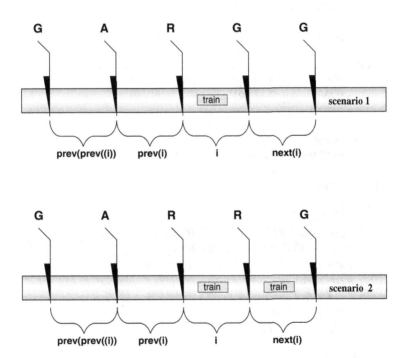

Fig. 1. Two different train-positioning scenarios

Static aspects of the system are described in the invariant. It is the point of the development where security and safety properties are expressed. The invariant alone is not enough to ensure safety. Therefore a cross check of the invariant with the events is needed. The most fundamental system properties here is that no two trains ever occupy the same section at the same time. The invariant is reasonably complex despite the fact that the example is small. When a train is occupying section i then we must ensure that $section(prev(i))$ has a red signal. Where a second train does not occupy section $prev(i)$, then the signal on section $prev(prev(i))$ is set to amber. All the other signals are set to green.

Dynamic aspects of the system are described in the Events clause. For instance the clear event specifies that when a train clears a section, the signal on that section changes to red and the signal on the previous section changes to amber -figure 1-.

All trains and signals are set at the initialisation stage. We set train positions, signal settings, and speeds. Two events are used: a train leaving a section to enter the next one, and when a train starting from a halt position (either being on a station or restricted by a red signal). A train will clear a section when the signal of the section is either amber or green. A train will start from a halt position, when the signal of the section is set to amber or green.

The GSL version of the global system follows:

System *Rlwy*

Sets
$$TRAIN = \{T1, T2\}$$
$$SIGNAL = \{G, A, R\}$$
$$SPEED = \{stop, slow, fast\}$$

Constants
$$SECTION, nsec, next, prev, stations$$

Properties
$$nsec = 8 \wedge$$
$$SECTION = 1 .. nsec \wedge$$
$$next \in SECTION \rightarrowtail SECTION \wedge$$
$$prev \in SECTION \rightarrowtail SECTION \wedge$$
$$\forall x \bullet x \in 1 .. nsec - 1 \Rightarrow next(x) = x + 1 \wedge$$
$$next(nsec) = 1 \wedge$$
$$\forall x \bullet x \in 2 .. nsec \Rightarrow prev(x) = x - 1 \wedge$$
$$prev(1) = nsec \wedge$$
$$stations = \{3, 7\}$$

Variables
$$posn, signal, speed, docked$$

Invariant
$$posn \in TRAIN \rightarrowtail SECTION \wedge$$
$$signal \in SECTION \rightarrow SIGNAL \wedge$$
$$speed \in TRAIN \rightarrow SPEED \wedge$$
$$docked \in \mathbb{P} \, TRAIN \wedge$$
$$(\forall i \bullet i \in SECTION \Rightarrow$$
$$\quad (occupied(i) \Rightarrow signal(prev(i)) = R) \wedge$$
$$\quad (occupied(i) \wedge \neg occupied(prev(i)) \Rightarrow signal(prev(prev(i))) = A)) \wedge$$
$$\quad (\neg occupied(next(i)) \wedge \neg occupied(next(next(i))) \Rightarrow signal(i) = G)$$
$$)$$
$$\wedge$$

$$(\forall t \bullet t \in TRAIN \Rightarrow$$
$$(t \in docked \Rightarrow speed(t) = stop) \land$$
$$signal(posn(t)) \neq R \land$$
$$(\neg (t \in docked) \land next(posn(t)) \in stations \Rightarrow speed(t) = slow) \land$$
$$(\neg (t \in docked) \land next(posn(t)) \notin stations \Rightarrow$$
$$(signal(posn(t)) = R \Rightarrow speed(t) = stop) \land$$
$$(signal(posn(t)) = A \Rightarrow speed(t) \neq fast)$$
$$)$$
$$)$$

Initialisation

$$posn := \{T1 \mapsto 1, T2 \mapsto 2\} \parallel$$
$$signal := AllGreen \Leftarrow \{1 \mapsto R, nsec \mapsto R, nsec - 1 \mapsto A\} \parallel$$
$$speed := \{T1 \mapsto stop, T2 \mapsto slow\} \parallel$$
$$docked := \{\}$$

Events

$clear(t : TRAIN, i : SECTION) \mathrel{\hat{=}}$
$\qquad posn(t) = s \land speed(t) \neq stop \Longrightarrow$
$\qquad\qquad (posn(t) := next(i) \parallel$
$\qquad\qquad signal(i) := R \parallel$
$\qquad\qquad signal(prev(i)) := A \parallel$
$\qquad\qquad \mathbf{if}\ (empty(prev(i)))\ \mathbf{then}\ signal(prev(prev(i))) := G\ \mathbf{endif}\ \parallel$
$\qquad\qquad \mathbf{if}\ (next(i) \in stations)\ \mathbf{then}$
$\qquad\qquad\qquad speed(t) := stop \parallel docked := docked \cup \{t\}$
$\qquad\quad \mathbf{else}$
$\qquad\qquad\qquad signal(next(i)) = R \Longrightarrow speed(t) := stop$
$\qquad\qquad\qquad\qquad \llbracket$
$\qquad\qquad\qquad signal(next(i)) = A \Longrightarrow speed(t) := slow$
$\qquad\qquad\qquad\qquad \llbracket$
$\qquad\qquad\qquad signal(next(i)) = G \Longrightarrow$
$\qquad\qquad\qquad\qquad \mathbf{if}\ (next(next(i))) \in stations\ \mathbf{then}\ speed(t) := slow$
$\qquad\qquad\qquad\qquad \mathbf{else}\ speed(t) := fast$
$\qquad\qquad\qquad\qquad \mathbf{endif}$
$\qquad\quad \mathbf{endif}$

$start(t : TRAIN) \mathrel{\hat{=}} signal(posn(t)) \neq R \Longrightarrow$
$\qquad (docked := docked \setminus \{t\}) \parallel$
$\qquad (signal(posn(t)) = G \Longrightarrow speed(t) := fast$
$\qquad\qquad \llbracket$
$\qquad signal(posn(t)) = A \Longrightarrow speed(t) := slow)$

Definitions

$\qquad occupied(i) == i \in ran(posn)$

$\qquad empty(i) == \neg (occupied(i))$

$\qquad AllGreen == SECTION \times \{G\}$

end

Note that, unlike Abrial, we do not hide event parameters with an "ANY" construct. Thus $clear(T1, 1)$ will be the event that train T1 clears section 1. This approach to event representation will enable us to express parameter passing when we come to formulate our communicating state machine model.

We also take a slightly more general approach to parallel composition than is given in the B Book, e.g. in:

$$signal(i) := R \parallel signal(prev(i)) := A$$

we violate the rule that parallel assignments should act on different variables, but the meaning of our parallel composition is hopefully obvious, i.e.

$$signal := signal \lhd\!\!\!- \{i \mapsto R, prev(i) \mapsto A\}$$

4   Distributed State Machine Model

The second approach to the formal description of the railway system is to split the model into small communicating parts. We use a diagrammatic state machine notation to describe various behavioural aspects [11], [10] of the distributed system. The key components of our model are sections, trains and signals.

We first express some global constants for the combined system in a machine specification:

Machine *Globals*

Sets
$TRAIN = \{T1, T2\}$
$SIGNAL = \{G, A, R\}$
$SPEED = \{stop, slow, fast\}$

Constants
$SECTION, nsec, next, prev, stations$

Properties
$nsec = 8 \wedge$
$SECTION = 1 .. nsec \wedge$
$next \in SECTION \rightarrowtail SECTION \wedge$
$prev \in SECTION \rightarrowtail SECTION \wedge$
$\forall x \bullet x \in 1 .. nsec - 1 \Rightarrow next(x) = x + 1 \wedge$
$next(nsec) = 1 \wedge$
$\forall x \bullet x \in 2 .. nsec \Rightarrow prev(x) = x - 1 \wedge$
$prev(1) = nsec \wedge$
$stations = \{3, 7\}$
end

4.1   Track Sections and Signals

We should first give an informal set of rules for the interaction of the system components:

  - An event can only occur when all machines that have that event in their repertoire are ready to take part in it.
  - When an event occurs each machine that has that event in its repertoire changes according to one of the possibilities offered by its next state relation. Other machines are unaffected.

Each clear event involves four sections of track, figure 2. Concider clear(train, 3, signal), i.e the event of clearing track section 3. The signal value is provided by section 4, the section being entered by the train. The signal aspect on section 3 will change to red, and on section 2 will change to amber. For section 1 there is a choice. If this section is red it stays red, (the case when another train is occupying section 2) otherwise it changes to green.

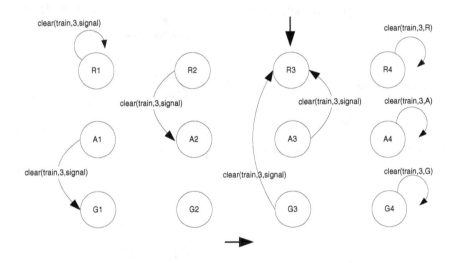

Fig. 2. Possible interactions between signal controllers: event clear(train,3,signal)

The state machine of figure 3 represents the signalling behaviour of a general section of the track. It shows all events which effect such a component, and which can take place in the context of the overall system. The state Ri represents a red signal for section i of the system. A red signal setting can change to amber when a train clears the next section. Equally a green signal can only change to red. In the signals diagram we also include the clear events for the previous section. A train approaching a red signal section is still allowed to clear the previous section only with the difference here that train can see the red signal ahead. A train can enter a red signaled section but it cannot clear it until the signal changes to green or amber. Thus the event $clear(train, prev(i), R)$ is permitted. Similarly events $clear(train, prev(i), A)$ and $clear(train, prev(i), G)$ are also allowed to happen and correspond to the cases that the train is approaching an amber signal and a green signal section respectively.

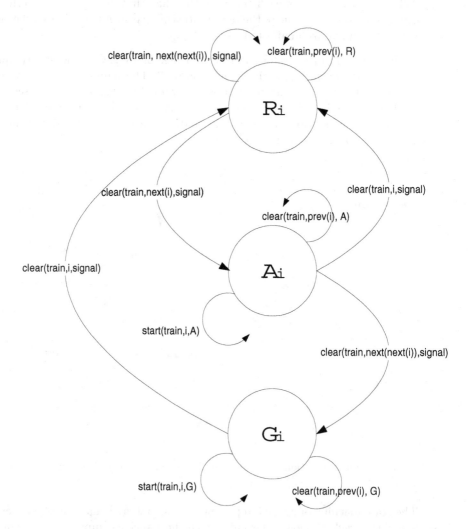

Fig. 3. Signal diagram with all possible events

Notice that when the signal is set to either amber or green and the train is stationery, the start event can happen. This is not the case when the signal of the section is set to red.

A diagram that shows all possible events <u>not prevented by the signal setting</u> is shown in figure 4. The dotted lines represent events that should never happen in the context of the overall system. However, they are shown here because it is not the job of the signal setting to prevent them; at least not directly. Rather,

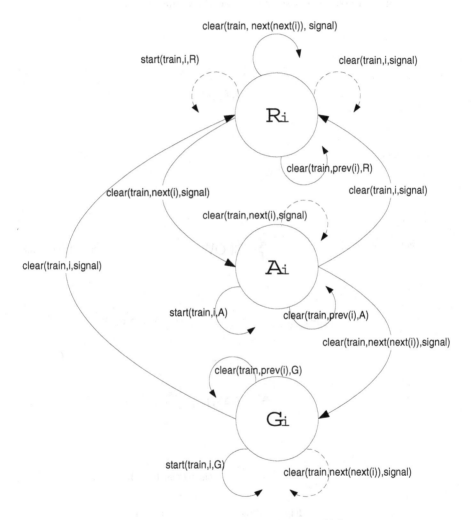

Fig. 4. Signal diagram including "impossible" events

they will be prevented by the interaction of the signal setting with other devices, namely the speed controller for each of the trains.

An event such as $clear(train, next(next(i)), signal)$ when the system is at a green signal state Gi will never be possible. That is because if a train occupies section $next(next(i))$ then the signal in section i cannot be green.

The "impossible" events will be checked to make sure that they will never occur by adding some extra checks to the invariant or the assertions clause of the combined system. Other examples of such "impossible" events are:

– A train cannot start when the signal is set to red and
– a train cannot clear a section when the signal is red.

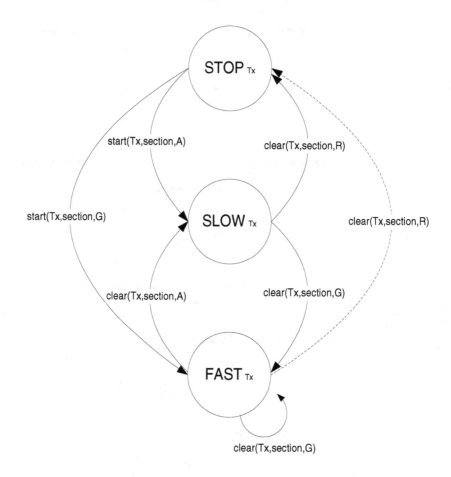

Fig. 5. Speed Controller diagram

Machine *Signal(i)*

Sees *Globals*

Constraints *i* : ℕ

Variables *state*

Invariant
   $state : SIGNAL$

Events
   $clear(t : TRAIN, sec : SECTION, sig : SIGNAL) \mathrel{\widehat{=}}$
      $state = G \land sec = i \implies state := R \;[\!]$
      $state = A \land sec = i \implies state := R \;[\!]$
      $state = A \land sec = next(next(i)) \implies state := G \;[\!]$
      $state = R \land sec = next(i) \implies state := A \;[\!]$
      $sec = prev(i) \land sig = state \implies skip \;[\!]$
      $\lnot\, sec \in \{prev(i), i, next(i), next(next(i))\} \implies skip$

   $start(...) \mathrel{\widehat{=}}$
      $...$

end

The penultimate choice in the clear event eliminates the possibility of a train not "seeing" the signals, because the only choice offered when $sec = prev(i)$ is one in which $state$ (the signal setting for this section) is equated to the signal parameter value sig.

   To complete our development we will define 8 signal machines, therefore $Signal(1)..Signal(8)$.

4.2   Speed Controllers

Each train has its own speed controller. When a train is about to clear a section the speed controller in cooperation with the signals part of the system will decide about the behaviour of the train. In other words it will set its speed restriction to STOP, SLOW or FAST.

   A speed controller - shown in figure 5 - will filter the events that breach the safety of the system. For the moment we ignore the existence of stations in our model. The subscript Tx shown in the states STOP, SLOW and FAST describes that the specific controller belongs to train x. In this case study x can be either 1 or 2, since we only have two trains.

   We convert the diagram of figure 5 into the following abstract system, which represents the speed controller of both trains. After some obvious simplification of GSL expression it can be written as follows:

Machine *Controllers*

Sees *Globals*

Variables *speed*

Invariant
   $speed : TRAIN \rightarrow SPEED$

Events

$clear(t : TRAIN, sec : SECTION, sig : SIGNAL) \ \widehat{=}$
$\qquad sig = R \implies speed(t) := stop \ []$
$\qquad sig = A \implies speed(t) := slow \ []$
$\qquad sig = G \implies speed(t) := fast$

$start(...) \ \widehat{=}$
$\qquad ...$

end

4.3   Trains

The speed controller part of the development does not have any information about the position of the train. It is therefore important to include an extra machine to represent this. It is not something we will later implement in software, it just represents the physical reality of the trains' position. The diagram shown at figure 6 is tailored for our 8 section railway system. The label of each state denotes the train and the section number. Tx,8 stands for train x occupying section 8.

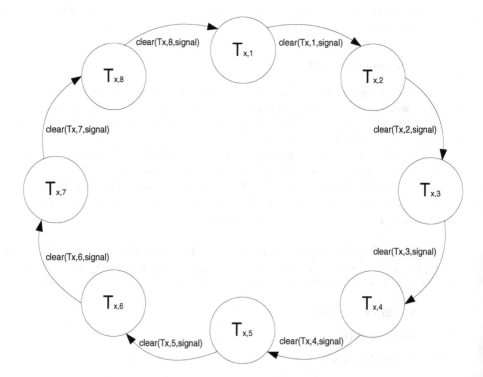

Fig. 6. The train position diagram

Since we want to allow the possibility of having more than one or two trains in a railway system we create the *Trains* machine. Instead of using an indexed machine to represent the trains, we use a single machine to represent the state of all components within a given class by means of a function. The train machine shown can be then translated into a B machine which we express in GSL form as follows:

Machine *Trains*

Sees *Globals*

Variables *posn*

Invariant
$posn : TRAIN \rightarrow SECTION$

Events
$clear(t : TRAIN, sec : SECTION, sig : SIGNAL) \,\widehat{=}$
$\qquad posn(t) = sec \Longrightarrow posn(t) := next(sec)$

Definitions
$occupied(i) == i \in \mathrm{ran}(posn)$

end

Note that the invariant of the *Trains* machine does not give the fundamental safety property of the railway, namely that no two trains should ever occupy the same section of track. In fact such a property could not be preserved by the *Trains* machine, which knows nothing about speeds and signals. We will only be able to assert this property once we have incorporated the *Trains* machine into an overall system machine.

4.4   Combining the State Machines into a Single Model

The system machine combines the Globals, the Signal, the Trains and the Controllers machines.

Machine *System*

Sees *Globals*

Includes
$Signal(1).Signal(1)...Signal(8).Signal(8),$
$controllers.Controllers,$
$trains.Trains$

Invariant
$(\forall\, t1, t2 : TRAIN \bullet t1 \neq t2 \Rightarrow trains.posn(t1) \neq trains.posn(t2))$
...

Assertions [1]

$\forall i : SECTION, t : TRAIN, sig : SIGNAL \bullet Signal(i).state = G \Longrightarrow$
$\neg \text{ fis } Signal(next(i)).clear(t, next(i), sig) \wedge$
$\neg \text{ fis } Signal(next(next(i))).clear(t, next(next(i)), sig)$

...

Initialisation

$(trains.posn(t1) := 1 \parallel$
$trains.posn(t2) := 2 \parallel$
$Signal(1).state := R \parallel$
$Signal(2).state := G...Signal(6).state := G \parallel$
$Signal(7).state := A \parallel$
$Signal(8).state := R \parallel$
$controllers.speed(1) := stop \parallel$
$controllers.speed(2) := slow)$

Events

$clear(t : TRAIN, sec : SECTION, sig : SIGNAL) \mathrel{\widehat{=}}$
$\quad (\parallel_{i:SECTION} Signal(i).clear(t, sec, sig)) \parallel$
$\quad controllers.clear(t, sec, sig) \parallel$
$\quad trains.clear(t, sec, sig)$

$start(...) \mathrel{\widehat{=}}$
$\quad ...$

end

In our system we need to express the parallel composition of the *clear* operation from each of the eight section machines. We introduce a "scaling notation"

$$\parallel_{i:SECTION} Signal(i).clear(t, sec, sig)$$

which we can use in place of writing

$$Signal(1).clear(t, sec, sig) \parallel$$
$$Signal(2).clear(t, sec, sig) \parallel$$
$$...$$
$$Signal(8).clear(t, sec, sig)$$

We now have a machine which models a railway with signals, trains, and speed controllers. We have expressed the individual components in terms of some simple state machines, and indicated how these are converted into a B model. So far, however, we have ignored the presence of stations. We now introduce them.

When stations are present, we suppose the speed controller of a train will know about them and control the train accordingly. This means the speed controller either needs to know where the train is, or should receive some communication as it is approaching a station. We will use the former technique, partly

---

[1] We take the liberty of considering fis(S) as part of our language, where fis(S) is defined as fis(S) = ¬ [S]false.

to show an approach we use when the complexity of our state machine diagram increases.

What if we could somehow combine the Trains machine and the speed Controllers machine into a single one? Since the speed controller really belongs to the train then it seems a good idea to try an integration. The diagram in figure 7 shows such integration. The diagram shows the possible change of the speed settings of train x when it moves from section to section.

Fig. 7. Combined Controller-Train diagram

Let us exploit the scenario of a train approaching a station section. The behaviour of that train will change. To clarify this point clearer let us zoom into figure 7 and cut off a part of it that consists of a station section and the two previous sections. This part of the railway is shown in figure 8. A station is located at section i: suppose that a train occupies section $prev(prev(i))$. While the train is moving to the next section - section $prev(i)$ - its speed can only be either slow or stop. Stop is the case when the second train is ahead and docked. Dotted states are never accessed. When the train then moves to section i it is clear that its speed will be set to stop. Figure 8 shows the changes on the speed of the train approaching the station. Event and state labels on this diagram are abbreviated.

5   Refinement

Since we have two specifications - global model and distributed state machine model - we are interested in comparing them and more specifically to show that the distributed state machine model refines the global model. The first problem we encounter here is that the distributed system is not a refinement machine in the B sense. It is just another B machine. One approach would be to re-express our state machine model as a refinement of the global model. Another approach would be to leave both machines as specifications but use the second machine to create an implementation machine for the first, following the approach used by Ken Robinson, [8]. Then the formal reconciliation can be achieved by generating the proof obligations for the implementation machine.

A second complication is that in the state machine model we have additional parameters in our events, and these are used to pass information between state machines. These arise because it is not the job of a single state machine to store all the information about our global model. There is a need for sharing information. This is done by passing this information from one machine to the other.

However we can hide certain information, for instance we could hide the signal parameter using the ANY construct as follows:

Events
$$clear(t : TRAIN, sec : SECTION) \; \widehat{=}$$
$$Any \; sig \in SIGNAL \; Then \; ($$
$$(\|_{i:SECTION} \; Signal(i).clear(t, sec, sig)) \; \|$$
$$controllers.clear(t, sec, sig) \; \|$$
$$trains.clear(t, sec, sig))$$

$$start(...) \; \widehat{=}$$
$$...$$

This gives us event parameters which allow us to directly compare our two styles of model. Note that this idea of hiding is rather different to the idea of hiding in process algebra where an entire event is hidden.

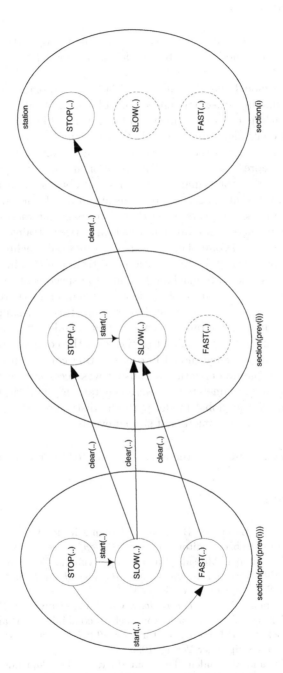

Fig. 8. Approaching a station scenario

6   Conclusion

The study of the small railway system gave us the opportunity to compare two forms of system model, a global model, and a model based on communicating components.

For the model based on communicating components, we identified three classes of component: signals, speed controllers, and trains. These were expressed in state machine diagrams which can be translated into B. In the case of trains and speed controllers, we were able to use a single B machine to represent all system components of the same class. The modelling technique used to achieve this was to represent the state of all components within a given class by means of a function. For the signal components this does not appear to be possible at least in a form that we can easily translate into B from our state machine diagrams. The reason appears to be that there is an interaction between different signals where there is no direct interaction between trains or speed controllers. To represent signal components we used an indexed machine.

When speed controller information and train position information were combined into a single state machine to allow the speed controller to automatically take account of the location of stations, our diagrammatic representation of the machine became reasonably complex, even for this very simple model. However from this complex diagram we can extract simpler diagrams which illustrate particular scenarios, and which together give the complete picture.

Partitioning a global model into interacting components gave us a different form of event parameterisation because it was necessary for the individual components to pass parameters that were generally available in the global machine. So to compare the global model with the distributed state machine model we need to hide some parameters of shared events.

Acknowledgments. We thank the referees for their valuable comments.

References

1. J R Abrial. Extending B without Changing it (for Developing Distributed Systems). In H Habrias, editor, *The First B Conference, ISBN : 2-906082-25-2*, 1996.
2. J R Abrial and L Mussat. Introducing Dynamic Constraints in B. In Bert D, editor, *B98: Recent Developments in the Use of the B Method.*, number 1393 in Lecture Notes in Computer Science, 1998.
3. Jean-Raymond Abrial. *The B Book.* Cambridge University Press, 1996.
4. M Butler. csp2B: A practical approach to combining CSP and B. In J M Wing, Woodcock J, and Davies J, editors, *FM99 vol 1*, Lecture Notes in Computer Science, no 1708. Springer Verlag, 1999.
5. M Butler and M Waldén. Distributed System Development in B. In H Habrias, editor, *The First B Conference, ISBN : 2-906082-25-2*, 1996.
6. M J Butler. An approach to the design of distributed systems with B AMN. In J P Bowen, M J Hinchey, and Till D, editors, *ZUM '97: The Z Formal Specification Notation*, number 1212 in Lecture Notes in Computer Science, 1997.

7. Treharne H and Schneider S. How to Drive a B Machine. In Bowen J P, Dunne S, Galloway A, and King S, editors, *ZB 2000: Formal Specification and Development in Z and B*, Lecture Notes in Computer Science, no 1878. Springer Verlag, 2000.

8. K Robinson. Reconciling Axiomatic and Model-Based Specifications Using the B Method. In Bowen J P, Dunne S, Galloway A, and King S, editors, *ZB 2000: Formal Specification and Development in Z and B*, Lecture Notes in Computer Science, no 1878. Springer Verlag, 2000.

9. W J Stoddart. An Introduction to the Event Calculus. In J P Bowen, M J Hinchey, and Till D, editors, *ZUM '97: The Z Formal Specification Notation*, number 1212 in Lecture Notes in Computer Science, 1997.

10. W J Stoddart, S E Dunne, Galloway A J, and Shore R. Abstract State Machines: Designing Distributed Systems with State Machines and B. In Bert D, editor, *B98: Recent Developments in the Use of the B Method.*, number 1393 in Lecture Notes in Computer Science, 1998.

11. W J Stoddart and P K Knaggs. The Event Calculus, (formal specification of real time systems by means of Z and diagrams). In H Habrias, editor, *5th International Conference on putting into practice methods and tools for information system design*. University of Nantes, 1992.

Verification of Dynamic Constraints for B Event Systems under Fairness Assumptions

Françoise Bellegarde, Samir Chouali, and Jacques Julliand

Université de Franche-Comté, Laboratoire d'Informatique de l'Université de Franche-Comté,
16 route de Gray, 25030 Besançon Cedex France
Ph:(33) 3 81 66 64 52, Fax:(33) 3 81 66 64 50
{bellegar, chouali, julliand}@lifc.univ-fcomte.fr,
http://lifc.univ-fcomte.fr

Abstract. A *B* event systems is supposed to specify a closed system, i.e., the system is meant to be specified in isolation. So, the specification includes the specification of the system of interest and of its environment. Often, the environment supposes fairness constraints. Therefore, classically in a *B* system approach, we express the fairness of the environment by the specification of fair scheduler together with the events of the system of interest. This leads to an infinite state model even when the system is finite state by nature. This does not facilitate *PLTL* properties verification by model checking which is only effective on finite state models. In this paper, we propose to keep separate the fairness of the environment from the specification of the system of interest by a *B* event system. Then, the fairness is expressed as events which have to be fairly fired. So, a finite state system of interest has a finite state model. The chosen model is a finite labeled transition system which allows the model checking of *PLTL* properties using the fair events as assumptions. In the paper, we make diverse proposals–some of them are proposed as perspectives–for a verification under fairness assumptions. We use the protocol *T=1* as a running example.

Keywords. *B* event systems [1], fairness hypotheses, specification, *PLTL* verification.

1 Motivations and Related Works

Abstract event systems were introduced by J. R. Abrial in [2] in the framework of the *B* method [1]. Events are guarded by conditions on the state and may only be fired when their guards are true, i.e. when they are enabled. The state of the system is changed according to the action of the fired event. So, events are not invoked as operations. An event system designs a *closed* system while a *B* machine is an *open* system.

---

[1] Submission to the B program committee

D. Bert et al. (Eds.): ZB 2002, LNCS 2272, pp. 477–496, 2002.
© Springer-Verlag Berlin Heidelberg 2002

Similarly to Action Systems [3] B may be used in the development of distributed and reactive systems [8,10,9]. Reactive systems, by definition, are engaged in an ongoing interaction with their environments. As a *closed* system, an event system specifies both the system of interest and its environment, i.e. the totality of the application. For example, a software which implements a communication protocol specifies the whole system – the protocol as well as the emission of the messages to be transmitted, i.e., the application environment. In such a case, the environment could be a bank distributor. Another example, a resource dispatcher specification includes the dispatcher and the processes requesting the resource which represents the application environment.

Consider the example of the dispatcher, the resource allocation to the many processes must be fair to guarantee request satisfactions. In other words, the satisfaction of a request is unavoidable. The problem arise when there are many choices. Then, we want to ensure that none of these choices will be infinitely delayed. This problem has been specified in the B event system methodology by J.R. Abrial and L. Mussat in [2] for a lift application and by D. Bert in [5] for the arbitration of the bus SCSI3. These B events systems includes the specification of a fair scheduler for the arbitration process. In these applications, it is necessary to specify the fairness which is intrinsic of the software for both verifying the system and implementing it.

Consider now the example of the communication protocol. The protocol is also concerned by fairness. However, the fairness is not intrinsic to the protocol but comes from environment constraints. For example, messages are always sent as a finite sequence of blocks. In a closed system, the emission of the messages are controlled by a fair scheduler which specifies that the sequence of blocks sent for one message is finite. In this application, it is necessary to take account of the fairness constraints of the environment to verify the liveness properties of the protocol. However, the software implementation of the protocol itself does not include the scheduler.

Notice that, the specification of a fair scheduler often leads to an infinite state model since it involves managing an *apriori* finite but undetermined number of resources. Less often, it does not change the finite nature of the model. This happens when ordering the events is enough to ensure fairness. For the protocol example, since the fair scheduler is a counter, it changes the finite nature of the specification. This is a serious shortcoming for the following reasons:

– Since the scheduler is part of the environment, it is not part of the implantation of the communication protocol.
– Since the model is infinite state, model checking is impossible. Liveness properties verification requires the fairness constraints, but notice that safety properties does not.

In many other approaches such as TLA [18], TLR [17], systems are specified within temporal logic. They are particular kinds of action systems. These systems are described as abstract fair transition systems. A fair transition system includes a set of compassionate transitions also called strongly fair transitions and a set of just transitions also called weakly fair transitions. Transitions, like events may be enabled and fired. A strongly fair transition requirement for a transition t disallows computations in which t is enabled infinitely many times but taken finitely many times.

These approaches have two advantages:

1. They avoid to specify a scheduler which is never implemented.
2. They often allows model checking liveness properties on a finite model.

In *TLR* the model checking of safety, response and reactivity properties –a hierarchy proposed by Z. Manna [20]– is based on a tableau method.

In this paper we intend to extend the B event systems with the possibility of specifying event systems which are submitted to fairness constraints. For that, we express fairness assumptions in the form of constrained events. We study different approaches for model checking *PLTL* properties of the system under fairness hypotheses. For the particular case of the verification of B modalities on event system submitted to fairness assumptions, we propose an approach based on Manna's proof rules [20]. We illustrate the approach on the example of the protocol $T=1$ [13] where the fairness comes from the environment of the system.

From our point of view, it is more convenient for the designer to propose hypotheses in the environment rather than to specify a scheduler.

Moreover, it is difficult to verify properties on a specification under fairness assumptions relying only on proof, i.e. not model-based, since fairness constraints are path properties expressed as follows by *PLTL* formulas:

- $\Box(\Diamond\Box e\text{-}enabled \Rightarrow \Diamond e\text{-}executed)$, for the weak fairness [20] meaning that *if the event e is continually enabled, it will be eventually executed.*
- $\Box(\Box\Diamond e\text{-}enabled \Rightarrow \Diamond e\text{-}executed)$ for the strong fairness [20], meaning that *if e is infinitely often enabled then it will be eventually executed.*

On the one hand, such constraints cannot be used by a first-order predicate prover such as the B prover [22]. However the B prover can prove modalities on an infinite state event system including a scheduler. On the other hand, model checking properties under fairness constraints is not a problem if the system is finite state and does not call forth a state explosion. Then, a *PLTL* model checker such as SPIN [15] or CTL with fairness [11] can be used for finite state specification under fairness constraints.

In the paper we propose various approaches for a verification under fairness assumptions, we use the protocol $T=1$ as a running example. In Section 2, we recall some preliminaries on the *PLTL* operators. In Section 3, we show how to introduce the specification of fairness hypotheses in the B events systems using the example of the protocol $T=1$. In Section 4, we propose an approach for a verification of *PLTL* formulas by model checking the fair transition systems we consider as a model for B event systems under fairness assumptions. In Section 5, we consider the particular case of verifying B modalities on event system under fairness assumptions. In the conclusion, in Section 6, we indicate what could be our future research directions.

2   Preliminaries on *PLTL*

For specifying the system properties we use the Propositional Linear Temporal Logic (*PLTL*). *PLTL* formulas are built up from atomic propositions, boolean connectives, and the following basic temporal operators:

- $\Box$ (Always)
- $\bigcirc$ (Next)
- $\Diamond$ (Eventually)
- $\mathcal{U}$ (Until)
- $\mathcal{W}$ (Unless)

A *PLTL* formula is interpreted over a model, which is an infinite sequence of state $\sigma : s_0, s_1, ...$, where each state s_j provides an interpretation for the variables mentioned in the formulas.

Given a model σ and a *PLTL* formula P, we present a definition for the notion of P holding at a position $j \geq 0$ in σ, denoted by $(\sigma, j) \models P$.

- for a state formula p,
  $(\sigma, j) \models p \Longleftrightarrow s_j \models p$,
  That is, we evaluate p locally, using the interpretation given by s_j.
- $(\sigma, j) \models \neg P \Longleftrightarrow (\sigma, j) \not\models P$,
- $(\sigma, j) \models P \vee Q \Longleftrightarrow (\sigma, j) \models P$ or $(\sigma, j) \models Q$,
- $(\sigma, j) \models \Box P \Longleftrightarrow$ for all $k \geq j$, $(\sigma, k) \models P$,
- $(\sigma, j) \models \bigcirc P \Longleftrightarrow (\sigma, j+1) \models P$,
- $(\sigma, j) \models \Diamond P \Longleftrightarrow (\sigma, k) \models P$ for some $k \geq j$,
- $(\sigma, j) \models P \mathcal{U} Q \Longleftrightarrow$ for some $k \geq j$, $(\sigma, j) \models Q$, and for every i such that $j \leq i < k, (\sigma, j) \models P$,
- $(\sigma, j) \models P \mathcal{W} Q \Longleftrightarrow (\sigma, j) \models P \mathcal{U} Q$ or $(\sigma, j) \models \Box P$.

Notice that $\Box\Diamond$ means infinitely often, and $\Diamond\Box$ means continually.

3   Fairness Hypotheses in the *B* Event Systems

In this section we use the example of the protocol *T=1* to illustrate how :

1. to specify fairness in the B events system,
2. to express fairness as assumptions without including its specification in the event system.

3.1   *B* Event System of the Protocol *T=1*

Figure 1 describes the *B* event system of a half duplex communication protocol between a chip integrated card and a card reader. The chip card and the reader exchange alternately messages. We view the messages as a sequence of blocks ended by a last block (value *lb*). A block sent (value *bl*) is acknowledged by an acknowledgement block (value *ackb*). We call frame these three types of exchanged information.

After a last block is sent by one of the devices, the other device answers with a sequence of blocks ending by a last block unless the card is ejected. These exchanges of messages alternate until the card is ejected.

The variables *CardF* and *ReaderF* describe the type of the last frame sent respectively by the chip card and the reader. The variable *SenderF* describes the device which will send the next frame. The variable *Cstatus* indicates the card status *in* or *out*. The variables *NbMes* and *NbBloc* show the number of remaining messages and the number of remaining blocks in the current message to be sent.

Here, the invariant express typing for the variables and the proposition I_1 and I_2: *When a device is sending a message the other device always acknowledge by a last block or by an acknowledgment block.*

We have 8 events :

- *Rsends:* the reader sends a message,
- *Csends:* the card sends a message,
- *Eject:* the card is ejected,
- *Cinsert:* inserts the card,
- *Rblocksends:* the reader sends a block,
- *Cblocksends:*the card sends a block,
- *Racksends:* the reader sends an acknowledgement,
- *Cacksends:* the card sends an acknowledgement.

Specification of the Environment Fairness. We consider that the transmission of the messages between the card and the reader must terminate by the ejection of the card.

The unavoidability of the card ejection is obtained by initializing the variable *NbMes*, decrementing *NbMes* when a message is sent, and activating *Eject* when *NbMes* is null. In the same way, the activations of *Csends* and *Rsends* become unavoidable by using the variable *NbBloc* which counts the number of remaining blocks to be sent. When this number is null, the emission of the last block of message is required.

3.2   *B* Specification under Fairness Assumptions

In this section we present a specification of the protocol *T=1* without specifying the fairness. However there are fairness hypotheses which must be satisfied by the protocol environment.

A fairness assumption is written (*e if p*), where *e* is the name of an event and *p* is a predicate characterizing the states in which *e* cannot be avoided when it is infinitely often enabled. When an event must always be fair, the fairness assumption is only the event name.

```
MACHINE teg1
SETS
     FRAME = {bl, lb, ackb}; SENDER = {chip_card, reader}; CARD_STATE = {in, out}
VARIABLES
     SenderF, Cstatus, CardF, ReaderF │ , NbMes, NbBloc │
INVARIANT
     SenderF ∈ SENDER ∧ CardF ∈ FRAME ∧ ReaderF ∈ FRAME ∧ Cstatus ∈ CARD_STATE ∧
     │ NbMes ∈ N ∧ NbBloc ∈ N │ ∧
     ((SenderF = chip_card ∧ ReaderF = bl) ⇒ (CardF = lb ∨ CardF = ackb))  ∧   (I₁)
     ((SenderF = reader ∧ CardF = bl) ⇒ (ReaderF = lb ∨ ReaderF = ackb))   (I₂)
INITIALIZATION
     SenderF := reader ║ Cstatus := in ║ CardF := lb ║ ReaderF := lb │ ║ NbMes :∈ N ║ NbBloc :∈ N │
EVENTS
     Rsends ≙ SELECT (SenderF = reader ∧ Cstatus = in │ ∧ NbMes > 0 ∧ NbBloc = 0 │
               ∧ (CardF = ackb ∨ CardF = lb))
               THEN SenderF := chip_card ║ ReaderF := lb │ ║ NbMes := NbMes − 1║ NbBloc :∈ N │ END;
     Csends ≙ SELECT (SenderF = chip_card ∧ Cstatus = in │ ∧ NbMes > 0 ∧ NbBloc = 0 │
               ∧ (ReaderF = ackb ∨ ReaderF = lb))
               THEN SenderF := reader ║ CardF := lb │ ║ NbMes := NbMes − 1 ║ NbBloc :∈ N │ END;
     Eject ≙ SELECT ((SenderF = chip_card ∧ ReaderF = lb) ∨ (SenderF = reader ∧ CardF = lb)) ∧
               Cstatus = in ∧ │ NbMes = 0 │
               THEN Cstatus := out END;
     Cinsert ≙ SELECT CStatus = out
               THEN SenderF := reader ║ Cstatus := in │ ║ NbMes :∈ N ║ NbBloc :∈ N │ END;
     Rblocksends ≙ SELECT (SenderF = reader ∧ Cstatus = in ∧ (CardF = ackb ∨ CardF = lb) │ ∧ NbBloc > 0 │)
               THEN SenderF := chip_card ║ ReaderF := bl │ ║ NbBloc := NbBloc − 1 │ END;
     Cblocksends ≙ SELECT (SenderF = chip_card ∧ Cstatus = in ∧ (ReaderF = ackb ∨ ReaderF = lb) │ ∧ NbBloc > 0 │)
               THEN SenderF := reader ║ CardF := bl │ ║ NbBloc := NbBloc − 1 │ END;
     Racksends ≙ SELECT (SenderF = reader ∧ Cstatus = in ∧ CardF = bl)
               THEN SenderF := chip_card ║ ReaderF := ackb END;
     Cacksends ≙ SELECT (SenderF = chip_card ∧ Cstatus = in ∧ ReaderF = bl)
               THEN SenderF := reader ║ CardF := ackb END;
END
```

Fig. 1. B event specification of the protocol $T=1$

Specification. The B event system under fairness assumptions is obtained by deleting the variables $NbBloc$ and $NbMes$, i.e., by deleting the boxed parts of Fig. 1. The fairness assumptions are defined by the declaration set $FAIR$-$EVENTS$={$Eject$, $Csends$ if $(CardF=bl)$, $Rsends$ if $(Reader =bl)$}, these assumptions are verified by all the possible environments of the protocol. The hypothesis about $Eject$ indicates that the end of the transmission is unavoidable. Assumptions about $Csends$ and $Rsends$ express that the messages contain a finite number of blocks. These assumptions are not concerned with one-block messages. So, the system cannot run infinitely along the simple cycles in Fig. 2. In this figure, each state is decorated by indicating the kind of frame sent by the reader and the card (see the key in the figure). Moreover the character '?' or '!' in the reader indicates respectively that the reader is the receiver or the sender device.

The transition system as shown in Fig. 2 has a finite number of states. Therefore, model checking becomes possible and property verification must be done

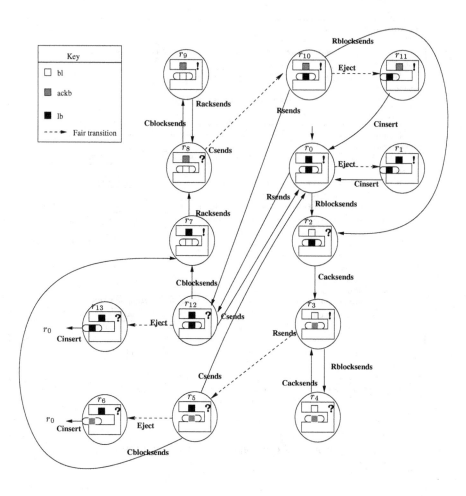

Fig. 2. Transition system of the protocol $T=1$ under fairness assumptions

under the three fairness hypotheses $\{Eject,\ Csends\ if\ (CardF=bl),\ Rsends\ if\ (ReaderF=bl)\}$. These assumptions are expressed in *PLTL* as follows:

- $\Box(\Box \Diamond Cstatus = in \Rightarrow \Diamond Cstatus = out)$ (F_1),
- $\Box(\Box \Diamond(SenderF = chip\_card \wedge CardF = bl) \Rightarrow \Diamond CardF = lb)$ (F_2),
- $\Box(\Box \Diamond(SenderF = reader \wedge ReaderF = bl) \Rightarrow \Diamond ReaderF = lb)$ (F_3).

We have presented two approaches for modelling the fairness of the environment of the protocol $T=1$. A first specification includes the fairness specification in the event system. A second specification does not since fairness hypotheses about the environment are expressed apart of the event system. In the next section, we examine how to verify liveness properties for the second specification.

4   Verification of the *PLTL* Formulas under Fairness Hypotheses

Model checking is a fully automatic approach to verify a *PLTL* formula. Model checking typically depends on a discrete model of a system. The event system behavior can be provided by a labeled transition system.

We now show how a labeled transition system (*TS*) represents the behavior of an event system and how to take into account fairness constraints H into the transition system (TS_H).

4.1   Fairness Assumptions and Labeled Transition Systems

The behavior of an event system is viewed as labeled transition system (see [6]). A labeled transition system TS is defined by $\langle S, Evt, \rightarrow, l \rangle$ where:

 - S is a set of states,
 - Evt is a set of the event labels,
 - $\rightarrow \subseteq S \times Evt \times S$ is a transition relation,
 - l is an interpretation of a state s as a predicate on the variables and their values.

Let $V \stackrel{\text{def}}{=} \{x_1, \ldots, x_n\}$ be a finite set of the *variables* of a component. Let s be a state. The semantics of s over V is a mapping that associates with each variable x_i its value v_i.

Usually, to describe *PLTL* formulas semantics, the set of all the propositions holding in a state s is considered to be a label of s. We consider one of these propositions $l(s)$ defined by $x_1 = v_1 \wedge \cdots \wedge x_n = v_n$. Then, we know that a predicate p holds on s by $l(s) \Rightarrow p$ holds. So, any *invariant* I, formulating a safety property, holds iff $\forall s \in S, l(s) \Rightarrow I$. The state space of the labeled transition system is the set of states which satisfy the invariant. Let s be a state, and let e be an event enabled in s, i.e. its guard $grd(e)$ is satisfied in s, or $l(s) \Rightarrow grd(e)$ holds.

Event names are labels of transitions in Evt. The events affect state variables. The transition $s \xrightarrow{e} s'$ is the result of the transformation of the state s into the state s' by the event e, i.e. $l(s) \Rightarrow \langle GS_e \rangle l(s')$ where GS_e is the generalized substitution of e, and $\langle GS_e \rangle p \equiv \neg [GS_e] \neg p$ (see [16,6]) represents the weakest precondition under which it is possible for GS_e to establish p.

The labeled transition system of the protocol is represented as usual by a graph structure–a reachability graph–where the nodes are the system's states and the edges are the transitions between states labeled by event names (see Fig. 2).

Such a transition system is not enough to model the behavior of a system under fairness assumptions. For that [20] uses a fair transition system. A fair transition system includes two more sets, the set of the strong fair transitions and the set of the weak fair transitions. A strong fair transition is taken infinitely many times when it is enabled infinitely many times. It is disallowed that a weak fair transition is continually enabled but taken only finitely many times. In this paper, we do not distinguish weak fairness from strong fairness, since weak fairness is a particular case of strong fairness. Moreover, we have a slightly different notion of strong fairness. In our approach, a fair transition, when enabled infinitely many times is eventually taken, i.e.

$$\Box(\Box\Diamond e\text{-}enabled \Rightarrow \Diamond e\text{-}executed)\quad(1)$$

while usually the *PLTL* formula for a strongly fair transition is as follows:

$$\Box\Diamond e\text{-}enabled \Rightarrow \Box\Diamond e\text{-}executed\quad(2).$$

Notice that (1) implies (2) by distributing $\Box$ w.r.t $\Rightarrow$ but the inverse implication is not valid.

We call TS_H the fair transition system which is a model for an event system under the fairness set of assumptions H.

In order to get TS_H, we can compute the set T_H of the (strong) fair transitions included into the transition set of TS as follows. Let $h = e$ *if* p a fairness hypothesis element of H. The set T_h of the fair transitions corresponding to h is the set of the transitions $s \xrightarrow{e} s'$ such that $l(s) \Rightarrow p$ and all of these transitions are cycles exiting transitions. We have $T_H = \bigcup_{h \in H} T_h$.

In the protocol *T=1*, the fair transitions corresponding to the hypothesis *Eject* are :

$$- t_1 = r_0 \xrightarrow{Eject} r_1$$
$$- t_2 = r_5 \xrightarrow{Eject} r_6$$
$$- t_3 = r_{10} \xrightarrow{Eject} r_{11}$$
$$- t_4 = r_{12} \xrightarrow{Eject} r_{13}$$

Figure 3 sketches an algorithm to compute the set T_h. For $h = e$ *if* p, T_h is the set of every transition which is enabled on a state where p holds, and which is an output transition of some cycle.

Let H be the set of the fair hypotheses. Let $TS = \langle S, Evt, \rightarrow, l \rangle$.
let $h \in H$, $h = e$ *if* p
let T_h be the set of the fair transitions corresponding to the hypothesis h.
$T_h = \emptyset$
FOR each $t \in \rightarrow$ s.t. $t = s_i \xrightarrow{e} s_j$ **DO**
    IF $(l(s_i) \Rightarrow p)$ **THEN**
        IF $\big(t$ *is an output transition of some cycle*$\big)$ **THEN**
            $T_h := T_h \cup t$
        ELSE
            Error in the specification of the fairness hypotheses

---

Fig. 3. Computation of the fair transitions

The fair transitions are exiting transitions of cycles. If a transition specified as fair by the hypotheses is not a transition exiting from a cycle, there is an error in the FAIR-EVENTS declaration set. We call such cycles *fair exiting cycles*.

Definition 1. *Let TS_H a transition system, h a fairness hypothesis and T_h the set of the fair transitions corresponding to h ($h = e$ if p). A cycle $c = s_i \xrightarrow{e_i} s_{i+1}$, $s_{i+1} \xrightarrow{e_{i+1}} s_{i+2}, \ldots, s_{i+n-1} \xrightarrow{e_{i+n-1}} s_i$ is a fair exiting cycle for the hypothesis h if there is a transition t ($t \in T_h$) s.t. $t = s_j \xrightarrow{e} s$, $i \leqslant j \leqslant i+n-1$, $l(s_j) \Rightarrow p$ and t is not a transition of the cycle c.*

For example, in Fig. 2 the cycle $r_3 \xrightarrow{Rblocksends} r_4 \xrightarrow{Casksends} r_3$ is a fair exiting cycle because it has a fair output transition $t = r_3 \xrightarrow{Rsends} r_5$. Figure 4 describes an algorithm to compute the fair exiting cycles.

In what follows c_f is the set of the fair exiting cycles

$C_f = \emptyset$
FOR $t \in T_H$, $t = s_i \xrightarrow{e} s_j$ **DO**

$\quad C_f := C_f \cup \{C$ s.t. C is a cycle in $TS_H \wedge s_i \in C \wedge s_j \notin C\}$.

Fig. 4. Fair exiting cycles computation

In the next section we present a classical approach for verifying a liveness *PLTL* property under fairness assumptions.

4.2   *PLTL* Model Checking on Labeled Transition Systems

Model checking views verification as checking whether the reachability graph TS satisfies the property P to be checked, i.e. $TS \models P$. The classical approach [26, 14] to solve this problem is to construct the Büchi automaton representing the negation of the property and to analyse if the language accepted by the synchronous product of the reachability graph with this automata is empty or not, i.e., to check if there is or not a cycle among accepting states.

The model checking is *PSPACE*-complete. In practice, for a given reachability graph, the complexity limit comes from the size of the Büchi automaton since the running time is proportional to the product of number of the nodes in the automaton and in the graph. The general approach to search for an accepting cycle uses linear algorithms either algorithmic search for Strongly Connected Components (see [24]) or simple algorithms for cycle detections. The space limit comes from the part of the synchronous product which has to be memorized to check for the accepting cycles. Also, the construction of the Büchi automaton generation is exponential in the number of temporal subformulas [27]. Recent works [12] propose optimizing the number of nodes of the constructed Büchi automaton for the SPIN model checker. But, the exponential complexity remains for the worst case. So, the *PLTL* formula P can be translated to a Büchi automaton in $\mathcal{O}(2^{|P|})$. The model checking complexity is $\mathcal{O}((|S|+|\rightarrow|)*2^{|P|+\Sigma_{h\in H}|h|+1})$.

Following this approach, to verifying a property P under fairness assumptions $H = \{h_1,\cdots,h_n\}$ for the transition system *TS*, we need to check if $TS \models (h_1 \wedge \cdots \wedge h_n) \Rightarrow P$

Notice that the conjunction of the $h_i, i = 1,n$ remains in the negation of the implication. This conjunction brings a complexity of $\mathcal{O}(2^{4n})$ since each of the fairness hypothesis expressed in *PLTL* brings 4 temporal operators.

In the next section, we propose an approach which checks the *PLTL* formula directly on the transition system TS_H.

4.3   Model Checking on a Fair Transitions System

We propose the following model checking approach:

- constructing the Büchi automaton BA corresponding to the *PLTL* formula $\neg P$ (see [25]),
- constructing the fair transition system TS_H from the transitions system *TS* and the set of the fair hypotheses H (see Section 4.1),
- constructing the synchronous product of TS_H and BA, written as $TS_H \boxtimes BA$ in Fig. 5
- searching for a non-fair exiting acceptation cycle.

Therefore, we verify the satisfaction of P for the fair transition system TS_H, i.e., $TS_H \models P$. Then, model checking complexity is in $\mathcal{O}((|S|+|\rightarrow|)*2^{|P|})$. Figure 5 shows a flow diagram for this approach.

The next section compares the two approaches in practice.

4.4   Comparison between the Two Above Model Checking Approaches under Fairness Hypotheses

For this purpose, we consider a liveness property P to be verified on the protocol $T=1$ under its fairness hypotheses $H=\{h_1, h_2, h_3\}$ where $h_1 = Eject$, $h_2 = Csends \ if \ (CardF = bl)$ and $h_3 = Rsends \ if \ (ReaderF = bl)$. It is enough to compare the size of the Büchi automaton of the *PLTL* formula $\neg((h_1 \wedge h_2 \wedge h_3) \Rightarrow P)$ for the first approach of the model checking with the size of $\neg P$ for the second approach.

Let $P_1 \hat{=} \Box((SenderF = chip\_card \wedge ReaderF = lb) \Rightarrow \Diamond(SenderF = reader \wedge CardF = lb))$ which expresses the alternation. The *PLTL* interpretation of the hypotheses are: the formula (F_1) for *Eject*, the formula (F_2) for *(Csends if (CardF= bl))* and the formula (F_3) for *(Rsends if (ReaderF= bl))* (see Section 4).

Figures 6 and 7 show the respective sizes of the Büchi automata. The size of the Büchi automaton for $\neg((h_1 \wedge ... \wedge h_n) \Rightarrow P_1)$ increases with the number n of fairness assumptions in H. Figure 8 describes the increase of the size of the

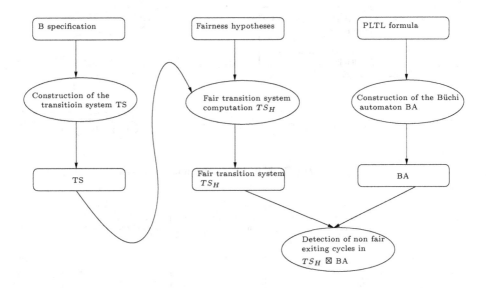

Fig. 5. *PLTL* model checking on TS_H

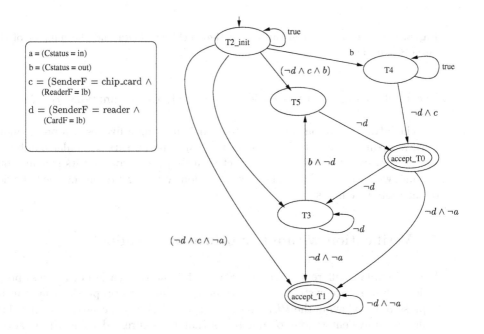

Fig. 6. The Büchi automaton for $\neg(F_1 \Rightarrow P_1)$

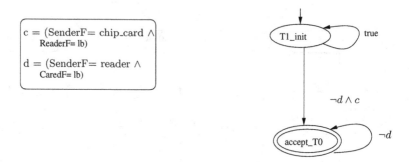

Fig. 7. The Büchi automaton for $\neg P_1$

Fairness hypotheses	Number of the Büchi automaton states	Number of the Büchi automaton transitions
F_1	6	13
$F_1 \wedge F_2$	19	51
$F_1 \wedge F_2 \wedge F_3$	50	130

Fig. 8. Comparison between the size of the Büchi automata and the number of the fairness hypotheses

Büchi automaton when we consider successively the assumptions h_1, $\{h_1,\ h_2\}$ and $\{h_1,\ h_2,\ h_3\ \}$.

The advantage of our approach to model checking a liveness property under fairness assumptions is that we use a Büchi automaton with a smaller number of states. However, the need to construct a Büchi automaton and its synchronous product with the automaton, i.e. the transition system, representing the behavior of the system remains.

5   Verification without Synchronous Product

We now explore different direct approaches for the verification of liveness properties of finite state systems. Here, we limit ourselves to properties that can be expressed by following the *PLTL* pattern $\square(p \Rightarrow \Diamond q)$, in other words, to the class which is equivalent to the *B* modalities that are expressed using the keyword *LEADS-TO*. For example, under the fairness assumptions derived from the specification of the fair-events *Eject*, *Csends if* ($CardF = bl$), *Rsends if* ($ReaderF = bl$), the verification of the formula $\square((Cstatus = in) \Rightarrow \Diamond(Cstatus = out))$ can

be done without proving a user provided decreasing variant or without constructing any Büchi automaton.

In this section, we first examine typical approaches that can be found in the
literature which uses a proof technique, or an algorithmic enumeration or both.
Second, as a research perspective, we propose a combination of proof techniques
with an algorithmic search for the fair exiting cycles.

5.1   Verification of the *B LEADS-TO* Modalities

The B approach to verification uses exclusively proof techniques. J. R. Abrial
and L. Mussat in [2] proposes a main *modality* of the form

$$SELECT\ p\ LEADS\text{-}TO\ q$$
$$INVARIANT\ J$$
$$VARIANT\ V$$

as an equivalent of the *PLTL* formula $\Box(p \Rightarrow \Diamond q)$.

The optional invariant predicate J, when present, must hold on a state where
p holds and must be maintained afterwards as long as $\neg q$ holds. The invariant J
might be a useful hint provided by the user to help the proof search. However, a
positive integer expression V is mandatory. The variant V has to decrease with
any activation of any event which is enabled when p or $\neg q$ holds. Let us recall
that such a modality holds when the following proof obligations hold:

1. The predicate J holds where p holds:

$$p \Rightarrow J\ \ (O_1)$$

2. The variant V is a positive integer expression:

$$p \Rightarrow \forall z.(I \wedge J \Rightarrow V \in \mathbb{N})\ \ (O_2)$$

   where z is the collection of the state variables that are affected by the events.
3. The invariant J is preserved by the events $e_i, i = 1, n$:

$$\forall i\ s.t.\ i \in [1..n].(p \Rightarrow (\forall z.(I \wedge J \wedge \neg q \Rightarrow [GS_i]J)))\ \ (O_3)$$

   where GS_i is the generalized substitution corresponding to the event e_i.
4. The variant V decreases with any activation of an event $e_i, i = 1, n$:

$$\forall i\ s.t.\ i \in [1..n].(p \Rightarrow (\forall z.(I \wedge J \wedge \neg q \Rightarrow [v := V][GS_i](V < v))))\ \ (O_4)$$

5. When the formula $\neg q$ is satisfied, at least one of the events $e_i, i \in [1..n]$ is
   enabled:

$$p \Rightarrow \forall z.(I \wedge J \wedge \neg q \Rightarrow grd(e_1) \vee \cdots \vee grd(e_n))\ \ (O_5)$$

   This means that contiguous activations of the events beginning in a state
   satisfying p never deadlock in a state which does not satisfy q.

Needless to say, these proof obligations have to be generated and proved for specifications of closed systems, i.e. specifications including the specification of the fairness provided by the environment. In our example, verifying the modality

$$SELECT \ Cstatus = in \ LEADSTO \ Cstatus = out \ VARIANT \ V$$

for the specification in Fig. 1 generates 12 proof obligations. We can specify the variant V as the following expression:

if $NbBloc = 0$ *then* $NbMes$
$$else \ 2 * NbBloc + g(SenderF, CardF, ReaderF) + f(Cstatus)$$
   where
   $g(reader, bl, ?) = 1$
   $g(chip\_card, ?, bl) = 1$
   $g(reader, ackb, ?) = 0$
   $g(reader, lb, ?) = 0$
   $g(chip\_card, ?, ackb) = 0$
   $g(chip\_card, ?, lb) = 0$
and
   $f(in) = 1$
   $f(out) = 0$

where ? means anything.

The variant decrease can be shown as follows:

- Activations of *Rsends* and *Csends* decrease *NbMes*.
- Activations of *Rblocksends* and *Cblocksends* do not affect *NbMes*, decrease *NbBloc* from one, and increase *g(SenderF, CardF, ReaderF)* from one.
- Activations of *Racksends* and *Cacksends* do not affect either *NbMes* or *NbBloc*, and decrease *g(SenderF, CardF, ReaderF)* from one.
- Activations of *Eject* decrease *f(Cstatus)*.

The proof obligation (O_4), holds for *Cinsert* since its guard *Cstatus=out* is not concerned with the decrease of the variant. Obviously , *Cinsert* is never enabled when p holds i.e. $Cstatus = in$ or when $\neg q$ holds i.e., $\neg(Cstatus = out)$. In the next section, we will see how the rules given by Manna and Pnueli for a direct model checking of $PLTL$ formulas are adapted to open system specifications.

5.2   Direct Model Checking Approach

Z. Manna and A. Pnueli have given rules for verifying $PLTL$ formulas (see [20, 19]). Here, a *fair* transition system similar to TS_H, represents the behavior of the system.

The rule for the verification of the $PLTL$ formula $\Box(p \Rightarrow \Diamond q)$ in [21] relies on a known and fair *helpful* transition τ_h. The fair transition τ_h leads to a state where q holds. Moreover, the rule uses an auxiliary predicate invariant Φ which must hold where p or q does not hold. Let $\mathcal{T}$ be the set of transitions. Let us examine the rule:

$$
\begin{array}{l}
(H_1) \ \Box(p \Rightarrow (q \vee \Phi)) \\
(H_2) \ \{\Phi\}\mathcal{T}\{q \vee \Phi\} \\
(H_3) \ \{\Phi\}\tau_h\{q\} \\
(H_4) \ \Box(\Phi \Rightarrow \Diamond(q \vee Enabled(\tau_h)))
\end{array}
$$

$$\Box(p \Rightarrow \Diamond q)$$

1. Premise (H_1) expresses that either q or Φ holds in *reachable* states where p holds.
2. Premise (H_2) expresses that any transition must begin in a state where Φ holds and ends in a state where either q or Φ holds.
3. Premise (H_3) expresses that τ_h is a *helpful* transition leading to a state where q holds.
4. Premise (H_4) expresses that any state where Φ holds leads to a state where either q holds or the helpful transition τ_h is enabled.

So, the conclusion relies on τ_h being a *strongly* fair transition.

The verification of the $PLTL$ formula is then a direct, i.e. no construction of a Büchi automaton and no construction of a synchronous product, verification on a finite state fair transition system.

The direct approach is not usual for $PLTL$ model checking. The Büchi automaton approach is linear in the size of the reachable state space and exponential in the size of the $PLTL$ formula.

Direct or not, model checking approaches for the verification of the B modalities on B event systems relies on a reachability analysis based on a transition system model. To end-up this state of the art, let us consider now an approach known as *assertion model checking* which is a direct model checking approach relying on a reachability analysis based on an annotated specification.

5.3   Assertion Model Checking

This approach, which can be found in [7] performs a depth-first search in the behaviors of the systems to verify assertions annotating the specification. For B event systems, it can be *pre* and *post* assertions for events.

Z. Manna and A. Pnueli rules can be used in this approach. Then, a *helpful* event replaces the *helpful* transition. This seems very realistic in practice. If we consider the example and the property $\Box((Cstatus = in) \Rightarrow \Diamond(Cstatus = out))$, the event *Eject* is obviously a *helpful* event. As for fairness hypotheses, only some of all the activations of an event can be *helpful*. So the specification of a *helpful* event may take the form of a fairness hypothesis: e *if* P.

We now can present a perspective for an approach combining direct verification and proof of B modalities under fairness assumptions without variant for finite B event systems.

5.4 Perspective about Verification

Here we present a perspective for verifying B modalities under fairness assumptions. Therefore we suppose a finite state specification by an event system which does not specify the fairness.

The fairness assumptions do not change the proof obligations (O_1), (O_3), and (O_5). Therefore, we can keep these proof obligations. The problem is the replacement of the role played by the variant, i.e. proof obligations (O_2) and (O_4), when the verification of the modality has to be done under fairness assumptions.

Notice that the specification of a variant is impossible without the variables provided by the fairness specification. We need to use the fair exiting cycles to verify that after a state where $p \wedge \neg q$ holds, any contiguous activations of the events leading to states where $\neg q$ holds lead to a state where q hold. This can obviously be done by an algorithmic exploration of the reachability graph. It is, for example enough to verify by a graph exploration that any cycle in the Strongly Connected Component of Fig. 2 is a fair exiting cycle.

This provides a perspective of a verification of B modalities under fairness assumptions which combine proof techniques with a linear algorithmic method.

6   Conclusion and Future Works

Using the B method and its event systems approach, the fairness of reactive systems is always specified whether for a fairness property of the system or a known property of its environment. In the first case, the property has to be specified and verified. In the second case, it is required to specify a scheduler modelling the fairness of the environment. However, this approach has the following drawbacks:

- The model is infinite state although, without the scheduler, it could be finite. This disallows using model checking.
- The specification of the fairness by the scheduler is not implemented since it is part of the environment specification. This is useless work for the specifier.

In this paper, we propose to extend the B event systems by allowing in B a specification of reactive systems under fairness hypotheses as it is currently done in other approaches [20,18]. For that, we propose a way to express fairness assumptions as events with optional additional guards so that some but not all of the activation of the events are fair. In this framework, we study and compare two ways to model check $PLTL$ properties of the event system. Our approach consists in model checking the $PLTL$ property on a model which takes account of the fairness assumptions by noticing the fair exiting cycles in the model. To implement this approach, we need the following:

- We need to build a finite labeled transition system from the B event system. For that, we use a formalism borrowed to constraint programming with a representation of states by a mapping between variables and set constraints [23, 6].

- We need to search for all fair exiting cycles in the reachability graph. For that we use an extension of Tarjan's algorithm.
- We need a model checker which uses the constrained model [23]. To take account of the fairness assumptions, it is enough to pay no attention to fair-exiting acceptation cycles, a minor extension to the already implemented constrained model checker.

Our goal is to compare the two approaches for industrial size applications. Moreover, we give other perspectives for the verification of B modalities on a specification assuming fairness.

In another paper [4], we proposed an algorithmic verification of the refinement between an abstract and a refined event system under fairness assumptions. This needs to be implemented and integrated with the verification tools so that we could observe a refinement development under fairness assumptions.

Furthermore, we would like to answer the two following questions:

- Assumed environment properties can be manyfold. It can be liveness assumptions which are more general than our fairness assumptions. Here again, the model authorizes more behaviors than necessary, but it is a problem to eliminate the useless behaviors from the transition system by a linear search algorithm. So, verifying a property P on the reduced model TS_L corresponding to the set of liveness hypotheses L=$\{l_1,...,l_n\}$ is certainly not interesting. However, the classical approach which consists in verifying that TS is a model for $(l_1 \wedge ... \wedge l_n) \Rightarrow P$ always works.
- There are often fairness properties which are intrinsic in the application. Then, they must be implemented. Often, the fairness properties arises in front of a nondeterministic choice. They guaranty that some choices are not infinitely delayed. It is the case for the lift problem studied by J. R. Abrial and L. Mussat in [2] and for the SCSI-3 bus studied by D. Bert in [5] in the framework of a B event system refinement development. We can notice that these so called fairness properties are expressed in *PLTL* as liveness properties, i.e. every active request is eventually satisfied $\Box$ (*request* $-$ *enabled* $\Rightarrow$ $\Diamond$ *request* $-$ *satisfied*). When this liveness property holds, fairness properties also hold. For example, the lift must not pass infinitely many times at a floor where a lift stop is requested. This is a fairness, i.e., if a request at a floor in infinitely enabled, then it must be infinitely satisfied. Our idea is to study an approach which begins at the abstract level assuming the fairness in the system so that we verify the abstract arbitration between the many choices. Then, refinement is data refinement of the representation of the arbitration process. There is no need to introduce a new event for such a refinement. Therefore, no variant is needed to prove refinement. For example, it can be verified by the presently available versions of the *Atelier B*.

References

1. J. R. Abrial. *The B Book.* Cambridge University Press – ISBN 0521-496195, 1996.
2. J. R. Abrial and L. Mussat. Introducing dynamic constraints in B. In *Second Conference on the B method, France*, volume 1393 of *LNCS*, pages 83–128. Springer Verlag, April 1998.
3. R. J. Back and R. Kurki-Sunio. Decentralisation of process nets with centralised control. In *2nd ACM SIGACT-SIGOPS Symposium on principles of distributed computing*, pages 131–142, 1983.
4. F. Bellegarde, S. Chouali, J. Julliand, and O. Kouchnarenko. Comment limiter la spécification de l'équité dans les systèmes d'événements B. In *Approches Formelles dans l'Assistance au Développement de Logiciels (AFADL'01)*, pages 205–219, Nancy, France, 2001.
5. D. Bert. Preuve de propriétés d'équité en B: étude du protocole du bus SCSI-3. In *Approches Formelles dans l'Assistance au Développement de Logiciels (AFADL'01)*, pages 221–241, Nancy, France, 2001.
6. D. Bert and F. Cave. Construction of finite labelled transition systems from B abstract systems. In *In proc. of IFM'2000*, volume 1945 of *LNCS*, pages 235–254. Springer Verlag, November 2000.
7. R. Bharadwaj and J. I. Zucker. Direct model checking of temporal properties (version 2). Technical Report CRL Report 317, Communications Research Laboratory, jan 1996.
8. M. J. Butler. Stepwise refinement of communicating systems. *Science of Computer Programming*, 27(2):139–173, 1996.
9. M. J. Butler. An approach to the design fo distributed systems with B amn. In J. P Brown and D. Till, editors, *10th International Conference of Z Users (ZUM'97)*, volume LNCS 1212, pages 223–241. Springer-Verlag, April 1997.
10. M. J. Butler and M. Walden. Distributed system development in B. In H. Habrias, editor, *First B Conference*, pages 155–168, November 1996.
11. E. A. Emerson and C. Lei. Modalities for model checking: Branching time logic strikes back. *Science of Computer Programming*, 8(3):275–306, 1987.
12. K. Etessami and G. Holzmann. Optimizing Büchi automata. In *Proceedings of 11th Int. Conf. on Concurrency Theory (CONCUR)*, pages 153–167, 2000.
13. Comité européen de Normalisation. En27816-3. European standard - identification cards - integrated circuit(s) card with contacts - electronic signal and transmission protocols. Technical Report ISO/CEI 7816-3, 1992.
14. R. Gerth, D. Peled, M. Vardi, and P. Wolper. Simple on-the-fly automatic verification of linear temporal logic. In *Proc. IFIP-WG6.1 Symposium On Protocols Specification, Testing and Verification (PSTV95)*, pages 2–21, Warsaw – Poland, 1995.
15. G. Holzmann. *Design and validation of protocols.* Prentice hall software series, 1991.
16. J. Julliand, P.A. Masson, and H. Mountassir. Vérification par model checking modulaire des propriétés dynamiques introduites en B. In *Technique et Science Informatiques*, 2001. to appear.
17. Y. Kesten, Z. Manna, and A. Pnueli. Temporal verification of simulation and refinement. In *REX Workshop A Decade of Concurrency*, volume 803 of *Lecture Notes in Computer Science*, pages 273–346. Springer Verlag, 1993.
18. L. Lamport. A temporal logic of actions. *ACM Transactions On Programming Languages And Systems, TOPLAS*, 16(3):872–923, May 1994.

19. Z. Manna and A. Pnueli. A hierarchy of temporal properties. In *Proceedings of the 9th ACM Symposium on Principles of Distributed Computing (PODC)*, pages 377–408, New York, NY, 1990. ACM Press.
20. Z. Manna and A. Pnueli. *The Temporal Logic of Reactive and Concurrent Systems: Specification.* Springer-Verlag - ISBN 0-387-97664-7, 1992.
21. Zohar Manna and Amir Pnueli. Models for reactivity. *Acta Informatica*, 30(7):609–678, 1993.
22. Steria Méditerranée. *Le langage B. Manuel de référence version 1.5.* S.A.V. Steria, BP 16000, 13791 Aix-en-Provence cedex 3, France.
23. B. Parreaux. *Vérification de systèmes d'événements B par model-checking PLTL.* PhD thesis, U.F.R. des Sciences et Techniques, Université de Franche-Comté, Décembre 2000.
24. R. E. Tarjan. Depth-first search and linear graph algorithms. *SIAM J. Comput*, 1:146–160, 1972.
25. M. Vardi and P. Wolper. An automata-theoric approach to automatic program verification. In *1 st IEEE Symp. Logic in Computer Science (LICS'86)*, pages 332–344, Cambridge, MA, USA, june 1986.
26. M. Y. Vardi and P. Wolper. Reasonning about infinite computations. *Information and Computation*, 115(1):1–37, 1994.
27. P. Wolper and V. Lovinfosse. Verifying properties of large sets of processes with network invariants. In *Int. Workshop on Automatic Verification Methods for Finite State Systems*, LNCS 407, pages 68–80, Grenoble, France, june 1989. Springer Verlag.

A Formal Model of the UML Metamodel: The UML State Machine and Its Integrity Constraints

Soon-Kyeong Kim and David Carrington

School of Computer Science and Electrical Engineering
The University of Queensland, Brisbane, 4072, Australia
soon@csee.uq.edu.au, davec@csee.uq.edu.au

Abstract. This paper presents a formal Object-Z model of the UML State Machine. We encapsulate the abstract syntax and the static and dynamic semantics for each individual model construct as a single Object-Z class. To formalize the dynamic semantics, a denotational semantics of the construct is given first ignoring detailed operational sequences. Based on this denotational semantics, an operational (execution) semantics is then defined in terms of (Object-Z) class operations and invariants constraining the operation sequences. The timed refinement calculus is used to define the operation sequences within Object-Z. Finally, integrity consistency constraints with other model constructs are formalized in terms of invariants defined in the state machine. Our approach not only enhances the precision of the UML state machine description but also overcomes the lack of modularity, extensibility and reusability of the current UML semantic representation.

1. Introduction

Among software engineers, the correctness of a software model has been emphasized as one of the major aspects for developing a correct software system. To develop such a correct software model, it is essential that the language used to develop the model itself should have a well-defined syntax and semantics for its notation. This is even more important when a language aims to be a standard such as the Unified Modeling Language (UML) [15].

UML is a widely accepted modeling language that can be used to visualize, specify, construct and document the artifacts of a software system and was accepted as a standard object-oriented modeling language by OMG in 1997 [15]. The syntax and semantics of notations provided in UML are defined in terms of its metamodel. The existence of such a metamodel for UML is beneficial in the following ways:

- it enhances understanding as both the syntax and the semantics of modeling constructs are defined in the single metamodel, and
- it allows the integration of syntactic and semantic consistency constraints within individual models (intra-view constraints) and integrity consistency constraints between different models (inter-view constraints).

D. Bert et al. (Eds.): ZB 2002, LNCS 2272, pp. 497-516, 2002.

Despite these advantages, there exist several drawbacks in the quality of the UML metamodel documentation. We explore the UML metamodel from two perspectives: structure and formality.

The structure: In the UML metamodel, the modeling constructs are defined using three distinct views: abstract syntax in UML class diagrams, static semantics in OCL [15], and dynamic semantics (specifying the meaning of the constructs) mainly in English. As a result, the abstract syntax, the static and the dynamic semantics are defined separately in distinct structures and representations, with consequently many redundancies and inconsistencies. For example, when a modeling construct is enhanced, each of these three structures must be extended. This lack of modularity and extendibility in the semantic presentation of UML results in the UML metamodel being not well-structured, despite the overall metamodel being structured in terms of metaclasses.

The level of formality: Another significant problem in the current UML metamodel is its lack of precision. None of the three meta-languages used to define UML have a precise basis yet. For example, the semantics of class diagrams is ambiguous [5, 7, 8]. OCL neither has a precise semantics [18] nor is it a consistent formal description language (e.g. OCL flattens set-valued attributes to single-valued attributes, which makes a consistent logic check over set properties impossible). Furthermore, the pre and post-condition style of OCL expressions is not appropriate to specify the execution sequences [16]. Finally, English descriptions are informal and do not provide much insight into the semantics of UML notations [4, 5].

The aim of our work is to develop a complete formal model of the UML metamodel using Object-Z [3, 20] as a meta-language. Previously, we presented an Object-Z specification for the three fundamental packages in the UML metamodel: the Core, DataTypes, and Common Behavior packages [11]. This paper extends our work to the State Machine package.

In our work, the abstract syntax and static semantics are formalized as follows. Since in the UML metamodel the abstract syntax is defined using UML class diagrams, we first formalize UML class constructs using Object-Z. In doing so, we define a metamodel of Object-Z [9] adopting the same architecture as that of the UML metamodel. Given the metamodels of both languages, we formalize the abstract syntax of model elements represented in UML class diagrams by translating the UML class constructs to the Object-Z class constructs. Since Object-Z is more expressive than UML class diagrams, we are able to encapsulate the static semantics of model elements with the abstract syntax in terms of invariants defined in the corresponding Object-Z classes (see Fig. 1). The dynamic semantics of the state machine are formalized in two ways. First, we give denotational semantics for the model constructs ignoring their operational semantics. Based on the denotational semantics given, then we define the operational (execution) semantics of the constructs considering the sequence of operations. Denotational semantics are defined by introducing additional variables (we call these *semantic variables* to distinguish them from the variables that formalize the abstract syntax), which are used to define the semantics and invariants in Object-Z classes. Operational semantics are specified in terms of class operations (we call these operations *meta-operations*) and invariants restricting the operation sequences. We use the timed trace notation of timed refinement calculus [19] to define these invariants with Object-Z. With this approach, we give a mutually consistent (formal) denotational and operational semantics for the UML state machine in a compositional way with respect to its syntax. Furthermore, we formalize integrity consistency constraints between the state machine and other UML models.

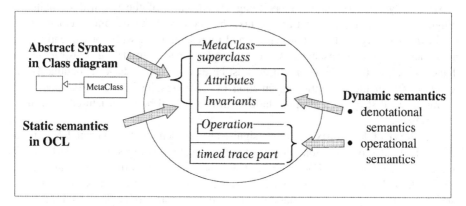

Fig. 1. A graphical description showing how to group the three different views of a construct into a single Object-Z class.

We believe our work is the first to integrate the syntax and the denotational and operational semantics of any variant of statecharts [2, 6] into a single model construct in a modular way. Several attempts [12, 13, 14, 17] have already been made to give a formal definition for a subset of the UML state machine. The major significance of our work is that it provides not only a formal definition for individual model concepts but also a *formal model* for the entire UML State Machine package, while maintaining the same structure as the UML metamodel in terms of classes. This feature enables us not only to enhance the precision of the UML state machine but also to make possible an intuitive cross-referencing between the two models. Furthermore, adopting an object-oriented approach for defining UML enables us to overcome the lack of modularity, extensibility and reusability of the current UML semantic representation. In our approach, extending an existing model element or adding a new model element can be readily achieved in a modular way using the inheritance mechanism in Object-Z. Since Object-Z has a precise semantics and reasoning techniques developed for it, the validity of any modification or extension to the existing UML metamodel can be systematically verified in a logical way. It should be also noticed that re-structuring the current UML metamodel does not affect our formal model. When the UML metamodel is re-structured into several parts (probably into three parts: the abstract syntax, the semantic domain and the semantic mapping between the two previous parts as suggested in [1]), for example, our formal model provides an integrated view of the parts.

The structure of the rest of this paper is as follows. Section 2 presents a brief description of how to model UML class constructs using Object-Z class constructs and the integrated use of the timed refinement calculus with Object-Z. Section 3 presents a formalization of the abstract syntax and static semantics of core model constructs in the State Machine package. Section 4 presents a formalization of the dynamic (execution) semantics of the model constructs. Section 5 discusses integrity consistency constraints from the state machine's point of view. Section 6 draws some conclusions and discusses future work.

2. Modeling UML Class Constructs Using Object-Z Class Constructs

Object-Z is an object-oriented extension to Z [20] designed to facilitate specification in an object-oriented style. Fig. 2 is a UML class diagram showing the abstract syntax of core modeling constructs in Object-Z. In this section, we summarize features of Object-Z needed in this paper (see [9] for details of the Object-Z metamodel and [9, 11] for detailed translation rules from UML class constructs to Object-Z class constructs).

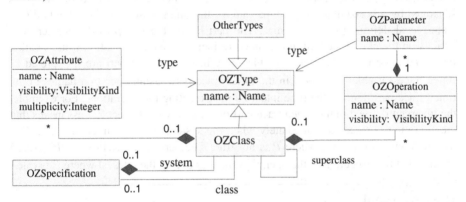

Fig. 2. A class diagram showing the structure of modeling constructs in Object-Z

Classes: A class in Object-Z specifies both the states that are possible for an object of that class and the operations that such an object can undergo [3]. That is, an Object-Z class is a description of all objects that share the same features: attributes (including relationships), and operations. Based on this semantic interpretation, we model each UML metaclass using a separate Object-Z class. When attributes appear in a metaclass, the attributes and their types are used as attributes and attribute types in its Object-Z class (see the Object-Z classes presented in Section 3 for examples).

Inheritance: Classes in Object-Z can be used to define other classes by inheritance. A class can inherit from several classes (multiple inheritance). We model UML inheritance using this inheritance mechanism in Object-Z. For each UML metaclass that inherits from other metaclasses, the names of the Object-Z classes corresponding to their superclasses are included within the Object-Z class corresponding to the metaclass. Since in UML generalization is a subtyping relationship, a down-arrow symbol (denoting subtyping in Object-Z) [3] is added whenever superclasses are used as a type within other classes.

Association: Classes in Object-Z can be instantiated in other classes as attributes. In Object-Z, instantiation is used as a mechanism for modeling relationships between objects. Objects that instantiate other classes as their attributes can refer to the objects of the instantiated classes. The values of these attributes are object-identities of the referenced objects. Consequently, we model an association in UML as an attribute of the corresponding Object-Z classes that it relates. For each associated class, the type of the attribute is the corresponding Object-Z class of its opposite class. For composition, the symbol © which represents unshared ownership in Object-Z [3] is added to the attribute name. When an association is directed, the association is modeled only as an attribute of the Object-Z class that can access. Multiplicity constraints and other constraints are formalized as invariants in Object-Z classes.

The timed refinement calculus [19] has been integrated with Object-Z for specifying time related properties for real-time systems (see [19] for details). The notation introduces a given type T for absolute time and an interval bracket $\langle \, \rangle$ to model a set of time intervals. It also has the concatenation operator ';' for joining intervals whose end points meet. Object-Z classes in the integrated notation are composed of two parts separated by a horizontal line. The part above the line is essentially the standard Object-Z local definitions and schemas. The part below the line provides further constraints on the class specified in the timed refinement calculus notation (this is called the timed trace part). All state variables $x: X$ in the standard part are interpreted as timed trace variables $x: T \rightarrow X$ in the timed trace part. Using operation names in the timed trace part denotes the time intervals for executing the operations. For example, the invariant defined in the timed trace part of the following Object-Z class means that the operation OP occurs immediately when the system enters an interval where P is true following an interval where P is not true. The part of the RHS ($\cup \, \langle OP \rangle;\langle true \rangle$) ensures that the execution of the operation finishes during the interval where P is true.

┌─*ExampleClass*────────────────────────────────────
│ [Standard part]
│ ────────────────────────────────
│ [timed trace part]
│ $\langle \neg P \rangle;\langle P \rangle \subseteq \langle \neg OP \rangle;(\langle OP \rangle \cup \langle OP \rangle;\langle true \rangle)$
└──

3. Formalizing the Abstract Syntax and Static Semantics

In this section, we formalize the abstract syntax of core model elements defined in the State Machine package (see Fig. 3) according to the translation rules described in the previous section. We also formalize the static semantics of the model elements as invariants in their corresponding Object-Z classes. In the Object-Z classes defined in this section, for brevity, we omit some constraints that formalize links between associated classes, and we assume that all types used in the Object-Z classes presented in this paper are defined as distinct Object-Z classes (see [11] for a full formal description of the data types).

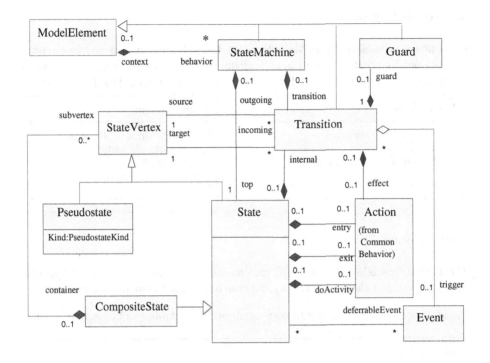

Fig. 3. State Machine -Main

3.1 State

Syntactically State inherits from StateVertex and has several associations with StateMachine, Action, Event and Transition: *stateMachine, entry, doActivity, exit, deferrableEvent,* and *internal* (see Fig. 2). The following Object-Z class is a formal description of State. StateVertex is included as a superclass (we assume that State-Vertex is also formalized as an Object-Z class with the same name) and the associations are formalized as attributes. No specific static semantics are given for states.

```
┌─State────────────────────────────────────────────────
│ StateVertex
│ ┌──────────────────────────────────────────────────
│ │ stateMachine : ℙ StateMachine
│ │ entry : ℙ↓Action ©
│ │ doActivity : ℙ↓ Action ©
│ │ exit : ℙ↓Action ©
│ │ deferrableEvent : ℙ↓Event
│ │ internal : ℙ Transition ©
│ ├──────────────────────────────────────────────────
│ │ {# stateMachine, # entry, # exit, # doActivity} ⊆ {0, 1}
│ │ ...
```

3.2 Event and Action

Event and Action inherit from ModelElement. Event has parameters and associations to Transition and State. Action has an attribute *isAsynchronous* indicating whether or not the action is asynchronous and several other attributes (refer to [10] for details).

```
┌─Event──────────────────────
│ ModelElement
│ ├──────────────────────────
│ │ transition : ℙ Transition
│ │ state : ℙ↓ State
│ │ parameters : seq Parameter ©
│ │ ...
│ └──────────────────────────
└────────────────────────────
```

```
┌─Action──────────────────────
│ ModelElement
│ ├────────────────────────────
│ │ isAsynchronous : Boolean
│ │ entryState : ℙ State
│ │ exitState : ℙ State
│ │ activityState : ℙ State
│ │ ...
│ └────────────────────────────
└──────────────────────────────
```

3.3 Pseudostate

The kinds of pseudo-states in UML are: *initial*, *deepHistory*, *shallowHistory*, *join*, *fork*, *junction*, and *choice*. For brevity, we omit detailed structure of pseudo-states.

$PseudostateKind := initial \mid deepHistory \mid shallowHistory \mid join \mid fork \mid junction \mid choice$

```
┌─Pseudostate────────────────────
│ StateVertex
│ ├───────────────────────────────
│ │ kind : PseudostateKind
│ │ ...
│ └───────────────────────────────
└─────────────────────────────────
```

3.4 CompositeState

CompositeState has two attributes *isConcurrent* and *isRegion* representing whether it is decomposed into two or more orthogonal regions and whether it is a substate of a concurrent state respectively [15]. It inherits from State and has a composition association to StateVertex. The static semantics are formalized as invariants in the predicate of the following Object-Z class (note that numbers are purely descriptive).

```
┌─CompositeState──────────────────────────────
│ State
│ ├────────────────────────────────────────────
│ │ isConcurrent : Boolean
│ │ isRegion : Boolean
│ │ subvertex : ℙ ↓ StateVertex ©
│ ├────────────────────────────────────────────
│ │ [1] # { s: subvertex | s ∈ Pseudostate ∧ s.kind = initial } ≤ 1
│ │ [2] # { s: subvertex | s ∈ Pseudostate ∧ s.kind = deepHistory } ≤ 1
│ │ [3] # { s: subvertex | s ∈ Pseudostate ∧ s.kind = shallowHistory } ≤ 1
│ │ [4] isConcurrent ⟹ #{ s: subvertex | s ∈ CompositeState } ≥ 2
│ │ [5] isConcurrent ⟹ ∀ s: subvertex • s ∈ CompositeState
│ └────────────────────────────────────────────
└──────────────────────────────────────────────
```

The static semantics given for composite states are:

[1] A composite state can have at most one initial vertex.

[2] A composite state can have at most one deep history vertex.

[3] A composite state can have at most one shallow history vertex.

[4] There must be at least two composite substates in a concurrent composite state.

[5] A concurrent state can only have composite states as substates.

[6] The substates of a composite state are part of only that composite state. Note that this rule is already defined by the composition relationship between the metaclass StateVertex and CompositeState (the symbol © added to the attribute *subvertex* formalizes this rule implicitly).

3.5 Transition

Transition has a source and a target state vertex. It also can have a guard, a trigger event, effect actions, and a state for internal transitions. The static semantics of transitions are formalized as invariants; [1] a fork or [2] a join segment should not have guards or triggers; [3] a fork segment should always target a state; [4] a join segment should always originate from a state; [5] transitions outgoing from pseudo-states may not have a trigger (this rule is inconsistent with rule [8] below so we restrict the rule to non-initial transitions); [6] join segments should originate from orthogonal states; [7] fork segments should target orthogonal states; and [8] an initial transition at the top-most level may have a trigger with the stereotype "create". An initial transition of a StateMachine modeling a behavioral feature has a CallEvent trigger associated with the BehavioralFeature. Apart from these cases, an initial transition never has a trigger.

┌─*Transition*───
│ *ModelElement*
│ ┌──
│ │ *stateMachine* : $\mathbb{P}$ *StateMachine*
│ │ *source* : $\downarrow$ *StateVertex*
│ │ *target* : $\downarrow$ *StateVertex*
│ │ *guard* : $\mathbb{P}$ *Guard* ©
│ │ *trigger* : $\mathbb{P} \downarrow$ *Event*
│ │ *effect* : $\mathbb{P} \downarrow$ *Action* ©
│ │ *state* : $\mathbb{P} \downarrow$ *State* ...
│ ├──
│ │ ...
│ │ [1] $source \in Pseudostate \land source.kind = fork \Rightarrow guard = \varnothing \land trigger = \varnothing$
│ │ [2] $target \in Pseudostate \land target.kind = join \Rightarrow guard = \varnothing \land trigger = \varnothing$
│ │ [3] $stateMachine \neq \varnothing \land source \in Pseudostate \land source.kind = fork \Rightarrow target \in State$
│ │ [4] $stateMachine \neq \varnothing \land target \in Pseudostate \land target.kind = join \Rightarrow source \in State$
│ │ [5] $source \in Pseudostate \land source.kind \neq initial \Rightarrow trigger = \varnothing$
│ │ [6] $target \in Pseudostate \land target.kind = join \Rightarrow \forall c : source.container \bullet c.isConcurrent$
│ │ [7] $source \in Pseudostate \land source.kind = fork \Rightarrow \forall c : target.container \bullet c.isConcurrent$
│ │ [8] $source \in Pseudostate \land source.kind = initial \Rightarrow trigger = \varnothing \lor$
│ │ $(\ source.container = \{s : stateMachine \bullet s.top\} \land$
│ │ $\cup\{s : \cup\{t : trigger \bullet t.stereotype\} \bullet s.name\}) \subseteq \{create\}) \lor$
│ │ $(\ \cup\{s : stateMachine \bullet s.context\} \subseteq BehavioralFeature \land trigger \subseteq CallEvent \land$
│ │ $\{t : trigger \bullet t.operation\} = \cup\{s : stateMachine \bullet s.context\})$
│ └──
└──

3.6 StateMachine

StateMachine has composition relationships to State and Transition, and an aggregation relationship to ModelElement which is the context of the state machine. The static semantics are formalized as invariants; [1] a StateMachine is aggregated within either a classifier or a behavioral feature; [2] a top state is always a composite (Note that this constraint could be removed by modifying the association to State as an association to CompositeState instead); [3] a top state cannot have any containing states; [4] the top state cannot be the source of a transition; and [5] if a StateMachine describes a behavioral feature, it contains no triggers of type CallEvent, apart from the trigger on the initial transition (also see the static semantics [8] of *Transition*).

┌*StateMachine*────────────────────────────────────
│ *ModelElement*
│
├──
│ *context* : $\mathbb{P} \downarrow ModelElement$
│ *top* : $\downarrow State$ ©
│ *transition* : $\mathbb{P}\ Transition$ ©
│
├──
│ $...$
│ $[1]\ \forall c : context \bullet c \in Classifier \lor c \in BehavioralFeature$
│ $[2]\ top \in CompositeState\ \land\ [3]\ top.container = \varnothing \land [4]\ top.outgoing = \varnothing$
│ $[5]\ \forall\ c: context \bullet c \in BehavioralFeature \Rightarrow$
│ $\{t: transition \mid \neg\ (t.source \in Pseudostate \land t.source.kind = initial)\ \land t.trigger \subseteq CallEvent\ \} = \varnothing$
└──

4. Formalizing the Dynamic Semantics

In the UML State Machine package, the semantics of state machines are specified in terms of the operations of a hypothetical machine that implements a state machine specification [15]. The hypothetical machine is composed of:

- an event queue holding incoming events
- an event dispatcher mechanism and
- an event processor that processes dispatched event instances according to the semantics of the state machine.

UML does not provide any specific requirements for the event queue and the event dispatcher mechanism leaving open the possibility of modeling different schemes [15]. For this reason, we do not provide a formal definition of the event queue and dispatcher mechanism concentrating on the event processor which refers to the state machine in the metamodel. In this section, we formalize the dynamic semantics of the model elements presented in the previous section.

4.1 Event

An event instance is the trigger of a transition. When an event instance arrives at the state machine, if the state machine is not blocked with another event, the event is referred to as the *current event*. When the event processing is completed, the event is *consumed*. To formalize these semantics, we define a (semantic) variable *consumed* of type Boolean. The variable is false until the event is completely processed by the state machine. Consequently, an event is said to be fully *consumed* when all transitions enabled by the event that are also selected to fire by the state machine are completed (see the invariant in the Object-Z class below and the Object-Z class *StateMachine*).

```
┌─Event──────────────────────────────────────────────────────────────
│ ...
│   ┌──────────────────────────────────────┐   ┌─INIT──────────────┐
│   │ transition : ℙ Transition            │   │ ¬ consumed        │
│   │ ...                                  │   └───────────────────┘
│   │ [Semantic variable]                  │
│   │ consumed : Boolean                   │
│   └──────────────────────────────────────┘
│   consumed ⟺ ∀ t: transition | ∃ sm: t.stateMachine | t ∈ sm.maximalSetTrans • t.completed
└──────────────────────────────────────────────────────────────────────
```

4.2 Action

An action is a specification of an executable statement that should be performed as a result of the execution of a transition or entering to or exiting from a state. The Object-Z class *Action* now has an additional semantic variable *completed* of type Boolean representing its state. For synchronous actions, this variable becomes true when the execution of the actions is completed. For asynchronous actions, the variable becomes true when the actions are executed, but without waiting for the completion of the actions. We also define a semantic variable *aborted* of type Boolean representing whether or not the action has been aborted. Based on this semantic extension, we define two abstract operations *execute* and *abort* to represent the necessary realization of the action execution and to cause an interrupt on the execution of the action, respectively (the action semantics is outside the scope of the semantics of the state machine). These operations are invoked by other model elements such as states and transitions, which need to execute the action in the context of their behaviors.

```
┌─Action─────────────────────────────────────────────────────────────
│ ...
│   ┌──────────────────────────────────────┐   ┌─INIT──────────────┐
│   │ ...                                  │   │ ¬ completed ∧ ¬ aborted │
│   │ [Semantic variable]                  │   └───────────────────┘
│   │ completed : Boolean                  │
│   │ aborted : Boolean                    │   ┌─abort─────────────┐
│   └──────────────────────────────────────┘   │ Δ(aborted)        │
│                                               ├───────────────────┤
│   ┌─execute──────────────────────────────┐   │ ¬ completed ∧ aborted' │
│   │                                      │   └───────────────────┘
│   └──────────────────────────────────────┘
└──────────────────────────────────────────────────────────────────────
```

4.3 State

The dynamic semantics of states are formalized in two ways: denotational and operational.

Formalizing the denotational semantics: A state becomes active when it is entered, and becomes inactive when it is exited. This semantics is formalized by introducing a semantic variable *active* of type Boolean, which becomes true when the state is entered, and becomes false when the state is exited. When a state is active, all enclosing states should also be active (see invariant [1] below). We also define several semantic variables *entryCompleted*, *activityCompleted*, *transitionCompleted*, and *exitCompleted* of type Boolean; each of them represents the completion of the behavior of its corresponding model element (see invariants, [2] to [5]). A state is said to be *completed* when its entry action, transition and activity are completed (see the semantic variable *completed* of type Boolean and invariant [6] constraining when this variable becomes true).

```
┌─State──────────────────────────────────────────────────────
│ ...
│ ┌────────────────────────────────────────────────────────
│ │ ...
│ │ [Semantic variables]                      ┌INIT──────────────
│ │  active : Boolean                          │ ¬ active
│ │  entryCompleted : Boolean                  │ ¬ entryCompleted
│ │  activityCompleted : Boolean               │ ¬ activityCompleted
│ │  transitionCompleted : Boolean             │ ¬ transitionCompleted
│ │  exitCompleted : Boolean                   │ ¬ exitCompleted
│ │  completed : Boolean                       │ ¬ completed
│ │
│ │ ...
│ │ [1] active ⟹ ∀ c: container • c.active
│ │ [2] entryCompleted ⟺ active ∧ ∀ e : entry • e.completed
│ │ [3] activityCompleted ⟺ active ∧ ∀ a: doActivity • a.completed
│ │ [4] transitionCompleted ⟺ active ∧ ∀ t : internal • t.completed
│ │ [5] exitCompleted ⟺ ¬ active ∧ ∀ e : exit • e.completed
│ │ [6] completed ⟺ active ∧ entryCompleted ∧ activityCompleted ∧ transitionCompleted
│ └────────────────────────────────────────────────────────
└────────────────────────────────────────────────────────────
```

Formalizing the operational semantics: To formalize the execution semantics of states, we define several operations. The operations *enter* and *exit* change the value of the variable *active* accordingly. The operation *OuterEnter* applies the operation itself to the all containing states recursively if they have not been entered yet, and enters the state. The operations *ExecuteEntry*, *ExecuteActivity* and *ExecuteExit* execute their corresponding actions by invoking the operation *execute* defined in the Object-Z class *Action*. The operation *AbortActivity* results in aborting the activity (if it is not completed) by invoking the operation *abort* defined in the Object-Z class *Action*. We also define two operations *Enter* and *Exit* which are used by transitions when they enter or exit a state. Since entering a state should start from the outermost ancestor state if it has not been entered, the operation *Enter* invokes the operation *OuterEnter*.

State

$\upharpoonright (Enter, Exit) \ldots$

enter

$\Delta \, (active)$

$\neg \, active \wedge \neg \, completed \wedge active\,'$

exit

$\Delta \, (active, completed)$

$active \wedge \neg \, active\,'$

$OuterEnter \; \hat{=} \; (\, \wedge \; c: container \mid \neg \, c.active \bullet c.OuterEnter \,) \wedge enter$
$ExecuteEntry \; \hat{=} \; \wedge \; e: entry \bullet e.execute$
$ExecuteActivity \; \hat{=} \; \wedge \; a: doActivity \bullet a.execute$
$AbortActivity \; \hat{=} \; \wedge \; a: doActivity \mid \neg \, a.completed \bullet a.abort$
$ExecuteExit \; \hat{=} \; \wedge \; x: exit \bullet x.execute$
$Enter \; \hat{=} \; OuterEnter$
$Exit \; \hat{=} \; exit$

Finally, we formalize the operation sequences of a state in terms of invariants defined in the timed trace part. An informal description of the invariants in the timed-trace part is given below.

- When a state becomes active, the entry action is executed prior to any other action. When a state is contained by a composite state, the entry action of the state is executed after the entry action of the composite state is executed. Invariant [1] formalizes this by ensuring that the operation *ExecuteEntry* occurs when the state is active and the entry action of its container state is executed.

- When the entry action is completed, activity defined in the state is executed. Invariant [2] formalizes this by ensuring that the operation *ExecuteActivity* occurs immediately when the entry action is completed.

- Invariant [3] formalizes the operation sequence when a state becomes inactive, but the activity is not completed. In this case, the activity is aborted by invoking the operation *AbortActivity*.

- Invariant [4] formalizes the operation sequence when a state is exited. The invariant ensures that the operation *ExecuteExit* occurs when both the state is not active and the activity has been completed or aborted. Since the execution sequence for exiting from composite states is slightly different, the invariant is restricted to non-composite states.

State

$\ldots$

$[1] \; \langle \neg \, (active \wedge (\forall \, c: container \bullet c.entry.completed)) \, \rangle ;$
$\qquad\qquad \langle \, active \wedge (\forall \, c: container \bullet c.entry.completed) \, \rangle \subseteq$
$\qquad\qquad\qquad \langle \neg \, ExecuteEntry \, \rangle ; (\langle ExecuteEntry \, \rangle \cup \langle ExecuteEntry \rangle ; \langle \, true \, \rangle)$

$[2] \; \langle \, active \wedge \neg \, entryCompleted \, \rangle ; \langle \, active \wedge entryCompleted \, \rangle \subseteq$
$\qquad\qquad\qquad \langle \neg \, ExecuteActivity \rangle ; (\langle ExecuteActivity \, \rangle \cup \langle ExecuteActivity \rangle ; \langle \, true \, \rangle)$

$[3] \; \langle \, active \wedge \neg \, activityCompleted \, \rangle ; \langle \neg \, active \wedge \neg \, activityCompleted \, \rangle \subseteq$
$\qquad\qquad\qquad \langle \neg \, AbortActivity \rangle ; (\langle AbortActivity \rangle \cup \langle AbortActivity \rangle ; \langle \, true \, \rangle)$

$[4] \; \forall \, s: {\downarrow} State \mid s \notin CompositeState \bullet \langle \, active \vee \neg \, (\forall \, a: doActivity \bullet a.completed \vee a.aborted) \, \rangle ;$
$\qquad\qquad \langle \neg \, active \wedge (\forall \, a: doActivity \bullet a.completed \vee a.aborted) \, \rangle \subseteq$
$\qquad\qquad\qquad \langle \neg \, ExecuteExit \rangle ; (\langle ExecuteExit \rangle \cup \langle ExecuteExit \rangle ; \langle \, true \, \rangle)$

4.4 CompositeState

A composite state is composed of a set of states which may have their own state hierarchies. UML refers to these state hierarchies as *state configurations*. Space does not allow us to present details, so we give only part of the Object-Z class formalizing composite states below (refer to [10] for details).

```
┌─CompositeState──────────────────────────────────────────────
│ ...
│ ┌──────────────────────────────────────────────────────────
│ │ ...
│ │ Δ
│ │ stateConfiguration : ℙ↓ StateVertex
│ │ ─────────────────────────────────────
│ │ stateConfiguration = subvertex ∪ (∪ { s: subvertex | s ∈ CompositeState •
│ │                                 s.stateConfiguration } )
```

Formalizing the operational semantics: We now formalize the operational semantics of the composite state in terms of class operations of the Object-Z class *CompositeState*. The informal description of the operations defined is as follows (for brevity, we omit the formal definition of these operations and refer readers to [10] for this):

- The operation *DefaultEnter* enters the target state using the default entry when the target state is not explicitly defined.
- The operation *InnerEnter* enters the target state. When the substates are not entered explicitly, the operation invokes the operation *DefaultEnter* for a non-concurrent composite state or the operation *Enter* of the substates for a concurrent composite state (causing the default entry in each of the substates).
- The operation *InnerExit* exits a composite state starting from the innermost active substate (or states) of the composite state.

Since the execution sequence for entering to or exiting from composite states is different from those for non-composite states, we redefine the operations *Enter* and *Exit* defined in the Object-Z class *State* as follows.

- The operation *Enter* formalizes the entering execution to a composite state by combining the two operations *OuterEnter* and *InnerEnter*. This guarantees that all the containing states of a composite state are always active when the composite state is active and also all the substates of the composite state are entered properly according to the entering rules for composite states.
- The operation *Exit* formalizes the exiting execution from a composite state by combing the two operations *InnerExit* and *exit*. This also guarantees that all the active substates of a composite state are exited (inactive) when the composite state is exited (inactive).

┌─ *CompositeState* ───
│ ...
│ *DefaultEnter* ≙ ..
│ *InnerEnter* ≙ ([¬ *isConcurrent*] ∧
│   ([{*s*: *subvertex* | *s.active* } = ∅] ∧ *DefaultEnter*
│   ☐
│   [{*s*: *subvertex* | *s.active* } ≠ ∅])
│   ☐
│   [*isConcurrent*] ∧ (∧ *s*: *subvertex* | ¬ *s.active* • *s.Enter*))
│ *InnerExit* ≙ ∧ *s*: *subvertex* | *s.active* • *s.Exit*
│ *Enter* ≙ *OuterEnter* ⨾ *InnerEnter*
│ *Exit* ≙ *InnerExit* ⨾ *exit*
└──

Finally, we formalize the execution sequence of the exit action of the composite state, which is that the exit entry of the composite state is executed after the exit actions of its substates are executed. We formalize this semantics by defining an additional invariant in the timed trace part as below. The invariant ensures that the operation causing the execution of the exit action of a composite state only can occur when the exit actions of its active substates are executed (completed) as well as the composite state is inactive and its activity is completed or aborted.

┌─ *CompositeState* ───
│ ...
│ ─────────────────────────────────
│ ⟨ *active* ∨ ¬ (∀ *a*: *doActivity* • *a.completed* ∨ *a.aborted*) ∨
│   ¬ (∀ *s*: *stateConfiguration* | *s* ∈ ↓*State* ∧ *s.entryCompleted* • *s.exit.completed*) ⟩;
│   ⟨ ¬ *active* ∧ (∀ *a*: *doActivity* • *a.completed* ∨ *a.aborted*) ∧
│   (∀ *s*: *stateConfiguration* | *s* ∈ ↓*State* ∧ *s.entryCompleted* • *s.exit.completed*) ⟩ ⊆
│   ⟨¬ *ExecuteExit*⟩;(⟨*ExecuteExit*⟩ ∪ ⟨ *ExecuteExit*⟩;⟨*true*⟩)
└──

4.5 Transition

The major semantics of transitions are the enabling mechanism for transitions and the execution sequences of actions associated.

Formalizing the denotational semantics: Prior to formalizing the semantics, we define several secondary variables formalizing the notions used in the StateMachine package [15]: *leastCommonAncestor, mainSourceState, mainTargetState, explicitSourceStates*, and *explicitTargetStates* representing their corresponding concepts in the state machine (see the variables defined below the Δ symbol in the following Object-Z class *Transition*). For brevity, we omit details of the variables (refer to [10]).

A transition is enabled if and only if all of its source states are in the active state configuration of the state machine; the trigger of the transition is satisfied by the current event; if a guard exists, it should be true; and there exists at least one full path from the source state to the target state [15]. The variable *enabled* becomes true when the transition satisfies all the conditions for enabling transitions described above (see invariant [1] defined in the following Object-Z class *Transition*). Note that we do not discuss transitions without triggers in this paper (see [10] for this).

We also define several semantic variables: *enabled, effectCompleted, exitCompleted, entryCompleted* and *completed* of type Boolean. These variables become true when their corresponding actions on the transition are completed (see invariants [2], [3] and [4]). Finally, the variable *completed* becomes true when these three variables become true (see invariant [5]). We refer readers to [10] for entering various pseudo states: join and choice states.

Formalizing the operational semantics: We first define several operations: *Fire, ExitSource, ExecuteEffect,* and *EnterTarget* (see the operations defined in the following Object-Z class *Transition*). The operation *Fire* is invoked by the state machine. When a transition is selected to fire, all the exit actions of the explicit source states are executed by invoking the operation *ExitSource*. The operations *ExitSource* and *EnterTarget* invoke the *Exit* operation in the main source state and the *Enter* operation in the target state respectively. The rest of the exit execution step (or the entry execution) is defined by the *Exit* operation (or the *Enter* operation). The operation *ExecuteEffect* invokes the execution of all actions defined on the transition in sequence.

Transition

...

...
Δ
$leastCommonAncestor : \downarrow CompositeState$
$mainSourceState : \downarrow StateVertex$
$mainTargetState : \downarrow StateVertex$
$explicitSourceStates : \mathbb{P} \downarrow StateVertex$
$explictTargetStates : \mathbb{P} \downarrow StateVertex$
[Semantic variables]
$enabled : Boolean$
$exitCompleted : Boolean$
$effectCompleted : Boolean$
$entryCompleted : Boolean$
$completed : Boolean$

INIT
$\neg\ enabled \wedge$
$\neg\ completed \wedge$
$\neg\ exitCompleted \wedge$
$\neg\ effectCompleted \wedge$
$\neg\ entryCompleted \wedge$

...

[1] $enabled \iff \forall s: explicitSourceStates \bullet s.active \wedge$
    $\forall g: guard \bullet g.expression \wedge$
    $trigger \neq \emptyset \wedge trigger = \cup\{s: stateMachine \bullet s.currentEvent\}\wedge$
    $\exists_1 s: \downarrow CompositeState \bullet \{source, target\} \subseteq s.stateConfiguration$
[2] $effectCompleted \iff enabled \wedge \forall e: effect \bullet e.completed$
[3] $exitCompleted \iff enabled \wedge \forall x: explicitSourceStates \mid x \in \downarrow State \bullet x.exitCompleted$
[4] $entryCompleted \iff enabled \wedge \forall e: explicitTargetStates \mid e \in \downarrow State \bullet e.entryCompleted$
[5] $completed \iff exitCompleted \wedge effectCompleted \wedge entryCompleted$

$Fire \cong [enabled] \wedge ExitSource$
$ExitSource \cong [\neg\ completed] \wedge mainSourceState.Exit$
$ExecuteEffect \cong [\neg\ completed] \wedge (\ {}^\circ_9 e: effect \bullet e.execute)$
$EnterTarget \cong [\neg\ completed] \wedge target.Enter$

Finally, we formalize the execution sequence of a transition using these operations in invariants [1] and [2] in the timed trace part.

- The invariant [1] ensures that when the exit execution is completed, the *ExecuteEffect* operation should occur, which results in executing actions on the transition.
- The invariant [2] ensures that when all the effect actions are completed, the operation *EnterTarget* should occur, which results in executing all the entry actions.

```
┌─Transition─────────────────────────────────────────────────────────
│ . . .
│ ┌──────────────────────────────────────────────
│ │
│ [1] ⟨¬ exitCompleted⟩;⟨exitCompleted⟩ ⊆
│                    ⟨¬ ExecuteEffect⟩;(⟨ExecuteEffect⟩ ∪ ⟨ExecuteEffect⟩;⟨true⟩)
│ [2] ⟨¬ effectCompleted⟩;⟨effectCompleted⟩ ⊆
│                    ⟨¬ EnterTarget⟩;(⟨EnterTarget⟩ ∪ ⟨EnterTarget⟩;⟨true⟩)
└─────────────────────────────────────────────────────────────────────
```

4.6 StateMachine

In the UML metamodel, the semantics of state machines are described in terms of the run-to-completion assumption and transition firing mechanisms.

Run-to-completion assumption: Under this assumption, an event can only be dequeued and dispatched for processing when the processing of the previous current event is fully completed [15]. When all transitions fired by the current event complete their steps as described in Section 4.4, the current event is fully *consumed* and the run-to-completion step is *completed* (the state machine is in a stable configuration).

Maximal set transitions: UML provides transition selection rules that exclude all conflicting transitions from the set of transitions that may fire. This set is called the *maximal set transitions* [15]. Transitions included in the maximal set transitions should satisfy the following properties: all transitions are enabled; there are no conflicting transitions in the set; and there is no transition outside of the set that has higher priority than a transition in the set. The firing priority is that a transition originating from a substate has higher priority than a conflicting transition originating from any of its containing states [15].

Formalizing the denotational semantics: Prior to formalizing the semantics, we define two secondary variables that formalize the notions introduced above: *maximalSetTrans* representing all transitions that will fire and *activeStateConfiguration* representing all active states in the state configuration (see [10] for details). We also define semantic variables *currentEvent* of type Event representing the current event in the state machine and *run-to-completion* of type Boolean representing the status of the run-to-completion step. Since a state machine dispatches events and processes the dispatched events [15], we define an abstract Object-Z class *EventQueue* without defining details (because UML does not give any semantics with event queue and the event dispatcher mechanism). The *EventQueue* class has abstract operations *dequeue* and *enqueue* handling events, and *empty* denoting whether or not the queue is empty. Finally, we define two event queues: *eventQueue* and *completionEventQueue* representing an event queue for external events and an event queue for completion events respectively. Invariants [1] and [2] below formalize the run-to-completion assumption.

---StateMachine---

```
                                          ┌─INIT─────────────────────
...                                       │ ¬ run–to–completion ∧
Δ                                         │ currentEvent.INIT
maximalSetTrans : ℙ Transition            └──────────────────────────
activeStateConfiguration: ℙ ↓StateVertex
...
[Semantic variables]
currentEvent : ↓Event
run–to–completion : Boolean
eventQueue : EventQueue
completionEventQueue : EventQueue
```

[1] currentEvent.consumed ⟺
                 ∀ t: maximalSetTrans • t.completed
[2] currentEvent.consumed ⟺ run–to–completion

Formalizing the operational semantics: The operational semantics of the state machine are formalized in terms of operations and invariants in the timed trace part:

[1] When the run-to-completion step is completed, the state machine dispatches a new event by invoking the operation *dispatchEvent* which in turn invokes the operation *dequeue* defined in the event queue. If the *completionEventQueue* is not empty (meaning that there exists a completion event generated), the operation *dispatchEvent* dispatches an event from the *completionEventQueue*. Otherwise, the operation dispatches a new event from the *eventQueue*. We treat completion transitions in the same way as other transitions.

[2] When a new event is delivered, the state machine processes the event by invoking the operation *fireTransitions*, which fires all states in the maximal set transitions by invoking the operation *Fire* defined in the individual transitions.

[3] When a state in the active state configuration is completed, a completion event is generated and enqueued into the *completionEventQueue* by invoking the operation *genCompletionEvent*.

---StateMachine---

```
...
dispatchEvent ≙ [ run–to–completion ] ∧
    ( [ ¬ completionEventQueue.empty] ∧ completionEventQueue.dequeue ▯
    [ completionEventQueue.empty] ∧ eventQueue.dequeue ) ⨟
                        [ Δ (currentEvent) e? : ↓Event | currentEvent' = e?]
fireTransitions ≙ ∧ t: maximalSetTrans • t.Fire
genCompletionEvent ≙ [comEvent !: ↓Event] ⨟ completionEventQueue.enqueue
```

[1] ⟨¬ run–to–completion⟩;⟨run–to–completion⟩ ⊆
        ⟨¬ dispatchEvent⟩;((⟨dispatchEvent⟩ ∪ ⟨dispatchEvent⟩;⟨true⟩)
[2] ∀ e:↓Event • ⟨currentEvent = e⟩;⟨currentEvent ≠ e⟩ ⊆
        ⟨¬ fireTransitions⟩;((⟨fireTransitions⟩ ∪ ⟨fireTransitions⟩;⟨true⟩)
[3] ∀ s: activeStateConfiguration • ⟨ ¬ s.completed ⟩;⟨ s.completed⟩ ⊆
        ⟨¬ genCompletionEvent ⟩;((⟨genCompletionEvent⟩ ∪ ⟨genCompletionEvent⟩;⟨true⟩)

5. Integrity Consistency Constraints

Unfortunately, the current UML metamodel does not define inter-model integrity consistency constraints precisely. In this paper, we demonstrate how integrity consistency constraints can be precisely defined similar to the way the static semantics is defined. When the context of the state machine is a classifier, for example,

- for all call or send events associated with the classifier, there should be transitions describing the detailed behavior of the events (see invariants [1] and [2]);
- for all action for which the sender instance and receiver instance are different, there should be a (navigable) communication link between the two instances (see invariant [3]);
- for all call actions, there should be operations of the classifiers corresponding to the call actions (see invariant [4]);
- for all send actions, there should be receptions of the classifiers corresponding to the send actions and also the subsequent effects should be described by the state machines that describe the behaviors of the classifiers (see invariant [5]);
- any invariants defined in the classifier should be preserved in the corresponding state machine of the classifier (see invariant [6]);
- if the guard condition attached to a transition (which invokes an operation) conflicts with the pre-condition of the operation, the operation cannot be performed resulting that the expected behavior of the transition cannot be achieved (see invariant [7]). In this case, the state machine is logically ill-formed. Consequently, this situation should be avoided (see [10] for the Object-Z class *Constraint*).

$$
\begin{array}{l}
\text{—StateMachine———————————————————————} \\
\quad \cdots \\
\hline
\quad context : \mathbb{P} \downarrow ModelElement \\
\quad transition : \mathbb{P}\ Transition\ \copyright \\
\quad \cdots \\
\hline
\quad \cdots \\
\quad [\textit{Integrity consistency constraints}\,] \\
\quad context \subseteq Classifeir \Rightarrow \\
\quad [1]\forall\ callEvent : CallEvent \mid context \in \cup \{o: callEvent.operation \bullet o.owner\} \Longleftrightarrow \\
\qquad\qquad \exists\ t: transition \bullet callEvent \in t.trigger \\
\quad [2]\forall\ sendEvent : SendEvent \mid context \in \cup \{r: sendEvent.reception \bullet r.owner\} \Longleftrightarrow \\
\qquad\qquad \exists\ t: transition \bullet sendEvent \in t.trigger \\
\quad [3]\forall\ a : \cup \{t: transition \bullet t.effect\ \} \bullet \forall\ s{:}a.stimulus \bullet \\
\qquad s.sender \neq s.receiver \Rightarrow (\exists\ l: s.communicationLink \bullet \exists\ le1, le2 : ran(l.connection) \bullet \\
\qquad le1.instance = s.sender \wedge le2.instance = s.receiver \wedge le2.associationEnd.isNavigable = true) \\
\quad [4]\quad a \in CallAction \Rightarrow \exists\ o: \cup\{c: s.receiver.classifier \bullet c.operation\} \bullet o = a.operation \\
\quad [5]\quad a \in SendAction \Rightarrow (\exists\ r: \{f: \cup\{c: s.receiver.classifier \bullet c.feature\} \mid f \in Reception\} \bullet \\
\qquad\qquad a.signal = r.signal) \wedge (\exists\ s: \cup\{c: s.receiver.classifier \bullet c.behavior\} \bullet \\
\qquad\qquad\qquad \exists\ t: s.transition \bullet a.signal.occurrence \in t.trigger) \\
\quad [6]\forall\ c: \cup \{c: context \bullet c.constraint\} \bullet c.stereotype = invariant \Rightarrow c.body.evaluate = true \\
\quad [7]\forall\ t: transition \bullet \forall\ g: \cup \{g: t.guard\} \bullet \\
\qquad g.expression.body.evaluate = true \Rightarrow \forall\ a : t.effect \mid a \in CallAction \bullet \\
\qquad\qquad \forall\ c : a.operation.constraint \mid c.stereotype = precondition \bullet c.body..evaluate = true
\end{array}
$$

Although we discuss the integrity consistency constraints from the state machine's point of view in this paper, the constraints can be defined from individual model elements' point of view. Once all consistency constraints are defined in a similar way, an integrity consistency check between different UML models can be achieved in a modular way by verifying whether or not the individual model elements composing the models conform to all the integrity consistency constraints defined in their corresponding metaclasses.

6. Conclusion and Future Work

In this paper, we have presented a formal denotational and operational semantics for the UML state machine. The abstract syntax and the static and dynamic semantics of each model construct are formalized within a single Object-Z class. In particular, the dynamic semantics of the state machine are first defined in a denotational way by introducing semantic variables and invariants ignoring the operation sequences. The operational semantics of the state machine is defined in terms of Object-Z class operations and invariants defined using the timed refinement calculus. With this approach, we provide a complete formal semantics of the UML state machine that combines the denotational and operational semantics in a fully compositional manner by composing the semantics of its components. Our work is also significant because we deal with the UML state machine as part of the entire UML metamodel, so that the formal model given for the state machine can be interpreted in the context of other model elements in the metamodel. This provides a consistent semantic basis, which can be used for checking inconsistencies between different UML models and for checking the completeness of the UML models.

Furthermore, the Object-Z expressions formalizing the static semantics introduced in this paper are designed to be as descriptive as those in OCL. The result demonstrates that using a well-defined object-oriented formal specification technique like Object-Z for defining the UML metamodel neither increases the complexity nor decreases the readability. Instead, the approach enables the current UML metamodel to be viewed in an alternative way which is unambiguous, but still descriptive.

References

[1] T. Clark, A. Evans, R. France, S. Kent, and B. Rumpe, Response to UML 2.0 Request for Information, 1999. available at
http://www.cs.york.ac.uk/puml/papers/RFIResponse.PDF

[2] B. P. Douglass, *Real -Time UML: Developing Efficient Objects for embedded systems*, Addison-Wesley, 1998.

[3] R. Duke and R. Gordon, *Formal Object-Oriented Specification Using Object-Z*, Macmillan, 2000.

[4] Evans and S. Kent, Core meta modelling semantics of UML: The pUML approach, *Proc. UML'99*, LNCS. No. 1723, pp. 140-155, 1999.

[5]   R. France, A. Evans, K. Lano, and B. Rumpe, Developing the UML as a Formal Modeling Notation, *Computer Standards and Interfaces*, No. 19, pp. 325-334, 1998.

[6]   D. Harel and A. Naamad, The STATEMATE semantics of Statecharts, *ACM Transaction on Software Engineering*, vol. 5, pp 293-333, 1996.

[7]   Henderson-Sellers and F. Barbier, Black and White Diamonds, *Proc UML'99*, LNCS, No. 1723, pp. 550-565, 1999.

[8]   S-K. Kim and D. Carrington, Formalizing the UML class diagram using Object-Z, *Proc. UML'99*, LNCS, No. 1723, pp. 83-98, 1999.

[9]   S-K. Kim and D. Carrington, A Formal Mapping between UML Models and Object-Z Specifications, *ZB2000*, LNCS, No. 1878, pp. 2-21, 2000.

[10]   S-K. Kim and D. Carrington, UML Metamodel Formalization with Object-Z: the State Machine package, *SVRC technical Report 00-29*, The University of Queensland, 2000.

[11]   S-K. Kim and D. Carrington, A Formal Denotational Semantics of UML in Object-Z, *the special issue of the journal of l'Objet*, Vol. 7(1), 2001.

[12]   G. Kwon, Rewrite Rules and Operational Semantics for Model Checking UML statecharts, *Proc. UML'2000*, LNCS, No. 1939, pp. 528-540, 2000.

[13]   Latella, I. Majzik, and M. Massink, Towards a formal operational semantics of UML statechart diagrams, *In 3rd International Conference on Formal Methods for Open Object-Oriented Distributed Systems(FMOODS)*, Kluwer, 1999.

[14]   J. Lilius and I. P. Paltor, Formalizing UML state machines for model checking, *Proc. UML'99*, LNCS, No. 1723, pp. 430-445, 1999.

[15]   OMG, *Unified Modeling Language Specification*, version 1.3, 1999, http://www.omg.org

[16]   OMG, Response to OMG RFP ad/98-11-01: Action Semantics for the UML, 2000. Available at ftp://ftp.omg.org/pub/docs/ad/00-08-03.pdf

[17]   G. Reggio, E. Astesiano, C. Choppy, and H. Hussmann, Analysing UML Active Classes and Associated State Machines-Lightweight Formal Approach, *Proc. FASE 2000*, LNCS. No.1783, pp. 127-146. Springer Verlag, 2000.

[18]   M. Richters and M. Gogolla, On formalizing the UML Object Constraint Language OCL, *Proc. 17th Int. Conf. Conceptual Modeling*, LNCS. No. 1507, pp. 449-464, 1998.

[19]   G. Smith and I. Hayes, Structuring Real-Time Object-Z Specifications, *IFM'2000*, LNCS, No. 1945, pp. 97-115, 2000.

[20]   G. Smith. *The Object-Z Specification Language. Advances in Formal Methods.* Kluwer Academic Publishers, 2000.

Coming and Going from UML to B: A Proposal to Support Traceability in Rigorous IS Development

Régine Laleau[1] and Fiona Polack[2]

[1] CEDRIC-IIE(CNAM),
18 allée Jean Rostand, 91025 Evry, France
laleau@iie.cnam.fr
[2] Department of Computer Science
University of York, York, YO10 5DD, UK
fiona@cs.york.ac.uk

Abstract. CEDRIC-IIE is researching rigorous information system (IS) development. Previous work includes translation rules for deriving a B specification from object-oriented diagrams, metamodels of IS UML structural and functional concepts, and a prototype translation tool. Here we outline the traceability needs for a tool to assist in rigorous IS development, and provide meta-structures for the required links among B and IS UML concepts, in the context of existing translation rules and IS UML metamodels.

1   Introduction

There is currently an overdue realisation that computer systems development in general must conform to good engineering practice: it is not just critical systems that must be optimally robust and reliable, and all computer systems should be developed within cost and time constraints. Computer systems development must become a verifiable industrial process not an art form.

Increasingly, information systems (IS) are at the centre of commercially-critical systems, and their development needs more rigour than has typically been applied. Light-weight or back-door formalism is a possible solution. In our work, we seek to extend diagrammatic tool support to provide rigour via a partly-invisible and largely machine-generated formalisation. The work derives from the extensive but under-exploited integrated methods work of the 1990s[26, 13], but exploits the increased use of and reliance on tool support for modelling, the increasing sophistication of formal methods support, and the increasing capacity of development support computers to manage the development process. The CEDRIC-IIE work has already published a guideline-based approach for deriving a formal specification of an IS (in B) from object-oriented class and behavioural models[22,21,10,11]. This allows the formal investigation (proof, etc) of the IS specification, but cannot guarantee either the correctness or the equivalence of the object-oriented models. To permit fully-rigorous IS specification,

D. Bert et al. (Eds.): ZB 2002, LNCS 2272, pp. 517–534, 2002.

the modelling notations, as well as the existing translation rules, are being rigorously defined for an IS-focused variant of UML. Ultimately, this will allow a support tool to comment not only on the syntactic correctness of each graphical model, but also on the semantic validity of the models, including cross-references between models. (Other research in this area is summarised in [17].)

In this paper, we explore issues of traceability between the UML and B specifications. We believe that tools for rigorous development must allow the designer to move between the UML and formal models in both directions. This helps to improve readability and allows evolution of the specification, for example as validation (including proofs of properties of the formal model) identifies errors in UML and/or formal models. To our knowledge, the existing tools deriving B specifications from UML specifications [20] do not support mechanisms that allow navigation in both direction.

2   Context

UML is widely used in commercial software engineering. However, the language is still evolving. One problem for a user of UML is to determine the subset of notations which is relevant to their particular project or level of abstraction. The Object Management Group (OMG), which oversees the development of the UML, has recognised the need to allow specialisation of the language for different types of system. The mechanisms of extension are not static or agreed[1]. For the IS UML, we follow the general approach of the pUML group (eg[4,5]), defining a small core language and extensions. Conventional work on UML semantics uses UML's diagrammatic notations and OCL constraints. Although OCL is intended for the expression of object constraints, it is not a mature language, and does not have a formally-defined semantics. A significant drawback of using OCL in language definition is the need to derive a definition of the language of expression at the same time as using it to express the syntax and semantics of the modelling concepts. A practical problem in the use of OCL is the current lack of commercial tools for checking OCL expressions. OCL provides more rigour than a natural language, but less rigour than most mature formal languages. The use of diagrams to specify system concepts and OCL for invariants makes the specifications hard to check or prove.

In our research, the full IS UML metamodel is written in B; UML class diagrams summarise the abstract syntax. B has a well-defined semantics that is used to give meaning to IS UML concepts. B provides the power of expression required to specify the extensive structural aspects of IS, and the functional specification is built directly on the structural specification. (It would be difficult to express dynamic properties such as liveness but this is not of concern here.) B also provides relational operators and types such as sets and sequences that facilitate precise, readable expression of data integrity constraints, and a well-

---

[1] see OMG's Requests for Proposals (RFPs),
   http://www.celigent.com/omg/adptf/wgs/uml2wg.htm

defined structuring mechanism, using an architecture of abstract machines. We have used the B proof facilities in IS specification[11].

We have already defined most of the structural aspects of UML required for IS specification[9,24], including a structure for IS constraints, which supports the construction of operations with complete, consistent preconditions. Because an IS manages large amounts of highly structured data, its specification depends on a well-defined structural model, with a large number of data constraints defining the integrity conditions of the IS. The structural aspect is an important determinant of the architecture of the final implementation. The structural metamodels are summarised in section 5.1.

More recently, we have established an abstract syntax and semantics for the IS functional aspect, comprising transactions that correspond to user requirements and enforce the integrity constraints in IS operations[17]. This is the principle component of the IS behavioural aspect. (In IS, dynamic features such as liveness and concurrency management are largely determined by the software used to implement the IS eg, database management system.) The operational aspect is summarised in sections 5.2 and 5.3.

The main contribution of this paper (section 6) is to model traceability links between IS UML and B concepts summarised in section 5. To set the context, we first describe the architecture of a prototype tool, indicating the traceability needs (section 3). The process of deriving a B specification from a UML class model is illustrated, and issues in the development of a tool assistant are noted (section 4).

3   A Tool to Support Rigorous IS UML Development

Conventional diagrammatic case tools can check specifications for syntactic precision, based on a built-in metamodel of syntax. However, it is hard to check diagram semantics, and to check that the components specified in one diagrammatic model are consistent with related components in other models. For example, a class model specifies the operations on each class, whereas the state model defines the events which cause a class to trigger an operation on a particular object. The preconditions of the operation bear a relationship to the guard conditions on the transition triggered by the event. We aim to provide these checks via an underlying formal metamodel and a formalisation of the specification which can be analysed by existing formal method tools.

Figures 1 and 2 outline the architecture of a tool to assist in rigorous IS specification, using a B specification derived from IS UML models. In the diagrams, solid arrows represent the current state of a prototype tool produced at CEDRIC-IIE [16,9]. The dashed lines show the needed traceability links which are the main contribution of this paper. Traceability between UML and B models (and vice versa) is crucial for rigorous tool support. Without bidirectional traceability, errors can be identified using the B method, but cannot be automatically associated with omissions or errors in specific features of the UML models.

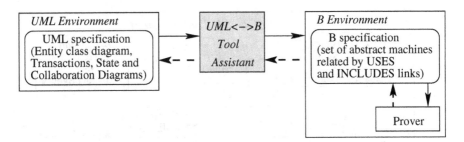

Fig. 1. Architecture of the Current Translation Tool

Figure 1 shows the context of the tool assistant. CEDRIC-IIE's prototype tool takes machine-readable UML models from the Rose case tool[6], and passes B machine structures to AtelierB[7], which performs formal analysis.

Figure 2 expands the architecture of the tool assistant. The machine-readable UML representation is verified against the metamodel definition of UML abstract syntax; it is then translated into B machines by the correspondence rules. The CEDRIC-IIE prototype performs these translations and verification using an OCaml implementation of the metamodel and rules. Figure 2 shows, as dashed arrows and boxes, the traceability mechanisms that are needed for the complete analysis of errors detected in translation, verification and AtelierB formal analysis. Thus, for example, if AtelierB finds an unprovable assertion, the traceability links are used to determine which parts of the UML model give rise to inadequate constraints or operations preconditions.

4   Deriving a B Model from a UML Class Model

This section illustrates the process of rigorous IS specification. Note that interaction between the system and the users would be modelled as operation calls (events), the signature of which are specified. The realisation of the user interface would be considered in design and implementation phases. The process is that automated in the CEDRIC-IIE prototype. The structure of the B specification that we produce mirrors the UML specification, to assist traceability. The process has five steps.

1. The IS UML entity class diagram is constructed. The diagram expresses the structure of data that is persistent over time. In IS, the class model is crucial, as it provides the basis of the final implementation structure.
2. The B specification corresponding to the diagram is derived. Primitive operations can be automatically generated once the B specification of the state has been defined. These low-level operations, generic to any IS, comprise update, insert, and delete operations for each class and association, changing the state of the object(s) and link(s). The operations are added to the IS

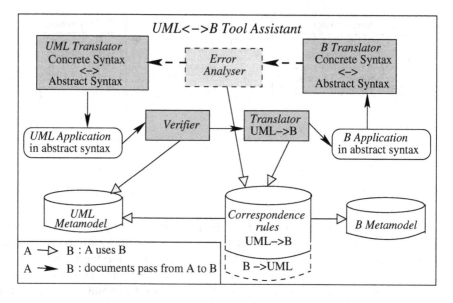

Fig. 2. Architecture of the Current Translation Tool

UML entity class diagram. The B specification is presented in an architecture of basic B machines that contain all the state variables of the system with the relevant primitive operations.

3. The user requirements and system business rules are analysed to identify the signatures of input/output events. In IS UML, the event effects can be specified in UML state and collaboration diagrams or can be written directly in B, as a transaction corresponding to the event. The latter approach is similar to Catalysis actions[8] or Meyer's contracts [19], in that the transaction must enforce all system integrity constraints.

4. The B machine architecture is completed to express the whole specification. New machines are defined to provide necessary state and operation structures. These machines have no variables, and must call generated operations from included basic machines in order to update the system state. A top-level B machine forms the interface of the B system specification. It contains one B operation for each event.

5. The properties of the B specification can be proved using a B tool. In order to "validate" the specification animators can also be used. If the proof or animation identifies problems with the specification, the B and UML specifications need to be modified accordingly. This is where traceability is a key issue.

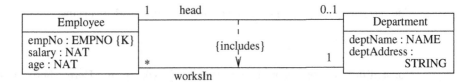

Fig. 3. Example of an IS UML Entity Class Diagram

4.1   Illustration of UML-to-B Translation on an Example

The first four steps of the translation process are demonstrated for a simple system of employees and departments. Steps 2 to 4 employ our existing translation rules[22,21,10], as implemented by the prototype tool assistant.

Step 1: Constructing the entity class diagram. Figure 3 shows the entity class diagram for the example. There are two associations between the *Employee* and *Department* classes. Every employee *worksIn* exactly one department, and, in each department, exactly one employee is designated as the *head* of the department. The links of the latter are included in the links of the former association. The keyword {*K*} against the *Employee* attribute *empNo* indicates that this is a key (unique identifier) of employees.

Step 2: The initial B representation. The B model represents an entity class as a B given set. A variable represents the set of instances of the class via a typing invariant. In the example, thus, the *Employee* class is represented by the given set, *EMPLOYEE*, the variable, *Employee*, and the typing invariant,

$Employee \subseteq EMPLOYEE$

The B representations of attributes and associations are isomorphic. Each is modelled by a B variable and a typing invariant. For the attributes, we use B given sets *EMPNO*, *NAME* and *STRING*, and the B type *NAT*. The invariant links the variable representing the attribute to its class. For *Employee*, this gives:

$empNo \in Employee \rightarrowtail EMPNO \;\wedge$
$salary \in Employee \rightarrow NAT \;\wedge$
$age \in Employee \rightarrow NAT$

For associations, the B model captures the whole association semantics. A B variable represents the named association. A typing invariant models the association end classes and their multiplicity:

$worksIn \in Employee \rightarrow Department \;\wedge$
$head \in Department \rightarrowtail Employee$

The inter-association constraint is represented in a B invariant,

$$head \subseteq worksIn$$

Constraints are represented in B invariant expressions. Here, there are requirements that all employees are aged at least 18, that employees under 30 years old earn under 30000 euros per annum, and that the salary of the head of a department is at least as great as that of any employee of the department:

$$C1 : \forall e \in Employee \bullet age(e) \geq 18 \wedge$$
$$C2 : \forall e \in Employee \bullet age(e) < 30 \Rightarrow salary(e) < 30000 \ euros \wedge$$
$$C3 : \forall e \in Employee \bullet salary(e) \leq salary(head(worksIn(e)))$$

There is one generated operation each for inserting and deleting an object (link) of each of class and association, and for updating each mandatory attribute of each class. The detail of each operations depends on the definition of keys, the integrity constraints, the multiplicity of the associations, and the machine in which an association is defined (for more details see [22]). Three operations are given here:

$UpdateSalary(emp, sal)$
  PRE
    $emp \in Employee \wedge sal \in Nat$
    $sal \in Nat$
  $THEN$
    $salary(emp) := sal$
  END

$InsertWorksIn(e, dep)$
  PRE
    $e \in EMPLOYEE - Employee \wedge$
    $dep \in DEPARTEMENT$
  $THEN$
    $WorksIn := WorksIn \cup \{e-> dep\}$
  END

$InsertEmployee(e, no_t, sal_t, age_t) \ PRE$
  $e \in EMPLOYEE - Employee \wedge no_t \in EMPNO - ran(empNo) \wedge$
  $sal_t \in NAT \wedge age_t \in NAT \wedge age_t \geq 18 \wedge$
  $THEN$
    $Employee := Employee \cup \{e\} \ || \ empNo := empNo \cup \{e-> no_t\} \ ||$
    $salary := salary \cup \{e-> sal_t\} \ || \ age := age \cup \{e-> age_t\}$
  END

The B specification of each entity class is presented in a B machine. A B machine is also created for any association that is subjected to independent operations (ie operations on links between existing objects or on the attributes of the association). **USES** links are defined from the association machine to the machines of the associated entity classes. If the association has no independent operations, it is defined in one of the entity class machine, according to the multiplicity, with a **USES** link to the other machine[9,22]. Generated operations are included in the relevant basic machine. Thus, *UpdateSalary()* and *InsertEmployee()* are defined in the machine *B_Employee*; *InsertWorksIn()* is defined in the machine *B_WorksIn*. The machine structure comprises the upper four machines in figure 4, below.

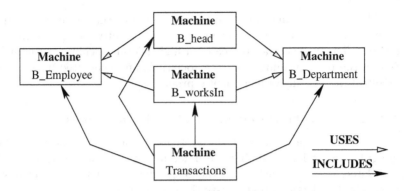

Fig. 4. Architecture of the final B specification

Step 3: Expressing input/output events. In the example, the users require a transaction to insert an employee object and link the employee to a department. Here, we express this directly in B syntax:

$newEmp < -AddNewEmployee(no_t, sal_t, age_t, dep_t)$
PRE
　　$no_t \in EMPNO \; - \; ran(empNo) \wedge sal_t \in NAT \wedge age_t \in NAT \wedge$
　　$age_t \; \geq \; 18 \wedge (age_t < 30 \Rightarrow sal_t < 3000) \wedge sal_t \; < \; salary(head(dep_t)) \wedge$
　　$dep_t \in Department \wedge \; Employee \subset \; EMPLOYEE$
THEN
　ANY e **WHERE** $e \in EMPLOYEE - Employee$ **THEN**
　　$InsertEmployee(e, no_t, sal_t, age_t)$
　　$|| \; InsertWorksIn(e, dep_t) \; || \; newEmp := e$
　END
END

Transactions such as *AddNewEmployee* are defined in a top-level *Transactions* B machine (figure 4). In general, it is possible to have an intermediate level of IS operation, between generated operations and transactions, *User Operations*, that define a context of use for generated operations. In essence, user operations differ from transactions only in their restriction to a single class or operation. A user operation can be used to specify additional preconditions dependent on the application, and to enforce the integrity constraints related to the class or association.

Step 4: Composing the complete B specification. The definitions from steps 2 and 3 give rise to the B architecture in figure 4. The B machine architecture is similar to the structure of the UML class model. The **USES** structures mirror the classes and associations; the **INCLUDES** structure builds the top-level transaction machine on the whole state.

The levels of B machine facilitate evolution of the specification. In IS, user requirements and associated business rules frequently emerge during the development. The changes can be made in higher-level machines, without modification of the base machine classes, associations and generated operations. The B architecture also allows incremental proof.

4.2   Issues in the Development of Assistance Tools

One of the characteristics of the systems we specify is the great number of integrity constraints that link the different entities and associations. They are translated as state invariants in the B specification. Thus it is not obvious how to specify transactions that satisfy the constraints.

Once the state invariant is complete, two approaches can offer assistance in the development of consistent operations[24,17]. Pro-active tool assistance would identify the parts of the invariant that are relevant to the operation, and then automatically generate preconditions and substitutions whenever possible, as in [18]. Subsequently, the specification would be presented to a B proof tool, which attempts to discharge the proof obligation. Assuming that the state invariant is sufficiently defined, the proof will fail if either the precondition or the substitution of an operation is insufficient. Error messages from the proof tool point to missing preconditions and/or substitutions. However, at present, these messages are often hard to understand. To be of value to a designer, the messages returned by the proof tool should be interpreted in terms of the UML models. For example, the omission of the call to *InsertWorksIn()* from the specification of the *AddNewEmployee()* transaction would produce a B prover error relating to the failure to prove the totality of the B function, *worksIn*. This needs converting into a comment on the failure to satisfy the multiplicity of the *Employee* association end of the *worksIn* association. This kind of assistance is called a reactive approach.

At present, for each UML construct, it is possible to identify all the corresponding B elements. To provide assistance as described below, we need to complete traceability between the textual (B) and the graphical (UML) specifications. As in the work of Freire[14], we use metamodels to unify the different views of the specification, as represented in the metamodels of the UML and B concepts.

5   Core IS UML Metamodels

Before defining the links between B and UML concepts, we summarise our IS UML metamodels[9,24,17]. These define the abstract syntax and semantics of core IS UML concepts, giving a precise definition of the relevant UML graphical notations. The metamodels are formally specified in B, and are used in the prototype tool.

In this section, graphical summaries (UML class model notations[25]) are presented, with the B representations for just a subset of the metamodels relevant to operation specification. The formal B metamodel is derived from the

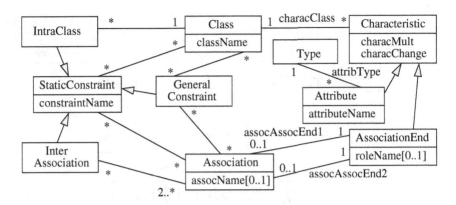

Fig. 5. IS UML Metamodel for Structural Concepts, from [9,24]

diagrammatic representation using our translation rules, as in the illustrative translation, section 4, above. As above, the B expresses additional constraints on the structure.

5.1 Metamodels for the Structural Aspect

Figure 5 summarises the abstract syntax of the main structural concepts of IS UML. *Class*es have many *Characteristics*, which are either *Attribute*s or *AssociationEnd*s. A *Characteristic* has attributes representing its multiplicity and whether its value can change; an *Attribute* also has a name, and is linked to an IS UML *Type*. An *Association* is linked to two distinct *AssociationEnd*s, each of which has a role name. *Class*es and *Association*s have many associated constraints (*IntraClass* and *InterAssociation* subclasses of *StaticConstraint*).

In [24], we define a hierarchical metamodel for constraint concepts, categorised by the parts of the structure on which they act. *IntraClass* constraints are the most common, and include *InterObject* constraints that relate the values of several objects of a class (or links of an association), and *IntraObject* constraints that relate to attributes within an object of a class (or a link of an association). *InterAssociation* constraints are less common. They capture subsetting and exclusion constraints on associations and association ends. *GeneralConstraints* represents business rules and other constraints requirements, and relate more than one system class and/or association.

5.2 Metamodels for the Functional Aspect

The metamodel of IS UML operations is summarised here; metamodels for state and collaboration diagrams[9] are omitted.

The hierarchy of operations (figure 6) inherits attributes from the superclass, *Operation*. These attributes together define the signature of an operation, ie its structural aspect.

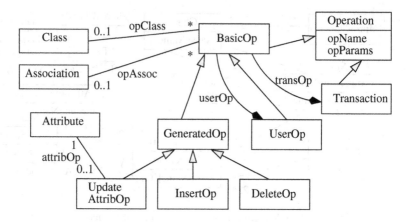

Fig. 6. IS UML Metamodel for Structural of Operations, from [17]

BasicOp and *Transaction* are subclasses of *Operation*. A *BasicOp* represents operations on one class or association, whereas a transaction, composed of basic operations, acts on many classes and/or associations. *BasicOp* is further specialised into *GeneratedOp* and *UserOp*. A *UserOp* is both a subclass of *BasicOp* and a composition of basic operations on a single class or association. *GeneratedOp*s are the generic, low-level operations of any IS, described above. The three subclasses, insert, delete, and attribute update operations, represent the only B operations that change the state; they cannot be modified by the designer.

5.3   B Representation of Abstract Syntax of Operation Concepts

The abstract syntax of the IS UML notations (figure 6) is represented in B according to step 2 of the process defined above.

The B metamodel requires a number of given sets, written in upper case. The given set, *TYPE*, is used to represent any relevant UML- and user-defined data types. The B given set, *NAME*, represents any class name in the metamodel.

Superclasses are represented in the same way as classes in section 4. Subclasses have the same type as their superclass; the instances are a subset of the superclass instances. The typing invariants for all these are given in table 1. An invariant is added to express the disjointness of subclasses. For example,

$$BasicOperation \cup Transaction = Operation \wedge$$
$$BasicOperation \cap Transaction = \varnothing$$

The *Operation* attributes are represented by total functions. These capture the fact that *opName* and *opParams* are both mandatory. Each could be associated with more than one operation, and a number of operations could have the same parameters:

Table 1. B Variables and Invariants for Sets of Known Instances of Operations

$Operation$	$\subseteq$	$OPERATION$	$GeneratedOp$	$\subseteq$	$BasicOp$
$Class$	$\subseteq$	$CLASS$	$UserOp$	$\subseteq$	$BasicOp$
$Association$	$\subseteq$	$ASSOCIATION$	$DeleteOp$	$\subseteq$	$GeneratedOp$
$BasicOp$	$\subseteq$	$Operation$	$InsertOp$	$\subseteq$	$GeneratedOp$
$Transaction$	$\subseteq$	$Operation$	$UpdateAttribOp$	$\subseteq$	$GeneratedOp$

Table 2. B Variables and their Typing Invariants Representing Associations in the IS Metamodel

$opClass$	$\in$	$BasicOp \nrightarrow Class$
$opAssociation$	$\in$	$BasicOp \nrightarrow Association$
$attribOp$	$\in$	$UpdateAttribOp \rightarrowtail Attribute$
$transOp$	$\in$	$Transaction \leftrightarrow BasicOp \wedge dom(transOp) = Transaction$
$userOp$	$\in$	$UserOp \leftrightarrow BasicOperation \wedge dom(userOp) = UserOp$

$$opName \in Operation \rightarrow NAME \wedge$$
$$opParams \in Operation \rightarrow seq(TPARAM)$$

The type of $opParams$, $seq(TPARAM)$, models an ordered list of parameters. $TPARAM$ is a triple of a name, a type, and an indicator of whether the parameter is input (i) or output (o):

$$TPARAM \in (NAME \times TYPE \times \{i, o\})$$

Metamodel associations are expressed as relations or functions, table 2. Composition is not well defined in UML. Here, we simply translate the two compositions ($transOp$ and $userOp$) as if they were simple associations. Each is a relation, whose domain is constrained to be all the known instances of the class ($Transaction$, $UserOp$).

Three of the additional B invariants are illustrated here.

A $BasicOp$ (and thus any of its subclasses) must reference only one class or one association. The partial functions in the B representation of the associations require that a $BasicOp$ can be associated with at most one class or association. To ensure that it is associated with exactly one, the associations are defined to be disjoint and total with respect to $BasicOp$:

$$dom(opClass) \cap dom(opAssoc) = \varnothing \wedge$$
$$dom(opClass) \cup dom(opAssoc) = BasicOp$$

The uniqueness of operation names within each class (or association) is an example of the internal and mutual consistency properties of the metamodel. The invariant states that, for any two operations of a particular class, if the operations have the same name then they are the same operation:

$$\forall\, c \,\in\, ran(opClass) \,\bullet$$
$$\forall\, o_1, o_2 \,\in\, opClass^{-1}[\{c\}] \,\bullet$$
$$opName(o_1) = opName(o_2) \Rightarrow o_1 = o_2$$

The third example requires that, for *UpdateAttribOp*s, the attribute updated must be of the class acted on by the superclass *BasicOp*:

$$\forall\, op \,\in\, UpdateAttribOp \,\bullet$$
$$attribOp(op) \in allClassAttributes(opClass(op))$$

where *allClassAttributes()* returns attributes of a class and its superclasses.

6   Modelling Links between IS UML and B

To navigate from UML diagrams to B specifications and *vice versa*, we need to formally establish the correspondence between B concepts and the UML meta-model. The correspondence metamodels do not represent a complete metamodel of B concepts - B is fully defined in [1].

6.1   Modelling with B Machines

The practicalities of building an B specification from IS UML models led us to develop guidelines for B machine structures [12,23], summarised in figure 7.

The instantiation of links from *BasicMachine* and its subtypes to the IS UML *Class*, *Association* and *GeneratedOp* are established during step 2 of the formalisation process. A *GeneratedOp* is always defined in a lowest-level basic machine, but cannot be accessed directly at this level.

The remaining links are established in step 4. All *BasicMachine*s are included in either *IntermediateMachine*s or *TransactionMachine*s. The *TransactionMachine* develops transactions required by users by combining (in parallel substitutions) the relevant operations defined on classes or associations.

In our operation metamodel, a *UserOp* is a combination of basic operations on one class or association. Figure 7 shows the ideal meta-structure, with *UserOp* represented in an B *IntermediateMachine* that **INCLUDES** the relevant basic machines. However, this is inhibited by the B rule that each operation of such a machine calls at most one operation of the included machine (because the parallel application of operations from an included machine may violate the invariant of the included machine[1, p317].) In effect, the including machine provides an interface and context for a single operation of the included machine. To take account of the B restriction, a further concept, a *BuiltOp* is added to the IS UML metamodel. This has no rationale other than to satisfy the needs of B specification.

A *BuiltOp* is also required when a transaction needs to call several *GeneratedOp*s and/or *UserOp*s from one *BasicMachine* or *IntermediateMachine*. Thus a *BuiltOp* can be composed of either *GeneratedOp*s or *UserOp*s (figure 8).

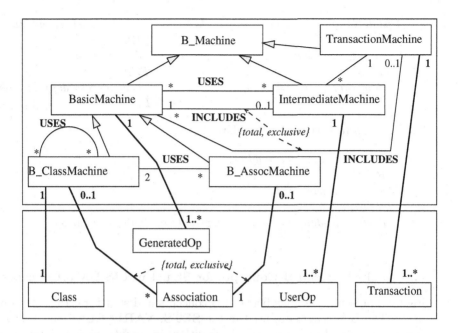

Fig. 7. B Machine Interrelationships and Links to IS UML Concepts

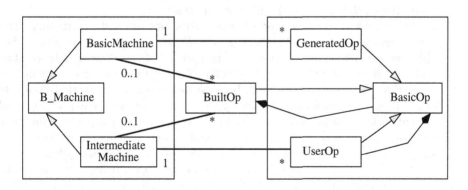

Fig. 8. B Machine and IS UML structures for User and Built Operations

A *BuiltOp* is automatically generated if the transaction requires such multiple operation calls. The *BuiltOp* is constructed by concatenation of the signatures of the required *BasicOp*s, conjunction of their preconditions and the parallel combination of their substitutions (resolving any duplication that occurs).

The use of the *BuiltOp* concept is illustrated for an operation that both updates the name of a department, and changes its address. The *GeneratedOp*s to achieve this are,

$$UpdateName(dep, name)$$
$$PRE$$
$$\quad dep \in Department$$
$$THEN$$
$$\quad deptName(dep) := name$$
$$END$$

$$UpdateAddress(dep, address)$$
$$PRE$$
$$\quad dep \in Department$$
$$THEN$$
$$\quad deptAddress(dep) := address$$
$$END$$

The *BuiltOp* to perform both updates is,

$$UpdateDept(dep, name, address)$$
$$PRE$$
$$\quad dep \in Department$$
$$THEN$$
$$\quad deptName(dep) := name \parallel deptAddress(dep) := address$$
$$END$$

6.2   Links between B Concepts for Structural IS UML Concepts

Figure 9 presents the links between B concepts and structural IS UML components. The B concepts are used in the **SETS**, **VARIABLES** and **INVARI-
ANT** clauses of basic machines. In addition, invariants defining integrity constraints appear in the **INVARIANT** of intermediate and transaction machines.
The diagram omits the components of B **OPERATION** clauses, namely preconditions (*B_Expressions*) and substitutions[17,9].

The links between B and IS UML concepts are self-explanatory. The B
structures require some explanation. A *B_Expression* (and its subclasses) is any
boolean expression, and is used to construct a *B_GlobalInvariant* by combining
relevant *B_Invariants*. The *B_ElementaryInvariant* is required to fully define association end multiplicities in B: because tools such as AtelierB[7] do not have
built-in proof rules for all function types, some functions are expressed by additional typing invariants on the domain, the range or the inverse of the function[7].
B_Variables represent class, attribute and association instances. The diagram
models the links between these subclasses and the subclasses of *B_Invariant*.
B_ClassVars and *B_AttribVars* are linked to *B_Types*; in IS specification, most
types are represented as *B_GivenSets*, but the *B_BasicTypes* are also available.

7   Conclusion

The work presented here brings together the tool development and metamodelling components of our research into rigorous IS development. Whilst much of
the work is published elsewhere, the traceability links and their part in a tool
architecture is new, and represents an important contribution to our long-term
goal of a tool assistant for rigorous IS development.

Much of our research is forward-looking. Although commercial tools for diagrammatic and formal modelling have progressed in recent years, they do not
yet provide all the features that we need. For example, tool interchange formats

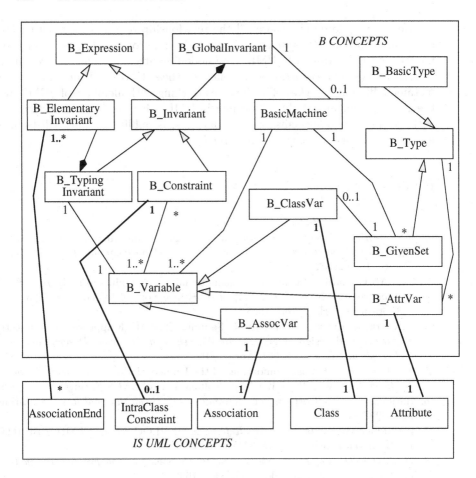

Fig. 9. B Concepts for Types and Invariants

allow us to import to the prototype tool assistant textual representations of UML models, but we cannot specialise the UML tool to our IS UML syntax. Formal methods tools provide a wide range of verification, proof and animation facilities, but their user interfaces are inadequate for general use by non-experts. Tool assistants need to be capable of use by conventional designers with no more than a basic knowledge in formal specifications.

In this paper, we have not looked at how to refine an IS specified in IS UML and B. CEDRIC-IIE work has shown how relational database implementations can be formally derived from the B specification[15,12]. Further research is planned into the connections between B refinement and the design process in UML. B is not ideal for the dynamic aspects, but there is ongoing research into event-oriented extensions of B[2,3].

We have explored elsewhere[17] the relationship of our work to other combined notations for IS specification, and the relationship between our metamodels and the "standard" UML metamodelling attempts. Ongoing research is looking at how our B metamodels relate to these UML-OCL metamodels and metamodelling approaches. This involves working with members of pUML, with the aim of producing a fully-conformant IS UML, when the UML 2.0 "standard", with extension and specialisation mechanisms, is published. This will facilitate use of our tool assistant with commercial UML tools.

References

1. J-R. Abrial. *The B-Book.* CUP, 1996.
2. J-R. Abrial. Event driven sequential program construction. In *AFADL01: Approches Formelles dans l'Assistance au Développement des Logiciels, Nancy, France,* June 2001.
3. J-R. Abrial and L. Mussat. Introducing dynamic constraints in B. In *B98: Second International B Conference, Montpellier, France,* volume 1393 of *LNCS,* pages 83–128. Springer Verlag, April 1998.
4. A. Clark, A. Evans, S. Kent, and P. Sammut. The MMF approach to engineering object-oriented design languages. In *Workshop on Language Descriptions, Tools and Applications, LTDA2001,* April 2001.
5. T. Clark and A. Evans. Foundations of the Unified Modeling Language. In *Second Northern Formal Methods Workshop,* volume 1241 of *LNCS.* Springer Verlag, 1997.
6. Rational Software Corporation. *Rational Rose - Using Rational Rose 98.* Rational Inc., 1998.
7. DIGILOG. *Atelier B, Manuel de Référence.* DIGILOG groupe STERIA, BP 16000, 13791 Aix-en-Provence Cedex 3, France, 1996.
8. D. F. D'Souza and A. C. Wills. *Objects, Components and Frameworks in UML: the Catalysis Approach.* Addison-Wesley, 1999.
9. P. Facon, R. Laleau, A. Mammar, and F. Polack. Formal specification of the UML metamodel for building rigorous CAiSE tools. Technical report, CEDRIC Laboratory, CNAM, September 1999.
10. P. Facon, R. Laleau, and H. P. Nguyen. Dérivation de spécifications formelles B à partir de spécifications semi-formelles de systèmes d'information. In *1st B Conference, Nantes, France,* 1996.
11. P. Facon, R. Laleau, and H. P. Nguyen. From OMT diagrams to B specifications. In M. Frappier and H. Habrias, editors, *Software Specification Methods. An Overview Using a Case Study,* pages 57–77. Springer, 2000.
12. P. Facon, R. Laleau, H. P. Nguyen, and A. Mammar. Combining UML with the B formal method for the specification of database applications. Technical report, CEDRIC Laboratory, CNAM, September 1999.
13. R. B. France and M. M. Larrondo-Petrie. A two-dimensional view of integrated formal and informal specification techniques. In *Ninth International Conference of Z Users, Limerick, Ireland, September 1995,* volume 967 of *LNCS,* pages 434–448. Springer Verlag, 1995.
14. J-C. Freire Junior. Pouvoir d'expression de modèles orientés objet. *Ingénierie des Systèmes d'Information,* 4(2):219–237, 1996.

15. R. Laleau and A. Mammar. A generic process to refine a B specification into a relational database implementation. In *Proceedings, ZB2000: Formal Specification and Development in Z and B, York, August-September 2000*, volume 1878 of *LNCS*, pages 22–41. Springer Verlag, 2000.

16. R. Laleau and A. Mammar. An overview of a method and its support tool for generating B specifications from UML notations. In *ASE: 15th IEEE Conference on Automated Software Engineering, Grenoble, France*. IEEE Computer Society Press, September 2000.

17. R. Laleau and F. Polack. Specification of integrity-preserving operations in information systems by using a formal UML-based language. Accepted for publication, Information and Software Technology, 2001.

18. Y. Ledru. Identifying preconditions with the Z/EVES theorem prover. In *ASE: 13th IEEE Conference on Automated Software Engineering*. IEEE Computer Society Press, October 1998.

19. B. Meyer. *Object-oriented software construction*. Prentice Hall, 1997.

20. E. Meyer and J. Souquières. A systematic approach to transform OMT diagrams to a B specification. In *FM99, Toulouse, France*, volume 1708 and 1709 of *LNCS*, pages 875–895. Springer Verlag, 1999.

21. F. Monge. Formalisation du méta modèle des méthodes graphiques d'analyse et conception orientées objet. Master's thesis, Institut d'Informatique d'Entreprise, Conservatoire National des Arts et Métiers, Evry, September 1997.

22. H. P. Nguyen. *Dérivation de spécifications formelles B à partir de spécifications semi-formelles*. PhD thesis, Laboratoire CEDRIC, Conservatoire National des Arts et Métiers, Evry, December 1998. Available from www.iie.cnam.fr/~laleau/thesePHNguyen.ps-gz.

23. F. Polack. Exploring the informal translations of OMT object models in B. Technical report, University of York, forthcoming, 2001.

24. F. Polack and R. Laleau. A rigorous metamodel for UML static conceptual modelling of information systems. In *CAiSE2001: 13th International Conference on Advanced Information Systems Engineering, Interlaken, Switzerland*, volume 2068 of *LNCS*. Springer Verlag, June 2001.

25. J. Rumbaugh, I. Jacobson, and G. Booch. *The Unified Modeling Language Reference Guide*. Addison-Wesley, 1998.

26. L. T. Semmens, R. B. France, and T. W. G. Docker. Integrated structured analysis and formal specification techniques. *The Computer Journal*, 35(6):600–610, 1992.

Author Index

Lecture Notes in Computer Science

For information about Vols. 1–2184
please contact your bookseller or Springer-Verlag

Vol. 2225: N. Abe, R. Khardon, T. Zeugmann (Eds.), Algorithmic Learning Theory. Proceedings, 2001. XI, 379 pages. 2001. (Subseries LNAI).

Vol. 2226: K.P. Jantke, A. Shinohara (Eds.), Discovery Science. Proceedings, 2001. XII, 494 pages. 2001. (Subseries LNAI).

Vol. 2227: S. Boztaş, I.E. Shparlinski (Eds.), Applied Algebra, Algebraic Algorithms and Error-Correcting Codes. Proceedings, 2001. XII, 398 pages. 2001.

Vol. 2228: B. Monien, V.K. Prasanna, S. Vajapeyam (Eds.), High Performance Computing – HiPC 2001. Proceedings, 2001. XVIII, 438 pages. 2001.

Vol. 2229: S. Qing, T. Okamoto, J. Zhou (Eds.), Information and Communications Security. Proceedings, 2001. XIV, 504 pages. 2001.

Vol. 2230: T. Katila, I.E. Magnin, P. Clarysse, J. Montagnat, J. Nenonen (Eds.), Functional Imaging and Modeling of the Heart. Proceedings, 2001. XI, 158 pages. 2001.

Vol. 2232: L. Fiege, G. Mühl, U. Wilhelm (Eds.), Electronic Commerce. Proceedings, 2001. X, 233 pages. 2001.

Vol. 2233: J. Crowcroft, M. Hofmann (Eds.), Networked Group Communication. Proceedings, 2001. X, 205 pages. 2001.

Vol. 2234: L. Pacholski, P. Ružička (Eds.), SOFSEM 2001: Theory and Practice of Informatics. Proceedings, 2001. XI, 347 pages. 2001.

Vol. 2235: C.S. Calude, G. Păun, G. Rozenberg, A. Salomaa (Eds.), Multiset Processing. VIII, 359 pages. 2001.

Vol. 2236: K. Drira, A. Martelli, T. Villemur (Eds.), Co-operative Environments for Distributed Systems Engineering. IX, 281 pages. 2001.

Vol. 2237: P. Codognet (Ed.), Logic Programming. Proceedings, 2001. XI, 365 pages. 2001.

Vol. 2239: T. Walsh (Ed.), Principles and Practice of Constraint Programming – CP 2001. Proceedings, 2001. XIV, 788 pages. 2001.

Vol. 2240: G.P. Picco (Ed.), Mobile Agents. Proceedings, 2001. XIII, 277 pages. 2001.

Vol. 2241: M. Jünger, D. Naddef (Eds.), Computational Combinatorial Optimization. IX, 305 pages. 2001.

Vol. 2242: C.A. Lee (Ed.), Grid Computing – GRID 2001. Proceedings, 2001. XII, 185 pages. 2001.

Vol. 2243: G. Bertrand, A. Imiya, R. Klette (Eds.), Digital and Image Geometry. VII, 455 pages. 2001.

Vol. 2244: D. Bjørner, M. Broy, A.V. Zamulin (Eds.), Perspectives of System Informatics. Proceedings, 2001. XIII, 548 pages. 2001.

Vol. 2245: R. Hariharan, M. Mukund, V. Vinay (Eds.), FST TCS 2001: Foundations of Software Technology and Theoretical Computer Science. Proceedings, 2001. XI, 347 pages. 2001.

Vol. 2246: R. Falcone, M. Singh, Y.-H. Tan (Eds.), Trust in Cyber-societies. VIII, 195 pages. 2001. (Subseries LNAI).

Vol. 2247: C. P. Rangan, C. Ding (Eds.), Progress in Cryptology – INDOCRYPT 2001. Proceedings, 2001. XIII, 351 pages. 2001.

Vol. 2248: C. Boyd (Ed.), Advances in Cryptology – ASIACRYPT 2001. Proceedings, 2001. XI, 603 pages. 2001.

Vol. 2249: K. Nagi, Transactional Agents. XVI, 205 pages. 2001.

Vol. 2250: R. Nieuwenhuis, A. Voronkov (Eds.), Logic for Programming, Artificial Intelligence, and Reasoning. Proceedings, 2001. XV, 738 pages. 2001. (Subseries LNAI).

Vol. 2251: Y.Y. Tang, V. Wickerhauser, P.C. Yuen, C.Li (Eds.), Wavelet Analysis and Its Applications. Proceedings, 2001. XIII, 450 pages. 2001.

Vol. 2252: J. Liu, P.C. Yuen, C. Li, J. Ng, T. Ishida (Eds.), Active Media Technology. Proceedings, 2001. XII, 402 pages. 2001.

Vol. 2253: T. Terano, T. Nishida, A. Namatame, S. Tsumoto, Y. Ohsawa, T. Washio (Eds.), New Frontiers in Artificial Intelligence. Proceedings, 2001. XXVII, 553 pages. 2001. (Subseries LNAI).

Vol. 2254: M.R. Little, L. Nigay (Eds.), Engineering for Human-Computer Interaction. Proceedings, 2001. XI, 359 pages. 2001.

Vol. 2255: J. Dean, A. Gravel (Eds.), COTS-Based Software Systems. Proceedings, 2002. XIV, 257 pages. 2002.

Vol. 2256: M. Stumptner, D. Corbett, M. Brooks (Eds.), AI 2001: Advances in Artificial Intelligence. Proceedings, 2001. XII, 666 pages. 2001. (Subseries LNAI).

Vol. 2257: S. Krishnamurthi, C.R. Ramakrishnan (Eds.), Practical Aspects of Declarative Languages. Proceedings, 2002. VIII, 351 pages. 2002.

Vol. 2258: P. Brazdil, A. Jorge (Eds.), Progress in Artificial Intelligence. Proceedings, 2001. XII, 418 pages. 2001. (Subseries LNAI).

Vol. 2259: S. Vaudenay, A.M. Youssef (Eds.), Selected Areas in Cryptography. Proceedings, 2001. XI, 359 pages. 2001.

Vol. 2260: B. Honary (Ed.), Cryptography and Coding. Proceedings, 2001. IX, 416 pages. 2001.

Vol. 2262: P. Müller, Modular Specification and Verification of Object-Oriented Programs. XIV, 292 pages. 2002.

Vol. 2264: K. Steinhöfel (Ed.), Stochastic Algorithms: Foundations and Applications. Proceedings, 2001. VIII, 203 pages. 2001.

Vol. 2267: M. Cerioli, G. Reggio (Eds.), Recent Trends in Algebraic Development Techniques. Proceedings, 2001. X, 345 pages. 2002.

Vol. 2272: D. Bert, J.P. Bowen, M.C. Henson, K. Robinson (Eds.), ZB 2002: Formal Specification and Development in Z and B. Proceedings, 2002. XII, 535 pages. 2002.

Vol. 2273: A.R. Coden, E.W. Brown, S. Srinivasan (Eds.), Information Retrieval Techniques for Speech Applications. XI, 109 pages. 2002.

Vol. 2275: N.R. Pal, M. Sugeno (Eds.), Advances in Soft Computing – AFSS 2002. Proceedings, 2002. XVI, 536 pages. 2002. (Subseries LNAI).